Lecture Notes in Computer Science   3562

Commenced Publication in 1973
Founding and Former Series Editors:
Gerhard Goos, Juris Hartmanis, and Jan van Leeuwen

Editorial Board

David Hutchison
  Lancaster University, UK
Takeo Kanade
  Carnegie Mellon University, Pittsburgh, PA, USA
Josef Kittler
  University of Surrey, Guildford, UK
Jon M. Kleinberg
  Cornell University, Ithaca, NY, USA
Friedemann Mattern
  ETH Zurich, Switzerland
John C. Mitchell
  Stanford University, CA, USA
Moni Naor
  Weizmann Institute of Science, Rehovot, Israel
Oscar Nierstrasz
  University of Bern, Switzerland
C. Pandu Rangan
  Indian Institute of Technology, Madras, India
Bernhard Steffen
  University of Dortmund, Germany
Madhu Sudan
  Massachusetts Institute of Technology, MA, USA
Demetri Terzopoulos
  New York University, NY, USA
Doug Tygar
  University of California, Berkeley, CA, USA
Moshe Y. Vardi
  Rice University, Houston, TX, USA
Gerhard Weikum
  Max-Planck Institute of Computer Science, Saarbruecken, Germany

T0189264

José Mira   José R. Álvarez (Eds.)

Artificial Intelligence and Knowledge Engineering Applications: A Bioinspired Approach

First International Work-Conference on the Interplay
Between Natural and Artificial Computation, IWINAC 2005
Las Palmas, Canary Islands, Spain, June 15-18, 2005
Proceedings, Part II

 Springer

Volume Editors

José Mira
José R. Álvarez
Universidad Nacional de Educación a Distancia
E.T.S. de Ingeniería Informática
Departamento de Inteligencia Artificial
Juan del Rosal, 16, 28040 Madrid, Spain
E-mail: {jmira, jras}@dia.uned.es

Library of Congress Control Number: Applied for

CR Subject Classification (1998): F.1, F.2, I.2, G.2, I.4, I.5, J.3, J.4, J.1

ISSN   0302-9743
ISBN-10   3-540-26319-5 Springer Berlin Heidelberg New York
ISBN-13   978-3-540-26319-7 Springer Berlin Heidelberg New York

Springer is a part of Springer Science+Business Media

springeronline.com

© Springer-Verlag Berlin Heidelberg 2005
Printed in Germany

Typesetting: Camera-ready by author, data conversion by Scientific Publishing Services, Chennai, India
Printed on acid-free paper   SPIN: 11499305   06/3142   5 4 3 2 1 0

Preface

The computational paradigm considered here is a conceptual, theoretical and formal framework situated above machines and living creatures (two instantiations), sufficiently solid, and still non-exclusive, that allows us:

1. to help neuroscientists to formulate intentions, questions, experiments, methods and explanation mechanisms assuming that neural circuits are the psychological support of calculus;
2. to help scientists and engineers from the fields of artificial intelligence (AI) and knowledge engineering (KE) to model, formalize and program the computable part of human knowledge;
3. to establish an interaction framework between natural system computation (NSC) and artificial system computation (ASC) in both directions, from ASC to NSC (in computational neuroscience), and from NSC to ASC (in bioinspired computation).

With these global purposes, we organized IWINAC 2005, the 1st International Work Conference on the Interplay Between Natural and Artificial Computation, which took place in Las Palmas de Gran Canaria, Canary Islands (Spain), during June 15–18, 2005, trying to contribute to both directions of the interplay:

I: From Artificial to Natural Computation. What can computation, artificial intelligence (AI) and knowledge engineering (KE) contribute to the understanding of the nervous system, cognitive processes and social behavior? This is the scope of computational neuroscience and cognition, which uses the computational paradigm to model and improve our understanding of natural science.

II: From Natural Sciences to Computation, AI and KE. How can computation, AI and KE find inspiration in the behavior and internal functioning of physical, biological and social systems to conceive, develop and build up new concepts, materials, mechanisms and algorithms of potential value in real-world applications? This is the scope of the new bionics, known as bioinspired engineering and computation, as well as of natural computing.

To address the two questions posed in the scope of IWINAC 2005, we made use of the "building of and for knowledge" concepts that distinguish three levels of description in each calculus: the physical level (PL), where the hardware lives, the symbol level (SL) where the programs live, and a third level, introduced by Newell and Marr, situated above the symbol level and named by Newell as "the knowledge level" (KL) and by Marr as the level of "the theory of calculus." We seek the interplay between the natural and the artificial at each one of these three levels (PL, SL, KL).

1. For the interplay at the **physical level** we consider:
   - **Computational Neuroscience**. *Tools*: Conceptual, formal and computational tools and methods in the modelling of neuronal processes and neural nets: individual and collective dynamics. *Mechanisms*: Computational modelling of neural mechanisms at the architectural level: oscillatory/regulatory feedback loops, lateral inhibition, reflex arches, connectivity and signal routing networks, distributed central-patterns generators. Contributions to the library of neural circuitry. *Plasticity*: Models of memory, adaptation, learning and other plasticity phenomena. Mechanisms of reinforcement, self-organization, anatomo-physiological coordination and structural coupling.
   - **Bio-inspired Circuits and Mechanisms**. *Electronics*: Bio-inspired electronics and computer architectures. Advanced models for ANN. Evolvable hardware (CPLDs, FPGAs, etc.). Adaptive cellular automata. Redundancy, parallelism and fault-tolerant computation. Retinotopic organizations. *Non-conventional (Natural) Computation:* Biomaterials for computational systems. DNA, cellular and membrane computing. *Sensory and Motor Prostheses:* Signal processing, artificial cochlea, audiotactile vision substitution. Artificial sensory and motor systems for handicapped people. Intersensory transfer and sensory plasticity.
2. For the interplay at the **symbol level** we consider:
   - **Neuro-informatics**. *Symbols:* From neurons to neurophysiological symbols (regularities, synchronization, resonance, dynamics binding and other potential mechanisms underlying neural coding). Neural data structures and neural "algorithms." *Brain Databases:* Neural data analysis, integration and sharing. Standardization, construction and use of databases in neuroscience and cognition. *Neurosimulators:* Development and use of biologically oriented neurosimulators. Contributions to the understanding of the relationships between structure and function in biology.
   - **Bio-inspired Programming Strategies**. *Behavior-Based Computational Methods:* Reactive mechanisms. Self-organizing optimization. Collective emergent behavior (ant colonies). Ethology and artificial life. *Evolutionary Computation*: Genetic algorithms, evolutionary strategies, evolutionary programming and genetic programming. Macroevolution and the interplay between evolution and learning. *Hybrid Approaches:* Neuro-symbolic integration. Knowledge-based ANN and connectionist KBS. Neuro-fuzzy systems. Hybrid adaptation and learning at the symbol level.
3. For the Interplay at the **knowledge level** we consider:
   - **Computational Approaches to Cognition**. *AI and KE:* Use of AI and KE concepts, tools and methods in the modelling of mental processes and behavior. Contribution to the AI debate on paradigms for knowledge representation and use: symbolic (representational), connectionist, situated, and hybrid. *Controversies:* Open questions and controversies in

AI and Cognition (semantics versus syntax, knowledge as mechanisms that know, cognition without computation, etc.). Minsky, Simon, Newell, Marr, Searle, Maturana, Clancey, Brooks, Pylyshyn, Fodor, and others. *Knowledge Modelling:* Reusability of components in knowledge modelling (libraries of tasks, methods, inferences and roles). Ontologies (generic, domain-specific, object-oriented, methods and tasks). Knowledge representation methodologies and knowledge edition tools.

- **Cognitive Inspiration in AI and KE.** *Synthetic Cognition:* Bio-inspired modelling of cognitive tasks. Perception, decision-making, planning and control. Biologically plausible (user-sensitive) man–machine interfaces. Natural language programming attempts. Social organizations, distributed AI, and multi-agent systems. *Bio-inspired Solutions to Engineering, Computational and Social Problems in Different Application Domains*: Medicine, image understanding, KBSs and ANNs for diagnoses, therapy planning, and patient follow-up. Telemedicine. Robotic paradigms. Dynamic vision. Path planning, map building, and behavior-based navigation methods. Anthropomorphic robots. Health biotechnology. Bio-inspired solutions for sustainable growth and development.

IWINAC 2005 was organized by the Universidad Nacional de Educación a Distancia (UNED) in cooperation with the Instituto Universitario de Ciencias y Tecnologías Cibernéticas de la Universidad de Las Palmas de Gran Canaria and the Las Palmas UNED Associated Center.

Sponsorship was obtained from the Spanish Ministerio de Ciencia y Tecnología and the organizing universities (UNED and Las Palmas de Gran Canaria).

The chapters of these two books of proceedings correspond to the talks delivered at the IWINAC 2005 conference. After the refereeing process, 117 papers were accepted for oral or poster presentation, according to the authors' preferences. We organized these papers into two volumes basically following the topics list previously mentioned. The first volume, entitled *"In Search of Mechanisms, Symbols, and Models Underlying Cognition,"* includes all the contributions mainly related to the methodological, conceptual, formal, and experimental developments in the fields of neurophysiology and cognitive science.

In the second volume, *"Artificial Intelligence and Knowledge Engineering Applications: A Bioinspired Approach,"* we have collected the papers related to bioinspired programming strategies and all the contributions related to the computational solutions to engineering problems in different application domains.

And now is the time for acknowledgements. A task like this, organizing a work conference with a well-defined scope, cannot be achieved without the active engagement of a broad set of colleagues who share with us the conference principles, foundations and objectives. First, let me express my sincere gratitude to all the scientific and organizing committees, in particular, the members of these committees who acted as effective and efficient referees and as promoters and managers of preorganized sessions on autonomous and relevant topics under the IWINAC global scope. Thanks also to the invited speakers, Joost N. Kok,

Dana Ballard and Juan Vicente Sánchez Andrés, for synthesizing in preparing the plenary lectures. Finally, thanks to all the authors for their interest in our call and their efforts in preparing the papers, condition sine qua non for these proceedings.

My debt of gratitude to José Ramón Alvarez and Félix de la Paz goes further the frontiers of a preface. Without their collaboration IWINAC 2005 would not have been possible, in the strictest sense. And the same is true concerning Springer and Alfred Hofmann, for being continuously receptive and for collaborating on all our editorial joint ventures on the interplay between neuroscience and computation, from the first IWANN in Granada (1991, LNCS 540), to the successive meetings in Sitges (1993, LNCS 686), Torremolinos (1995, LNCS 930), Lanzarote (1997, LNCS 1240), Alicante (1999, LNCS 1606 and 1607), again in Granada (2001, LNCS 2084 and 2085), then in Maó (Menorca) (2003, LNCS 2686 and 2687) and, now, the first IWINAC in Las Palmas.

June 2005   José Mira

Organization

General Chairman

José Mira, UNED (Spain)

Organizing Committee

José Ramón Álvarez Sánchez, UNED (Spain)
Félix de la Paz López, UNED (Spain)

Local Organizing Committee

Roberto Moreno-Díaz, Jr., Univ. Las Palmas de Gran Canaria (Spain)
Alexis Quesada, Univ. Las Palmas de Gran Canaria (Spain)
José Carlos Rodriguez, Univ. Las Palmas de Gran Canaria (Spain)
Cristobal García Blairsy, UNED (Spain)
José Antonio Muñoz, Univ. Las Palmas de Gran Canaria (Spain)

Invited Speakers

Joost N. Kok, Leiden University (The Netherlands)
Dana Ballard, University of Rochester (USA)
Juan Vicente Sánchez Andrés, University of La Laguna (Spain)

Field Editors

Eris Chinellato, Universitat Jaume I (Spain)
Carlos Cotta, University of Málaga (Spain)
Angel P. del Pobil, Universitat Jaume I (Spain)
Antonio Fernández-Caballero, Universidad de Castilla-La Mancha (Spain)
Oscar Herreras, Hospital Ramón y Cajal (Spain)
Heinz Hügli, University of Neuchâtel (Switzerland)
Roque Marín, Universidad de Murcia (Spain)
Carlos Martin-Vide, Rovira i Virgili University of Tarragona (Spain)
Victor Mitrana, Rovira i Virgili University of Tarragona (Spain)
José T. Palma Méndez, University of Murcia (Spain)
Miguel Angel Patricio Guisado, Universidad de Alcalá (Spain)
Eduardo Sánchez Vila, Universidad de Santiago de Compostela (Spain)
Ramiro Varela Arias, Universidad de Oviedo (Spain)

Scientific Committee (Referees)

Ajith Abraham, Chung Ang University (South Korea)
Igor Aleksander, Imperial College London (UK)
José Ramón Álvarez Sánchez, UNED (Spain)
Margarita Bachiller Mayoral, UNED (Spain)
Antonio Bahamonde, Universidad de Oviedo (Spain)
Emilia I. Barakova, RIKEN (Japan)
Alvaro Barreiro, Univ. A Coruña (Spain)
Senen Barro Ameneiro, University of Santiago de Compostela (Spain)
Luc Berthouze, AIST (Japan)
Joanna J. Bryson, University of Bath (UK)
Lola Cañamero, University of Hertfordshire (UK)
Joaquín Cerdá Boluda, Univ. Politécnica de Valencia (Spain)
Enric Cervera Mateu, Universitat Jaume I (Spain)
Eris Chinellato, Universitat Jaume I (Spain)
Carlos Cotta, University of Málaga (Spain)
Paul Cull, Oregon State University (USA)
Kerstin Dautenhahn, University of Hertfordshire (UK)
Félix de la Paz López, UNED (Spain)
Ana E. Delgado García, UNED (Spain)
Javier de Lope, Universidad Politécnica de Madrid (Spain)
Angel P. del Pobil, Universitat Jaume I (Spain)
Jose Dorronsoro, Universidad Autónoma de Madrid (Spain)
Richard Duro, Universidade da Coruña (Spain)
Juan Pedro Febles Rodriguez, Centro Nacional de Bioinformática (Cuba)
Antonio Fernández-Caballero, Universidad de Castilla-La Mancha (Spain)
Jose Manuel Ferrández, Universitöt Politécnica de Cartagena (Spain)
Nicolas Franceschini, Université de la Méditerranée (France)
Marian Gheorghe, University of Sheffield (UK)
Karl Goser, universitöt Dortmund (Germany)
Carlos G. Puntonet, Universidad de Granada (Spain)
Manuel Graña Romay, Universidad Pais Vasco (Spain)
John Hallam, University of Southern Denmark (Denmark)
Denise Y.P. Henriques, York University (Canada)
Oscar Herreras, Hospital Ramón y Cajal (Spain)
Juan Carlos Herrero, (Spain)
Heinz Hügli, University of Neuchâtel (Switzerland)
Shahla Keyvan, University of Missouri, Columbia (USA)
Kostadin Koroutchev, Universidad Autónoma de Madrid (Spain)
Elka Korutcheva, UNED (Spain)
Max Lungarella, University of Tokyo (Japan)
Francisco Maciá Pérez, Universidad de Alicante (Spain)
george Maistros, University of Edinburgh (UK)
Dario Maravall, Universidad Politécnica de Madrid (Spain)
Roque Marín, Universidad de Murcia (Spain)

Table of Contents

Evolutionary Computation

Electronics and Robotics

Other Applications

Cultural Operators for a Quantum-Inspired Evolutionary Algorithm Applied to Numerical Optimization Problems

André V. Abs da Cruz, Marco Aurélio C. Pacheco,
Marley Vellasco, and Carlos R. Hall Barbosa

ICA — Applied Computational Intelligence Lab,
Electrical Engineering Department,
Pontifícia Universidade Católica do Rio de Janeiro
{andrev, marco, marley, hall}@ele.puc-rio.br

Abstract. This work presents the application of cultural algorithms operators to a new quantum-inspired evolutionary algorithm with numerical representation. These operators (fission, fusion, generalization and specialization) are used in order to provide better control over the quantum-inspired evolutionary algorithm. We also show that the quantum-inspired evolutionary algorithm with numerical representation behaves in a very similar manner to a pure cultural algorithm and we propose further investigations concerning this aspect.

1 Introduction

Many research efforts in the field of quantum computing have been made since 1990, after the demonstration that computers based on principles of quantum mechanics can offer more processing power for some classes of problems. The principle of superposition, which states that a particle can be in two different states simultaneously, suggests that a high degree of parallelism can be achieved using this kind of computers. Its superiority was shown with few algorithms such as the Shor's algorithm [1, 2] for factoring large numbers, and the Grover's algorithm [3] for searching databases. Shor's algorithm finds the prime factors of a n-digit number in polynomial time, while the best known classical algorithm has a complexity of $O(2^{n^{1/3}} log(n)^{2/3})$. On the other hand, Grover's algorithm searches for an item in a non-ordered database with n items with a complexity of $O(\sqrt{n})$ while the best classical algorithm has a complexity of $O(n)$.

Research on merging evolutionary algorithms with quantum computing has been developed since the end of the 90's. This research can be divided in two different groups: one that, motivated by the lack of quantum algorithms, focus on developing new ones by using techniques for automatically generating programs [4]; and another which focus on developing quantum–inspired evolutionary algorithms with binary [5, 6, 7] and real [8] representations which can be executed on classical computers.

J. Mira and J.R. Álvarez (Eds.): IWINAC 2005, LNCS 3562, pp. 1–10, 2005.

This work is an extension of [8], where new operators are proposed for the quantum–inspired algorithm with real representation. Those new operators, based on another class of algorithms known as Cultural Algorithms [9], are used in order to provide a more consistent set of operators and to avoid some problems regarding premature convergence to local minima, providing a more consistent set of operators.

This paper is organized as follows: section 2 describes the proposed quantum-inspired evolutionary algorithm; section 3 describes the cultural operators proposed and implemented; section 4 describes the experiments that were carried out; section 5 presents the results obtained; and finally section 6 draws some conclusions regarding this work.

2   Quantum-Inspired Evolutionary Algorithm

Quantum-inspired evolutionary algorithms rely on the concepts of "quantum bits", or qubits, and on superposition of states from quantum mechanics [5, 6]. The state of a quantum bit can be represented, using the Dirac notation, as:

$$|\varphi >= |\alpha > +|\beta >$$
(1)

Where α and β are complex numbers that represent probability amplitudes of the corresponding states. $|\alpha|^2$ and $|\beta|^2$ give the probability of the qubit to be in state 0 and in state 1, respectively, when observed. The amplitude normalization guarantees that:

$$|\alpha|^2 + |\beta|^2 = 1$$
(2)

The quantum-inspired evolutionary algorithm with binary representation [5, 6] works properly in problems where this kind of representation is more suitable. However, in some specific situations, real numbers representation is more adequate such as in function optimization, where a maximum or minimum must be found by adjusting some variables).

The proposed algorithm using real representation works as following: initially, a set of rectangular pulses is generated for each variable that must be optimized. This set of pulses will be used to represent probability in replacement to the α and β values of the binary representation. The lower (l) and upper (u) bounds of these pulses are the same as the bounds of the problem domain. The height (h) of the pulses is such that their areas sum up to 1 or, as presented in equation 3:

$$h = \frac{1}{(l - u)}$$
(3)

The algorithm's initialization procedure begins with the definition of a value N that indicates how many pulses will be used to represent each variable's probability distribution function. Then, for each single pulse used in each variable, it must be defined:

- The pulse centre in the mean point of the variable domain;
- The pulse height as the inverse of the domain length divided by N.

At the end of this process, the sum of the N pulses related to a variable will have a total area of 1.

Suppose, for instance, that one wishes to initialize a variable with an universe of discourse equals to the interval $[-50, 50]$ and wants to use 4 rectangular pulses to represent the probability distribution function for this variable; in this case, each pulse would have a width equal to 100, centered at zero and height equals to $1/100/4 = 0.0025$.

The resultant set of probability distribution functions creates a superposition $Q_i(t)$ for each variable i of the problem. From this $Q_i(t)$ distribution, a set of n points are randomly drawn, which will form the population $P(t)$.

After choosing the individuals that will form the population $P(t)$, it is necessary to update the probability distribution $Q_i(t)$, in order to converge to the optimal or sub-optimal solution, similarly to the conventional crossover from classical genetic algorithms. The method employed in this work consists of choosing randomly m individuals from the population $P(t)$ using a roulette method identical to the one used in classical genetic algorithms. Then, the central point of the first pulse is redefined as the mean value of those m chosen individuals. This process is repeated for each one of the N pulses that define the distribution $Q_i(t)$. The value m is given by:

$$m = \frac{n}{N} \tag{4}$$

Where N is the number of pulses used to represent the probability distribution function and n is size of the population $P(t)$.

In addition, after each generation, the pulses' width is contracted symmetrically related to its center. This contraction is performed by an exponential decay function, according to the following formula:

$$\sigma = (u - l)^{(1 - \frac{t}{T})^{\lambda}} - 1 \tag{5}$$

Where σ is the pulse width, u is the domain's upper limit, l is the lower limit, t is the current algorithm generation, T is the total number of generations and λ is a parameter that defines the decay rate for the pulse width.

It is important to notice that as the pulses have their widths contracted and their mid-points changed, their sums will look less like a rectangular signal and will start to have several different shapes.

More details on how the basic proposed quantum–inspired algorithm works can be found in [8].

3   Cultural Operators

In the original quantum–inspired algorithm with real representation the operator responsible for contracting the pulses' width is a very restrictive operator in the sense that it can lead the optimization to a local minima (maxima). This premature convergence is irreversible, since contraction is the only operator allowed. This work proposes the use of 4 new operators (where one of them is very similar to the contraction operator) which are inspired by similar operator that are present in the field of Cultural Algorithms. Cultural Algorithms were introduced in [9] and a detailed description of the operators can be found in [10].

The proposed operators are based on the concept of density: the density is calculated by counting the number of genes inside a pulse and then dividing the total by the width of the pulse (which is the upper bound minus the lower bound of the pulse). The pulses' density cannot be lower than a specified threshold. The operators introduced are (the examples represent the pulses' boundaries with the '[' and ']' signals and the genes with the '+' signals):

- *Specialization* - this operator is similar to the cotraction operator. If the density is lower than the acceptable threshold, the pulse is contracted.

  Before using the operator: [++++++++++   ]
  After using the operator: [++++++++++]

- *Generalization* - If there is an evidence of acceptable individuals outside of the current pulse, then expand it to include them.

  Before using the operator: +++   [++++++++++] ++
  After using the operator: [+++   ++++++++++ ++]

- *Fusion* - this is a special case of generalization where two disjoint pulses are merged when there exist acceptable individuals between them.

  Before using the operator: [++++++] [++++++++++]
  After using the operator: [++++++   ++++++++++]

- *Fission* - this is a special case of specialization where the interior of the current pulse is removed to produce two separate pulses.

  Before using the operator: [++++++   ++++++++++]
  After using the operator: [++++++] [++++++++++]

The operators were used without restrictions. The algorithm is allowed to use each one of them, one time per individual per generation.

4   Experiments

To assess the new proposed algorithm, several functions, each of them with different characteristics, were used. A subset of real benchmark functions from [11] were chosen, more specifically:

- The Sphere Model

$$f(x) = \sum_{i=1}^{n} (x_i - 1)^2 \qquad (6)$$

where $x_i \in [-5, 5]$

– The Griewank's Function

$$f(x) = \frac{1}{d} \sum_{i=1}^{n} (x_i - 100)^2 - \prod_{i=1}^{n} \cos(\frac{x_i - 100}{\sqrt{i}}) + 1 \qquad (7)$$

where $d = 4000$ and $x_i \in [-600, 600]$

– The Michalewicz' Function

$$f(x) = - \sum_{i=1}^{n} \sin(x_i) \sin^{2m}(\frac{ix_i^2}{\pi}) \qquad (8)$$

where $m = 10$ and $x_i \in [0, \pi]$

The n value in each of the above equations represents the number of dimensions used (the number of variables). In this work, the functions were optimized with two different values of n: $n = 5$ and $n = 10$. To compare the results, a classical genetic algorithm was used with the parameters shown in Table 1.

Table 1. Parameters for the classical genetic algorithm

Mutation Rate	10%
Crossover Rate	90%
Gap	40%
Population Size	100
Generations	40
Number of Evaluations	4000
Genetic Operators	Arithmetical Crossover, Uniform and Creep Mutation
Selection Method	Roulette with steady state

For the quantum-inspired evolutionary algorithm the set of parameters presented in Table 2 was employed.

Table 2. Parameters for the quantum–inspired evolutionary algorithm

Pulses per Variable	3
Number of Observations $P(t)$	100
Generations	40
Number of Evaluations	4000

These parameters have provided the best results and were obtained after systematic experiments, with several different configurations. For each experiment 20 rounds were executed and the mean value for the evaluation was calculated.

5   Results

The results obtained for the selected functions are presented in figures 1 to 6. Figures 1 and 2 shows the results for the Sphere function with $n = 5$ and $n = 10$ respectively. Each figure depicts the average value of the best individual in each generation for 20 experiments.

Figures 3 and 4 show the results for the Griewank's function with $n = 5$ and $n = 10$ respectively. Again, each figure depicts the average value of the best individual in each generation for 20 experiments.

Those figures shows that the quantum–inspired algorithm performs better in both configurations of the problem. Also, it is clear that the algorithm reaches lower values and converges towards the minimum faster than its counterpart.

Finally, figures 5 and 6 present the results for the Michalewicz's function with $n = 5$ and $n = 10$ respectively. Each figure presents the average value of the best individual in each generation for 20 experiments as in the case of the sphere and Griewank functions.

In this particular case, the quantum–inspired algorithm does not reach a value as good as the classical algorithm for the 5 dimension problem but it performs better for the 10 dimension one. This suggests that the quantum–inspired algorithm may perform better for more functions of high dimensionality than the classical genetic algorithm.

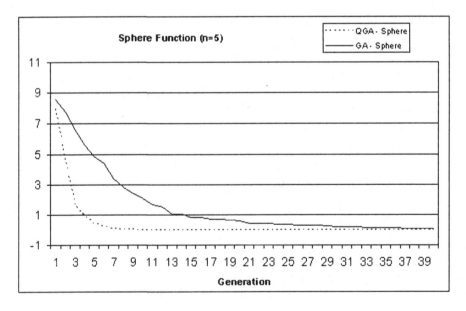

Fig. 1. Comparison between the quantum–inspired(QGA) and the classical(GA) evolutionary algorithms for the sphere function with 5 dimensions

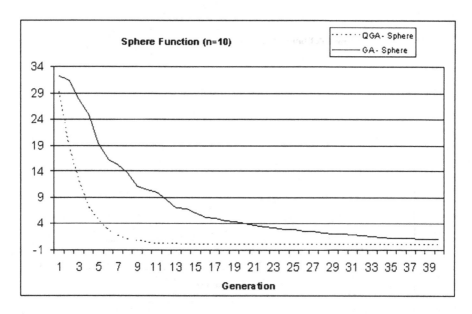

Fig. 2. Comparison between the quantum–inspired(QGA) and the classical(GA) evolutionary algorithms for the sphere function with 10 dimensions

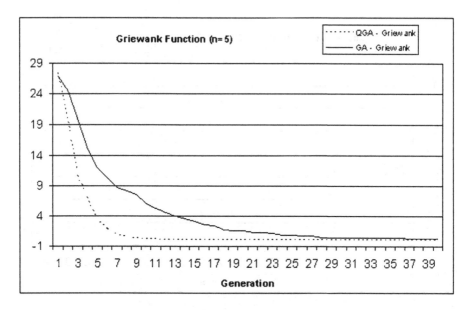

Fig. 3. Comparison between the quantum–inspired(QGA) and the classical(GA) evolutionary algorithms for Griewank's function with 5 dimensions

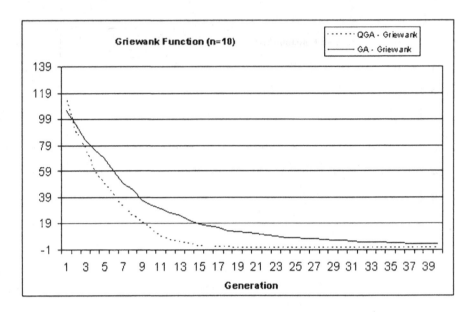

Fig. 4. Comparison between the quantum–inspired(QGA) and the classical(GA) evolutionary algorithms for Griewank's function with 10 dimensions

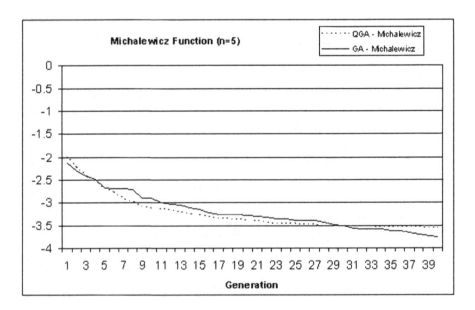

Fig. 5. Comparison between the quantum–inspired(QGA) and the classical(GA) evolutionary algorithms for Michalewicz's function with 5 dimensions

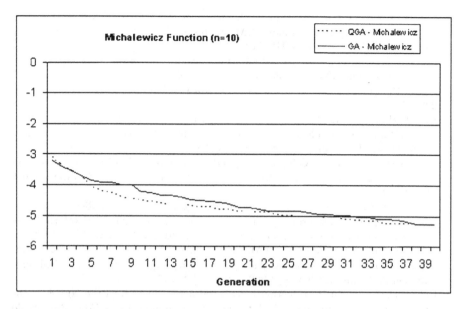

Fig. 6. Comparison between the quantum–inspired(QGA) and the classical(GA) evolutionary algorithms for Michalewicz's function with 10 dimensions

6 Conclusions

This paper described new operators for the quantum–inspired evolutionary algorithm. They are based on operators once developed for Cultural Algorithms. They were chosen in order to provide a more consistent way of updating the algorithm's state and to avoid premature convergence to local minima. The results demonstrated that the use of these operators improved the results and maintained the desired characteristics of the algorithm such as robustness and faster convergence.

Three different functions were used to compare the results between the classical and the quantum–inspired algorithms. Each of these functions have a different characteristic which makes optimization difficult.

The results showed that the quantum–inspired algorithm performed better for most functions (the only exception being the 5-dimensional version of the Michalewicz's function). This shows that this kind of algorithm might be a good option for real-function optimization. However, more research must be carried out, mainly over real world problems instead of benchmark problems.

References

1. Shor, P.W.: Algorithms for quantum computation: Discrete log and factoring. In: Foundations of Computer Science, Proc. 35th Ann. Symp., IEEE Computer Society Press (1994) 124–134
2. Shor, P.W.: Quantum computing. Documenta Mathematica (1998) 467–486

3. Grover, L.K.: A fast quantum mechanical algorithm for database search. In: Proceedings of the 28th Annual ACM Symposium on the Theory of Computing (STOC), ACM Press (1996) 212–219
4. Spector, L., Barnum, H., Bernstein, H.J., Swami, N.: Finding a better-than-classical quantum AND/OR algorithm using genetic programming. In: Proceedings of the Congress on Evolutionary Computation. Volume 3., IEEE Press (1999) 2239–2246
5. Han, K.H., Kim, J.H.: Genetic quantum algorithm and its application to combinatorial optimization problem. In: Proceedings of the 2000 Congress on Evolutionary Computation, IEEE Press (2000) 1354–1360
6. Han, K.H., Kim, J.H.: Quantum-inspired evolutionary algorithm for a class of combinatorial optimization. IEEE Transactions on Evolutionary Computation **6** (2002) 580–593
7. Narayanan, A., Moore, M.: Genetic quantum algorithm and its application to combinatorial optimization problem. In: Proceedings of the 1996 IEEE International Conference on Evolutionary Computation (ICEC96), IEEE Press (1996) 61–66
8. Abs da Cruz, A.V., Vellasco, M.M.B.R., Pacheco, M.A.C., Hall Barbosa, C.R.: Quantum-inspired evolutionary algorithms and its application to numerical optimization problems. In: ICONIP 2004. (2004) 212–217
9. Reynolds, R.G.: An introduction to cultural algorithms. In: Proceedings of the Third Annual Conference on Evolutionary Programming, River Edge, NJ: World Scientific (1994) 131–139
10. Chung, C., Reynolds, R.G.: A testbed for solving optimization problems using cultural algorithms. In: Proceedings of EP 96. (1996)
11. Bersini, H., Dorigo, M., Langerman, S., Seront, G., Gambardella, L.: Results of the first international contest on evolutionary optimisation (1st iceo). In: Proceedings of the 1996 IEEE International Conference on Evolutionary Computation (ICEC96), IEEE Press (1996) 622–627

New Codification Schemas for Scheduling with Genetic Algorithms*

Ramiro Varela, David Serrano, and María Sierra

Dep. of Computer Science, University of Oviedo,
Artificial Intelligence Center,
Campus de Viesques, 33271 Gijón, Spain
{ramiro, mariasierra}@aic.uniovi.es
http://www.aic.uniovi.es/Tc

Abstract. Codification is a very important issue when a Genetic Algorithm is designed to dealing with a combinatorial problem. In this paper we introduce new codification schemas for the Job Shop Scheduling problem which are extensions of two schemas of common use, and are worked out from the concept of underlying probabilistic model. Someway the underlying probabilistic model of a codification schema accounts for a tendency of the schema to represent solutions in some region of the search space. We report results from an experimental study showing that in many cases any of the new schemas results to be much more efficient than conventional ones due to the new schema tends to represent more promising solutions than the others. Unfortunately the selection in advance of the best schema for a given problem instance is not an easy problem and remains still open.

1 Introduction

Genetic Algorithms (GAs) are a flexible search paradigm for dealing with complex optimization and combinatorial problems. Even though a conventional GA often produce moderate results, it is well-known that its efficiency can be improved by incorporating heuristic knowledge from the problem domain in any of the genetic operators, or by combining the GA with a local search procedure as done, for example, by D. Mattfeld in [5] and by T. Yamada and R. Nakano in [8] for the Job Shop Scheduling (JSS) problem.

In this paper we consider the issue of codification and propose new codifications schemas. These new schemas are worked out from the study of two schemas commonly used in scheduling problems: conventional permutations (CP) and permutations with repetition (PR). Through the concept of underlying probabilistic model proposed in [7] we developed two new schemas termed as partial PR (PPR) and extended PR (EPR) respectively, as extensions of PR and CP.

---

* This work has been supported by project FEDER-MCYT TIC2003-04153 and by FICYT under grant BP04-021.

J. Mira and J.R. Álvarez (Eds.): IWINAC 2005, LNCS 3562, pp. 11–20, 2005.

These new codifications improve the capacity of CP and PR to represent good schedules and consequently in many cases are able to improve the GA performance as well.

The rest of the paper is organized as follows. In section 2 we formulate the JSS problem and describe the search space of active schedules together with the G&T algorithm that allows for greedy search over this space. In section 3 we review two common codifications for scheduling problems: CP and PR, and propose the new models: PPR and EPR. Then in section 5 we report results from an experimental study over a subset of problem instances taken from a standard repository. Finally in section 6 we summarize the main conclusions and propose a number of ideas for future work.

2 Problem Formulation and Search Space for the JSS Problem

The JSS problem requires scheduling a set of N jobs $\{J_0, \ldots, J_{N-1}\}$ on a set of M physical resources or machines $\{R_0, \ldots, R_{M-1}\}$. Each job J_i consists of a set of task or operations $\{\theta_{i0}, \ldots, \theta_{i(M-1)}\}$ to be sequentially scheduled. Each task θ_{il} having a single resource requirement, a duration $du\theta_{il}$ and a start time $st\theta_{il}$ whose value should be determined. The JSS has two binary constraints: *precedence constraints* and *capacity constrains*. Precedence constraints defined by the sequential routings of the tasks within a job translate into linear inequalities of the type: $st\theta_{il} + du\theta_{il} \leq st\theta_{i(l+1)}$ (i.e. θ_{il} before $\theta_{i(l+1)}$). Capacity constraints that restrict the use of each resource to only one task at a time translate into disjunctive constraints of the form: $st\theta_{il} + du\theta_{il} \leq st\theta_{jk} \vee st\theta_{jk} + du\theta_{jk} \leq st\theta_{il}$. The objective is to come up with a feasible schedule such that the completion time, i.e. the *makespan*, is minimized.

The JSS problem has interested to many researches over the last three decades. In [4] Jain and Meeran review the most interesting approaches to this problem. One of the first efficient approaches is the well-known algorithm proposed by Giffler and Thomson in [3]. Here we consider a variant termed as hybrid G&T (see Algorithm 1). The hybrid G&T algorithm is an active schedule builder. A schedule is active if to starting earlier any operation, at least another one must be delayed. Active schedules are good in average and at the same time this space contains at least one optimal schedule. For these reasons it is worth to restrict the search to this space. Moreover the search space can be reduced by means of the parameter $\delta \in [0, 1]$ (see Algorithm 1, step 7). When $\delta < 1$ the search space gets narrowed so that it may contain none of the optimal schedules. At the extreme $\delta = 0$ the search is constrained to non-delay schedules: in such a schedule a resource is never idle if an operation that requires the resource is available. The experience demonstrates that as long as parameter δ decreases, in general, the mean value of solutions within the search space improves.

Algorithm 1 Hybrid G&T

---

1. Let $A = \{\theta_{j0}, 0 \leq j < N\}$;
while $A \neq \emptyset$ **do**
   2. $\forall \theta_i \in A$ let $st\theta_i$ be the lowest start time of i if scheduled at this stage;
   3. Let $\theta_1 \in A$ such that $st\theta_1 + du\theta_1 \leq st\theta + du\theta, \forall \theta \in B$;
   4. Let $R = MR(\theta_1)$; $\{MR(\theta)$ *is the machine required by operation* $\theta\}$
   5. Let $B = \{\theta \in A; MR(\theta) = R, st\theta < st\theta_1 + du\theta_1\}$
   6. Let $\theta_2 \in B$ such that $st\theta_2 \leq st\theta, \forall \theta \in B$;
     $\{$*the earliest starting time of every operation in B,if it is selected next,is a value*
     of the interval $[st\theta_2, st\theta_1 + du\theta_1[\}$
   7. Reduce the set B such that
     $B = \{\theta \in B : st\theta \leq st\theta_2 + \delta((st\theta_1 + du\theta_1) - st\theta_2), \delta \in [0,1]\}$;
     $\{$*now the interval is reduced to* $[st\theta_2, st\theta_2 + \delta((st\theta_1 + du\theta_1) - st\theta_2)]\}$
   8. Select $\theta^* \in B$ at random and schedule it at time $st\theta^*$;
   9. Let $A = A\backslash\{\theta^*\} \cup \{SUC(\theta^*)\}$;
     $\{SUC(\theta)$ *is the next operation to* θ *in its job if any exists*$\}$
end while

---

3   Codification Schemas for JSS with GAs

In this work we consider a standard GA such as the one showed in Algorithm 2, and for chromosome codification in principle we consider CP and PR schemas. In both cases a chromosome expresses a total ordering among all operations of the problem. For example, if we have a problem with $N = 3$ jobs and $M = 4$ machines, one possible ordering is given by the permutation $(\theta_{10}\ \theta_{00}\ \theta_{01}\ \theta_{20}\ \theta_{21}$ $\theta_{11}\ \theta_{21}\ \theta_{02}\ \theta_{12}\ \theta_{13}\ \theta_{03}\ \theta_{22})$, where θ_{ij} represents the operation $j, 0 \leq j < M$, of job $i, 0 \leq i < N$. In the CP schema, the operations are codified by the numbers $0, \ldots, N \times M - 1$, starting from the first job, so that the previous ordering would be codified by the chromosome $(4\ 0\ 1\ 8\ 9\ 5\ 10\ 2\ 6\ 7\ 3\ 11)$. Whereas in the PR schema an operation is codified by just its job number, so that the previous order would be given by $(1\ 0\ 0\ 2\ 2\ 1\ 2\ 0\ 1\ 1\ 0\ 2)$. PR schema was proposed by C. Bierwirth in [1]; and CP were also used by C. Bierwirth and D. Mattfeld

---

Algorithm 2 Genetic Algorithm

---

1. Generate the Initial Population;
2. Evaluate the Population
while No termination criterion is satisfied **do**
   3. Select chromosomes from the current population;
   4. Apply the Crossover and Mutation operators to the chromosomes selected at step 1. to generate new ones;
   5. Evaluate the chromosomes generated at step 4.;
   6. Apply de Acceptation criterion to the set of chromosomes selected at step 3. together with the chromosomes generated at step 4.;
end while
7. Return the best chromosome evaluated so far;

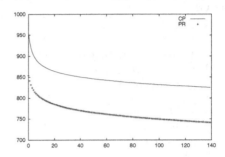

Fig. 1. Mean makespan evolution, over 140 generations, with CP and PR codifications for the problem instance ABZ8. In either case the GA was run 30 times starting from random initial populations

in [2]. In any case we chose the G&T algorithm as decoder. This only requires modify the selection criteria a step 8 (see Algorithm 1) by a deterministic one; in this case

8. Select $\theta^* \in B$ such that θ^* is the leftmost operation of B in the chromosome and schedule it at time $st\theta^*$

As genetic operators we consider generalized crossover (GOX) and mutation by swapping two consecutive operations, as described in [2]. Regarding selection and acceptation criteria we consider 2-2 tournament acceptation after organizing all chromosomes of the previous generation in pairs and apply crossover and mutation to every pair accordingly to crossover and mutation probabilities.

In [7] we have demonstrated that PR codification is better than CP. By simple experimentation it is easy to demonstrate that an initial population composed by random PRs is, in most of the cases, much better than a random population of CPs. Moreover a typical GA converges towards better solutions when PRs are used. Figure 1 shows the mean makespan convergence of the GA to solve the problem instance ABZ8 using PR and CP codifications. Furthermore we have also provided some explanation for such a behavior: a PR tends to represent *natural orders*. We explain this by means of an example. Let us consider that operations θ_{12} and θ_{20} require the same resource. As θ_{12} is the third operation of job 0 and θ_{20} is the first operation of job 2, the most natural or probable order among these two operations within an optimal (or at least a good) schedule can be considered in principle as $(\theta_{20}\ \theta_{12})$. The intuition behind this assumption is that the operation θ_{12} has to wait for at least two operations, θ_{10} and θ_{11}, to be processed, while the operation θ_{20} could be processed with no waiting at all. Now if we consider the probability that the operation θ_{20} appears before the operation θ_{12}, in a random PR this value is 0.95 whereas it is 0.5 in a random CP. In general, the probability that operation θ_{li} falls in a previous position to operation θ_{mj} in a random PR depends on the value of M and the positions i and j and is calculated by

$$P_{PR}(i,j) = (j+1) * \binom{M}{j+1} * \sum_{k=i+1}^{M} \frac{\binom{M}{k}}{\binom{2M}{k+j+1} * (k+j+1)} \tag{1}$$

whereas for random CPs we have $P_{CP}(i,j) = 0.5$. We call these probability distributions *underlying probabilistic models* of the corresponding schemas PR and CP respectively. Figure 2a shows the PR underlying probabilistic model for a problem with 15 machines.

From PR and CP codifications, in particular from their probabilistic models, the following question raises: would be it possible to improve the PR schema?, If so, how should the underlying probabilistic model be? We conjecture that the essential of a given probabilistic model is the slope of the probability curves, in particular the way the slope varies as long as j varies from 0 to $M-1$ for a given i. This way we look for schemas with probabilistic models having slopes raising lower than PR, or to the contrary slopes raising more quickly than PR. In the first case we would have an intermediate schema between CP and PR, and in the second we would have an extension of PR in the opposite direction to CP. From this rationale we propose two extensions of the CP and PR schemas. The first one is termed Partial PR (PPR) and the second is Extended PR (EPR).

PPR consists on using a number of K different numbers to codify the set of operations of each job, M being the number of machines and K being an integer number that divides to M. In PPR for a given K the operations of job 0 are represented by numbers $0, 1, \ldots, K-1$; the operations of job 1 by $K, \ldots, 2(K-1)$, and so on. As every job has M operations each number should be repeated M/K times. For example operations of job number 0 are numbered by $0, 1, \ldots, K-1, 0, 1, \ldots, K-1, \ldots$ This way the PR chromosome *(1 0 0 2 2 1 2 0 1 1 0 2)* is equivalent to the $PPR(K=2)$ chromosome *(2 0 1 4 5 3 4 0 2 3 1 5)*. As we can observe CP and PR are limit cases of PPR with $K = M$ and $K = 1$ respectively.

On the other hand EPR is an extension of PR that consists on representing each operation by P numbers, instead of just 1 as in PR, taking into account that the last of the P numbers is the one that actually represents the operation when the decoding algorithm is applied, whereas the previous $P-1$ are only for the purpose of modify the probabilistic model in the way mentioned above. For example, the PR chromosome *(1 0 0 2 2 1 2 0 1 1 0 2)* is equivalent to the EPR($P=2$) chromosome *(0 1 1 0 0 2 0 2 2 2 2 1 1 2 1 0 0 1 2 0 1 1 0 2)*.

Figures 2b and 2c show probability distributions corresponding to PPR and EPR schemas, with $K = 3$ and $P = 2$ respectively, for a problem with 15 machines. As we can observe PPR curves raise lower than PR curves, whereas EPR curves raise more quickly. In these cases the curves are calculated experimentally from a random population of 5000 chromosomes. However equations analogous to expression (2) for PR can be derived from PPR and EPR schemas as well. PPR curves are composed by steps of size K due to for every group of K consecutive operations with different numbers in a job all the relative orders among them in a random chromosome are equally probable. Moreover, for the same reason, every group of curves corresponding to K consecutive values of i, starting from $i = 0$, degenerates to the same curve.

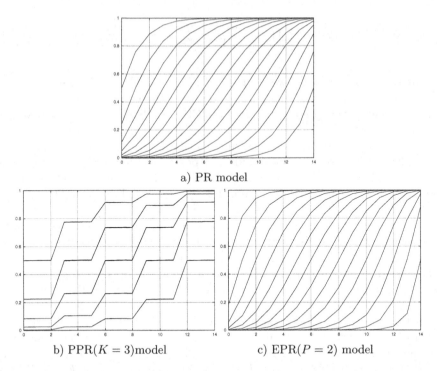

a) PR model

b) PPR($K = 3$)model   c) EPR($P = 2$) model

Fig. 2. Probability profiles for a problem with $M = 15$ for RP , PPR and EPR models. In any case each of the curves represents the values of $P(i, j)$ for a fixed i whereas j ranges from 0 to $M - 1$. The curves from up to bottom correspond to values of i ranging from 0 to $M - 1$. PR model is obtained by evaluation of expression (2), whereas PRP and EPR are obtained empirically from random populations of 5000 chromosomes

In any case it can be observed that for PPR, the larger the value of parameter K, the lower the rise of the curves. Whereas for EPR, the larger the value of P, the larger the rise. Here it is important to remark that for EPR the chromosome gets larger with increasing values of P so that some genetic operators get more time consuming.

4   Experimental Study

In this study we have experimented with a set of selected problems taken from a well-known repository: the OR library. These selected problems are recognized as hard to solve for a number of researches such as D. Mattfeld [5]. Table 1 reports results about the mean values of makespan of random initial populations regarding various codifications. As we can observe, with the only exception of FT20 instance, PR populations are much better than CP populations. Moreover PPR populations gets better and better as long as the value of parameter K augments, but in neither case PPR is better than PR. Regarding EPR, popula-

Table 1. Summary of mean values from random populations of 100 chromosomes with different codifications. In PPR a "-" means that the corresponding value of K is not applicable (it does not divide to M). Chromosomes are evaluated by means of the G&T algorithm with $\delta = 0.5$

| Problem Instance | CP | PPR | | | PR | EPR | | |
Name(size)	BS	K=5	K=3	K=2		P=2	P=3	P=4
$abz7(20 \times 15)$	665	940	876	854	-	818	801	795	**792**
$abz8(20 \times 15)$	670	957	905	882	-	854	828	819	**815**
$abz9(20 \times 15)$	686	950	915	903	-	895	883	882	**881**
$ft10(10 \times 10)$	930	1265	1242	-	1218	1200	1185	**1181**	**1181**
$ft20(20 \times 5)$	1165	**1510**	-	-	-	1531	1571	1590	1603
$la21(15 \times 10)$	1046	1432	1380	-	1364	1324	1315	1315	**1314**
$la24(15 \times 10)$	935	1321	1260	-	1216	1192	1164	1150	**1141**
$la25(15 \times 10)$	977	1400	1357	-	1277	1238	1204	1193	**1185**
$la27(20 \times 10)$	1235	1691	1624	-	1572	1555	**1541**	1542	1545
$la29(20 \times 10)$	1153	1685	1595	-	1521	1481	1444	1431	**1422**
$la38(15 \times 15)$	1196	1660	1591	1576	-	1530	1519	1512	**1509**
$la40(15 \times 15)$	1222	1675	1599	1581	-	1520	1491	1479	**1473**

Table 2. Summary of results from the GA with various codification schemas starting from random initial populations. The GA was ran 30 times for each problem instance with crossover probability 0.7, mutation probability 0.2, population size of 100 chromosomes, 200 generations and parameter $\delta = 0.5$ in decoding algorithm G&T. For each one of the codifications the mean error in percent of the best solutions reached in the 30 trials is represented

| Problem Instance | CP | PPR | | | PR | EPR | | |
Name(size)	BS	K=5	K=3	K=2		P=2	P=3	P=4
$abz7(20 \times 15)$	665	15.6	7.7	5.6	-	2.4	**1.7**	1.8	1.8
$abz8(20 \times 15)$	670	17.0	8.3	11.1	-	8.3	**7.8**	**7.8**	8.0
$abz9(20 \times 15)$	686	18.7	15.9	14.1	-	**11.5**	12.1	12.7	13.2
$ft10(10 \times 10)$	930	6.5	5.1	-	**3.7**	**3.7**	4.3	4.7	5.6
$ft20(20 \times 5)$	1165	**1.4**	-	-	-	3.2	5.7	6.8	7.4
$la21(15 \times 10)$	1046	10.0	8.5	-	5.8	**4.4**	4.6	4.7	4.9
$la24(15 \times 10)$	935	9.8	8.3	-	5.6	4.9	**4.3**	4.5	4.8
$la25(15 \times 10)$	977	8.4	6.2	-	3.7	**3.1**	3.3	4.1	4.3
$la27(20 \times 10)$	1235	12.9	10.7	-	7.0	4.9	4.8	**4.7**	4.9
$la29(20 \times 10)$	1153	15.2	12.0	-	7.3	**6.7**	7.1	7.8	8.2
$la38(15 \times 15)$	1196	10.1	7.9	7.7	-	**7.3**	9.1	10.3	10.9
$la40(15 \times 15)$	1222	9.1	7.0	5.6	-	**4.5**	**4.5**	5.2	5.6
Average	11.2	8.9	8.8	5.5	**5.4**	5.8	6.3	6.6

tions slightly improve as long as parameter P augments, but we have to take into account that in this case the chromosome size is proportional to the value of P.

In the next experiment we have run the GA with the codifications considered in previous experiment; in any case starting from random initial populations.

Table 2 reports results about the mean makespan in percent of the solutions reached by the GA calculated as

$$((Mean - Best)/Best) * 100, \tag{2}$$

Mean being the mean of the best solutions reached in the 30 trials of each version of the GA, and *Best* being the makespan of the optimal solution to the problem instance.

As we can observe the original PR schema produces the best results in average, even though PR initial populations are not the best ones, as shown in Table 1. However EPR produces the best results for 5 of the 12 problem instances. Again problem instance FT20 is an exception, in this case CP is the best schema, not only for initial population but also regarding the GA evolution.

From comparison of Tables 1 and 2 we can observe that in principle the final values of the mean error reached by the successive GA versions improves as long as the mean makespan of initial populations gets lower. However the mean error reaches a minimum around PR schema, and beyond this point the mean error values augment in spite of the fact that the initial populations have a lower mean makespan. This behavior can be explained by taking into account the degree of diversity of the population. As showed in Table 3, as the initial populations get lower values of mean makespan, these populations get also a lower diversity that translates into a premature convergence of the corresponding GA version. Therefore we could conclude that, regarding a conventional GA such as the one proposed in section 3, PR schema presents the best tradeoff between quality and diversity, and therefore in general the GA reaches better solutions when starting from random initial populations. However in a remarkable number of cases EPR performs better than PR. This fact suggest us that EPR is a suitable codification that can be considered as comparable to PR. Regarding the remaining codifications, mainly EPR($P > 2$), a control mechanism should

Table 3. Summary of standard deviation values of makespan from random populations of 100 chromosomes with different codifications

Problem Instance		CP	PPR			PR	EPR		
Name(size)	BS		K=5	K=3	K=2		P=2	P=3	P=4
$abz7(20 \times 15)$	665	54.3	39.8	33.1	-	26.7	21.8	20.3	19.6
$abz8(20 \times 15)$	670	54.3	40.2	36.8	-	30.1	24.8	22.8	21.0
$abz9(20 \times 15)$	686	46.1	35.1	31.9	-	32.7	29.1	27.1	25.5
$ft10(10 \times 10)$	930	76.5	70.6	-	64.8	59.7	51.6	46.0	44.1
$ft20(20 \times 5)$	1165	72.6	-	-	-	62.7	54.6	48.7	44.6
$la21(15 \times 10)$	1046	87.9	70.5	-	67.4	57.6	49.9	45.5	43.2
$la24(15 \times 10)$	935	98.5	75.3	-	67.3	60.0	51.1	44.7	43.2
$la25(15 \times 10)$	977	95.7	94.8	-	74.3	65.1	51.5	45.9	43.7
$la27(20 \times 10)$	1235	92.0	70.5	-	63.2	59.9	58.9	59.5	58.2
$la29(20 \times 10)$	1153	91.1	80.8	-	66.3	57.7	50.7	44.9	44.0
$la38(15 \times 15)$	1196	104.4	79.7	71.5	-	59.3	49.7	45.3	44.4
$la40(15 \times 15)$	1222	108.5	88.8	85.2	-	67.2	60.7	53.8	50.8

be introduced into the GA to maintain an acceptable degree of diversity; not only in the initial population, but also along the GA evolution. Maybe in this case EPR($P > 2$) could be competitive. At the same time PPR and CP in principle seem to be not competitive with the remaining ones due to the GA converges to much worse values in spite of the high diversity degree that these codifications produce into the initial populations. We have also conducted experiments over a number of 1000 generations with similar conclusions.

Regarding time consumption, CP, PPR and PR schemas needs approximately the same values. For example an execution in the conditions reported in Table 2, that is over 200 generations, takes about 1.9 secs. for a problem of size 10×10 and about 7.5 secs. for a problem of size 20×15. However for EPR schema the time consumption augments with the value of parameter P so that the times required for the 10×10 instance are about 2, 2.3 and 2.5 secs. for values of P of 2, 3 and 4 respectively. Whereas these times are about 8.5, 9 and 10 secs. for the 20×15 instance. The experiments were conducted on a Pentium IV processor at 1.7 Ghz. and 512 Mbytes of RAM, under Windows XP operating system.

5   Concluding Remarks

In this paper we have proposed two new codification schemas to solve scheduling problems by means of genetic algorithms: PPR and EPR. These new schemas are in fact extensions of two codifications of common use in scheduling and related problems: CP and PR. By simple experimentation we have demosntrated that PR is in general much better than CP. We have observed that populations of randomly generated PRs has a mean value of makespan much lower than random populations of CPs. Moreover the convergence of a conventional GA is much better when PR is used. In [7] we have provided an explanation to these facts: PRs tends to represent natural orders among the operations requiring the same resource. In other words, operations that are more likely to appear at the beginning of a good schedule have a larger probability of appearing at the first positions of the chromosome as well. Moreover we have formalized the explanation by means of the underlying probabilistic model of a codification. This is a probability distribution that accounts for the probability that operation at position i in its job sequence appears before than operation at position j of another job. By observation of the probabilistic models of CP and PR codifications we have worked out two extensions: EPR and PPR. EPR generalized both CP and PR and is parameterized by a value K ranging from M to 1. In principle only values of K that divides to M are considered in order to simplify the genetic operators. PPR($K = M$) is the same as CP and PPR($K = 1$) is the same as PR. On the other hand, EPR is an extension of PR but in the opposite direction to CP. EPR is parameterized as well by a value $P \geq 1$ that indicates the number of digits we use to represent an operation. EPR($P = 1$) is the same as PR and for larger values of P has the inconvenient of requiring a chromosome length proportional to P, what in practice restricts this value to small values as 2, 3 or 4.

The reported experimental results shown that the performance of the GA depends on the codification chosen. Moreover in average PR is the best schema, however for a significative number of problem instances other schemas, generally close to PR such as PPR with a small value of K or EPR with a small value of P, are better. Moreover schemas far from PR such as CP and EPR with larger values of P, are in general the worse ones. This fact suggest that it is worth to consider schemas other than PR. However as it does not seem easy to envisage a method to select in advance the best schema for a given problem instance, in principle the only way is trying various schemas at the same time and take the value provided for the best one. It would also be possible to allow the GA consider in the initial population chromosomes codified with different schemas and let the evolution process selecting the best one by itself.

References

1. Bierwirth, C.: A Generalized Permutation Approach to Jobshop Scheduling with Genetic Algorithms. OR Spectrum **17** (1995) 87-92.
2. Bierwirth, C, Mattfeld, D.: Production Scheduling and Rescheduling with Genetic Algorithms. Evolutionary Computation **7(1)** (1999) 1-17.
3. Giffler, B. Thomson, G. L.: Algorithms for Solving Production Scheduling Problems. Operations Reseach **8** (1960) 487-503.
4. Jain, A. S. and Meeran, S.: Deterministic job-shop scheduling: Past, present and future. European Journal of Operational Research **113** (1999) 390-434.
5. Mattfeld, D. C.: Evolutionary Search and the Job Shop. Investigations on Genetic Algorithms for Production Scheduling. Springer-Verlag, November 1995.
6. Varela, R., Vela, C. R., Puente, J., Gmez A.: A knowledge-based evolutionary strategy for scheduling problems with bottlenecks. European Journal of Operational Research **145** (2003) 57-71.
7. Varela, R., Puente, J. and Vela, C. R.: Some Issues in Chromosome Codification for Scheduling Problems with Genetic Algorithms. ECAI 2004, Workshop on Constraint Satisfaction Techniques for Planning and Scheduling Problems (2004) 7-16.
8. Yamada, T. and R. Nakano.: Scheduling by Genetic Local Search with multi-step crossover. Fourth Int. Conf. On Parallel Problem Solving from Nature (PPSN IV), Berlin, Germany, (1996) 960-969.

Solving the Multidimensional Knapsack Problem Using an Evolutionary Algorithm Hybridized with Branch and Bound

José E. Gallardo, Carlos Cotta, and Antonio J. Fernández

Dept. Lenguajes y Ciencias de la Computación, ETSI Informática,
University of Málaga, Campus de Teatinos, 29071 - Málaga, Spain
{pepeg, ccottap, afdez}@lcc.uma.es

Abstract. A hybridization of an evolutionary algorithm (EA) with the branch and bound method (B&B) is presented in this paper. Both techniques cooperate by exchanging information, namely lower bounds in the case of the EA, and partial promising solutions in the case of the B&B. The multidimensional knapsack problem has been chosen as a benchmark. To be precise, the algorithms have been tested on large problems instances from the OR-library. As it will be shown, the hybrid approach can provide high quality results, better than those obtained by the EA and the B&B on their own.

1 Introduction

Branch and Bound (B&B) [1] is an algorithm for finding optimal solutions to combinatorial problems. Basically, the method produces convergent lower and upper bounds for the optimal solution using an implicit enumeration scheme. The algorithm starts from the original problem, and proceeds iteratively. In each stage, the problem is split into subproblems such that the union of feasible solutions for these subproblems gives the whole set of feasible solutions for the current problem. Subproblems are further divided until they are solved, or their upper bounds are below the best feasible solution found so far (maximization is assumed here). Thus, the approach produces a branching tree in which each node corresponds to a problem and the children of the node represent the subproblems into which it is split. Several strategies can be used to traverse the search tree. The most efficient one consists of expanding more promising (according to the attainable solution, i.e., the upper bound) problems first, but memory resources may be exhausted. A depth-first expansion requires less memory, but will likely expand much more nodes than the previous strategy.

A different approach to optimization is provided by evolutionary algorithms [2, 3] (EAs). These are powerful heuristics for optimization problems based on principles of natural evolution, namely adaptation and survival of the fittest. Starting from a *population* of randomly generated *individuals* (representing solutions), a process consisting of *selection*, (promising solutions are chosen from the population) *reproduction* (new solutions are created by combining selected

J. Mira and J.R. Álvarez (Eds.): IWINAC 2005, LNCS 3562, pp. 21–30, 2005.

ones) and *replacement* (some solutions are replaced by the new ones) is repeated. A *fitness* function measuring the quality of the solution is used to guide the process.

A key aspect of EAs is robustness, meaning that they can be deployed on a wide range of problems. However, it has been shown that some kind of domain knowledge has to be incorporated into EAs for them to be competitive with other domain specific optimization techniques [4, 5, 6]. A promising approach to achieve this knowledge-augmentation is the hybridization with other (domain-specific) heuristics for the optimization problem to be solved. In this paper a hybridization of an EA with B&B is presented. This hybridization is aimed to combining their search capabilities in a synergetic way.

The remainder of the article is organized as follows: Sect. 2 presents the *multidimensional knapsack problem* (MKP) –the benchmark used to test our hybrid model– and describes both an efficient evolutionary algorithm and two different B&B implementations that have been successfully applied to solve the MKP. Then, Sect. 3 discusses related work regarding the hybridization of evolutionary algorithms and B&B models; a novel proposal for this hybridization is described here too. Subsequently, Sect. 4 shows and analyzes the empirical results obtained by the application of each of the described approaches (i.e., the EA, pure B&B models and the hybrid model) on different instances of the benchmark. Finally, Sect. 5 provides the conclusions and outlines ideas for future work.

2   The Multidimensional Knapsack Problem

Let us firstly describe the target problem, and several approaches –both meta-heuristic and exact– used for solving it.

2.1   Description of the Problem

The *Multidimensional Knapsack Problem* (MKP) is a generalization of the classical *knapsack problem*, so it is worth starting with the description of the latter. There is a knapsack with an upper weight limit b, and a collection of n items with different values p_j and weights r_j. The problem is to choose the collection of items which gives the highest total value without exceeding the weight limit of the knapsack.

In the MKP, m knapsacks with different weight limits b_i must be filled with the same items. Furthermore, these items have a different weight r_{ij} for each knapsack i. Formally, the problem can be formulated as:

$$\text{maximise} \quad \sum_{j=1}^{n} p_j x_j, \tag{1}$$

$$\text{subject to} \quad \sum_{j=1}^{n} r_{ij} x_j \leq b_i, \quad i = 1, \ldots, m, \tag{2}$$

$$x_j \in \{0, 1\}, \quad j = 1, \ldots, n. \tag{3}$$

Each of the m constraints in Eq. (2) is called a knapsack constraint, and vector $\boldsymbol{x}$ describes which objects are chosen in the solution. The problem is NP-hard [7], and can be seen as a general statement of any zero-one integer programming problem with non-negative coefficients. Many practical problems can be formulated as an instance of the MKP, for example, the capital budgeting problem, project selection and capital investment, budget control, and numerous loading problems (see e.g. [8]).

2.2 An Evolutionary Algorithm

EAs have been used in several works for solving the MKP, e.g., [9, 10, 11, 12, 13] among others. To the best of our knowledge, the EA developed by Chu and Beasley in [11] represents the state-of-the-art in solving the MKP with EAs. This particular algorithm has an additional advantage: it uses the *natural* representation of solutions, i.e., n-bit binary strings, where n is the number of items in the MKP. For this representation, a value of 0 or 1 in the j-th bit indicates the value of x_j in the solution.

Since this representation allows infeasible solutions, a repair operator is used to correct them. In order to implement this operator, a preprocessing routine is first applied to each problem to sort variables according to the decreasing order of their *pseudo-utility* ratios u_j's (the greater the ratio, the higher the chance that the corresponding variable will be set to one in the solution, see [11] for details). Then, an algorithm consisting in two phases (see Fig. 1) is applied to every solution. In the first phase, variables are examined in increasing order of u_j and set to zero if feasibility is violated. In the second phase, variables are examined in reverse order and set to one as long as feasibility is not violated. The aim of the first phase is to obtain a feasible solution, whereas the second phase seeks to improve its fitness.

By restricting the EA to search only the feasible region of the solution space, the simple fitness function $f(\boldsymbol{x}) = \sum_{j=1}^{n} p_j x_j$ can be considered.

```
1:   initialize R_i = ∑ⁿ_{j=1} r_{ij}x_j, ∀i ∈ {1, ···, n};
2:   for j = n down to 1 do      /* DROP phase */
3:       if (x_j = 1) and (∃i ∈ {1, ···, n} : R_i > b_i) then
4:           set x_j ← 0;
5:           set R_i ← R_i - r_{ij}, ∀i ∈ {1, ···, n}
6:       end if
7:   end for
8:   for j = 1 up to n do      /* ADD phase */
9:       if (x_j = 0) and (∀i ∈ {1, ···, n} : R_i + r_{ij} ≤ b_i) then
10:          set x_j ← 1;
11:          set R_i ← R_i + r_{ij}, ∀i ∈ {1, ···, n}
12:      end if
13:  end for
```

Fig. 1. Repair operator for the MKP

2.3   Branch and Bound Algorithms

Two B&B algorithms have been evaluated. The first one is a simple implementation that expands the search tree by introducing or excluding an arbitrary item in the knapsack until a complete solution is generated. When an item j is included, the lower bound for the problem is increased with the corresponding profit p_j (and the remaining available space is decreased by r_{ij} in each knapsack i), whereas the upper bound is decreased by p_j when the item is excluded. Of course, infeasible solutions are pruned during the process. The problem queue is examined in a depth-first way in order to avoid memory exhaustion when solving large problems. Although this method is very naive, it can be very efficiently implemented and may be the only one available for other problems for which no sophisticated heuristics have been developed.

The second implementation performs a standard linear programming (LP)-based tree search. The algorithm solves the linear relaxation of the current problem (that is, allowing fractional values for decision variables), and considers the problem solved if all variables have integral values in the relaxed solution. Otherwise, it expands two new problems by introducing or excluding an item with an associated fractional value (the one whose value in the LP-relaxed solution is closest to 0.5). This method is more accurate, but is not very fast when large problems are considered, as their relaxation may take some time to be solved.

3   Hybrid Models

In this section we present a hybrid model that integrates an EA with B&B. Our aim is to combine the advantages of both approaches and, at the same time, avoid (or at least minimize) their drawbacks working alone. Firstly, in the following subsection, we briefly discuss some related works existing in the literature regarding the hybridization of B&B techniques and EAs.

3.1   Related Work

Cotta *et al.* [14] used a problem-specific B&B approach for the traveling salesman problem based on 1-trees and the Lagrangean relaxation [15], and made use of an EA to provide bounds in order to guide the B&B search. More specifically, they analyzed two different approaches for the integration. In the first model, the genetic algorithm plays the role of master and the B&B is incorporated as a tool of it. The primary idea was to build a hybrid genetic operator based in the B&B philosophy. The second model proposed consisted of executing in parallel the B&B algorithm with a certain number of EAs which generate a number of solutions of different structure. The diversity provided by the independent EAs contributed to make that edges suitable to be part of the optimal solution were likely included in some individuals, and non-suitable edges were unlikely taken into account. Despite these approaches showed encouraging results, the work in [14] described only preliminary results.

Another relevant research was developed by Nagard *et al.* [16], combining a B&B tree search and an EA which was used to provide bounds for solving flowshop scheduling problems. Later, a hybrid algorithm, combining genetic algorithms and integer programming B&B approaches to solve MAX-SAT problems was described by French *et al.* [17]. This hybrid algorithm gathers information during the run of a linear programming based B&B algorithm, and uses it to build an EA population. The EA is eventually activated, and the best solution found is used to inject new nodes in the B&B search tree. The hybrid algorithm is run until the search tree is exhausted, and hence it is an exact approach. However, in some cases it can expand more nodes than the B&B alone.

More recently, Cotta and Troya [18] presented a framework for the hybridization based on using B&B as an operator embedded in the EA. This hybrid operator is used for recombination: it intelligently explores the possible children of solutions being recombined, providing the best possible outcome. The resulting hybrid metaheuristic provides better results than pure EAs in problems where a full B&B exploration is unpractical on its own.

3.2   Our Hybrid Algorithm

One way to do the integration of evolutionary techniques and B&B models is via a *direct collaboration* that consists of letting both techniques work alone in parallel (i.e., let both processes perform independently), that is, at the same level. Both processes will share the solution. There are two ways of obtaining a benefit of this parallel execution:

- The B&B can use the lower bound provided by the EA to purge the problem queue, deleting those problems whose upper bound is smaller than the one obtained by the EA.
- The B&B can inject information about more promising regions of the search space into the EA population in order to guide the EA search.

In our hybrid approach (see Fig. 2), a single solution is shared among the EA and B&B algorithms that are executed in an interleaved way. Whenever one of the algorithms finds a better approximation, it updates the solution and yields control to the other algorithm.

The hybrid algorithm starts by running the EA in order to obtain a first approximation to the solution. In this initial phase, the population is randomly initialized and the EA executed until the solution is not improved for a certain number of iterations. This approximation can be later used by the B&B algorithm to purge the problem queue. No information from the B&B algorithm is incorporated in this initial phase of the EA, in order to avoid the injection of high-valued building blocks that could affect diversity, polarizing further evolution.

Afterwards, the B&B algorithm is executed. Whenever a new solution is found, it is incorporated into the EA population (replacing the worst individual), the B&B phase is paused and the EA is run to stabilization. Periodically, pending

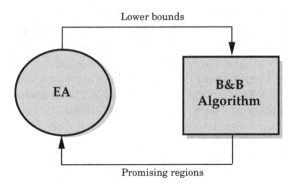

Fig. 2. The hybrid algorithm

nodes in the B&B queue are incorporated into the EA population. Since these are partial solutions and the EA population consists of full solutions, they are completed and corrected using the repair operator. The intention of this transfer is to direct the EA to these regions of the search space. Recall that the nodes in the queue represent the subset of the search space still unexplored. Hence, the EA is used for finding probably good solutions in this region. Upon finding an improved lower bound (or upon stabilization of the EA, in case no improvement is found), control is returned to the B&B, hopefully with an improved lower bound. This process is repeated until the search tree is exhausted, or a time limit is reached. The hybrid is then an anytime algorithm that provides both a quasi-optimal solution, and an indication of the maximum distance to the optimum.

4   Experimental Results

We have tested our algorithms with problems available at the OR-library [19] maintained by Beasley. We took two instances per problem set. Each problem set is characterized by a number, m, of constraints (or knapsacks), a number, n, of items and a *tightness ratio*, $0 \leq \alpha \leq 1$. The closer to 0 the tightness ratio the more constrained the instance.

We solved these problems on a Pentium IV PC (1700MHz and 256MB of main memory) using the EA, the B&B and the hybrid algorithms (all of them coded in C). A single execution for each instance was performed for the B&B method whereas ten runs were carried out for the EA and hybrid algorithms. The algorithms were run for 600 seconds in all cases. For the EA and the hybrid algorithm, the size of the population was fixed at 100 individuals that were initialized with random feasible solutions. The probability of mutation was set to 2 bits per string, recombination probability was set to 0.9, the binary tournament selection method was used, and a standard uniform crossover operator was chosen.

Table 1. Results (averaged for ten runs) of the B&B algorithm, the EA, and the hybrid thereof for problem instances of different number of items (n), knapsacks (m), and tightness ratio (α)

α	m	n	B&B	GA best	GA mean ± std.dev	B&B-GA best	B&B-GA mean ± std.dev
		100	24373	24381	24381.0 ± 0.0	24381	24381.0 ± 0.0
	5	250	59243	59243	59211.7 ± 18.0	59312	59305.1 ± 20.7
		500	120082	120095	120054.0 ± 25.1	120148	120122.0 ± 14.0
		100	23064	23064	23050.2 ± 19.2	23064	23059.1 ± 3.2
0.25	10	250	59071	59133	59068.7 ± 29.1	59164	59146.3 ± 11.6
		500	117632	117711	117627.3 ± 64.7	117741	117702.4 ± 20.5
		100	21516	21946	21856.1 ± 112.5	21946	21946.0 ± 0.0
	30	250	56277	56796	56606.9 ± 126.6	56796	56796.0 ± 0.0
		500	115154	115763	115619.9 ± 79.7	115820	115779.6 ± 18.6
		100	59960	59960	59960.0 ± 0.0	59965	59965.0 ± 0.0
	5	250	154654	154668	154626.2 ± 31.7	154668	154668.0 ± 0.0
		500	299904	299885	299842.7 ± 26.9	299904	299902.3 ± 5.1
		100	60633	60633	60629.7 ± 4.9	60633	60633.0 ± 0.0
0.75	10	250	149641	149686	149622.7 ± 39.6	149704	149685.3 ± 15.1
		500	306949	306976	306893.7 ± 56.0	307027	307002.7 ± 8.4
		100	60574	60593	60560.9 ± 32.1	60603	60603.0 ± 0.0
	30	250	149514	149514	149462.8 ± 44.4	149595	149528.6 ± 24.4
		500	300309	300351	300218.8 ± 94.5	300387	300359.0 ± 21.9

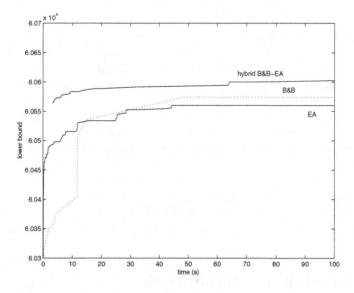

Fig. 3. Evolution of the lower bound in the three algorithms during the first 100 seconds of execution for an problem instance with $\alpha = .75$, $m = 30$, $n = 100$. Curves are averaged for the ten runs in the case of the EA and the hybrid algorithm

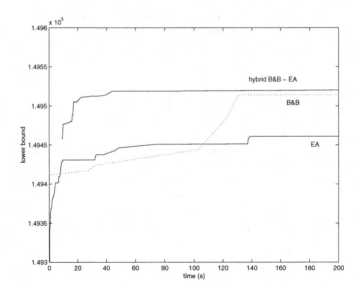

Fig. 4. Evolution of the lower bound in the three algorithms during the first 100 seconds of execution for an problem instance with $\alpha = .75$, $m = 30$, $n = 250$. Curves are averaged for the ten runs in the case of the EA and the hybrid algorithm

The results are shown in Table 1. The first three columns indicate the sizes (m and n) and the tightness ratio (α) for a particular instance. The next column reports results for the B&B algorithm, whereas the last two columns report the best and average solutions over 10 runs for the EA and the hybrid algorithm. As it can be seen, the hybrid algorithm always outperforms the original algorithms. Notice also that the difference in the mean values is notably larger than the corresponding standard deviations, thus reinforcing the significance of the results.

Figs. 3 and 4 show the on-line evolution of the lower bound for the three algorithms. Notice how the hybrid algorithm yields consistently better results all over the run. This confirms the goodness of the hybrid model as an anytime algorithm.

We are currently testing the second implementation of the hybrid algorithm that solves LP-relaxation of the problems. The preliminary results indicate that the lower bounds obtained by this algorithm are not better than the ones reported in this paper, although more accurate upper bounds can be achieved.

5   Conclusions and Future Work

We have presented a hybridization of an EA with a B&B algorithm. The EA provides lower bounds that the B&B can use to purge the problem queue, whereas the B&B guides the EA to look into promising regions of the search space.

The resulting hybrid algorithm has been tested on large instances of the MKP problem with encouraging results: the hybrid EA produces better results than the constituent algorithms at the same computational cost. This indicates the synergy of this combination, thus supporting the idea that this is a profitable approach for tackling difficult combinatorial problems. In this sense, further work will be directed to confirm these findings on different combinatorial problems, as well as to study alternative models for the hybridization of the B&B with EAs.

Acknowledgements

This work is partially supported by Spanish MCyT and FEDER under contracts TIC2002-04498-C05-02 and TIN2004-7943-C04-01.

References

1. Lawler, E., Wood, D.: Branch and bounds methods: A survey. Operations Research **4** (1966) 669–719
2. Bäck, T.: Evolutionary Algorithms in Theory and Practice. Oxford University Press, New York NY (1996)
3. Bäck, T., Fogel, D., Michalewicz, Z.: Handbook of Evolutionary Computation. Oxford University Press, New York NY (1997)
4. Davis, L.: Handbook of Genetic Algorithms. Van Nostrand Reinhold, New York NY (1991)
5. Wolpert, D., Macready, W.: No free lunch theorems for optimization. IEEE Transactions on Evolutionary Computation **1** (1997) 67–82
6. Culberson, J.: On the futility of blind search: An algorithmic view of "no free lunch". Evolutionary Computation **6** (1998) 109–128
7. Garey, M., Johnson, D.: Computers and Intractability: A Guide to the Theory of NP-Completeness. Freeman and Co., San Francisco CA (1979)
8. Salkin, H., Mathur, K.: Foundations of Integer Programming. North Holland (1989)
9. Khuri, S., Bäck, T., Heitkötter, J.: The zero/one multiple knapsack problem and genetic algorithms. In Deaton, E., Oppenheim, D., Urban, J., Berghel, H., eds.: Proceedings of the 1994 ACM Symposium on Applied Computation, ACM Press (1994) 188–193
10. Cotta, C., Troya, J.: A hybrid genetic algorithm for the 0-1 multiple knapsack problem. In Smith, G., Steele, N., Albrecht, R., eds.: Artificial Neural Nets and Genetic Algorithms 3, Wien New York, Springer-Verlag (1998) 251–255
11. Chu, P., Beasley, J.: A genetic algorithm for the multidimensional knapsack problem. Journal of Heuristics **4** (1998) 63–86
12. Gottlieb, J.: Permutation-based evolutionary algorithms for multidimensional knapsack problems. In Carroll, J., Damiani, E., Haddad, H., Oppenheim, D., eds.: ACM Symposium on Applied Computing 2000, ACM Press (2000) 408–414
13. Raidl, G., Gottlieb, J.: Empirical analysis of locality, heritability and heuristic bias in evolutionary algorithms: A case study for the multidimensional knapsack problem. Technical Report TR 186-1-04-05, Institute of Computer Graphics and Algorithms, Vienna University of Technology (2004)

14. Cotta, C., Aldana, J., Nebro, A., Troya, J.: Hybridizing genetic algorithms with branch and bound techniques for the resolution of the TSP. In Pearson, D., Steele, N., Albrecht, R., eds.: Artificial Neural Nets and Genetic Algorithms 2, Wien New York, Springer-verlag (1995) 277–280

15. Volgenant, A., Jonker, R.: A branch and bound algorithm for the symmetric traveling salesman problem based on the 1-tree relaxation. European Journal of Operational Research **9** (1982) 83–88

16. Nagard, A., Heragu, S., Haddock, J.: A combined branch and bound and genetic algorithm based for a flowshop scheduling algorithm. Annals of Operation Research **63** (1996) 397–414

17. French, A., Robinson, A., Wilson, J.: Using a hybrid genetic-algorithm/branch and bound approach to solve feasibility and optimization integer programming problems. Journal of Heuristics **7** (2001) 551–564

18. Cotta, C., Troya, J.: Embedding branch and bound within evolutionary algorithms. Applied Intelligence **18** (2003) 137–153

19. Beasley, J.: Or-library: distributing test problems by electronic mail. Journal of the Operational Research Society **41** (1990) 1069–1072

Cryptanalysis of Substitution Ciphers Using Scatter Search

Mohamed Amine Garici[1] and Habiba Drias[2]

[1] Electronics and Computer Science Faculty,
Research Laboratory in Artificial Intelligence, LRIA-USTHB,
BP 32 El-Alia Bab-Ezzouar, 16111, Algiers, Algeria
mgarici@yahoo.fr
[2] Computer Science National Institute I.N.I.,
BP 68m Oued Smar El Harrach, Algiers, Algeria
drias@wissal.dz

Abstract. This paper presents an approach for the automated cryptanalysis of substitution ciphers based on a recent evolutionary metaheuristic called Scatter Search. It is a population-based metaheuristic founded on a formulation proposed two decades ago by Fred Glover. It uses linear combinations on a population subsets to create new solutions while other evolutionary approaches like genetic algorithms resort to randomization. First, we implement the procedures of the scatter search for the cryptanalysis of substitution ciphers. This implementation can be used as a framework for solving permutation problems with scatter search. Then, we test the algorithm and show the importance of the improvement method and the contribution of subset types. Finally, we compare its performances with those of a genetic algorithm.

Keywords: automated cryptanalysis, substitution ciphers, scatter search, evolutionary approach, heuristic search, genetic algorithm, optimization problem.

1 Introduction

Simple ciphers were first used hundreds years ago. A particular interest is carried to this kind of systems because most of the modern cryptosystems use operations of the simple ciphers as their building blocks. Many ciphers have a finite key space and, hence, are vulnerable to an exhaustive key search attack. Yet, these systems remain secure from such an attack because the key space size is such that the time and resources for a search are not available. Thus, automated reasoning tools can be used to perform attack against this systems. Many researches showed that a range of modern-day cryptological problems can be attacked successfully using metaheuristic search[3].

Many automated attacks have been proposed in the literature for cryptanalysing classical ciphers. Previously, Spillman and al. [11] have published an attack on the simple substitution cipher using a genetic algorithm, Forsyth and

J. Mira and J.R. Álvarez (Eds.): IWINAC 2005, LNCS 3562, pp. 31–40, 2005.

Safavi-Naini[5] presented an attack using simulated annealing. Tabu search was also used in[2]; and recently, Russell and al.[10] used ants (ACO) to attack this ciphers.

The evolutionary population based approach, the scatter search has been introduced recently as a metaheuristic for solving complex optimization problems[8]. It is based on a formulation for integer programming developed in 1977 by Fred Glover[6] and uses linear combination of a population subset to create new solutions. It could be viewed as a bridge between taboo search and genetic algorithms[7]. It has been recently applied with success to a number of combinatorial optimization problems, for instance, the linear ordering problem[9] and the satisfiability problem (SAT)[4].

In this paper, scatter search is used to attack substitution ciphers. First, we implement the procedures of the scatter search for the cryptanalysis of this class of ciphers. Since cryptanalysis of simple ciphers is a permutation problem, then this implementation can be used as a framework for solving permutation problems with scatter search. After, we test the method and we show the importance of the improvement method and the contribution of subset types. Finally, we compare its performances with those of a genetic algorithm.

1.1 Substitution Ciphers

There are several variants of substitution ciphers, the one used here is the most general form (mono-alphabetic substitution). A detailed description of these ciphers is given in[2]. In simple substitution ciphers, each symbol in the plaintext is replaced by another symbol in the ciphertext. A substitution cipher key can be represented as a permutation of the plaintext alphabet symbols. The main propriety of this kind of ciphers is that the $n - grams$ statistics are unchanged by the encryption procedure.

2   A General Overview of the Scatter Search

Basically, the scatter search method starts with a population of good and scattered solutions. At each step, some of the best solutions are extracted from the collection to be combined and included in a set called the reference set. A new solution is then obtained as a result of applying a linear combination on the extracted solutions. The quality of the new solution is then enhanced by an improvement technique such as a local search. The final solution will be included in the reference set if it presents interesting characteristics with regards to the solution quality and dispersion.

Although it belongs to the population-based procedures family, scatter search differs mainly from genetic algorithms by its dynamic aspect that does not involve randomization at all. Scatter search allows the combination of more than two solutions, it gets thus at each step more information. By combining a large number of solutions, different sub-regions of the search space are implicated to build a solution. Besides, the reference set is modified each time a good solution is encountered and not at the combination process termination. Furthermore, since this process considers at least all pairs of solutions in the reference set, there is a practical need for keeping the cardinality of the set small ($<= 20$). The scatter search can be summarized in a concise manner as follows:

1: Generate an initial population P
2: **while** not *Stop-Condition* **do**
3:    Initialize the reference set with the solutions selected to be combined processing
4:    Generate new solutions by applying the combination process
5:    Improve new solutions quality
6:    Insert new solutions in population with respect to quality and dispersion criteria
7: **end while**

The procedure stops as in many metaheuristics, when during a small number of iterations no improvement in solutions quality is recorded or when we reach a certain number of iterations limited by physical constraints.

The fact that the mechanisms within scatter search are not restricted to a single uniform design allows the exploration of strategic possibilities that may prove effective in a particular implementation. These observations and principles lead to the following template for implementing scatter search[8]:

- A *Diversification Generation Method* to generate a collection of diverse trial solutions, using an arbitrary trial solution (or seed solution) as an input.
- An *Improvement Method* to transform a trial solution into one or more enhanced trial solutions.
- A *Reference Set Update Method* to build and maintain a *reference set* consisting of the b "best" solutions found (where the value of b is typically small, e.g., no more than 20), organized to provide efficient accessing by other parts of the method. Solutions gain membership to the reference set according to their quality or their diversity.
- A *Subset Generation Method* to operate on the reference set, to produce a subset of its solutions as a basis for creating combined solutions.
- A *Solution Combination Method* to transform a given subset of solutions produced by the subset generation method into one or more combined solution vectors.

2.1   The Reference Set

The utility of the reference set $RefSet$ consists in maintaining the b best solutions found in terms of quality or diversity, where b is an empirical parameter. $RefSet$ is partitioned into $RefSet_1$ and $RefSet_2$, where $RefSet_1$ contains the b_1 best solutions and $RefSet_2$ contains the b_2 solutions chosen to augment the diversity. The distance between two solutions is defined to measure the solutions diversity. We compute the solution that is not currently in the reference set and that maximizes the distance to all this solutions currently in this set.

2.2   The Subset Generation Method

The solution combination procedure starts by constituting subsets from the reference set that have useful properties, while avoiding the duplication of subsets previously generated. The approach for doing this, consists in constructing four different collections of subsets, with the following characteristics:

- Subset-Type 1: all 2-element subsets.
- Subset-Type 2: all 3-element subsets derived from the 2-element subsets by augmenting each 2-element subset to include the best solution not in this subset.
- Subset-Type 3: all 4-element subsets derived from the 3-element subsets by augmenting each 3-element subset to include the best solution not in this subset.
- Subset-Type 4: the subsets consisting of the best i elements, for $i = 5$ to b.

The experiments described in[1] showed that at least 80% of the solutions that were admitted to the reference set came from combinations of type-1 subsets, but this should not be interpreted as a justification for completely disregarding the use of combinations other than those from type-1 subsets.

2.3   The Solution Combination

Scatter search generates new solutions by combining solutions of $RefSet$. Specifically, the design of a combination method considers the solutions to combine and the objective function. A new solution replaces the worst one in $RefSet_1$ if its quality is better. In the negative, the distances between the new solution and the solutions in $RefSet$ are computed. If diversification is improved, the new solution replaces the element of $RefSet_2$ that has the smallest distance. Otherwise, it is discarded.

3   The Design of Scatter Search for the Cryptanalysis of Substitution Ciphers

In our case, a solution is a cipher key. A key is a permutation of the plaintext's alphabet. The alphabet's characters are ordered according to the decreasing order of their standard frequency; e.g., in English this order is $(\_, e, t, a, o, n, h, i, s, r, d, l, u, m, w, g, y, c, f, b, p, k, v, x, j, q, z)$. The reason for this ordering will become apparent when generation method and combination method are presented.

Before implementing scatter search's methods, we must define two basic notions of the scatter search: how to estimate solution's fitness and distance between two given solutions. This last measure is a typical characteristic of the scatter search.

3.1   The Fitness Function

To estimate the fitness of a given solution, this solution is used to decrypt the intercepted ciphertext, then we calculate the difference between $n-gram$ statistics of the decrypted text with those of the language assumed known.

$$f(K) = \sum_{j=1}^{MaxNgram} \left(\alpha_j \cdot \sum_{i_1,...,i_j \in A} |P(i_1,...,i_j) - C(i_1,...,i_j)| \right) \qquad (1)$$

where:

- A: The plaintext's alphabet.
- α_j : constants which allow assigning of different weights to each $n-gram$, and $\sum \alpha_j = 1$.
- $P(i_1,...,i_n)$: standard frequency of the $n-gram$ $(i_1,...,i_n)$.
- $C(i_1,...,i_n)$: frequency of the $n-gram$ $(i_1,...,i_n)$ in the decrypted message.

All attacks on classical ciphers use at most $MaxNgram = 3$, i.e. the $n-grams$ are restricted to unigrams, bigrams and trigrams. Function (1) provides an estimation for the distance between frequencies of decrypted text's $n-grams$ and frequencies of the plaintext's language, many keys can provide the optimum value, or the authentic key don't give the optimum value. For this, we'll use another heuristic called '*Word*', which is more time-consuming. It estimates the number of correct words in the decrypted text.

$$Word(P) = \frac{1}{L} \sum_{M_P^l \in P} \left(RecognizedWord(M_P^l) \right) \qquad (2)$$

$$RecognizedWord(M_P^l) = \begin{cases} 1; \text{if } M_P^l \in \text{Dictionary} \\ 0; \text{else} \end{cases} \qquad (3)$$

where:

- M_P^l : a word belonging to the text P whose length is l.
- L : length of the text P.

Formula (2) estimates the ratio of the sum of recognized word's lengths on the total text's length. The use of this function is restricted to evaluate solutions newly inserted in *RefSet* at each iteration, and therefore to stop the search if a suitable value is reached. Formula (1) will be used when an evaluation of the solution's quality is required in the scatter search's methods.

3.2   The Distance Measure

The way to evaluate the distance between two solutions is an important element of the scatter search, because the diversification aspect is essentially based on this measure. We defined the distance between two given solutions $p = (p_1, \ldots, p_n)$ and $q = (q_1, \ldots, q_n)$ as follow:

$$d(p,q) = \text{number of permutations of } p_i \text{ and } p_{i+1} \text{ to obtain } p = q \qquad (4)$$

For example: the distance between $s_1 = (1, 2, 3, 4)$ and $s_2 = (2, 1, 4, 3)$ is 2.
$s_2 = (\underline{2, 1}, 4, 3) \longrightarrow (1, 2, \underline{4, 3}) \longrightarrow s_1 = (1, 2, 3, 4) \Longrightarrow d(s_1, s_2) = 2$

3.3   The Improvement Method

This method consists of a simple local search procedure exploring the solution's neighbourhood. In this context, a neighbouring solution is a solution obtained by permuting two neighbouring elements of the current solution. The research of the best improvement for all solutions is very expensive, therefore this implementation is restricted to explore the first improvement and stops when a local optimum is found or after a fixed number of iterations.

3.4   The Diversification Generation Method

Our method generates the initial population by two different approaches, where each one generates a part of the population, all generated solutions are improved by the previous method before being inserted in the initial set P.

- The first generator uses an existing solution of good quality (seed solution) and browses its neighbourhood (solutions being to a small distance of this solution but with avoiding the immediate neighbours since those will be browsed by the improvement method). The seed solution can be reached by a previous resolution tentative, or according to a heuristic like this one: order characters according to the decreasing order of their apparition frequencies in the ciphertext, it permits to minimize the difference of unigrams frequencies.
- The second generator employs controlled randomised process drawing upon frequency memory to generate a set of diverse solutions.

3.5   The Solution Combination Method

This method –like the improvement method– is a problem-specific mechanism, since it is directly related to the solution representation. The adopted method uses a vote mechanism, it browses each solution to combine in a left to right direction, and the new solution is constructed element by element: at each step the vote mechanism determines the following element to add. The number of voices granted by a solution to its element depends on the position of this element in this solution. For example, An element being in the first position of a solution, and after 3 iterations not appearing again in the constructed solution will receive 3 voices of its solution.

- $Vote$: contains the vote scores.
- F : contains for each element, the maximal fitness value obtained by its solutions.
- $OldElement$: contains for each solution, a list of its not elected elements.
- $(a_1, \ldots, a_N)$: the alphabet.
- $S_j[i]$: denotes the ith element of the jth solution.

Algorithm 1. *CombinationMethod* $(S_1, ..., S_p :$ Solution) : Solution;

1: **for** $i = 1$ to N **do**
2:    **for** $j = 1$ to p **do**
3:       **if** $S_j[i] \notin \{S_{New}[1], ..., S_{New}[i-1]\}$ **then**
4:          Find k such $S_j[i] = a_k$; { $S_j[i]$ is the kth alphabet character}
5:          $Vote[k] + +$;
6:          **if** $f(S_j) > F[k]$ **then**
7:             $F[k] = f(S_j)$;
8:          **end if**
9:       **end if**
10:    **end for**
11:    find m such $Vote[m] = \underset{j=1}{\overset{N}{Max}}(Vote[j])$ and $F[m] = \underset{j=1}{\overset{N}{Max}}(F[j])$;
12:    $S_{New}[i] = a_m$;
13:    $Vote[m] = 0$; $F[m] = 0$; { //don't consider this element at the next votes}
14:    **for** $j = 1$ to p **do**
15:       Delete a_m from the list $OldElement[j]$;
16:       **for** (each element c in $OldElement[j]$) **do** $Vote[k] + +$; \ $c = a_k$
17:       **if** $S_j[i] \notin \{S_{New}[1], ..., S_{New}[i]\}$ **then**
18:          Add $S_j[i]$ in the list $OldElement[j]$;
19:       **end if**
20:    **end for**
21: **end for**
22: **Return** (S_{New}) :

4   Experiments

The cryptanalysis procedure has been implemented in Pascal on a personal computer. First, numerical tests were carried out to set the parameters of the scatter search algorithm. In a second steps, we performed experiments in order to evaluate method's performances and to compare them with those of a genetic algorithm. The used plaintexts become from various texts (articles, classics) chosen at random and of total size adjoining 10 Millions of characters. Standard frequencies have been calculated from these texts.

4.1   Setting the Algorithm's Parameters

The Effectiveness of the n-Grams Types. The aim of this experiment is to evaluate the effectiveness of each one of the $n - grams$ types. We evaluated the average number of key elements correct with varying values of the constants α_j in equation (1). We applied the following restriction to the n-gram's weights followed in order to keep the number of combinations of the constants α_1, a_2 and α_3 workable.

- $\alpha_1, \alpha_2, \alpha_3 \in \{0, 0.1, 0.2, 0.3, 0.4, 0.5, 0.6, 0.7, 0.8, 0.9, 1\}$
- $\alpha_1 + \alpha_2 + \alpha_3 = 1$

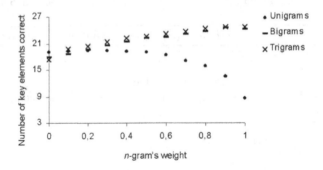

Fig. 1. The search's results with varying weights of $n - grams$

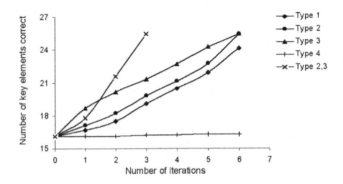

Fig. 2. Number of necessary iterations to find the best solution for each subset type

Figure 1 shows that a fitness function using bigrams or trigrams have better results than the one using unigrams, with a small advantage for the trigrams. But the profit obtained of the grams doesn't compensate the necessary resources for their use. A similar result is found in[2].

Contribution of the Subset's Types. In this experiment(Figure 2), we determine types of subsets that contribute best in the generation of reference solutions, and thereafter, to eliminate types of subsets that seem inert. Figure 2 shows that most reference solutions are generated by the combination of solutions of subsets of type 2 and type 3, contrary to most of the scatter search's implementations for other problems[1]. This difference can be explained by the combination method mechanism: high quality solutions are often close, and in most cases the combination of those solutions returns one of the input solutions. But when combination method have 3 or 4 input solutions, it works better and returns new solutions of high quality. It's clear that using all subset types will give the best quality, but using only type 2 and 3 will give almost the same quality and reduce significantly necessary time for the execution.

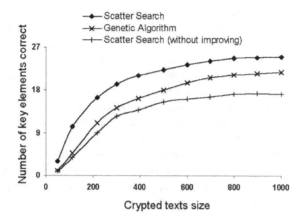

Fig. 3. A comparison between scatter search (with and without improving) and a genetic algorithm

4.2   Computational Results

In this section, we present results concerning performances of the cryptanalysis procedure and a comparison with a genetic algorithm. For the genetic algorithm implementation, we used parameters values presented in [11]: population size = 1000 and mutation rate = 0,1. The crossover and mutation operators are similar to those presented in [2].

Figure 3 shows clearly that scatter search returns solutions of better quality than the genetic algorithm (approximately 15%). The good performance of the scatter search procedure is -for the most part- the result of the improvement method which permits to explore better the neighbourhood of every considered solution. But in return, it needs more time to converge to the returned solution (approximately 75%).

To clearly show the importance of the improvement method; we remade the previous experimentation, but without using the improvement method for the research procedure. The curve (fig.3) shows that solutions returned by the scatter search (without improving method) are of very lower quality than those returned previously.The elimination of the improvement method from the scatter search decreased the rate of correct elements, because the scatter search mechanism is in this case equivalent to the genetic algorithm's one; but the fact that scatter search operates on a small population sees to it that the set of references converges prematurely and often toward middle quality solutions.

5   Conclusion

In this paper, scatter search is used to perform an automated attack against classical ciphers. First, we presented an implementation of the scatter search's procedures for the cryptanalysis of this class of ciphers. Since cryptanalysing simple ciphers is a permutation problem, then this implementation can be used as a framework for solving permutation problems with scatter search. Then, we performed tests, and the algorithm gave good results. We showed the contribution of each subset type and stressed the difference of contribution of certain subset types between this problem and other problems. We showed also that the robustness of the algorithm relies essentially on the improvement method.

It is clear that the heuristic methods have an important role to play in cryptanalysis. The next step is the cryptanalysis of more complex systems which use bit as an encoding unity. In this case, it's impossible to perform attack based on linguistic characteristics or frequencies analysis, but a 'known plaintext' attack or a 'chosen plaintext attack' are used. Another perspective is to use the metaheuristic methods in conjunction with classical cryptanalysis techniques (like the differential analysis) to improve their efficacy

References

1. Campos V., Glover F., Laguna M., Martí R.: An Experimental Evaluation of a Scatter Search for the Linear Ordering Problem. *Journal of Global Optimization*, vol. 21, 2001.
2. Clark A. J.: *Optimisation Heuristics for Cryptology*. PhD Thesis, Queensland University of Technology, 1998.
3. Clark J. A.: *Metaheuristic Search as a Cryptological Tool*. PhD Thesis, University of York, 2001.
4. Drias H., Azi N.: Scatter search for SAT and MAX-W-SAT problems. *Proceedings of the IEEE SSST*, Ohio, USA, 2001..
5. Forsyth W. S., Safavi-Naini R.: Automated cryptanalysis of substitution ciphers. *Cryptologia*, n°17, vol. 4, 1991.
6. Glover F.: Heuristics for Integer Programming Using Surrogate Constraints. *Decision Sciences*, vol 8, n° 1, 1977.
7. Glover F., Kelly J.P., Laguna M.: Genetic Algorithms and Taboo Search: Hybrids for Optimization. *Computers and Operation Reseach*, vol 22, n° 1, 1995.
8. Glover F.: A Template for Scatter Search and Path Relinking. *Notes in Computer Sciences*, Spinger-Verlag, 1998.
9. Laguna M., Armentano V.: Lessons from Applying and Experimenting with Scatter Search. Adaptive Memory and Evolution: Tabu Search and Scatter Search, *Cesar Rego and Bahram Alidaee Editions*, 2003.
10. Russell M., Clark J. A., Stepney S.: Using Ants to Attack a Classical Cipher. *Proceedings of the Genetic and Evolutionary Computation Conference (GECCO)*, 2003.
11. Spillman R., Janssen M., Nelson B., Kepner M.: Use of a genetic algorithm in the cryptanalysis of simple substitution ciphers. *Cryptologia*, n°17, vol. 1, 1993.

Combining Metaheuristics and Exact Algorithms in Combinatorial Optimization: A Survey and Classification

Jakob Puchinger and Günther R. Raidl

Institute of Computer Graphics and Algorithms,
Vienna University of Technology, Vienna, Austria
{puchinger, raidl}@ads.tuwien.ac.at

Abstract. In this survey we discuss different state-of-the-art approaches of combining exact algorithms and metaheuristics to solve combinatorial optimization problems. Some of these hybrids mainly aim at providing optimal solutions in shorter time, while others primarily focus on getting better heuristic solutions. The two main categories in which we divide the approaches are collaborative versus integrative combinations. We further classify the different techniques in a hierarchical way. Altogether, the surveyed work on combinations of exact algorithms and metaheuristics documents the usefulness and strong potential of this research direction.

1 Introduction

Hard combinatorial optimization problems (COPs) appear in a multitude of real-world applications, such as routing, assignment, scheduling, cutting and packing, network design, protein alignment, and many other fields of utmost economic, industrial and scientific importance. The available techniques for COPs can roughly be classified into two main categories: *exact* and *heuristic* methods. Exact algorithms are guaranteed to find an optimal solution and to prove its optimality for every instance of a COP. The run-time, however, often increases dramatically with the instance size, and often only small or moderately-sized instances can be practically solved to provable optimality. In this case, the only possibility for larger instances is to trade optimality for run-time, yielding heuristic algorithms. In other words, the guarantee of finding optimal solutions is sacrificed for the sake of getting good solutions in a limited time.

Two independent heterogeneous streams, coming from very different scientific communities, had significant success in solving COPs:

- *Integer Programming (IP)* as an exact approach, coming from the operations research community and based on the concepts of linear programming [11].
- Local search with various extensions and independently developed variants, in the following called *metaheuristics*, as a heuristic approach.

Among the exact methods are branch-and-bound (B&B), dynamic programming, Lagrangian relaxation based methods, and linear and integer programming

J. Mira and J.R. Álvarez (Eds.): IWINAC 2005, LNCS 3562, pp. 41–53, 2005.
© Springer-Verlag Berlin Heidelberg 2005

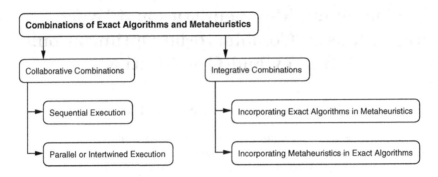

Fig. 1. Major classification of exact/metaheuristic combinations

based methods, such as branch-and-cut, branch-and-price, and branch-and-cut-and-price [30].

Metaheuristics include, among others, simulated annealing [21], tabu search [18], iterated local search [26], variable neighborhood search [20], and various population-based models such as evolutionary algorithms [3], scatter search [19], memetic algorithms [28], and various estimation of distribution algorithms [24].

Recently there have been very different attempts to combine ideas and methods from these two scientific streams. Dumitrescu and Stützle [13] describe existing combinations, focusing on local search approaches that are strengthened by the use of exact algorithms. In their survey they concentrate on integration and exclude obvious combinations such as preprocessing.

Here, we present a more general classification of existing approaches combining exact and metaheuristic algorithms for combinatorial optimization. We distinguish the following two main categories:

– *Collaborative Combinations*: By collaboration we mean that the algorithms exchange information, but are not part of each other. Exact and heuristic algorithms may be executed sequentially, intertwined or in parallel.
– *Integrative Combinations*: By integration we mean that one technique is a subordinate embedded component of another technique. Thus, there is a distinguished master algorithm, which can be either an exact or a metaheuristic algorithm, and at least one integrated slave.

In the following sections this classification is further refined and examples from the literature are presented, reflecting the current state-of-the-art. Figure 1 gives an overview of this classification.

2   Collaborative Combinations

The different algorithms and approaches described in this section have in common that they are top-level combinations of metaheuristics and exact techniques; no algorithm is contained in another. We further distinguish whether the algorithms are executed sequentially or in an intertwined or even parallel way.

2.1   Sequential Execution

Either the exact method is executed as a kind of preprocessing before the metaheuristic, or vice-versa. Sometimes, it is difficult to say if the first technique is used as initialization of the second, or if the second is a postprocessing of the solution(s) generated by the first.

Clements et al. [7] propose a column generation approach in order to solve a production-line scheduling problem. Each feasible solution of the problem consists of a line-schedule for each production line. First, the squeaky wheel optimization (SWO) heuristic is used to generate feasible solutions to the problem. SWO is a heuristic using a greedy algorithm to construct a solution, which is then analyzed in order to find the problematic elements. Higher priorities, such that these elements are considered earlier by the greedy algorithm, are assigned to them, and the process restarts until a termination condition is reached. SWO is called several times in a randomized way in order to generate a set of diverse solutions. In the second phase, the line-schedules contained in these solutions are used as columns of a set partitioning formulation for the problem, which is solved using MINTO[1]. This process always provides a solution which is at least as good as, but usually better than the best solution devised by SWO. Reported results indicate that SWO performs better than a tabu-search algorithm.

Applegate et al. [2] propose an approach for finding near-optimal solutions to the traveling salesman problem. They derive a set of diverse solutions by multiple runs of an iterated local search algorithm. The edge-sets of these solutions are merged and the traveling salesman problem is finally solved to optimality on this strongly restricted graph. In this way a solution is achieved that is typically superior to the best solution of the iterated local search.

Klau et al. [22] follow a similar idea and combine a memetic algorithm with integer programming to heuristically solve the prize-collecting Steiner tree problem. The proposed algorithmic framework consists of three parts: extensive preprocessing, a memetic algorithm, and an exact branch-and-cut algorithm applied as post-optimization procedure to the merged final solutions of the memetic algorithm.

Plateau et al. [31] combine interior point methods and metaheuristics for solving the multiconstrained knapsack problem. The first part is an interior point method with early termination. By rounding and applying several different ascent heuristics, a population of different feasible candidate solutions is generated. This set of solutions is then used as initial population for a path-relinking (scatter search) algorithm. Extensive computational experiments are performed on standard multiconstrained knapsack benchmark instances. Obtained results show that the presented combination is a promising research direction.

---

[1] http://www.isye.gatech.edu/faculty/Martin_Savelsbergh/software

Sometimes, a relaxation of the original problem is solved to optimality and the obtained solution is repaired to act as a promising starting point for a subsequent metaheuristic. Often, the linear programming (LP) relaxation is used for this purpose, and only a simple rounding scheme is needed. For example, Feltl and Raidl [36] solve the generalized assignment problem using a hybrid genetic algorithm (GA). The LP-relaxation of the problem is solved using CPLEX[2] and its solution is used by a randomized rounding procedure to create a population of promising integral solutions. These solutions are, however, often infeasible; therefore, randomized repair and improvement operators are additionally applied, yielding an even more meaningful initial population for the GA. Reported computational experiments suggest that this type of LP-based initialization is effective.

Vasquez and Hao [43] heuristically solve the multiconstrained knapsack problem by reducing and partitioning the search space via additional constraints that fix the total number of items to be packed. The bounds for these constraints are calculated by solving a modified LP-relaxation of the multiconstrained knapsack problem. For each remaining part of the search space, parallel tabu-search is finally performed starting with a solution derived from the LP-relaxation of the partial problem. This hybrid algorithm yields excellent results also for large benchmark instances with up to $2\,500$ items and 100 constraints.

Lin et al. [25] describe an exact algorithm for generating the minimal set of affine functions that describes the value function of the finite horizon partially observed Markov decision process. In the first step a GA is used to generate a set Γ of witness points, which is as large as possible. In the second step a component-wise domination procedure is performed in order to eliminate redundant points in Γ. The set generated so far does, in general, not fully describe the value function. Therefore, a Mixed Integer Program (MIP) is solved to generate the missing points in the final third step of the algorithm. Reported results indicate that this approach requires less time than some other numerical procedures.

Another kind of sequential combination of B&B and a GA is described by Nagar et al. [29] for a two-machine flowshop scheduling problem in which solution candidates are represented as permutations of jobs. Prior to running the GA B&B is executed down to a predetermined depth k and suitable bounds are calculated and recorded at each node of the explicitly stored B&B tree. During the execution of the GA the partial solutions up to position k are mapped onto the correct tree node. If the bounds indicate that no path below this node can lead to an optimal solution, the permutation is subjected to a mutation operator that has been specifically designed to change the early part of the permutation in a favorable way.

Tamura et al. [40] tackle a job-shop scheduling problem and start from its IP formulation. For each variable, they take the range of possible values and partition it into a set of subranges, which are then indexed. The chromosomes

---

[2] http://www.ilog.com

of the GA are defined so that each position represents a variable, and its value corresponds to the index of one of the subranges. The fitness of a chromosome is calculated using Lagrangian relaxation to obtain a bound on the optimal solution subject to the constraints that the values of the variables fall within the correct ranges. When the GA terminates, an exhaustive search of the region identified as the most promising is carried out to produce the final solution.

2.2   Parallel or Intertwined Execution

Instead of a strictly sequential batch approach, exact and heuristic algorithms may also be executed in a parallel or intertwined way. Such peer-to-peer combinations of exact/heuristic techniques are less frequent. An interesting framework for this purpose was proposed by Talukdar et al. [38, 39] with the so-called *asynchronous teams* (A-Teams). An A-Team is a problem solving architecture consisting of a collection of agents and memories connected into a strongly cyclic directed network. Each of these agents is an optimization algorithm and can work on the target problem, on a relaxation—i.e., a superclass—of it, or on a subclass of the problem. The basic idea of A-Teams is having these agents work asynchronously and autonomously on a set of shared memories. These shared memories consist of trial solutions for some problem (the target problem, a superclass, or a subclass as mentioned before), and the action of an agent consists of modifying the memory by adding a solution, deleting a solution, or altering a solution. A-Teams have been successfully utilized in a variety of combinatorial optimization problems, see e.g. [5, 39].

Denzinger and Offerman [12] present a similar multi-agent based approach for achieving cooperation between search-systems with different search paradigms. The TECHS (TEams for Cooperative Heterogenous Search) approach consists of teams of one or more agents using the same search paradigm. The communication between the agents is controlled by so-called send- and receive-referees, in order to filter the exchanged data. Each agent is in a cycle between searching and processing received information. In order to demonstrate the usefulness of TECHS, a GA and a B&B based system for job-shop scheduling is described. The GA and B&B agents exchange only positive information (solutions), whereas the B&B agents can also exchange negative information (closed subtrees). Computational experiments show that the cooperation results in finding better solutions given a fixed time-limit and in finding solutions comparable to the ones of the best individual system alone in less time.

3   Integrative Combinations

In this section we discuss approaches of combining exact algorithms and metaheuristics in an integrative way such that one technique is a subordinate embedded component of another technique.

3.1   Incorporating Exact Algorithms in Metaheuristics

We start by considering techniques where exact algorithms are incorporated into metaheuristics.

Exactly Solving Relaxed Problems. The usefulness of solutions to relaxations of an original problem has already been mentioned in Section 2.1. Besides exploiting them to derive promising initial solutions for a subsequent algorithm, they can be of great benefit for heuristically guiding neighborhood search, recombination, mutation, repair and/or local improvement. Examples where the solution of the LP-relaxation and its dual were exploited in such ways are the hybrid genetic algorithms for the multiconstrained knapsack problem from Chu and Beasley [6] and Raidl [35].

Exactly Searching Large Neighborhoods. A common approach is to search neighborhoods in local search based metaheuristics by means of exact algorithms. If the neighborhoods are chosen appropriately, they can be relatively large and nevertheless an efficient search for the best neighbor is still reasonable. Such techniques are known as Very Large-Scale Neighborhood (VLSN) search [1].

Burke et al. [4] present an effective local and variable neighborhood search heuristic for the asymmetric traveling salesman problem in which they have embedded an exact algorithm in the local search part, called HyperOpt, in order to exhaustively search relatively large promising regions of the solution space. Moreover, they propose a hybrid of HyperOpt and 3-opt which allows to benefit from the advantages of both approaches and gain better tours overall. Using this hybrid within the variable neighborhood search metaheuristic framework also allows to overcome local optima and to create tours of high quality.

Dynasearch [8] is another example where exponentially large neighborhoods are explored. The neighborhood where the search is performed consists of all possible combinations of mutually independent simple search steps and one Dynasearch move consists of a set of independent moves that are executed in parallel in a single local search iteration. Independence in the context of Dynasearch means that the individual moves do not interfere with each other; in this case, dynamic programming can be used to find the best combination of independent moves. Dynasearch is restricted to problems where the single search steps are independent, and it has so far only been applied to problems, where solutions are represented as permutations.

For the class of partitioning problems, Thompson et al. [41, 42] defined the concept of a cyclic exchange neighborhood, which is the transfer of single elements between several subsets in a cyclic manner; for example, a 2–exchange move can be seen as a cyclic exchange of length two. Thompson et al. showed that for any current solution to a partitioning problem a new, edge-weighted graph can be constructed, where the set of nodes is split into subsets according to a partition induced by the current solution of the partitioning problem. A cyclic exchange for the original problem corresponds to a cycle in this new graph that uses at most one node of each subset. Exact and heuristic methods that solve the problem of finding the most negative-cost subset-disjoint cycle (which corresponds to the best improving neighbor of the current solution) have been developed.

Puchinger et al. [34] describe a combined GA/B&B approach for solving a real-world glass cutting problem. The GA uses an order-based representation, which is decoded using a greedy heuristic. The B&B algorithm is applied with a certain probability enhancing the decoding phase by generating locally optimal subpatterns. Reported results indicate that the approach of occasionally solving subpatterns to optimality may increase the overall solution quality.

The work of Klau et al. [22] has already been mentioned in Section 2.1 in the context of collaborative sequential combinations. When looking at the memetic algorithm we encounter another kind of exact/heuristic algorithm combination. An exact subroutine for the price-collecting Steiner tree problem on trees is used to locally improve candidate solutions.

Merging Solutions. Subspaces defined by the merged attributes of two or more solutions can, like the neighborhoods of single solutions, also be searched by exact techniques. The algorithms by Clements et al. [7], Applegate et al. [2], and Klau et al. [22], which were already discussed in Section 2.1, also follow this idea, but are of sequential collaborative nature. Here, we consider approaches where merging is iteratively applied within a metaheuristic.

Cotta and Troya [9] present a framework for hybridizing B&B with evolutionary algorithms. B&B is used as an operator embedded in the evolutionary algorithm. The authors recall the necessary theoretical concepts on forma analysis (formae are generalized schemata), such as the dynastic potential of two chromosomes x and y, which is the set of individuals that only carry information contained in x and y. Based on these concepts the idea of dynastically optimal recombination is developed. This results in an operator exploring the potential of the recombined solutions using B&B, providing the best possible combination of the ancestors' features that can be attained without introducing implicit mutation. Extensive computational experiments on different benchmark sets comparing different crossover operators with the new hybrid one show the usefulness of the presented approach.

Marino et al. [27] present an approach where a GA is combined with an exact method for the Linear Assignment Problem (LAP) to solve the graph coloring problem. The LAP algorithm is incorporated into the crossover operator and generates the optimal permutation of colors within a cluster of nodes, hereby preventing the offspring to be less fit than its parents. The algorithm does not outperform other approaches, but provides comparable results. The main conclusion is that solving the LAP in the crossover operator strongly improves the performance of the GA compared to the GA using crossover without LAP.

Exact Algorithms as Decoders. In evolutionary algorithms, candidate solutions are sometimes only incompletely represented in the chromosome, and an exact algorithm is used as decoder for determining the missing parts in an optimal way.

Staggemeier et al. [37], for example, present a hybrid genetic algorithm to solve a lot-sizing and scheduling problem minimizing inventory and backlog costs of multiple products on parallel machines. Solutions are represented as product subsets for each machine at each period. Corresponding optimal lot sizes are determined when the solution is decoded by solving a linear program. The approach outperforms a MIP formulation of the problem solved using CPLEX.

3.2   Incorporating Metaheuristics in Exact Algorithms

We now turn to techniques where metaheuristics are embedded within exact algorithms.

Metaheuristics for Obtaining Incumbent Solutions and Bounds. In general, heuristics and metaheuristics are often used to determine bounds and incumbent solutions in B&B approaches. For example, Woodruff [44] describes a chunking-based selection strategy to decide at each node of the B&B tree whether or not reactive tabu search is called in order to eventually find a better incumbent solution. The chunking-based strategy measures a distance between the current node and nodes already explored by the metaheuristic in order to bias the selection toward distant points. Reported computational results indicate that adding the metaheuristic improves the B&B performance.

Metaheuristics for Column and Cut Generation. In branch-and-cut and branch-and-price algorithms, the dynamic separation of cutting-planes and the pricing of columns, respectively, is sometimes done by means of heuristics including metaheuristics in order to speed up the whole optimization process.

Filho and Lorena [14] apply a heuristic column generation approach to graph coloring. They describe the principles of their constructive genetic algorithm and give a column generation formulation of the problem. The GA is used to generate the initial columns and to solve the slave problem (the weighted maximum independent set problem) at every iteration. Column generation is performed as long as the GA finds columns with negative reduced costs. The master problem is solved using CPLEX. Some encouraging results are presented.

Puchinger and Raidl [32, 33] propose new integer linear programming formulations for the three-stage two-dimensional bin packing problem. Based on these formulations, a branch-and-price algorithm was developed in which fast column generation is performed by applying a hierarchy of four methods: (a) a greedy heuristic, (b) an evolutionary algorithm, (c) solving a restricted form of the pricing problem using CPLEX, and finally (d) solving the complete pricing problem using CPLEX. Computational experiments on standard benchmark instances document the benefits of the new approach. The combination of all four pricing algorithms in the proposed branch-and-price framework yields the best results in terms of the average objective value, the average run-time, and the number of instances solved to proven optimality.

Metaheuristics for Strategic Guidance of Exact Search. French et al. [16] present a GA/B&B hybrid to solve feasibility and optimization IP problems. Their hybrid algorithm combines the generic B&B of the MIP-solver XPRESS-MP[3] with a steady-state GA. It starts by traversing the B&B tree. During this phase, information from nodes is collected in order to suggest chromosomes to be added to the originally randomly initialized GA-population. When a certain criterion is fulfilled, the GA is started using the augmented initial population. When the GA terminates, its fittest solution is passed back and grafted onto the B&B tree. Full control is given back to the B&B-engine, after the newly added nodes were examined to a certain degree. Reported results on MAX-SAT instances show that this hybrid approach yields better solutions than B&B or the GA alone.

Kotsikas and Fragakis [23] determine improved node selection strategies within B&B for solving MIPs by using genetic programming (GP). After running B&B for a certain amount of time, information is collected from the B&B tree and used as training set for GP, which is performed to find a node selection strategy more appropriate for the specific problem at hand. The following second B&B phase then uses this new node selection strategy. Reported results show that this approach has potential, but needs to be enhanced in order to be able to compete with today's state-of-the-art node selection strategies.

Applying the Spirit of Metaheuristics. Last but not least, there are a few approaches where it is tried to bring the spirit of local search based techniques into B&B. The main idea is to first search some neighborhood of incumbent solutions more intensively before turning to a classical node selection strategy. However, there is no explicit metaheuristic, but B&B itself is used for doing the local search. The metaheuristic may also be seen to be executed in a "virtual" way.

Fischetti and Lodi [15] introduced local branching, an exact approach combining the spirit of local search metaheuristics with a generic MIP-solver (CPLEX). They consider general MIPs with 0-1 variables. The idea is to iteratively solve a local subproblem corresponding to a classical k-OPT neighborhood using the MIP-solver. This is achieved by introducing a local branching constraint based on an incumbent solution $\bar{x}$, which partitions the search space into the k-OPT neighborhood and the rest: $\Delta(x, \bar{x}) \leq k$ and $\Delta(x, \bar{x}) \geq k + 1$, respectively, with Δ being the Hamming distance of the 0-1 variables. The first subproblem is solved, and if an improved solution could be found, a new subproblem is devised and solved; this is repeated as long as an improved solution is found. If the process stops, the rest of the problem is solved in a standard way. This basic mechanism is extended by introducing time limits, automatically modifying the neighborhood size k and adding diversification strategies in order to improve the performance. Reported results are promising.

Danna et al. [10] present an approach called Relaxation Induced Neighborhood Search (RINS) in order to explore the neighborhoods of promising MIP

---

[3] http://www.dashoptimization.com/

solutions more intensively. The main idea is to occasionally devise a sub-MIP at a node of the B&B tree that corresponds to a certain neighborhood of an incumbent solution: First, variables having the same values in the incumbent and in the current solution of the LP-relaxation are fixed. Second, an objective cutoff based on the objective value of the incumbent is set. Third, a sub-MIP is solved on the remaining variables. The time for solving this sub-MIP is limited. If a better incumbent could be found during this process, it is passed to the global MIP-search which is resumed after the sub-MIP termination. CPLEX is used as MIP-solver. The authors experimentally compare RINS to standard CPLEX, local branching, combinations of RINS and local branching, and guided dives. Results indicate that RINS often performs best.

4   Conclusions

We gave a survey on very different, existing approaches for combining exact algorithms and metaheuristics. The two main categories in which we divided these techniques are collaborative and integrative combinations. Some of the combinations are dedicated to very specific combinatorial optimization problems, whereas others are designed to be more generally useful. Altogether, the existing work documents that both, exact optimization techniques and metaheuristics, have specific advantages which complement each other. Suitable combinations of exact algorithms and metaheuristics can benefit much from synergy and often exhibit significantly higher performance with respect to solution quality and time. Some of the presented techniques are mature, whereas others are still in their infancy and need substantial further research in order to make them fully developed. Future work on such hybrid systems is highly promising.

References

1. R. K. Ahuja, Ö. Ergun, J. B. Orlin, and A. P. Punnen. A survey of very large-scale neighborhood search techniques. *Discrete Applied Mathematics*, 123(1-3):75–102, 2002.
2. D. Applegate, R. Bixby, V. Chvátal, and W. Cook. On the solution of the traveling salesman problem. *Documenta Mathematica*, Extra Volume ICM III:645–656, 1998.
3. T. Bäck, D. Fogel, and Z. Michalewicz. *Handbook of Evolutionary Computation*. Oxford University Press, New York NY, 1997.
4. E. K. Burke, P. I. Cowling, and R. Keuthen. Effective local and guided variable neighborhood search methods for the asymmetric travelling salesman problem. In E. Boers et al., editors, *Applications of Evolutionary Computing: EvoWorkshops 2001*, volume 2037 of *LNCS*, pages 203–212. Springer, 2001.
5. S. Chen, S. Talukdar, and N. Sadeh. Job-shop-scheduling by a team of asynchronous agents. In *IJCAI-93 Workshop on Knowledge-Based Production, Scheduling and Control*, Chambery, France, 1993.
6. P. C. Chu and J. E. Beasley. A genetic algorithm for the multidimensional knapsack problem. *Journal of Heuristics*, 4:63–86, 1998.

7. D. Clements, J. Crawford, D. Joslin, G. Nemhauser, M. Puttlitz, and M. Savelsbergh. Heuristic optimization: A hybrid AI/OR approach. In *Proceedings of the Workshop on Industrial Constraint-Directed Scheduling*, 1997. In conjunction with the Third International Conference on Principles and Practice of Constraint Programming (CP97).

8. R. K. Congram. *Polynomially Searchable Exponential Neighbourhoods for Sequencing Problems in Combinatorial Optimisation*. PhD thesis, University of Southampton, Faculty of Mathematical Studies, UK, 2000.

9. C. Cotta and J. M. Troya. Embedding branch and bound within evolutionary algorithms. *Applied Intelligence*, 18:137–153, 2003.

10. E. Danna, E. Rothberg, and C. Le Pape. Exploring relaxation induced neighbourhoods to improve MIP solutions. Technical report, ILOG, 2003.

11. G. B. Dantzig. *Linear Programming and Extensions*. Princeton University Press, 1963.

12. J. Denzinger and T. Offermann. On cooperation between evolutionary algorithms and other search paradigms. In *Proceedings of the 1999 Congress on Evolutionary Computation (CEC)*. IEEE Press, 1999.

13. I. Dumitrescu and T. Stuetzle. Combinations of local search and exact algorithms. In G. R. Raidl et al., editors, *Applications of Evolutionary Computation*, volume 2611 of *LNCS*, pages 211–223. Springer, 2003.

14. G. R. Filho and L. A. N. Lorena. Constructive genetic algorithm and column generation: an application to graph coloring. In *Proceedings of APORS 2000 - The Fifth Conference of the Association of Asian-Pacific Operations Research Societies within IFORS*, 2000.

15. M. Fischetti and A. Lodi. Local Branching. *Mathematical Programming Series B*, 98:23–47, 2003.

16. A. P. French, A. C. Robinson, and J. M.Wilson. Using a hybrid genetic-algorithm/branch and bound approach to solve feasibility and optimization integer programming problems. *Journal of Heuristics*, 7:551–564, 2001.

17. F. Glover and G. Kochenberger, editors. *Handbook of Metaheuristics*, volume 57 of *International Series in Operations Research & Management Science*. Kluwer Academic Publishers, 2003.

18. F. Glover and M. Laguna. *Tabu Search*. Kluwer Academic Publishers, 1997.

19. F. Glover, M. Laguna, and R. Martí. Fundamentals of scatter search and path relinking. *Control and Cybernetics*, 39(3):653–684, 2000.

20. P. Hansen and N. Mladenović. An introduction to variable neighborhood search. In S. Voß, S. Martello, I. Osman, and C. Roucairol, editors, *Meta-heuristics: advances and trends in local search paradigms for optimization*, pages 433–438. Kluwer Academic Publishers, 1999.

21. S. Kirkpatrick, C. Gellat, and M. Vecchi. Optimization by simulated annealing. *Science*, 220:671–680, 1983.

22. G. Klau, I. Ljubić, A. Moser, P. Mutzel, P. Neuner, U. Pferschy, G. Raidl, and R. Weiskircher. Combining a memetic algorithm with integer programming to solve the prize-collecting Steiner tree problem. In K. Deb et al., editors, *Genetic and Evolutionary Computation - GECCO 2004*, volume 3102 of *LNCS*, pages 1304–1315. Springer, 2004.

23. K. Kostikas and C. Fragakis. Genetic programming applied to mixed integer programming. In M. Keijzer et al., editors, *Genetic Programming - EuroGP 2004*, volume 3003 of *LNCS*, pages 113–124. Springer, 2004.

24. P. Larrañaga and J. Lozano. *Estimation of Distribution Algorithms. A New Tool for Evolutionary Computation*. Kluwer Academic Publishers, 2001.

25. A. Z.-Z. Lin, J. Bean, and I. C. C. White. A hybrid genetic/optimization algorithm for finite horizon partially observed markov decision processes. *Journal on Computing*, 16(1):27–38, 2004.

26. H. R. Lourenço, O. Martin, and T. Stützle. Iterated local search. In Glover and Kochenberger [17], pages 321–353.

27. A. Marino, A. Prügel-Bennett, and C. A. Glass. Improving graph colouring with linear programming and genetic algorithms. In *Proceedings of EUROGEN 99*, pages 113–118, Jyväskyiä, Finland, 1999.

28. P. Moscato and C. Cotta. A gentle introduction to memetic algorithms. In Glover and Kochenberger [17], pages 105–144.

29. A. Nagar, S. S. Heragu, and J. Haddock. A meta-heuristic algorithm for a bicriteria scheduling problem. *Annals of Operations Research*, 63:397–414, 1995.

30. G. Nemhauser and L. Wolsey. *Integer and Combinatorial Optimization*. John Wiley & Sons, 1988.

31. A. Plateau, D. Tachat, and P. Tolla. A hybrid search combining interior point methods and metaheuristics for 0-1 programming. *International Transactions in Operational Research*, 9:731–746, 2002.

32. J. Puchinger and G. R. Raidl. An evolutionary algorithm for column generation in integer programming: an effective approach for 2D bin packing. In X. Yao et al., editors, *Parallel Problem Solving from Nature – PPSN VIII*, volume 3242 of *LNCS*, pages 642–651. Springer, 2004.

33. J. Puchinger and G. R. Raidl. Models and algorithms for three-stage two-dimensional bin packing. Technical Report TR 186-1-04-04, Institute of Computer Graphics and Algorithms, Vienna University of Technology, 2004. submitted to the European Journal of Operations Research.

34. J. Puchinger, G. R. Raidl, and G. Koller. Solving a real-world glass cutting problem. In J. Gottlieb and G. R. Raidl, editors, *Evolutionary Computation in Combinatorial Optimization – EvoCOP 2004*, volume 3004 of *LNCS*, pages 162–173. Springer, 2004.

35. G. R. Raidl. An improved genetic algorithm for the multiconstrained 0–1 knapsack problem. In D. B. Fogel, editor, *Proceedings of the 1998 IEEE International Conference on Evolutionary Computation*, pages 207–211. IEEE Press, 1998.

36. G. R. Raidl and H. Feltl. An improved hybrid genetic algorithm for the generalized assignment problem. In H. M. Haddadd et al., editors, *Proceedings of the 2003 ACM Symposium on Applied Computing*, pages 990–995. ACM Press, 2004.

37. A. T. Staggemeier, A. R. Clark, U. Aickelin, and J. Smith. A hybrid genetic algorithm to solve a lot-sizing and scheduling problem. In *Proceedings of the 16th triannual Conference of the International Federation of Operational Research Societies*, Edinburgh, U.K., 2002.

38. S. Talukdar, L. Baeretzen, A. Gove, and P. de Souza. Asynchronous teams: Cooperation schemes for autonomous agents. *Journal of Heuristics*, 4:295–321, 1998.

39. S. Talukdar, S. Murty, and R. Akkiraju. Asynchronous teams. In Glover and Kochenberger [17], pages 537–556.

40. H. Tamura, A. Hirahara, I. Hatono, and M. Umano. An approximate solution method for combinatorial optimisation. *Transactions of the Society of Instrument and Control Engineers*, 130:329–336, 1994.

41. P. Thompson and J. Orlin. The theory of cycle transfers. Technical Report OR-200-89, MIT Operations Research Center, Boston, MA, 1989.

42. P. Thompson and H. Psaraftis. Cycle transfer algorithm for multivehicle routing and scheduling problems. *Operations Research*, 41:935–946, 1993.
43. M. Vasquez and J.-K. Hao. A hybrid approach for the 0–1 multidimensional knapsack problem. In *Proceedings of the International Joint Conference on Artificial Intelligence 2001*, pages 328–333, 2001.
44. D. L. Woodruff. A chunking based selection strategy for integrating meta-heuristics with branch and bound. In S. Voss et al., editors, *Metaheuristics: Advances and Trends in Local Search Paradigms for Optimization*, pages 499–511. Kluwer Academic Publishers, 1999.

Convergence Analysis of a GA-ICA Algorithm

J.M. Górriz, C.G. Puntonet, F. Rojas, and E.G. Medialdea

Facultad de Ciencias, Universidad de Granada,
Fuentenueva s/n, 18071 Granada, Spain
gorriz@ugr.es

Abstract. In this work we consider the extension of Genetic-Independent Component Analysis Algorithms (GA-ICA) with guiding operators and prove their convergence to the optimum. This novel method for Blindly Separating unobservable independent component Sources (BSS) consists of novel guiding genetic operators (GGA) and finds the separation matrix which minimizes a contrast function. The convergence is shown under little restrictive conditions for the guiding operator: its effect must disappear in time like the simulated annealing.

1   Introduction

The starting point in the ICA research can be found in the 60's where a principle of redundancy reduction as a coding strategy in neurons, using statistically independent features, was suggested by Barlow. ICA algorithms have been applied successfully to several fields such as biomedicine, speech, sonar and radar, signal processing, etc. [1] and more recently also to time series forecasting [2]. Any abstract task to be accomplished can be viewed as a search through a space of potential solutions and whenever we work with large spaces, GAs are suitable artificial intelligence techniques for developing this optimization.

The extensive use of ICA as the statistical technique for solving BSS, may have lead in some situations to the erroneous utilization of both concepts as equivalent. In any case, ICA is just the technique which in certain situations can be sufficient to solve a given problem, that of BSS. In fact, statistical independence insures separation of sources in linear mixtures, up to the known indeterminacies of scale and permutation. However, generalizing to the situation in which mixtures are the result of an unknown transformation (linear or not) of the sources, independence alone is not a sufficient condition in order to accomplish BSS successfully.

In this work we prove how guided GA-ICA algorithms converge to the optimum. We organize the essay as follows. Section 3 introduces the linear ICA model (the post-nonlinear model as an alternative to the unconstrained pure nonlinear model is straight-forward). The proof of the convergence is detailed in the rest of the paper. In section 7, we complete the work with some experiments, using image signals. Finally we state some conclusions in section 8.

J. Mira and J.R. Álvarez (Eds.): IWINAC 2005, LNCS 3562, pp. 54–62, 2005.

2   Definition of ICA

We define ICA using a statistical latent variables model (Jutten & Herault, 1991). Assuming the number of sources n is equal to the number of mixtures, the linear model can be expressed as:

$$x_j(t) = b_{j1}s_1 + b_{j2}s_2 + \ldots + b_{jn}s_n \quad \forall j = 1 \ldots n \ , \tag{1}$$

where we explicitly emphasize the time dependence of the samples of the random variables and assume that both the mixture variables and the original sources have zero mean without loss of generality. Using matrix notation instead of sums and including additive noise, the latter mixing model can be written as:

$$\mathbf{x}(t) = \mathbf{B} \cdot \mathbf{s}(t) + \mathbf{b}(t) \ , \ or \tag{2}$$

$$\mathbf{s}(t) = \mathbf{A} \cdot \mathbf{x}(t) + \mathbf{c}(t) \ , \ where \ \mathbf{A} = \mathbf{B}^{-1}, \ \mathbf{c}(t) = -\mathbf{B}^{-1} \cdot \mathbf{b}(t) \ . \tag{3}$$

Due to the nature of the mixing model we are able to estimate the original sources $\tilde{s}_i$ and the de-mixing weights b_{ij} applying i.e. ICA algorithms based on higher order statistics like cumulants.

$$\tilde{s}_i = \sum_{i=1}^{N} b_{ij} x_j \tag{4}$$

Using vector-matrix notation and defining a time series vector $\mathbf{x} = (x_1, \ldots, x_n)^T$, $\mathbf{s}$, $\tilde{\mathbf{s}}$ and the matrix $\mathbf{A} = \{a_{ij}\}$ and $\mathbf{B} = \{b_{ij}\}$ we can write the overall process as:

$$\tilde{\mathbf{s}} = \mathbf{B}\mathbf{x} = \mathbf{B}\mathbf{A}\mathbf{s} = \mathbf{G}\mathbf{s} \tag{5}$$

where we define G as the overall transfer matrix. The estimated original sources will be, under some conditions included in the Darmois-Skitovich theorem [?], a permuted and scaled version of the original ones. Thus, in general, it is only possible to find G such that $\mathbf{G} = \mathbf{PD}$ where $\mathbf{P}$ is a permutation matrix and $\mathbf{D}$ is a diagonal scaling matrix.

2.1   Post-Non-linear Model

The linear assumption is an approximation of nonlinear phenomena in many real world situations. Thus, the linear assumption may lead to incorrect solutions. Hence, researchers in BSS have started addressing the nonlinear mixing models, however a fundamental difficulty in nonlinear ICA is that it is highly non-unique without some extra constraints, therefore finding independent components does not lead us necessarily to the original sources.

Blind source separation in the nonlinear case is, in general, impossible. Taleb and Jutten added some extra constraints to the nonlinear mixture so that the nonlinearities are independently applied in each channel after a linear mixture (see Fig. 2). In this way, the indeterminacies are the same as for the basic linear instantaneous mixing model: invertible scaling and permutation.

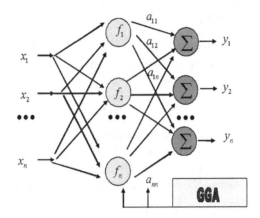

Fig. 1. Schematic Representation of the Separation System in ICA-GA

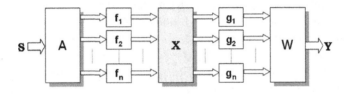

Fig. 2. Post-nonlinear model

The mixture model can be described by the following equation:

$$\mathbf{x}(t) = \mathbf{F}\left(\mathbf{A} \cdot \mathbf{s}(t)\right) \tag{6}$$

The unmixing stage, which will be performed by the algorithm here proposed is expressed by Equation 7:

$$\mathbf{y}(t) = \mathbf{W} \cdot \mathcal{G}\left(\mathbf{x}(t)\right) \tag{7}$$

The post-nonlinearity assumption is reasonable in many signal processing applications where the nonlinearities are introduced by sensors and preamplifiers, as usually happens in speech processing. In this case, the nonlinearity is assumed to be introduced by the signal acquisition system.

3   ICA and Statistical Independence Criterion

We define ICA using a statistical latent variables model (Jutten & Herault, 1991). Assuming the number of sources n is equal to the number of mixtures, the linear model can be expressed, using vector-matrix notation and defining a time series vector $\mathbf{x} = (x_1, \ldots, x_n)^T$, $\mathbf{s}$, $\tilde{\mathbf{s}}$ and the matrix $\mathbf{A} = \{a_{ij}\}$ and

$\mathbf{B} = \{b_{ij}\}$ as $\tilde{\mathbf{s}} = \mathbf{Bx} = \mathbf{BAs} = \mathbf{Gs}$ where we define $\mathbf{G}$ as the overall transfer matrix. The estimated original sources $\tilde{\mathbf{s}}$ will be a permuted and scaled version of the original ones $\mathbf{s}$ [4]. The statistical independence of a set of random variables can be described in terms of their joint and individual probability distribution. This is equivalent to the minimization of [3]:

$$\Pi = \sum_{\{\lambda, \lambda^*\}} \beta_\lambda \beta_{\lambda^*}^* \cdot \mathbf{\Gamma}_{\lambda, \lambda^*} \qquad |\lambda| + |\lambda^*| < \tilde{\lambda} \qquad (8)$$

where the expression defines a summation of cross cumulants [3] and is used as a fitness function in the GA (to seek for optimal matrices). The latter function satisfies the definition of a contrast function $\Psi(\tilde{\mathbf{s}} = \mathbf{Bx})$ defined in [4].

4   GAs, a Theoretical Background

Let $\mathcal{C}$ be the set of all possible creatures in a given world and a function $f : \mathcal{C} \to R^+$ be called fitness function. Let $\Xi : \mathcal{C} \to \mathcal{V}_{\mathbf{C}}$ a bijection from the creature space onto the free vector space over $\mathcal{A}^\ell$, where $\mathcal{A} = \{\overline{a}(i), \quad 0 \leq i \leq a - 1\}$ is the alphabet which can be identified by $\mathcal{V}_1$ the free vector space over $\mathcal{A}$. Then we can establish $\mathcal{V}_{\mathbf{C}} = \otimes_{\lambda=1}^{\ell} \mathcal{V}_1$ and define the free vector space over populations $\mathcal{V}_{\mathcal{P}} = \otimes_{\sigma=1}^{N} \mathcal{V}_{\mathbf{C}}$ with dimension $L = \ell \cdot N$ and a^L elements. Finally let $S \subset \mathcal{V}_{\mathcal{P}}$ be the set of probability distributions over $\mathcal{P}_{\mathbf{N}}$, that is the state which identifies populations with their probability value. A Genetic Algorithm is a product of stochastic matrices (mutation, selection, crossover, etc..) act by matrix multiplication from the left:

$$\mathcal{G}^{\mathbf{n}} = \mathbf{F_n} \cdot \mathbf{C^k_{P^n_c}} \cdot \mathbf{M_{P^n_m}} \qquad (9)$$

where $\mathbf{F_n}$ is the selection operator, $\mathbf{C^k_{P^n_c}}$ is the simple crossover operator and $\mathbf{M_{P^n_m}}$ is the local mutation operator (see [3] and [6] for definitions of the canonical operators)

5   Guided GAs

In order to include statistical information into the algorithm we define an hybrid statistical genetic operator as follows. The value of the probability to go from individual p_i to q_i depends on contrast functions (i.e. based on cumulants) as: $P(\xi_{n+1} = p_i | \xi_n = q_i) = \frac{1}{\aleph(T_n)} \exp\left(-\frac{\Psi(p_i) + \Psi(q_i)}{T_n}\right)$; $p_i, q_i \in \mathbf{C}$ where $\aleph(T_n)$ is the normalization constant depending on iteration n; temperature follows a variation decreasing schedule, that is $T_{n+1} < T_n$ converging to zero, and $\Psi(q_i)$ is the value of the selected contrast function over the individual (an encoded separation matrix). This sampling (simulated annealing law) is applied to the population and offspring emerging from the canonical genetic procedure.

Proposition 1. *The guiding operator can be described using its associated transition probability function (t.p.f.) by column stochastic matrices* $\mathbf{M_G^n}$, $n \in \mathcal{N}$ *acting on populations.*

1. *The components are determined as follows: Let p and $q \in \wp_N$, then we have*

$$\langle q, \mathbf{M_G^n} p \rangle = \frac{N!}{z_{0q}! z_{1q}! \ldots z_{a^L-1q}!} \prod_{i=0}^{a^L-1} \{P(i)\}^{z_iq}; \qquad p, q \in \mathcal{P_N} \qquad (10)$$

where z_{iq} is the number of occurrences of individual i on population q and $P(i)$ is the probability of producing individual i from population p given above. The value of the guiding probability $P(i) = P(i, \Psi)$ depends on the fitness

function used:[1] $P(i) = \dfrac{z_{ip} \exp\left(-\frac{\Psi(p_i) + \Psi(q_i)}{T_n}\right)}{\sum_{i=0}^{a^L-1} z_{ip} \exp\left(-\frac{\Psi(p_i) + \Psi(q_i)}{T_n}\right)}$

2. *For every permutation $\pi \in \Pi_N$, we have $\pi \mathbf{M_G^n} = \mathbf{M_G^n} = \mathbf{M_G^n} \pi$.*
3. *$\mathbf{M_G^n}$ is an identity map on $\mathbf{U}$ in the optimum, that is $\langle p, \mathbf{M_G^n} p \rangle = 1$; and has strictly positive diagonals since $\langle p, \mathbf{M_G^n} p \rangle > 0 \quad \forall p \in \mathcal{P_N}$.*
4. *All the coefficients of a GA consisting of the product of stochastic matrices: the simple crossover $\mathbf{C_{P_c}^k}$, the local multiple mutation $\mathbf{M_{Pm}^n}$ and the guiding operator $\mathbf{M_G^n}$ for all $n, k \in \mathcal{N}$ are uniformly bounded away from 0.*

Proof: (1) follows from the transition probability between states. (2) is obvious and (3) follows from [4] and checking how matrices act on populations. (4) follows from the fact that $\mathbf{M_{Pm}^n}$ is fully positive acting on any stochastic matrix $\mathbf{S}$.

It can be viewed as a suitable fitness selection and as a certain Reduction Operator, since it preserves the best individuals into the next generation using a non heuristic rule, unlike the majority of GAs used.

6 Convergence Analysis of GGAs

A markov chain (MC) modelling a canonical GA (CGA) has been proved to be strongly ergodic (hence weak ergodic, see [6]). So we have to focus our attention on the transition probability matrix that emerges when we apply the guiding operator. The overall transition matrix can be written as:

$$\langle q, \mathcal{G}^n p \rangle = \sum_{v \in \wp_N} \langle q, \mathbf{M_G^n} v \rangle \langle v, \mathcal{C}^n p \rangle \qquad (11)$$

where $\mathcal{C}^n$ is the product of stochastic matrices associated to the CGA and $\mathbf{M_G^n}$ is given by equation 10.

---

[1] The condition that must be satisfied the transition probability matrix $P(i, f)$ is that it must converge to a positive constant as $n \to \infty$ (since we can always define a suitable normalization constant). The fitness function or selection method of individuals used in it must be injective.

Proposition 2. Weak Ergodicity

A MC modelling a GGA satisfies weak ergodicity if the t.p.f associated to guiding operators converges to uniform populations (populations with the same individual).

Proof: The ergodicity properties depend on the new operator since CGAs satisfy them as we said before. We just have to check the convergence of the t.p.f. of the guiding operator on uniform populations. The following condition is satisfied:

$$\langle u, \mathcal{G}^n p \rangle \to 1 \quad u \in \mathbf{U} \tag{12}$$

Then we can find a series of numbers: $\sum_{n=1}^{\infty} \min_{n,p}(\langle u, \mathcal{G}^n p \rangle) = \infty \le \sum_{n=1}^{\infty} \min_{q,p} \sum_{v \in \wp_N} \min(\langle v, \mathbf{M}_{\mathbf{G}}^n p \rangle \langle v, \mathcal{C}^n q \rangle)$ which is equivalent to weak ergodicity.

Proposition 3. Strong Ergodicity

Let $\mathbf{M}_{\mathbf{P}_m}^n$ describes multiple local mutation, $\mathbf{C}_{\mathbf{P}_m}^k$ describes a model for crossover and $\mathbf{F}^n$ describes the fitness selection. Let $(P_m^n, \hat{P}_c^n)_n \in \mathcal{N}$ be a variation schedule and $(\phi_n)_{n \in \mathcal{N}}$ a fitness scaling sequence associated to $\mathbf{M}_{\mathbf{G}}^n$ describing the guiding operator according to this scaling. [2] Let $\mathcal{C}^n = \mathbf{F}^n \cdot \mathbf{M}_{\mathbf{P}_m}^n \cdot \mathbf{C}_{\mathbf{P}_c}^k$ represent the first n steps of a CGA. In this situation,

$$v_\infty = \lim_{n \to \infty} \mathcal{G}^n v_0 = \lim_{n \to \infty} (\mathbf{M}_{\mathbf{G}}^\infty \mathcal{C}^\infty)^n v_0 \tag{13}$$

exists and is independent of the choice of v_0, the initial probability distribution. Furthermore, the coefficients $\langle v_\infty, p \rangle$ of the limit probability distribution are strictly positive for every population $p \in \wp_N$.

Proof: This is a consequence of Theorem 16 in [6] and point 4 in Proposition 1. We only have to replace the canonical selection operator $\mathbf{F}_n$ with our guiding selection operator $\mathbf{M}_{\mathbf{G}}^n$ which has the same essential properties.

Proposition 4. Convergence to the Optimum

Under the same conditions of propositions 2 and 3 the GGA algorithm converges to the optimum.

Proof: To reach this result, one has to prove that the probability to go from any uniform population to the population p^* containing only the optimum is equal to 1 when $n \to \infty$:

$$\lim_{n \to \infty} \langle p^*, \mathbf{G}^n u \rangle = 1 \tag{14}$$

since the GGA is an strongly ergodic MC, hence any population tends to uniform in time. If we check this expression we finally have the equation 14. In addition we use point 3 in Proposition 1 to make sure the optimum is the convergence point.

---

[2] A scaling sequence $\phi_n : (\mathcal{R}^+)^N \to (\mathcal{R}^+)^N$ is a sequence of functions connected with a injective fitness criterion f as $f_n(p) = \phi_n(f(p))$ $p \in \wp_N$ such that $\mathbf{M}_{\mathbf{G}}^\infty = \lim_{n \to \infty} \mathbf{M}_{\mathbf{G}}^n$ exists.

7   Experiments

This section illustrates the validity of the genetic algorithms here proposed and investigates the accuracy of the method. The Computer used in these simulations was a PC 2 GHz, 256 MB RAN and the software used is an extension of ICATOOLBOX2.0 in MatLab code, protected by the Spanish law N° CA-235/04. In order to measure the accuracy of the algorithm, we evaluate using the Mean Square Error (MSE) and Normalized Round MSE (NRMSE) dividing by standard deviation) and the Crosstalk in decibels (Ct):

$$MSE_i = \frac{\sum\limits_{t=1}^{N} (s_i(t) - y_i(t))^2}{N}; \quad Ct_i = 10\log\left(\frac{\sum\limits_{t=1}^{N} (s_i(t) - y_i(t))^2}{\sum\limits_{t=1}^{N} (s_i(t))^2}\right) \quad (15)$$

We propose a two dimensional mixing problem with the input signals plotted in the Figures 3 and 4. These images are mixed using a linear mixing matrix $A = [0.697194 \quad -0.313446; 0.408292 \quad 0.382734]$ with $cond(A) = 1.67686$ and $det(A) = 0.39481$. Then we add some noise to mixtures, i.e. (gaussian noise $20dB$ on 50% pixel points. We compare the results with ERICA algorithm [5] with regard to number of iterations to reach convergence using the same number of individuals ($N = 6$). The GGA uses a guiding operator based on cumulants (see section 3). ERICA evolution is six deterministic trajectories in a 4 dimensional space while GGA is an stochastic process around the best trajectory (see Table 4). Of course in this low dimensional problem (only two sensors) ERICA is faster unlike in high dimensional scenario where GGA method prevails [3].

8   Conclusions

A GGA-based ICA method has been developed to solve BSS problem from mixtures of independent sources. Extensive simulation results proved the potential of the proposed method [3],[7] but no explanation about convergence was given. In this paper we introduced demonstrations of the GA-ICA methods used till now [3],[7], etc. proving the convergence to the optimum unlike the ICA algorithms which usually suffer of local minima and non-convergent cases. Any injective contrast function can be used to build a guiding operator, as an elitist strategy.

Finally, in the experimental section, we have checked the potential of the proposed algorithms versus ERICA algorithm [5]. Experimental results showed promising results, although in low dimensional scenario deterministic algorithms prevail. Future research will focus on the adaptation of the algorithm for higher dimensionality and nonlinearities.

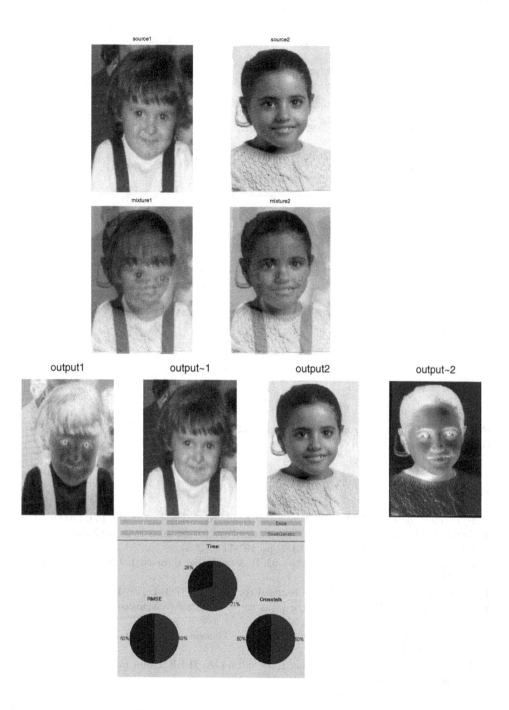

Fig. 3. Sources, Mixtures, Estimated Sources and CPU time (ERICA vs GGA)

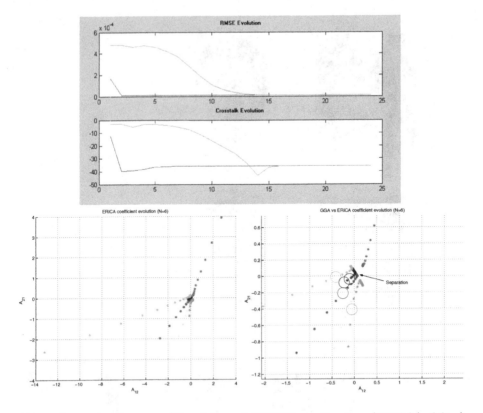

Fig. 4. Separation features: (Crosstalk and NRMSE vs. iteration (ERICA(red line) Guided Genetic Algorithm (green line) ; matrix coefficient evolution (a_{12} vs a_{22})

References

1. Hyvärinen, A. et al. ICA: Algorithms and Appl. Neural Net., 13: 411-430, (2000).
2. Górriz, J.M. et al. New Model for Time Series Forecasting using rbfs and Exogenous Data. Neural Comp. and Appl., 13/2, pp 101-111, (2004).
3. Górriz J.M. et al. Hybridizing GAs with ICA in Higher dimension LNCS 3195, pp 414-421, (2004).
4. Comon, P., ICA, a new concept? Signal Proc. 36, pp 287-314, (1994).
5. Cruces, S. et al. Robust BSS algorithms using cumulants. Neurocomp. 49 (2002) 87-118
6. Schmitt, L.M. et al. *Linear Analysis of GAs*, Theoretical Computer Science, 200, pp 101-134, (1998).
7. Tan, Y. et al. Nonlinear BSS Using HOS and a GA. IEEE Trans. on Evol. Comp., 5/6, pp 600-612 (2001).

An Evolutionary Strategy for the Multidimensional 0-1 Knapsack Problem Based on Genetic Computation of Surrogate Multipliers

César L. Alonso[1,*], Fernando Caro[2], and José Luis Montaña[3,**]

[1] Centro de Inteligencia Artificial, Universidad de Oviedo,
Campus de Viesques, 33271 Gijón, Spain
calonso@aic.uniovi.es
[2] Departamento de Informática,
Universidad de Oviedo
[3] Departamento de Matemáticas,
Estadística y Computación,
Universidad de Cantabria
montana@matesco.unican.es

Abstract. In this paper we present an evolutionary algorithm for the multidimensional 0-1 knapsack problem. Our algorithm incorporates a heuristic operator which computes problem-specific knowledge. The design of this operator is based on the general technique used to design greedy-like heuristics for this problem, that is, the surrogate multipliers approach of Pirkul (see [7]). The main difference with work previously done is that our heuristic operator is computed following a genetic strategy -suggested by the greedy solution of the one dimensional knapsack problem- instead of the commonly used simplex method. Experimental results show that our evolutionary algorithm is capable of obtaining high quality solutions for large size problems requiring less amount of computational effort than other evolutionary strategies supported by heuristics founded on linear programming calculation of surrogate multipliers.

Keywords: Evolutionary computation, genetic algorithms, knapsack problem, 0-1 integer programming, combinatorial optimization.

1 Introduction

The multidimensional 0-1 knapsack problem (MKP) is a NP-complete combinatorial optimization problem which captures the essence of linear 0-1 integer programming problems. It can be formulated as follows. We are given a set of n objects where each object yields p_j units of profit and requires a_{ij} units

---

* Partially supported by the Spanish MCyT and FEDER grant TIC2003-04153.
** Partially supported by the Spanish MCyT, under project MTM2004-01167.

J. Mira and J.R. Álvarez (Eds.): IWINAC 2005, LNCS 3562, pp. 63–73, 2005.

of resource consumption in the i-th knapsack constraint. The goal is to find a subset of the objects such that the overall profit is maximized without exceeding the resource capacities of the knapsacks. More precisely, given $A = (a_{ij})$, $\bar{p} = (p_1, \ldots, p_n)$, and $\bar{b} = (b_1, \ldots, b_m)$, an instance of MKP can be described as:

$$\text{maximize } f(x_1, \ldots, x_n) = \sum_{j=1}^{n} p_j x_j \tag{1}$$

$$\text{subject to constraint } C_i : \quad \sum_{j=1}^{n} a_{ij} x_j \leq b_i \quad i = 1, \ldots, m \tag{2}$$

The MKP is one of the most popular constrained integer programming problems with a large domain of applications. Many practical problems can be formulated as a MKP instance: the capital budgeting problem in economy; the allocation of databases and processors in a distributed computer system ([6]) or, more recently, the problem of the daily management of a remote sensing satellite ([15]). Most of the research on knapsack problems deals with the simpler one-dimensional case ($m = 1$). For this single constraint case, the problem is not strongly NP-hard and some exact algorithms and also very efficient approximation algorithms have been developed for obtaining near-optimal solutions. A good review of the one-dimensional knapsack problem can be found in ([10]).

In the multidimensional case several exact algorithms that compute different upper bounds for the optimal solutions are known. For example that in ([7]). But this method, based on the computation of optimal surrogate multipliers, becomes no applicable for large values of m and n and other strategies must be introduced. In this context heuristic approaches for the MKP have appeared during the last decades following different ideas: greedy-like assignment ([5], [11]); LP-based search ([1]); surrogate duality information ([11]) are some examples. Also an important number of papers using genetic algorithms and other evolutionary strategies have emerged. The genetic approach has shown to be well suited for solving large MKP instances. In ([9]) a genetic algorithm is presented where infeasible individuals are allowed to participate in the search and a simple fitness function with a penalty term is used. Thiel and Voss (see [14]) presented a hybrid genetic algorithm with a tabu search heuristic. Chu and Beasley (see [4]) have developed a genetic algorithm that searches only into the feasible search space. They use a repair operator based on the surrogate multipliers of some suitable surrogate problem. Surrogate multipliers are calculated using linear programming. A review of exact and heuristic techniques for MKP can be found in ([4]).

In the present paper we propose an evolutionary strategy for MKP which takes as starting point the surrogate multipliers approach appeared first in ([11]) and later in ([4]) but using a genetic algorithm for finding an approximation of some optimal set of weights. In this approach the m-knapsack constraints of equation 2 are reduced to a single constraint $\sum_{j=1}^{n} (\sum_{i=1}^{m} \omega_i a_{ij}) x_j \leq \sum_{i=1}^{m} \omega_i b_i$ computing a suitable weight ω_i for each restriction i. These weights are called the surrogate multipliers and the single knapsack constraint is called the surrogate constraint. Any optimal solution for the surrogate problem generates an upper bound for the optimal solution of the initial MKP. Ideally the best surrogate multipliers are those that generate the smallest upper bound for the solution of the initial problem. But computing this optimal set of multipliers is a computationally difficult task. The aforementioned genetic strategy to compute an approximate set of surrogate multipliers is the main novelty in this paper.

To our knowledge, one of the best evolutionary strategies for MKP based on surrogate constraints is CHUGA algorithm (see [4]). Our experimental results show that our heuristic based upon genetic algorithm is competitive -respect to the quality of the generated solutions- with CHUGA but requiring less amount of computational effort. The rest of the paper is organized as follows. Section 2 deals with the surrogate problem associated to an instance of the MKP and describes the mathematical background needed to justify the election of the fitness function of our genetic algorithm to approximate the optimal surrogate multipliers. In section 3 we present our evolutionary algorithm for solving the MKP that incorporates as heuristic operator the genetic procedure developed in section 2. In section 4 experimental results of our evolutionary algorithm and comparisons with that of Chu and Beasley ([4]) over problems taken from the OR-Library ([3]) are included. Finally, section 5 contains some conclusive remarks.

2   The MKP and the Surrogate Multipliers

This section resumes the mathematical background and the detailed formulation of our genetic heuristic for computing surrogate multipliers.

Definition 1. *An instance of the MKP is a 5-tuple:*

$$K = (n, m, \overline{p}, A, \overline{b}) \tag{3}$$

where n, $m \in \mathbb{N}$ are both natural numbers representing (respectively) the number of objects and the number of constraints; $\overline{p} \in (\mathbb{R}^+)^n$ is a vector of positive real numbers representing the profits; $A \in M_{m \times n}(\mathbb{R}^+ \cup \{0\})$ is $m \times n$-matrix of non-negative real numbers corresponding to the resource consumptions and $\overline{b} \in (\mathbb{R}^+)^m$ is a vector of positive real numbers representing the knapsack capacities.

Remark 2. Given an instance of the MKP $K = (n, m, \overline{p}, A, \overline{b})$ the objective is

$$\text{maximize } f(\overline{x}) = \overline{p} \cdot \overline{x}' \tag{4}$$

$$\text{subject to } A \cdot \overline{x}' \leq \overline{b}' \tag{5}$$

where $\overline{x} = (x_1, \ldots, x_n)$ is a vector of variables that takes values in $\{0, 1\}^n$. Notation $\overline{v}'$ stands for the transposition of the vector $\overline{v}$.

Definition 3. *Let* $K = (n, m, \overline{p}, A, \overline{b})$ *be an instance of the MKP. A bit-vector* $\alpha \in \{0,1\}^n$ *is a feasible solution of instance K if it verifies the constraint given by equation 5. A bit-vector* $\alpha_K^{opt} \in \{0,1\}^n$ *is a solution of instance K if it is a feasible solution and for all feasible solution* $\alpha \in \{0,1\}^n$ *of instance K it holds*

$$f(\alpha) \le f(\alpha_K^{opt}) = \sum_{j=1}^{n} p_j \alpha_K^{opt}[j] \tag{6}$$

Here $\alpha[j]$ stands for the value of variable x_j.

Remark 4. We say that an instance $K = (n, m, \overline{p}, A, \overline{b})$ of MKP is well stated if it satisfies

$$a_{ij} \le b_i < \sum_{j=1}^{n} a_{ij} \quad \forall i = 1, \ldots, m;\ j = 1, \ldots, n \tag{7}$$

Well stated MKP instances can always be assumed since otherwise some or all of the variables could be fixed to 0 or 1.

Remark 5. (LP-relaxation) Along this section we shall deal with a relaxed version of the MKP where the vector of variables $\overline{x}$ can take values in the whole interval $[0,1]^n$. We will refer to this case as the LP-relaxed MKP. An instance of the LP-relaxed MKP will be denoted by K^{LP}. A solution of some instance K^{LP} of the LP-relaxed MKP will be denoted by $\alpha_{K^{LP}}^{opt}$.

Definition 6. *Let $K = (n, m, \overline{p}, A, \overline{b})$ be an instance of MKP and let $\overline{\omega} \in (\mathbb{R}^+)^m$ be a vector of positive real numbers. The surrogate constraint for K associated to $\overline{\omega}$, denoted by $Sc(K, \overline{\omega})$, is defined as follows.*

$$\sum_{j=1}^{n} (\sum_{i=1}^{m} \omega_i a_{ij}) x_j \le \sum_{i=1}^{m} \omega_i b_i \tag{8}$$

The vector $\overline{\omega} = (\omega_1, \ldots, \omega_m)$ is called the vector of surrogate multipliers.
The surrogate instance for K, denoted by $SR(K, \overline{\omega})$, is defined as the 0–1 one-dimensional knapsack problem given by:

$$SR(K, \overline{\omega}) = (n, 1, \overline{p}, \overline{\omega} \cdot A, \overline{\omega} \cdot \overline{b}) \tag{9}$$

Remark 7. Given a surrogate instance $SR(K, \overline{\omega})$ the surrogate problem is

$$\text{maximize } f(\overline{x}) = \overline{p} \cdot \overline{x}' \tag{10}$$

$$\text{subject to } Sc(K, \overline{\omega})$$

where $\overline{x} = (x_1, \ldots, x_n)$ takes values in $\{0,1\}^n$
We shall denote by $\alpha_{SR(K, \overline{\omega})}^{opt}$ the solution of the surrogate instance $SR(K, \overline{\omega})$.

The key result concerning solutions of surrogate problems is stated in the next lemma. The proof of this useful fact is an easy consequence of the definition of the surrogate problem (see [7]).

Lemma 8. *Let $K = (n, m, \overline{p}, A, \overline{b})$ be any instance of MKP and $\overline{\omega} = (\omega_1, \ldots, \omega_m)$ any vector of surrogate multipliers. Then*

$$f(\alpha_{SR(K,\overline{\omega})}^{opt}) \geq f(\alpha_K^{opt}) \tag{11}$$

Remark 9. The best possible upper bound for $f(\alpha_K^{opt})$ using lemma 8 is computed by finding the minimum value $min\{f(\alpha_{SR(K,\overline{\omega})}^{opt}) : \overline{\omega} \in (\mathbb{R}^+)^m\}$. We shall call optimal any vector of surrogate multipliers that provides the above minimum. Following Gavish and Pirkul ([7]) we point out that, from this optimal set of surrogate multipliers, the generated upper bound on $f(\alpha_K^{opt})$ is less than or equal to bounds generated by LP-relaxation. Unfortunately , in terms of time complexity, computing the optimal surrogate multipliers is a computationally hard problem ([7]). Next proposition motivates our genetic strategy to approximate the optimal values of the surrogate multipliers $\overline{\omega}$.

Proposition 10. *Let $K = (n, m, \overline{p}, A, \overline{b})$ be any instance of MKP. Then the following inequalities are satisfied.*

$$min\{f(\alpha_{SR(K,\overline{\omega})^{LP}}^{opt}) : \overline{\omega} \in (0,1]^m\} = min\{f(\alpha_{SR(K,\overline{\omega})^{LP}}^{opt}) : \overline{\omega} \in (\mathbb{R}^+)^m\} \tag{12}$$

$$min\{f(\alpha_{SR(K,\overline{\omega})^{LP}}^{opt}) : \overline{\omega} \in (0,1]^m\} \geq f(\alpha_K^{opt}), \tag{13}$$

Here $\alpha_{SR(K,\overline{\omega})^{LP}}^{opt}$ denotes the solution of the relaxed surrogate problem $SR(K,\overline{\omega})^{LP}$.

Proof. First note that, although $\overline{\omega}$ has been defined in $(\mathbb{R}^+)^m$, we can normalize the constraint 8 assuming without lost of generality that each ω_i belongs to $(0, 1]$. This justifies equality 12. By definition of LP-relaxed problem (see remark 5 above) any instance I of MKP satisfies

$$f(\alpha_{I^{LP}}^{opt}) \geq f(\alpha_I^{opt}), \tag{14}$$

Previous inequality holds, in particular, if instance I is a surrogate instance, that is $I := SR(K, \overline{\omega})$. This immediately gives:

$$\forall \overline{\omega} \in (0,1]^m, \quad f(\alpha_{SR(K,\overline{\omega})^{LP}}^{opt}) \geq f(\alpha_{SR(K,\overline{\omega})}^{opt}) \tag{15}$$

Now the result follows from inequality 11 in lemma 8.

2.1   A Genetic Algorithm for Computing the Surrogate Multipliers

We propose a simple genetic algorithm (GA) to obtain approximate values for $\overline{\omega}$. In this GA the individuals will be binary 0–1 strings, representing $\overline{\omega} = (\omega_1, \ldots, \omega_m)$. Each ω_i will be represented as a q-bit binary substring, where q determines the desired precision of $\omega_i \in (0, 1]$.

Definition 11. *Given an MKP instance $K = (n, m, \overline{p}, A, \overline{b})$, $q \in \mathbb{N}$, and a representation $\gamma \in \{0, 1\}^{qm}$ of a candidate vector of surrogate multipliers $\overline{\omega} = (\omega_1, \ldots, \omega_m)$ the fitness value of γ is defined as follows:*

$$fitness(\gamma) = f(\alpha^{opt}_{SR(K,\overline{\omega})^{LP}}) = \sum_{j=1}^{n} p_j \alpha^{opt}_{SR(K,\overline{\omega})^{LP}}[j] \tag{16}$$

Remark 12. (Computing the fitness) The solution $\alpha^{opt}_{SR(K,\overline{\omega})^{LP}}$ is obtained by means of the well known greedy procedure for one-dimensional instances of the knapsack problem that we describe next. Let $I = (n, 1, \overline{p}, \overline{a}, b)$ be any instance of the LP relaxed one-dimensional knapsack problem. Assume that the items are sorted according to non-increasing efficiencies: $u_j = p_j/a_j$, $j = 1, \ldots, n$. Define $s = min\{k \in \{1, \ldots, n\} : \sum_{j=1}^{k} a_j > b\}$, which is usually known as the *split term*. Then, the solution α^{opt}_{ILP} is given by:

$$\alpha^{opt}_{ILP}[j] = 1 \text{ for } 1 \leq j \leq s - 1; \quad \alpha^{opt}_{ILP}[j] = 0 \text{ for } s + 1 \leq j \leq n; \tag{17}$$

$$\alpha^{opt}_{ILP}[s] = \frac{(b - \sum_{j=1}^{s-1} a_j)}{a_s} \tag{18}$$

The split term s and also the solution α^{opt}_{ILP} can be found in optimal $O(n)$ time using the median search algorithm in [2]. Finally, given an instance of MKP $K = (n, m, \overline{p}, A, \overline{b})$, $q \in \mathbb{N}$, and a representation $\gamma \in \{0, 1\}^{qm}$ of an individual $\overline{\omega} = (\omega_1, \ldots, \omega_m)$, set $I := SR(K, \overline{\omega})^{LP}$ and apply the previous procedure to compute $\alpha^{opt}_{SR(K,\overline{\omega})^{LP}}$ and then the value $fittness(\gamma)$ according to equation 16.

Note that, according to proposition 10, the objective of our GA is to minimize the *fitness* function defined by equation 16. The remainder of the parameters of our genetic algorithm have been chosen as follows: the roulette wheel rule as selection procedure, uniform crossover as recombination operator and bitwise mutation according to a given probability p_m (see [8] for a detailed description of these operators).

3   An Evolutionary Algorithm for the MKP

Given an instance of MKP, $K = (n, m, \overline{p}, A, \overline{b})$, and a set of surrogate multipliers, $\overline{\omega}$, computed by the genetic algorithm described in the previous section, we will run an steady state evolutionary algorithm for solving K with a repair operator for the infeasible individuals -that uses as heuristic operator the set of surrogate multipliers- and also a local improvement technique. The management of infeasible individuals and their improve will be done following ([4]).

3.1   Individual Representation and Fitness Function

We choose the standard 0–1 binary representation since it represents the underlying 0–1 integer values. A chromosome α representing a candidate solution for $K = (n, m, \overline{p}, A, \overline{b})$ is an n-bit binary string. A value $\alpha[j] = 0$ or 1 in the j-bit means that variable $x_j = 0$ or 1 in the represented solution. The fitness of chromosome $\alpha \in \{0,1\}^n$ is defined by

$$f(\alpha) = \sum_{j=1}^{n} p_j \alpha[i] \tag{19}$$

3.2   Repair Operator and Local Improve Procedure

The repair operator lies on the notion of utility ratios. Let $K = (n, m, \overline{p}, A, \overline{b})$ be an instance of MKP and let $\overline{\omega} \in (0,1]^m$ a vector of surrogate multipliers. The utility ratio for the variable x_j is defined by $u_j = \frac{p_j}{\sum_{i=1}^{m} \omega_i a_{ij}}$. Given an infeasible individual $\alpha \in \{0,1\}^n$ we apply the following procedure to repair it.

Procedure DROP
input: $K = (n, m, \overline{p}, A, \overline{b})$; $\overline{\omega} \in (0,1]^m$ and a chromosome $\alpha \in \{0,1\}^n$

```
begin
    for j=1 to n compute u_j
    P:=permutation of (1,...,n) with u_P[j] <= u_P[j+1]
    for j=1 to n do
        if (alpha[P[j]]=1 and infeasible(alpha)) then
            alpha[P[j]]:=0
end
```

Once we have transformed α into a feasible individual α', a second phase is applied in order to improve α'. This second phase is denominated the ADD phase.

Procedure ADD
input: a chromosome $\alpha \in \{0,1\}^n$

```
begin
    P:=permutation of (1,...,n) with u_P[j] >= u_P[j+1]
    for j=1 to n do
        if alpha[P[j]]=0 then alpha[P[j]]:=1
        if infeasible(alpha) then alpha[P[j]]:=0
end
```

3.3   Genetic Operators and Initial Population

We use the roulette wheel rule as selection procedure, the uniform crossover as replacement operator and bitwise mutation with a given probability p. So when mutation must be applied to an individual α, a bit j from α is randomly selected and flipped from its value $\alpha[j] \in \{0,1\}$ to $1 - \alpha[j]$.

We construct the individuals of the initial population generating random permutations from $(1, \ldots, n)$ and applying the DROP procedure to get feasibility. Then we apply to these permutations the ADD procedure to improve their fitness.

4   Experimental Results

We have executed our evolutionary algorithm on the problems included in the OR-Library proposed in [4] [1]. This library contains randomly generated instances of MKP with number of constraints $m \in \{5, 10, 30\}$, number of variables $n \in \{100, 250, 500\}$ and tightness ratios $r \in \{0.25, 0.5, 0.75\}$. The tightness ratio r fixes the capacity of i-th knapsack to $r \sum_{j=1}^{n} a_{ij}$, $1 \leq i \leq m$. There are 10 instances for each combination of m, n and r giving a total of 270 test problems. We have run our double genetic algorithm (DGA) setting the parameters to the following values. The GA computing the surrogate multipliers uses population size 75, precision $q = 10$, probability of mutation 0.1 and 15000 generations to finish. The steady state GA solving the MKP instances uses population size 100, probability of mutation equal to 0.1 and 10^6 non-duplicate evaluations to finish. Since the optimal solution values for most of these problems are unknown the quality of a solution α is measured by the percentage gap of its fitness value with respect to the fitness value of the optimal solution of the LP-relaxed problem: $\%gap = 100 \frac{f(\alpha_{KLP}^{opt}) - f(\alpha)}{f(\alpha_{KLP}^{opt})}$. We measure the computational time complexity by means of the number of evaluations required to find the best obtained solution. We have executed our evolutionary algorithm on a Pentium IV; 3GHz. Over this platform the execution time for a single run of our DGA ranges from 3 minutes, for the simplest problems, to 25 minutes, for the most complex ones.

Experimentation has been very extensive but due to lack of space reasons we only report on comparisons with other well known algorithms. The results of our experiments are displayed in Tables 1 and 2 based on 10 independent executions for each instance. In table 1, our algorithm is compared with that of Chu et. al (CHUGA) ([4]). To our knowledge this is the best known genetic algorithm for MKP based on surrogate constraints. The first two columns identify the 30 instances that corresponds to each combination of m, n. In the remainder columns we show for the two compared algorithms the average $\%gap$ and the average number of evaluations required until the best individual was encountered (A.E.B.S). We have taken the values corresponding to CHUGA from ([4]) and ([12]). As the authors have pointed out in their work these results are based on only one run for each problem instance whereas in the case of our double genetic algorithm DGA, 10 runs were executed. In table 2 we display results that measure the quality of our surrogate multipliers. From this table we conclude that in all

---

[1] Public available on line in http://www.brunel.ac.uk/depts/ma/research/jeb/info.html

Table 1. Computational results for CHUGA and DGA. Values for DGA are based on 10 runs for each instance. We also include the average % speed-up of DGA with respect to CHUGA, attending on the respective A.E.B.S

Problem		GHUGA		DGA		
m n	A. %gap	A.E.B.S	A. %gap	A.E.B.S	A. %Speed-up
5 100	0.59	24136	0.58	58045	–
5 250	0.16	218304	0.15	140902	35.46
5 500	0.05	491573	0.06	185606	62.24
10 100	0.94	318764	0.98	70765	77.80
10 250	0.35	475643	0.32	153475	67.73
10 500	0.14	645250	0.15	179047	72.25
30 100	1.74	197855	1.71	106542	46.15
30 250	0.73	369894	0.71	184446	50.14
30 500	0.40	587472	0.44	233452	60.26

Table 2. Average percentage gap of $f(\alpha^{opt}_{SR(K,\overline{\omega})LP})$ for the obtained surrogate multipliers $\overline{\omega}$, with respect to the value of the optimal solution of the LP-relaxed MKP instance (α^{opt}_{KLP})

Problem	DGA
m n	A. %gap
5 100	0.003
5 250	0.000
5 500	0.000
10 100	0.03
10 250	0.009
10 500	0.004
30 100	0.33
30 250	0.11
30 500	0.05

tested cases they are almost optimal respect to their best possible lower bound. From our experimentation we deduce that our DGA performs as well as the CHUGA algorithm in terms of the average %gaps but using less amount of computational effort because the best solution is reached within a considerably less number of individual evaluations in almost all tested cases: last column in table 1 shows the the average % speed-up obtained by DGA respect to CHUGA algorithm.

5 Conclusive Remarks

In this paper we have presented a heuristic algorithm based on GAs for solving multidimensional knapsack problems (DGA). The core component of DGA is that used in a standard GA for MKP. Our approach differs from previous GA based techniques in the way that a heuristic is calculated incorporating another genetic algorithm. On a large set of randomly generated problems, we have shown that the GA heuristic is capable of obtaining high-quality solutions for problems of various characteristics, whilst requiring less amount of computational effort than other GA techniques based on linear programming heuristics. Clearly, the effectiveness of the greedy heuristics based on the surrogate problem strongly depends on the ability of the surrogate constraint to grasp the aggregate weighted consumption level of resources for each variable, and this in turn relies on the determination of a good set of weights. In this sense, for the large-sized test data used, this new DGA converged most of the time much faster to high quality solutions than the comparable GA from Chu et al. and this feature can only be explained by the better quality of the surrogate multipliers obtained using our heuristic method. In particular this supports the conjecture that genetic computation is also well suited when tackling optimization problems of parameters with variables in continuous domains as it is the case of computing surrogate multipliers.

References

1. Balas, E., Martin, C. H.: Pivot and Complement–A Heuristic for 0–1 Programming. Management Science 26 (1) (1980) 86–96
2. Balas, E., Zemel, E.: An Algorithm for Large zero–one Knapsack Problems. Operations Research 28 (1980) 1130–1145
3. Beasley, J. E.: Obtaining Test Problems via Internet. Journal of Global Optimization 8 (1996) 429–433
4. Chu, P. C., Beasley, J. E.: A Genetic Algorithm for the Multidimensional Knapsack Problem. Journal of Heuristics 4 (1998) 63–86
5. Freville, A., Plateau, G.: Heuristics and Reduction Methods for Multiple Constraints 0–1 Linear Programming Problems. Europena Journal of Operationa Research 24 (1986) 206–215
6. Gavish, B., Pirkul, H.: Allocation of Databases and Processors in a Distributed Computing System. J. Akoka ed. Management od Distributed Data Processing, North-Holland (1982) 215–231
7. Gavish, B., Pirkul, H.: Efficient Algorithms for Solving Multiconstraint Zero–One Knapsack Problems to Optimality. Mathematical Programming 31 (1985) 78–105
8. Goldberg, D. E.: Genetic Algorithms in Search, Optimization and Machine Learning. Addison-Wesley (1989)
9. Khuri, S., Bäck, T., Heitkötter, J.: The Zero/One Multiple Knapsack Problem and Genetic Algorithms. Proceedings of the 1994 ACM Symposium on Applied Computing (SAC'94), ACM Press (1994) 188–193
10. Martello, S., Toth, P.: Knapsack Problems: Algorithms and Computer Implementations. John Wiley & Sons (1990)
11. Pirkul, H.: A Heuristic Solution Procedure for the Multiconstraint Zero–One Knapsack Problem. Naval Research Logistics 34 (1987) 161–172
12. Raidl, G. R.: An Improved Genetic Algorithm for the Multiconstraint Knapsack Problem. Proceedings of the 5th IEEE International Conference on Evolutionary Computation (1998) 207–211

13. Rinnooy Kan, A. H. G., Stougie, L., Vercellis, C.: A Class of Generalized Greedy Algorithms for the Multi-knapsack Problem. Discrete Applied Mathematics 42 (1993) 279–290

14. Thiel, J., Voss, S.: Some Experiences on Solving Multiconstraint Zero–One Knapsack Problems with Genetic Algorithms. INFOR 32 (1994) 226–242

15. Vasquez, M., Hao, J. K.: A Logic-constrained Knapsack Formulation and a Tabu Algorithm for the Daily Photograph Scheduling of an Earth Observation Satellite. Computational Optimization and Applications 20 (2001) 137–157

An Evolutionary Approach to Designing and Solving Fuzzy Job-Shop Problems

Inés González-Rodríguez, Camino R. Vela, and Jorge Puente*

Dept. of Computer Science, University of Oviedo (Spain)
inesgr@uniovi.es, camino@aic.uniovi.es, puente@aic.uniovi.es

Abstract. In the sequel we shall consider the *fuzzy job-shop problem*, a variation of the job-shop problem where the duration of tasks may be uncertain and where due-date constraints are flexible. Our aim is to provide a semantics for this problem and fix some criteria to analyse solutions obtained by Evolutionary Algorithms.

1   Introduction

In the last decades, scheduling problems have been subject to intensive research due to their multiple applications in areas of industry, finance and science [1]. More recently, fuzzy scheduling problems have tried to model the uncertainty and vagueness pervading real-life situations [2]. Scheduling problems are highly complex and practical approaches to solving them usually involve heuristic strategies. In particular, genetic algorithms have proved to be a powerful tool for solving scheduling problems, due to their ability to cope with huge search spaces involved in optimizing schedules [3]. This motivates our description of a fuzzy job-shop problem and the proposal of a genetic algorithm in order to solve it.

2   Description of the Problem

The *job shop scheduling problem,* also denoted *JSSP*, consists in scheduling a set of jobs $\{J_1, \ldots, J_n\}$ on a set of physical resources or machines $\{M_1, \ldots, M_m\}$, subject to a set of constraints. Each job J_i, $i = 1, \ldots, n$, consists of m tasks $\{\theta_{i1}, \ldots, \theta_{im}\}$ to be sequentially scheduled. Also, each task θ_{ij} requires the uninterrupted and exclusive use of one of the machines for its whole processing time, du_{ij}. Finally, we suppose that for each job there is a minimum starting time and a maximum finishing time, so that all its tasks must be scheduled within this time interval. Hence, there are three types of constraints: *precedence constraints, capacity constraints* and *release and due-date constraints*. The goal is twofold: we need to find a *feasible* schedule, so that all constraints hold and, at the same time, we want this schedule to be *optimal*, in the sense that its *makespan* (i.e., the time it takes to finish all jobs) is minimal.

---

* All authors supported by MCYT-FEDER Grant TIC2003-04153.

J. Mira and J.R. Álvarez (Eds.): IWINAC 2005, LNCS 3562, pp. 74–83, 2005.
© Springer-Verlag Berlin Heidelberg 2005

In real-life applications, it is often the case that the exact duration of a task is not known a priori. For instance, in ship-building processes, some tasks related to piece cutting and welding are performed by a worker and, depending on his/her level of expertise, the task will take a different time to be processed. Here, it is impossible to know a priori the exact duration of this task, even if an expert is able to estimate, for instance, the minimum time needed to process the task, the most likely processing time or an upper bound for the processing time. Clearly, classical job shop problems, are not adequate to deal with this type of situations. Instead, it is necessary to somehow model uncertain processing times and thus take advantage of the expert's knowledge.

In the last years, fuzzy sets have become a very popular tool to model uncertainty. Indeed, it is possible to find many examples in the literature where fuzzy sets are used to represent uncertain processing times (cf. [2]). In particular, we propose to represent each task's processing time by a *triangular fuzzy number* or *TFN*. For such a fuzzy set, the membership function takes a triangular shape, completely determined by three real numbers, $a^1 \leq a^2 \leq a^3$, as follows:

$$\mu_A(x) = \begin{cases} 0 & : x < a^1 \\ \frac{x-a^1}{a^2-a^1} & : a^1 \leq x \leq a^2 \\ \frac{x-a^3}{a^2-a^3} & : a^2 < x \leq a^3 \\ 0 & : a^3 < x \end{cases} \tag{1}$$

Every TFN A may be identified with the tuple formed by the three numbers a^1, a^2, a^3, using the notation $A = (a^1, a^2, a^3)$. Imprecise durations modelled using TFNs have a clear interpretation. a^1 can be interpreted as the shortest possible duration, a^2 as the most likely duration and a^3 as the longest possible duration. In the case that the exact duration of a task is known a priori and no uncertainty is present, the TFN model is still valid. The reason is that any real number $r \in \mathbb{R}$ can be seen as a special case of TFN $R = (r, r, r)$, with full membership at r and null membership anywhere else.

Once a task's starting time is known, its completion time will be calculated by adding the task's processing time to the given starting time. This can be done using *fuzzy number addition*, which in the case of TFNs $A = (a^1, a^2, a^3)$ and $B = (b^1, b^2, b^3)$ is reduced to adding three pairs of real numbers as follows:

$$A + B = (a^1 + b^1, a^2 + b^2, a^3 + b^3) \tag{2}$$

A consequence of this operation is that completion times are TFNs as well. Notice that this definition is coherent with traditional real-number addition.

Another situation where the need of fuzzy number arithmetic arises is when the starting time for a given task θ must be found. Here, it is necessary to find the maximum between two TFNs, the completion time of the task preceding θ in its job J and that preceding θ in its resource M. Now, given two TFNs $A = (a^1, a^2, a^3)$ and $B = (b^1, b^2, b^3)$, the *maximum* $A \vee B$ is obtained by extending the lattice operation max on real numbers using the Extension Principle. However, the resulting membership function might be quite complex to compute. Also, the result of such an operation, while still being a fuzzy number, is not guaranteed to be a TFN. For these reasons, we approximate $A \vee B$ by a TFN, $A \sqcup B$, as follows:

$$A \vee B \approx A \sqcup B = (a^1 \vee b^1, a^2 \vee b^2, a^3 \vee b^3) \tag{3}$$

Notice that in some cases the approximation $\sqcup$ coincides with the maximum $\vee$. Even if this is not the case, the support of both fuzzy sets $A \vee B$ and $A \sqcup B$ is always equal and the unique point x with full membership in $A \sqcup B$ also has full membership in $A \vee B$. Finally, notice that $\sqcup$ is coherent with the maximum of two real numbers.

Using the addition and the maximum $\sqcup$, it is possible to find the completion time for each job. The fuzzy makespan C_{max} would then correspond to the greatest of these TFNs. Unfortunately, the maximum $\vee$ and its approximation $\sqcup$ cannot be used to find such TFN, because they do not define a total ordering in the set of TFNs. Instead, it is necessary to use a method for *fuzzy number ranking* [4]. The chosen method consists in obtaining three real numbers $C_1(A), C_2(A), C_3(A)$ from each TFN A and then use real number comparisons. These three numbers are defined as follows:

$$C_1(A) = \frac{a^1 + 2a^2 + a^3}{4}, \quad C_2(A) = a^2, \quad C_3(A) = a^3 - a^1 \tag{4}$$

They can be used to rank a set of $n \geq 1$ TFNs according to Algorithm 1. This

1: order the *TFNs* according to the value of C_1
2: **if** there are TFNs with identical value of C_1 **then**
3:    order these TFNs using the real value C_2
4:    **if** there are TFNs with identical value of C_1 and C_2 **then**
5:       rank them using C_3

Algorithm 1. Ranking Method for TFNs

process establishes a total ordering in any set $A_1, \ldots, A_n$ of TFNs and, in the particular case of real numbers, yields exactly the same total ordering as the classical maximum for real numbers.

In practice, due-date constraints are often flexible. For instance, a customer may have a preferred delivery date d^1, but some delay will be allowed until a later date d^2, after which the order will be cancelled. Similarly, an early due-date d^1 exists, after which business profit starts to decline until a later date d^2, after which losses begin. We would then be completely satisfied if the job finishes before d^1 and after this time our level of satisfaction would decrease, until the job surpasses the later date d^2, after which date we will be clearly dissatisfied. The satisfaction of a due-date constraint becomes a matter of degree, our degree of satisfaction that a job is finished on a certain date. A common approach to modelling such satisfaction levels is to use a fuzzy set D with linear decreasing membership function:

$$\mu_D(x) = \begin{cases} 1 & : x \le d^1 \\ \frac{x-d^2}{d^1-d^2} & : d^1 < x \le d^2 \\ 0 & : d^2 < x \end{cases} \tag{5}$$

According to Dubois *et alt.* [2], such membership function expresses a flexible threshold "less than" and expresses the satisfaction level $sat(t) = \mu_D(t)$ for the ending date t of the job. However, when dealing with uncertain task durations, the job's completion time is no longer a real number t, but a TFN C. In this case, Sakawa and Kubota [5] propose to measure the degree to which a completion time C satisfies the due-date constraint D using the following *agreement index*:

$$AI(C,D) = \frac{area(D \cap C)}{area(C)} \tag{6}$$

Notice the similarity of this definition to that of degree of subsethood for discrete fuzzy sets.

Once we have established a means of modelling uncertain duration times and flexible due-dates, we can find a schedule for a given problem. For every job J_i, $i = 1, \ldots, n$, we may decide to what degree its completion time C_i satisfies the flexible due-date D_i, as given by $AI_i = AI_i(C_i, D_i)$ and we may use these completion times to obtain a fuzzy makespan, C_{max}. Now, we need to decide on the quality of this schedule, based on its feasibility and makespan.

To decide on the feasibility of the given schedule, we simply combine the satisfaction degrees AI_i, $i = 1, \ldots, n$. We may use the average aggregation operator, so the degree to which a schedule is feasible is given by:

$$z_1 = \frac{1}{n} \sum_{i=1}^{n} AI_i \tag{7}$$

A more restrictive approach is to combine the satisfaction degrees using the minimum aggregation operator as follows:

$$z_2 = AI_{min} = \min_{i=1..n} AI_i \tag{8}$$

The value of z_1 can be seen as the probability $Pr(F)$ of the fuzzy event F: "*the schedule is feasible*" over the finite domain of jobs $D = \{J_1, \ldots, J_n\}$, provided that the membership of job J_i in F is $\mu_F(J_i) = AI_i$, $i = 1, \ldots, n$. Similarly, z_2 corresponds to the necessity measure $N(F)$ of the fuzzy event F over the finite domain D. Both $z_1 \in [0,1]$ and $z_2 \in [0,1]$ measure the schedule's *degree of feasibility*, that is, the degree to which due-date constraints are satisfied. Obviously, the bigger this degree of feasibility, the better the schedule, so our goal should be to maximise these degrees z_1 and z_2.

Regarding the makespan C_{max}, the "smaller" it is, the better the schedule. Now, because C_{max} is a TFN, it is not totally clear what is meant by "smaller". If we consider the total ordering defined by the ranking method, it would mean a smaller $C_1(C_{max})$ and our goal should be to minimise:

$$z_3 = C_1(C_{max}) \tag{9}$$

Overall, we aim at maximising z_1 and z_2 (maximise feasibility) and, simultaneously, minimising z_3 (minimise makespan). Hence, given a set of possible schedules S, we need to decide which schedule $s \in S$ satisfies the following goals: G_1, maximise z_1; G_2, maximise z_2; and G_3, minimise z_3. If we consider this problem in the framework of fuzzy decision making, according to the model proposed by Bellman and Zadeh [6], the degree to which a given schedule $s \in S$ satisfies the three goals G_1, G_2, G_3 is given by $\min(\mu_{G_1}(s), \mu_{G_2}(s), \mu_{G_3}(s))$, where μ_{G_i} represents the degree to which s satisfies goal G_i, $i = 1, 2, 3$. Obviously, our aim is to find a schedule $s \in S$ maximising this satisfaction degree $\mu_D(s)$.

For the problem to be well-posed, the satisfaction degrees μ_{G_i} must be defined. For the first two goals G_1 and G_2 the satisfaction degree $\mu_{G_i}(s)$ should be given by $\mu_i(z_i)$ where $\mu_i : [0,1] \to [0,1]$ is an increasing function such that $\mu_i(0) = 0, \mu_i(1) = 1$, $i = 1, 2$. For the third goal G_3 the satisfaction degree $\mu_{G_3}(s)$ is given by $\mu_3(z_3)$, where $\mu_3 : [0, \infty] \to [0,1]$ is some decreasing function. Clearly, their exact definition should be dependent on the nature of the scheduling problem and should ideally be elicited by an expert. In practice, such an expert might not be available; in this case, we propose to use simple linear functions of the form:

$$\mu_i(z_i) = \begin{cases} 0 & : z_i \leq z_i^0, \\ \frac{z_i - z_i^0}{z_i^1 - z_i^0} & : z_i^0 < z_i < z_i^1, \\ 1 & : z_i \geq z_i^1 \end{cases} \quad \mu_3(z_3) = \begin{cases} 1 & : z_3 \leq z_3^0, \\ \frac{z_3 - z_3^0}{z_3^1 - z_3^0} & : z_3^0 < z_3 < z_3^1, \\ 0 & : z_3 \geq z_3^1 \end{cases} \quad (10)$$

where $i = 1, 2$ and the adequate values of z_j^0, z_j^1, $j = 1, 2, 3$ are determined in the experimentation process. The *objective function* for the FJSSP will then be given by:

$$f(s) = \min\{\mu_1(z_1), \mu_2(z_2), \mu_3(z_3)\} \quad (11)$$

where s is a possible schedule, and the solution to the FJSSP will be a schedule maximising the value of such objective function.

In summary, the *Fuzzy Job Shop Scheduling Problem* or *FJSSP* consists in maximising the objective function f subject to precedence and capacity constraints.

3   Using Genetic Algorithms to Solve FJSSP

In classical JSSP, the search for an optimal schedule is usually limited to the space of active schedules. The best-known algorithm to find active schedules is the *G&T Algorithm* [7], which allows to use complementary techniques to reduce the search space [8]. A possible extension of this algorithm for the FJSSP is proposed in Algorithm 2. Also, based on this algorithm and inspired in the work of Sakawa and Kubota [5], we propose a GA to solve the FJSSP.

In the proposed GA, *chromosomes* are a direct codification of schedules. If there are n jobs and m machines, each individual will be represented by a $n \times m$ matrix, where element (i, j) represents the completion time for the task in job

1: $A = \{\theta_{i1}, i = 1, \ldots, n\}$; /*first task of each job*/
2: **while** $A \neq \emptyset$ **do**
3:     Find the task $\theta' \in A$ with minimum earliest completion time /*$CT(\theta)^1$*/;
4:     Let M' be the machine required by θ' and B the subset of tasks in A requiring machine M';
5:     Delete from B any task that cannot overlap with θ'; /*$ST(\theta)^1 > CT(\theta')^3$*/
6:     Select $\theta^* \in B$ **according to some criteria** to be scheduled;
7:     Remove θ^* from A and insert in A the task following θ^* in the job if θ^* is not the last task of its job;

Algorithm 2. Fuzzy G&T

J_i requiring resource M_j. Each row then represents the schedule of a job's tasks over the corresponding resources.

Each chromosome in the *initial population* for the GA can be generated with fuzzy G&T algorithm, choosing a task at random from the Conflict Set B. To prevent premature convergence, it is advisable that the initial population be diverse enough. For this reason, a new individual will only be incorporated to the population if its similarity to the other members of the population is less than a given threshold σ, where similarity is measured using phenotype distance. For a given individual I, the *similarity* between two individuals I_1 and I_2 is defined as:

$$Sim(I_i, I_2) = \frac{\sum_{i=1}^{n} \sum_{j=1}^{m} |Pr_{I_1}(\theta_{ij}) \cap Pr_{I_2}(\theta_{ij})| + |Su_{I_1}(\theta_{ij}) \cap Su_{I_2}(\theta_{ij})|}{n \cdot m \cdot (m-1)}$$

(12)

where $Pr_I(\theta)$ is the set of tasks preceding θ in its machine according to the ordering induced by individual I and $Su_I(\theta)$ is the set of tasks following θ in its machine. Clearly, this method can become computationally very expensive for large populations. A possible solution to this problem is to divide the population into N sub-populations of a reasonable size and ensure diversity within each sub-population.

The value of the *fitness function* for a chromosome is simply the value of the objective function for the corresponding schedule, as given by (11).

The *crossover operator* consists in performing the fuzzy G&T algorithm and solving non-determinism situation using the information provided by the parents. Every time the conflict set B has more than one task, the criterion followed to select one of them is to choose that task with earliest completion time in the parents, according the ranking algorithm. The *mutation operator* is embedded in the crossover operator, so that, with a given probability pm, the task from the conflict set is selected at random.

The *general scheme* of the GA is designed to avoid premature convergence to local optima by using a niche-based system. Thus, the population is initially divided in N sub-populations, containing K individuals each. Each initial sub-population is generated ensuring enough diversity by means of similarity. These N sub-populations evolve separately, until a certain convergence is obtained. At this stage, these sub-populations are merged into a single population of NK

1: Generate initial population divided in N groups $P_1, \ldots, P_N$ containing K individuals each;
2: **while** terminating condition T_1 is not satisfied **do**
3:    **for** $i = 1; i \leq N; i++$ **do**
4:       **repeat**
5:          obtain 3 children by crossover and mutation in P_i;
6:          select the best of 3 children and the best of remaining children and parents for the new population NP_i;
7:       **until** a new population NP_i is complete
8:       Replace the worst individual in NP_i with the best of P_i.
9: Join $P_1, \ldots, P_N$ into a single population P;
10: **while** Terminating condition T_2 is not satisfied **do**
11:    Obtain a new population from P following the scheme above;

Algorithm 3. Genetic Algorithm for FJSSP

individuals, which will again evolve until some terminating condition holds. A pseudo-code description of the resulting GA can be seen in Algorithm 3.

4   Experimental Results

The results shown correspond to the benchmark problems from [5], three problems of size 6×6 and three problems of size 10×10. For each problem, we are given the duration of each task as a TFN, $Du = (t^1, t^2, t^3)$, and the flexible due-date for each job, $D = (d^1, d^2)$. Notice that the problem definitions are not complete, as no objective function is given in each case.

We shall now provide some semantics for FJSSP and its solutions. This will help us to find some criteria to analyse the quality of solutions and a heuristic method to fully determine the objective function in the absence of an expert. The underlying concept is that solutions to the FJSSP are *a-priori solutions*, found when the duration of tasks is not exactly known. It is then hard to know which schedule will be optimal, because it depends on the realisation of the task's durations, which is not known. Each schedule provides an ordering of tasks and, it is not until tasks are executed according to this ordering that we know their real duration and, hence, obtain a crisp schedule, the *a-posteriori solution*. Similar ideas, in a somewhat different framework, are proposed in [9].

For any experimental analysis of the problem to be performed, the objective function, i.e. $\mu_i, i = 1, 2, 3$, must be completely defined. This means that the exact values of $z_i^0, z_i^1, i = 1, 2, 3$ must be known. Hence, we need a methodology to automatically define these values in the absence of an expert who elicits them. According to the proposed semantics, the optimality of a solution to the FJSSP lies in the ordering of tasks it provides a priori, when the information about the problem is incomplete. This ordering should yield good schedules when put in practice and tasks have exact durations. For this reason, for a given FJSSP we shall generate a set of N crisp JSSP, which can be interpreted as realisations of the fuzzy problem. Random durations for each task are generated according to a probability distribution which is coherent with the fuzzy durations (namely, the TFNs membership functions normalised so that the additivity axiom holds). Crisp due-date restrictions are given by the latest possible date d^2. In addition to these N problems, we also consider three *limit problems* those where each task has its minimum possible, most likely and maximum possible duration.

The quality of a fuzzy solution should be measured based on feasibility and makespan for these crisp problems. In particular, we define two measures for a given crisp problem: the *Feasibility Error*, FE, as the proportion of due-date restrictions that do not hold for the given ordering, and the *Relative Makespan Error*, RME, as the makespan error with respect to a solution obtained with the Branch & Bound Method [1]. It should be noticed that the implementation used [10], does not take into account due-date restrictions, so the obtained makespan value is a lower bound for the makespan, but may not correspond to a feasible solution.

We then try different values of $z_i^0, z_i^1, i = 1, 2, 3$ and for each of the resulting objective functions we run the fuzzy GA M times. Each of the M executions provides an ordering of tasks ord_k, $k = 1, \ldots, M$. Given an ordering ord_k, for each crisp problem JSSP$_l$, $l = 1, \ldots, N$ we apply an algorithm of semiactive schedule building and obtain the values of FE and RME. We may now calculate the average value of RME, $\bar{E}$, and the average value of FE, $\bar{F}$, across the M executions and the N crisp problems. The optimal configuration would be that which simultaneously minimises the average relative makespan error $\bar{E}$ and the average feasibility error $\bar{F}$. However, it is not usually the case that a configuration exists complying with both criteria. A compromise solution is to take as optimal the configuration that minimises $\max(\bar{E}, \bar{F})$.

Interestingly, from experiments with all six problems, the best configuration achieved is $z_1^0 = 0.3, z_1^1 = 0.8, z_2^0 = 0, z_2^1 = 0.3$. Notice that this is semantically sound, because z_2 is more restrictive than z_1. The values for z_3^0 and z_3^1 correspond to the lower bound obtained by Branch&Bound for the limit JSSP with maximum task durations, t^3, and the maximum of the due-date upper bounds $\max_{i=1,\ldots,n} d_i^1$ respectively. Furthermore, even in the case where these values do not provide the optimal solutions, the difference in terms of $\bar{E}$ and $\bar{F}$ is very small. We may conclude that a heuristic method to automatically determine the objective function from the task durations and due-dates is to set $z_i^0, z_i^1, i = 1, 2, 3$ as above.

Once the objective function is defined, let us analyse the performance of the proposed GA, both in terms of the fitness value of a solution and its meaning as a *a-priori solution*. In Table 1 we present the results obtained with the proposed GA (labelled as FGA). We also present the fitness values obtained in [5] (denoted by S) for the sake of completeness, even if such value is only meaningful when the definition of the satisfaction levels $\mu_i, i = 1, 2, 3$ is known.

The average fitness value and its variance across several executions of the GA in Table 1 suggest that this algorithm converges adequately, with an average fitness value in all executions has a value which is close to 1. This is

Table 1. Results for 20 executions of the GA, with population size $N = 100$ for 6×6 and $N = 200$ for 10×10; similarity threshold $\sigma = 0.8$, number of generations with niche-evolution $I_{min} = 50$ for 6×6, $I_{min} = 100$ for 10×10; total number of generations $I_{max} = 100$ for 6×6, $I_{max} = 200$ for 10×10, crossover probability $p_c = 0.9$ and mutation probability $p_m = 0.03$. NB stands for the number of trials with best result. Errors have been estimated using 50 crisp problems. Average execution time on a Pentium 4 at 3Ghz is 9 seconds (0.88 to generate the initial population) for 6×6 and 97 seconds (27 to generate the initial population) for 10×10

Problem	Ver.	NB	Best	Avg.	Worst	Var.	$\%\bar{F}$	$\%\bar{E}$
6x6-1	S.	18	0.775	0.761	0.628	0.002		
	FGA	20	1	1	1	0	9.7	20
6x6-2	S.	19	0.792	0.779	0.542	0.003		
	FGA	20	1	1	1	0	1.7	13.6
6x6-3	S.	20	0.580	0.580	0.580	0		
	FGA	20	1	1	1	0	3.1	9.5
10x10-1	S.	1	0.714	0.574	0.439	0.009		
	FGA	9	1	0.964	0.832	0.002	17	18.74
10x10-2	S.	8	0.818	0.722	0.545	0.008		
	FGA	20	1	1	1	0	10.72	19.92
10x10-3	S.	1	0.560	0.525	0.475	0.003		
	FGA	1	0.989	0.815	0.597	0.007	19.92	25.85

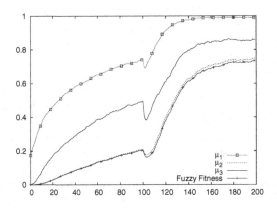

Fig. 1. Evolution of the average value of $\mu_i, i = 1, 2, 3$ in problem $10 \times 10 - 3$ for the configuration minimising $\max(\bar{E}, \bar{F})$ on a random sample of 50 crisp problems

further confirmed by running a t-test on problems $10 \times 10 - 1$ and $10 \times 10 - 3$. The obtained p-value indicates that the fitness value does not converge to a value lower than 0.97 and 0.81 respectively. Figure 1 further illustrates, as example, the convergence of the GA for problem $10 \times 10 - 3$. The fitness value decreases when niches are merged, but few generations are needed to recover previous fitness values, with a bigger slope in the evolution curves. We may conclude that the algorithm converges to a similar fitness value for all the solutions. However, it remains to see if all solutions are equally good a posteriori. Unfortunately, a variance analysis shows that this is not the case, concluding that differences in makespan and feasibility errors across different solutions are indeed significant.

5   Conclusions and Future Work

We have proposed a semantics for the FJSSP and a GA for solving it. Using both, we have obtained a heuristic method to completely define the objective function in the absence of expert's knowledge. Once the problem is well-specified, the semantics have also been used to analyse the solutions obtained with the GA, with good results.We believe that further analysis is needed, in order to select the best fuzzy schedule in terms of a posteriori performance. Furthermore, the proposed semantics could be used to develop and compare different methods for solving FJSSP. It would also be interesting to propose more benchmark problems, based on those available for the classical JSSP.

References

1. Brucker, P.: Scheduling Algorithms. 4th edn. Springer (2004)
2. Dubois, D., Fargier, H., Fortemps, P.: Fuzzy scheduling: Modelling flexible constraints vs. coping with incomplete knowledge. European Journal of Operational Research **147** (2003) 231–252
3. Mattfeld, D.C.: Evolutionary Search and the Job Shop Investigations on Genetic Algorithms for Production Scheduling. Springer-Verlag (1995)
4. Bortolan, G., Degani, R.: A review of some methods for ranking fuzzy subsets. In Dubois, D., Prade, H., Yager, R., eds.: Readings in Fuzzy Sets for Intelligence Systems. Morgan Kaufmann, Amsterdam (NL) (1993) 149–158
5. Sakawa, M., Kubota, R.: Fuzzy programming for multiobjective job shop scheduling with fuzzy processing time and fuzzy duedate through genetic algorithms. European Journal of Operational Research **120** (2000) 393–407
6. Bellman, R.E., Zadeh, L.A.: Decision-making in a fuzzy environment. Management Science **17** (1970) 141–164
7. Giffler, B., Thomson, G.L.: Algorithms for solving production scheduling problems. Operations Research **8** (1960) 487–503
8. Varela, R., Vela, C.R., Puente, J., Gómez, A.: A knowledge-based evolutionary strategy for scheduling problems with bottlenecks. European Journal of Operational Research **145** (2003) 57–71
9. Chanas, S., Kasperski, A.: Possible and necessary optimality of solutions in the single machine scheduling problem with fuzzy parameters. Fuzzy Sets and Systems **142** (2004) 359–371
10. Brucker, P., Jurisch, B., Sievers, B.: Code of a branch & bound algorithm for the job shop problem. `ftp://ftp.mathematik.uni-osnabrueck.de/pub/osm/-preprints/software/JOB-SHOP.tar.gz` (1994)

Memetic Algorithms with Partial Lamarckism for the Shortest Common Supersequence Problem

Carlos Cotta

Dept. Lenguajes y Ciencias de la Computación,
ETSI Informática, University of Málaga,
Campus de Teatinos, 29071 - Málaga, Spain
ccottap@lcc.uma.es

Abstract. The Shortest Common Supersequence problem is a hard combinatorial optimization problem with numerous practical applications. We consider the use of memetic algorithms (MAs) for solving this problem. A specialized local-improvement operator based on character removal and heuristic repairing plays a central role in the MA. The trade-off between the improvement achieved by this operator and its computational cost is analyzed. Empirical results indicate that strategies based on partial lamarckism (i.e., moderate use of the improvement operator) are slightly superior to full-lamarckism and no-lamarckism.

1 Introduction

The Shortest Common Supersequence Problem (SCSP) is a classical problem from the realm of string analysis. In essence, the SCSP consists of finding a minimal-length sequence S of symbols such that all strings in a certain set L are *embedded* in S (a more detailed description of the problem and the notion of embedding will be provided in next section). The SCSP provides a "clean" combinatorial optimization problem of great interest from the point of view of Theoretical Computer Science. In this sense, the SCSP has been studied in depth, and we now have accurate characterizations of its computational complexity. These characterizations range from the classical complexity paradigm (i.e., uni-dimensional complexity) to the more recent parameterized complexity paradigm (i.e., multidimensional complexity). We will survey some of these results in next section as well, but it can be anticipated that the SCSP is intrinsically hard [1, 2, 3] under many formulations and/or restrictions.

These hardness results would not be critical were the SCSP a mere academic problem. However, the SCSP turns out to be also a very important problem from an applied standpoint: it has applications in planning, data compression, and bioinformatics among other fields [4, 5, 6]. Thus, the practical impossibility of utilizing exact approaches for tackling this problem in general motivates attention be re-directed to heuristic approaches. Such heuristic approaches are aimed to producing *probably-* (yet not *provably-*) optimal solutions to the SCSP. Examples of such heuristics are the MAJORITY MERGE (MM) algorithm, and related

J. Mira and J.R. Álvarez (Eds.): IWINAC 2005, LNCS 3562, pp. 84–91, 2005.

variants [7], or the ALPHABET-LEFTMOST (AL) algorithm [8]. More sophisticated heuristics have been also proposed, for instance, evolutionary algorithms (EAs) [7, 9].

This work will follow the way paved by previous EA approaches to this problem. To be precise, the use of memetic algorithms (MAs) will be considered. The main feature of this MA is the utilization of a twofold local-improvement strategy: on one hand, a repair mechanism is used to restore feasibility of solutions, shortening them if possible; on the other hand, an iterated local-search strategy is used to further improve solution quality. The computational impact of this latter component will be here analyzed, and confronted with the quality improvement attainable.

2   The Shortest Common Supersequence Problem

First of all, let us define the notion of supersequence. Let s and r be two strings of symbols taken from an alphabet Σ. String s can be said to be a supersequence of r (denoted as $s \succ r$) using the following recursive definition:

$$s \succ r \triangleq \begin{cases} \text{TRUE} & \text{if } r = \epsilon \\ \text{FALSE} & \text{if } r \neq \epsilon \text{ and } s = \epsilon \\ s' \succ r' & \text{if } r = \alpha r' \text{ and } s = \alpha s', \ \alpha \in \Sigma \\ s' \succ r & \text{if } r = \alpha r' \text{ and } s = \beta s' \text{ and } \alpha \neq \beta, \ \alpha, \beta \in \Sigma \end{cases} \quad (1)$$

Plainly, the definition above implies that r can be embedded in s, meaning that all symbols in r are present in s in the very same order (although not necessarily consecutive). We can now formally define the decisional version of the SCSP:

SHORTEST COMMON SUPERSEQUENCE PROBLEM
Instance: A set L of m strings $\{s_1, \cdots, s_m\}$, $s_i \in \Sigma^*$ (where Σ is a certain alphabet), and a positive integer k.
Question: Does there exist a string $s \in \Sigma^*$, $|s| \leqslant k$, such that $s \succ s_i$ for all $s_i \in L$?

Obviously, associated with this decisional problem, we have its optimization version in which the smallest k is sought such that the corresponding instance is a yes-instance. Let us now consider the computational complexity of the SCSP.

The SCS problem can be shown to be NP-hard, even if strong constraints are posed on L, or on Σ. For example, it is NP-hard in general when all s_i have length two [5], or when the alphabet size $|\Sigma|$ is two [2]. At any rate, it must be noted that NP-hard results are usually over-stressed; in fact, the NP-characterization is a worst-case scenario, and such worst cases may be unlikely (for example, 3-SAT is NP-hard, yet most instances are easily solvable; only those located at the phase transition between satisfiability and non-satisfiability are hard to solve). A more sensible characterization of hardness is required in order to deal with these issues, and parameterized complexity is the key.

Parameterized complexity [10] approaches problems from a multidimensional perspective, realizing its internal structure, and isolating some *parameters*. If hardness (that is, non-polynomial behavior) can be isolated within these parameters, the problem can be *efficiently* solved for fixed values of them. Here, efficiently means in time $O(f(k)n^c)$, where k is the parameter value, n is the problem size, f is an arbitrary function of k only, and c is a constant independent of k and n. A paradigmatic example of this situation is provided by VERTEX COVER: it is NP-hard in general, but it can be solved in time $O(1.271^k+n)$, where n is the number of vertices, and k is the maximum size of the vertex cover sought [11,12]. Problems such as VERTEX COVER for which this hardness-isolation is possible are termed *fixed-parameter tractable* (FPT). Non-FPT problem will fall under some class in the W−hierarchy. Hardness for class $W[1]$ is the current measure of intractability.

Several parameterizations are possible for the SCSP. Firstly, the maximum length k of the supersequence sought can be taken as a parameter. If the alphabet size is constant, or another parameter, then the problem turns in this case to be FPT, since there are at most $|\Sigma|^k$ supersequences, and these can be exhaustively checked. However, this is not very useful in practice because $k \geqslant \max|s_i|$. If the number of strings m is used as a parameter, then SCS is $W[1]$−hard, and remains so even if $|\Sigma|$ is taken as another parameter [6], or is constant [3]. Failure of finding FPT results in this latter scenario is particularly relevant since the alphabet size in biological problems is fixed (e.g., there are just four nucleotides in DNA). Furthermore, the absence of FPT algorithms implies the absence of fully polynomial time approximation schemes (FPTAS) for the corresponding problem, that is, there does not exist an algorithm returning solutions within factor $1 + \epsilon$ from the optimum in time which is polynomial in n and $1/\epsilon$.

3   Heuristics for the SCSP

The hardness results mentioned in the previous subsection motivate the utilization of heuristic approaches for tackling the SCSP. One of the most popular algorithms for this purpose is MAJORITY MERGE (MM). This is a greedy algorithm that constructs a supersequence incrementally by adding the symbol most frequently found at the front of the strings in L, and removing these symbols from the corresponding strings. More precisely:

Heuristic MM $(L = \{s_1 \cdots, s_m\})$
1:    let $s \leftarrow \epsilon$
2:    do
3:        for $\alpha \in \Sigma$ do let $\nu(\alpha) \leftarrow \sum_{s_i=\alpha s_i'} 1$
4:        let $\beta \leftarrow \max^{-1}\{\nu(\alpha) \mid \alpha \in \Sigma\}$
5:        for $s_i \in L, s_i = \beta s_i'$ do let $s_i \leftarrow s_i'$
6:        let $s \leftarrow s\beta$
7:    until $\sum_{s_i \in L} |s_i| = 0$
8:    return s

The myopic functioning of MM makes it incapable of grasping the global structure of strings in L. In particular, MM misses the fact that the strings can have different lengths [7]. This implies that symbols at the front of short strings will have more chances to be removed, since the algorithm has still to scan the longer strings. For this reason, it is less urgent to remove those symbols. In other words, it is better to concentrate in shortening longer strings first. This can be done by assigning a weight to each symbol, depending of the length of the string in whose front is located. Branke *et al.* [7] propose to use precisely this string length as weight, i.e., step 3 in the previous pseudocode would be modified to have $\nu(\alpha) \leftarrow \sum_{s_i = \alpha s'_i} |s'_i|$.

Another heuristic has been proposed by Rahmann [8] in the context of the application of the SCSP to a microarray production setting. This algorithm is termed ALPHABET-LEFTMOST (AL), and its functioning can be described as follows:

Heuristic AL $(L = \{s_1 \cdots, s_m\}, \Pi = \langle \pi_1 \cdots \pi_{|\Sigma|} \rangle)$
1:      **let** $s \leftarrow \epsilon$
2:      **let** $i \leftarrow 1$
3:      **do**
4:          **if** $\exists s_j \in L : s_j = \pi_i s'_j$ **then**
5:              **for** $s_j \in L, s_j = \pi_i s'_j$ **do let** $s_j \leftarrow s'_j$
6:              **let** $s \leftarrow s\pi_i$
7:          **end if**
8:          **let** $i \leftarrow (i \; MOD \; |\Sigma|) + 1$
9:      **until** $\sum_{s_i \in L} |s_i| = 0$
10:     **return** s

As it can be seen, AL takes as input the list of strings whose supersequence is sought, and a permutation of symbols in the alphabet. The algorithm then proceeds with successive repetitions of this pattern until the all strings in L are embedded. Obviously, unproductive steps (i.e., when the next symbol in row does not appear at the front of any string in L) are ignored. Such a simple algorithm can provide very good results for some SCSP instances.

4   Memetic Algorithms for the SCSP

One of the difficulties faced by an EA (or by a MA) when applied to the SCSP is the existence of feasibility constraints, i.e., an arbitrary string $s \in \Sigma^*$, no matter its length, is not necessarily a supersequence of strings in L. Typically, these situations can be solved in three ways: (i) allowing the generation of infeasible solutions and penalizing accordingly, (ii) using a repairing mechanism for mapping infeasible solutions to feasible solutions, and (iii) defining appropriate operators and/or problem representation to avoid the generation of infeasible solutions. We have analyzed these three approaches elsewhere, and we have found that option (ii) provided better results than option (i) and (iii) (in this latter case, we considered an EA that used ideas from GRASP [13] as suggested in [14]). We will thus elaborate on this option.

Our MA evolves sequences in $|\Sigma|^\lambda$, where $\lambda = \sum_{s_i \in L} |s_i|$. Before being submitted for evaluation, these sequences are repaired using a function $\rho : \Sigma^* \times (\Sigma^*)^m \to \Sigma^*$ whose behavior can be described as follows:

$$\rho(s, L) = \begin{cases} s & \text{if } \forall i : s_i = \epsilon \\ \rho(s', L) & \text{if } \exists i : s_i \neq \epsilon \text{ and } \nexists i : s_i = \alpha s_i' \text{ and } s = \alpha s' \\ \alpha \rho(s', L|_\alpha) & \text{if } \exists i : s_i = \alpha s_i' \text{ and } s = \alpha s' \\ MM(L) & \text{if } \exists i : s_i \neq \epsilon \text{ and } s = \epsilon \end{cases} \quad (2)$$

where $L|_\alpha = \{s_1|_\alpha, \cdots, s_m|_\alpha\}$, and $s|_\alpha$ equals s' when $s = \alpha s'$, being s otherwise. As it can be seen, this repairing function not only completes s in order to have a valid supersequence, but also removes unproductive steps, as it is done in AL. Thus, it also serves the purpose of local improver to some extent. After this repairing, raw fitness (to be minimized) is simply computed as:

$$fitness(s, L) = \begin{cases} 0 & \text{if } \forall i : s_i = \epsilon \\ 1 + fitness(s', L|_\alpha) & \text{if } \exists i : s_i \neq \epsilon \text{ and } s = \alpha s' \end{cases} \quad (3)$$

As mentioned in Sect. 1, an additional local-improvement level is considered. To do so, we have considered the neighborhood define by the DELETE$_k$: $\Sigma^* \times (\Sigma^*)^m \to \Sigma^*$ operation [8]. The functioning of this operation is as follows:

$$\text{DELETE}_k(s, L) = \begin{cases} \rho(s', L) & \text{if } k = 1 \text{ and } s = \alpha s' \\ \alpha \text{DELETE}_{k-1}(s', L|_\alpha) & \text{if } k > 1 \text{ and } s = \alpha s' \end{cases} \quad (4)$$

This operation thus removes the k-th symbol from s, and then submits it to the repair function so that all all strings in L can be embedded. Notice that the repairing function can actually find that the sequence is feasible, hence resulting in a reduction of length by 1 symbol. A full local-search scheme is defined by iterating this operation until no single deletion results in length reduction:

Heuristic LS $(s \in \Sigma^*, L = \{s_1 \cdots, s_m\})$
```
1:   let k ← 0
2:   while k < |s| do
3:       let r ← DELETE_k(s, L)
4:       if fitness(r) < fitness(s) then
5:           let s ← r
6:           let k ← 0
7:       else
8:           let k ← k + 1
9:       end if
10:  end while
11:  return s
```

The application of this LS operator has a computational cost that we measure as the number of partial evaluations in step 3 above. More precisely, since the application of the repairing function starts at position k, we compute each application of DELETE_k to s as $(|r| - k)/|r|$ fitness evaluations. This is accumulated during the run of the MA to have a more sensible estimation of the search cost.

5   Experimental Results

The experiments have been done with a steady-state MA ($popsize = 100$, $p_X = .9$, $p_m = 1/n$, $maxevals = 100,000$), using binary tournament selection, uniform crossover, and random-substitution mutation. In order to analyze the impact of local search, the LS operation is not always applied, but randomly with probability p. The values $p \in \{0, 0.01, 0.1, 0.5, 1\}$ have been considered. We denote by $\text{MA}^{x\%}$ the use of $p = x/100$. Notice that $\text{MA}^{0\%}$ would then be a plain repair-based EA.

Two different sets of problem instances have been used in the experimentation. The first one is composed of random strings with different lengths. To be precise, each instance is composed of eight strings, four of them with 40 symbols, and the remaining four with 80 symbols. Each of these strings is randomly built, using an alphabet Σ. Four subsets of instances have been defined using different alphabet sizes, namely $|\Sigma| =2$, 4, 8, and 16. For each alphabet size, five different instances have been generated.

Secondly, a more realistic benchmark consisting of strings with a common source has been considered. A DNA sequence from a SARS coronavirus strain has been retrieved from a genomic database[1], and has been taken as supersequence; then, different sequences are obtained from this supersequence by scanning it from left to right, and skipping nucleotides with a certain fixed probability. In these experiments, the length of the supersequence is 158, the gap probability is 10%, 15%, or 20% and the number of so-generated sequences is 10.

First of all, the results for the random strings are shown in Table 1. All MAs perform notably better than AL. The results for MM (not shown) are similar to those of AL (more precisely, they are between 2.5% and 10% better, still far worse than the MAs). Regarding the different MAs, performance differences tend to be higher for increasing alphabet sizes. In general, MAs with $p > 0$ are better than $\text{MA}^{0\%}$ (the differences are statistically significant according to a Wilcoxon ranksum test [15] in above 90% of the problem instances). $\text{MA}^{1\%}$ provides somewhat better results, although the improvement with respect to the other MAs ($p > 0$) is only significant in less than 20% of the problem instances.

The results for the strings from the SARS DNA sequence are shown in Table 2. Again, AL performs quite poorly here. Unlike the previous set of instances, MM (not shown) does perform notably better than AL. Actually, it matches the performance of $\text{MA}^{0\%}$ for low gap probability (10% and 15%), and yields an average 227.8 for the larger gap probability. In this latter problem instance, the MAs with $p > 0$ seem to perform marginally better. $\text{MA}^{100\%}$ and $\text{MA}^{1\%}$ provide

---

[1] http://gel.ym.edu.tw/sars/genomes.html, accession AY271716.

Table 1. Results of the different heuristics on 8 random strings (4 of length 40, and 4 of length 80), for different alphabet sizes $|\Sigma|$. The results of AL are averaged over all permutations of the alphabet (or a maximum 100,000 permutations for $|\Sigma| = 16$), and the results of the EAs are averaged over 30 runs. In all cases, the results are further averaged over five different problem instances

	AL		$MA^{0\%}$		$MA^{1\%}$	
$|\Sigma|$	best	mean ± std.dev.	best	mean ± std.dev.	best	mean ± std.dev.
2	121.4	123.4 ± 2.0	111.2	112.6 ± 0.8	110.4	112.8 ± 1.0
4	183.0	191.2 ± 4.7	151.6	155.2 ± 1.9	149.4	152.7 ± 1.7
8	252.2	276.8 ± 6.4	205.4	213.5 ± 4.0	201.8	207.3 ± 2.2
16	320.6	352.9 ± 7.4	267.0	281.8 ± 5.9	266.2	274.2 ± 3.0
	$MA^{10\%}$		$MA^{50\%}$		$MA^{100\%}$	
$|\Sigma|$	best	mean ± std.dev.	best	mean ± std.dev.	best	mean ± std.dev.
2	111.6	113.1 ± 0.8	111.4	113.2 ± 0.8	111.2	113.1 ± 0.8
4	149.4	153.5 ± 1.5	150.0	153.3 ± 1.4	149.2	153.3 ± 1.6
8	202.0	207.9 ± 2.2	204.0	208.2 ± 2.0	203.0	208.2 ± 2.1
16	266.6	274.7 ± 3.0	268.4	275.0 ± 2.8	267.2	275.0 ± 3.2

Table 2. Results of the different heuristics on the strings from the SARS DNA sequence. The results of AL are averaged over all permutations of the alphabet (or a maximum 100,000 permutations for $|\Sigma| = 16$), and the results of the EAs are averaged over 30 runs

	AL		$MA^{0\%}$		$MA^{1\%}$	
gap%	best	mean ± std.dev.	best	mean ± std.dev.	best	mean ± std.dev.
10%	307	315.2 ± 6.8	158	158.0 ± 0.0	158	158.0 ± 0.0
15%	293	304.3 ± 8.8	158	158.0 ± 0.0	158	159.0 ± 2.8
20%	274	288.3 ± 8.6	165	180.8 ± 15.7	159	177.0 ± 9.3
	$MA^{10\%}$		$MA^{50\%}$		$MA^{100\%}$	
gap%	best	mean ± std.dev.	best	mean ± std.dev.	best	mean ± std.dev.
10%	158	158.0 ± 0.0	158	158.0 ± 0.0	158	158.0 ± 0.0
15%	158	159.8 ± 3.7	158	159.8 ± 3.0	158	159.1 ± 2.1
20%	163	179.4 ± 9.2	159	178.1 ± 9.9	161	176.5 ± 9.0

the best and second best mean results (no statistical difference between them). A Wilcoxon ranksum test indicates that the difference with respect to $MA^{0\%}$ is significant (at the standard 5% significance).

6   Conclusions

We have studied the deployment of MAs on the SCSP. The main goal has been to determine the way that local search affects the global performance of the algorithm. The experimental results seem to indicate that performance differences are small but significant with respect to a plain repair-based EA (i.e., no local search). Using partial lamarckism ($0 < p < 1$) provides in some problem instances better results, and does not seem to be harmful on any of the remaining instances. Hence, it can offer the best tradeoff between quality improvement and computational cost. Future work will be directed to confirm these results on other neighborhood structures for local search. In this sense, alternatives based on symbol insertions or symbol swaps can be considered [8].

Acknowledgements. This work is partially supported by Spanish MCyT and FEDER under contract TIC2002-04498-C05-02.

References

1. Bodlaender, H., Downey, R., Fellows, M., Wareham, H.: The parameterized complexity of sequence alignment and consensus. Theoretical Computer Science **147** (1994) 31–54
2. Middendorf, M.: More on the complexity of common superstring and supersequence problems. Theoretical Computer Science **125** (1994) 205–228
3. Pietrzak, K.: On the parameterized complexity of the fixed alphabet shortest common supersequence and longest common subsequence problems. Journal of Computer and System Sciences **67** (2003) 757–771
4. Foulser, D., Li, M., Yang, Q.: Theory and algorithms for plan merging. Artificial Intelligence **57** (1992) 143–181
5. Timkovsky, V.: Complexity of common subsequence and supersequence problems and related problems. Cybernetics **25** (1990) 565–580
6. Hallet, M.: An integrated complexity analysis of problems from computational biology. PhD thesis, University of Victoria (1996)
7. Branke, J., Middendorf, M., Schneider, F.: Improved heuristics and a genetic algorithm for finding short supersequences. OR-Spektrum **20** (1998) 39–45
8. Rahmann, S.: The shortest common supersequence problem in a microarray production setting. Bioinformatics **19** (2003) ii156–ii161
9. Branke, J., Middendorf, M.: Searching for shortest common supersequences by means of a heuristic based genetic algorithm. In: Proceedings of the Second Nordic Workshop on Genetic Algorithms and their Applications, Finnish Artificial Intelligence Society (1996) 105–114
10. Downey, R., Fellows, M.: Parameterized Complexity. Springer-Verlag (1998)
11. Chen, J., Kanj, I., Jia, W.: Vertex cover: further observations and further improvements. In: Proceedings of the 25th International Workshop on Graph-Theoretic Concepts in Computer Science. Number 1665 in Lecture Notes in Computer Science, Berlin Heidelberg, Springer-Verlag (1999) 313–324
12. Niedermeier, R., Rossmanith, P.: A general method to speed up fixed-parameter-tractable algorithms. Information Processing Letters **73** (2000) 125–129
13. Feo, T., Resende, M.: Greedy randomized adaptive search procedures. Journal of Global Optimization **6** (1995) 109–133
14. Cotta, C., Fernández, A.: A hybrid GRASP-evolutionary algorithm approach to golomb ruler search. In Yao, X., et al., eds.: Parallel Problem Solving From Nature VIII. Volume 3242 of Lecture Notes in Computer Science., Berlin, Springer-Verlag (2004) 481–490
15. Lehmann, E.: Nonparametric Statistical Methods Based on Ranks. McGraw-Hill, New York NY (1975)

2D and 3D Pictural Networks of Evolutionary Processors

K.S. Dersanambika[1,*], K.G. Subramanian[2], and A. Roslin Sagaya Mary[3]

[1] Department of Computer Science and Engineering,
Indian Institute of Technology, Madras
Chennai - 600 036, India
dersanapdf@yahoo.com
[2] Department of Mathematics,
Madras Christian College, Chennai - 600 059, India
kgsmani1948@yahoo.com
[3] Research Group on Mathematical Linguistics,
Rovira i Virgili University,
43005 Tarragona, Spain
writetoroslin@yahoo.co.uk

Abstract. Networks of Evolutionary Processors constitute an interesting computing model motivated by cell biology. Pictural networks of evolutionary processors have been introduced to handle generation of pictures of rectangular arrays of symbols. Here we introduce an extension of this model to generate pictures that are arrays, not necessarily rectangular, and compare these with array generating Puzzle grammars. We also provide an extension to generate 3D rectangular arrays.

1   Introduction

An interesting Computing model inspired by cell biology, called Network of Evolutionary processors, was introduced by Castellanos et al [1] and the investigation of this model continued in Castellanos et al [2] and Martin-Vide et al [6]. This notion has been carried over to pictures and Pictural networks of evolutionary processors (PNEP) have been considered by Mitrana et al [7] . A PNEP has nodes that are very simple processors able to perform just one type of operation, namely insertion or deletion of a row or substitution of a symbol in rectangular arrays. These nodes are endowed with filters defined by some membership or random context conditions.

On the other hand a variety of two-dimensional grammars generating picture languages of rectangular or non-rectangular arrays have been introduced and studied extensively in the literature [8, 4, 5]. Certain grammars for description of 3D pictures have also been introduced [11].

---

* This work is partially supported by University Grants Commission, India.

J. Mira and J.R. Álvarez (Eds.): IWINAC 2005, LNCS 3562, pp. 92–101, 2005.

In this paper networks of evolutionary processors generating pictures that are arrays of symbols, not necessarily rectangular, are introduced. Comparison with Array generating Puzzle grammars [8, 9] is made. Also extending the study in [7], generation of 3D rectangular pictures of symbols using Networks of Evolutionary Processors is considered.

2   Preliminaries

We refer to [4] for notions relating to two dimensional arrays. We recall the definition of a puzzle grammar introduced in [8, 9].

Definition 1. *A basic puzzle grammar* (BPG) *is a structure* $G = (N, T, R, S)$ *where* N *and* T *are finite sets of symbols;* $N \cap T = \phi$. *Elements of* N *are called non-terminals and elements of* T, *terminals.* $S \epsilon N$ *is the start symbol or the axiom.* R *consists of rules of the following forms:*
$A \rightarrow \textcircled{a}B, \; A \rightarrow a\,\textcircled{B}, \; A \rightarrow B\,\textcircled{a}, \; A \rightarrow \textcircled{B}\,a$

$$A \rightarrow \begin{matrix} \textcircled{a} \\ B \end{matrix}, \; A \rightarrow \begin{matrix} a \\ \textcircled{B} \end{matrix}, \; A \rightarrow \begin{matrix} B \\ \textcircled{a} \end{matrix}, \; A \rightarrow \begin{matrix} \textcircled{B} \\ a \end{matrix}$$

$A \rightarrow \textcircled{a}$ *where* $A, \; B, \; \in N$ *and* $a \in T$

Derivations begin with S written in a unit cell in the two-dimensional plane, with all the other cells containing the blank symbol $\#$, not in $N \cup T$. In a derivation step, denoted $\Rightarrow$, a non-terminal A in a cell is replaced by the right-hand member of a rule whose left-hand side is A. In this replacement, the circled symbol of the right-hand side of the rule used, occupies the cell to the right or the left or above or below the cell of the replaced symbol depending on the type of rule used. The replacement is possible only if the cell to be filled in by the non-circled symbol contains a blank symbol.

The set of pictures or figures generated by G, denoted $L(G)$, is the set of connected, digitized finite arrays over T, derivable in one or more steps from the axiom.

Definition 2. *A context-free puzzle grammar* $(CFPG)$ *is a structure* $G = (N, T, R, S)$ *where* N, T, S *are as in Definition 1 and* R *is the set of rewriting rules of the form* $A \rightarrow \alpha$, *where* α *is a finite, connected array of one or more cells, each cell containing either a nonterminal or a terminal symbol, with a symbol in one of the cells of* α *being circled.*

The set of pictures generated by a context-free puzzle grammar G is defined analogous to Definition 1

Example 1. The BPG $G_1 = (N, T, R, S)$ where $N = \{S, A, B\}$, $T = \{a\}$.
$R = \{S \rightarrow \textcircled{a}S, \; S \rightarrow \textcircled{a}A, \; B \rightarrow \textcircled{a}S, \; B \rightarrow a, A \rightarrow \begin{matrix} A \\ \textcircled{a} \end{matrix}, \; A \rightarrow \begin{matrix} B \\ \textcircled{a} \end{matrix}\}$

This BPG generates the picture language of pictures describing 'staircases' shown below.

$$
\begin{array}{ccc}
 & & a \\
 & & a \\
a & a & a \\
 & a & \\
 & a & \\
a & a & a
\end{array}
$$

Example 2. Consider the context free puzzle grammar, $G_2 = (N, T, R, S)$ where $N = \{S, C, D, B\}, T = \{a\}$

$$R = \{S \rightarrow \begin{array}{c} C\ @D \\ B \end{array}, B \rightarrow \begin{array}{c} @ \\ B \end{array}, C \rightarrow C\ @, D \rightarrow @D, C \rightarrow a, D \rightarrow a, B \rightarrow a\}$$

This CFPG generates the picture language of pictures describing 'token T' shown below.

$$
\begin{array}{cccccc}
a & a & a & a & a & a \\
 & & a & & & \\
 & & a & & & \\
 & & a & & &
\end{array}
$$

3   Contextual Pictural Networks of Evolutionary Processors(CPNEP)

We now introduce the notion of contextual pictural networks of evolutionary processors using contextual insertion and deletion rules. These rules are a special case of contextual insertion/deletion studied by Mitrana [10].

A *contextual pictural network of evolutionary processors* (CPNEP) of size n is a construct

$$\Gamma = (V, N_1, N_2, \ldots, N_n, G, N_{i_0}),$$

where:

- V is an alphabet,
- for each $1 \le i \le n$, $N_i = (A_i, M_i, PI_i, FI_i, PO_i, FO_i)$ is the i-th evolutionary node processor of the network. The parameters of every processor are:
  - A_i is a multiset of finite support of 2D pictures over V. This multiset represents the 2D pictures existing in the i-th node at the beginning of any computation. Actually, in what follows, we consider that each 2D picture appearing in any node at any step has an arbitrarily large number of copies in that node, so that we identify multisets by their supports. Therefore, the set A_i is the set of initial pictures in the i-th node.
  - M_i is a finite set of contextual evolution rules of one of the following forms:
    $A \rightarrow B$(substitution rules),
    $(a, \varepsilon) \rightarrow (a, A)(r)$(right cell insertion rules),
    $(\varepsilon, a) \rightarrow (A, a)(l)$(left cell insertion rules),

$$\begin{pmatrix} \varepsilon \\ a \end{pmatrix} \rightarrow \begin{pmatrix} A \\ a \end{pmatrix} (u)(\text{up cell insertion rules})$$

$$\begin{pmatrix} a \\ \varepsilon \end{pmatrix} \rightarrow \begin{pmatrix} a \\ A \end{pmatrix} (d)(\text{down cell insertion rules})$$

$(a, A) \rightarrow (a, \varepsilon)(r)(\text{right cell deletion rules}),$
$(A, a) \rightarrow (\varepsilon, a)(l)(\text{left cell deletion rules}),$

$$\begin{pmatrix} A \\ a \end{pmatrix} \rightarrow \begin{pmatrix} \varepsilon \\ a \end{pmatrix} (u)(\text{up cell deletion rules})$$

$$\begin{pmatrix} a \\ A \end{pmatrix} \rightarrow \begin{pmatrix} a \\ \varepsilon \end{pmatrix} (d)(\text{down cell deletion rules})$$

More clearly, the set of evolution rules of any processor contains either substitution, or deletion or insertion rules.

- PI_i and FI_i are subsets of V representing the input filter. This filter, as well as the output filter, is defined by random context conditions; PI_i forms the enforcing context condition and FI_i forms the forbidding context condition. A 2D picture $w \in V^{*2}$ can pass the input filter of the node processor i, if w contains each element of PI_i irrespective of the direction in which it appears, but w can contain no element of FI_i. Note that any of the random context conditions may be empty, in which case the corresponding context check is omitted.

With respect to the input filter we define the predicate

$$\rho_i(w) : w \ can \ pass \ the \ input \ filter \ of \ the \ node \ processor \ i.$$

- PO_i and FO_i are subsets of V representing the output filter. Analogously, a 2D picture can pass the output filter of a node processor if it satisfies the random context conditions associated with that node. Similarly, we define the predicate

$$\tau_i(w) : w \ can \ pass \ the \ output \ filter \ of \ the \ node \ processor \ i.$$

- $G = (\{N_1, N_2, \ldots, N_n\}, E)$ is an undirected graph called the *underlying graph* of the network. The nodes of G correspond to the evolutionary processors of the CPNEP. The edges of G, that is, the elements of E, are given in the form of sets of two nodes.
- N_{i_0} is the *output node* of the network.

By a configuration (state) of an CPNEP as above we mean an n-tuple $C = (L_1, L_2, \ldots, L_n)$, with $L_i \subseteq V^{*2}$ for all $1 \leq i \leq n$. A configuration represents the sets of 2D pictures (remember that each 2D picture appears in an arbitrarily large number of copies) which are present in any node at a given moment; clearly the initial configuration of the network is $C_0 = (A_1, A_2, \ldots, A_n)$. A configuration can change either by an *evolutionary* step or by a *communicating* step. When changing by an evolutionary step, each component L_i of the configuration is changed in accordance with the evolutionary rules associated with the node i.

Formally, we say that the configuration $C_1 = (L_1, L_2, \ldots, L_n)$ directly changes for the configuration $C_2 = (L_1', L_2', \ldots, L_n')$ by an evolutionary step, written as $C_1 \to C_2$ if L_i' is the set of 2D pictures obtained by applying the rules of R_i to the 2D pictures in L_i as follows (we present one of the cases of contextual insertion/deletion, the other cases being similar):

- A node having substitution rules performs a substitution as follows: one occurrence of the lefthand side of a substitution rule is replaced by the righthand side of that rule. If a letter can be replaced by more than one new letter (there are more than one substitution rules with the same lefthand side), then this replacement will be done in different copies of the original 2D picture, thus resulting in a multiset of new pictures, in which each 2D picture appears in an arbitrary number of copies.
  If the procedure above is applicable to more than one occurrence of the same letter, then each such occurrence will be replaced accordingly, thus resulting again in an even larger multiset of new 2D pictures, in which each 2D picture appears in an arbitrary number of copies.
- A node having a left cell insertion rule
  $(\varepsilon, a) \to (A, a)(l)$ performs a cell insertion as follows: A is inserted on the left of the cell containing a. Similarly for the other insertion rules.
- A node having a left cell deletion rule
  $(A, a) \to (\varepsilon, a)(l)$ performs a cell deletion as follows: A is deleted on the left of the cell containing a. Similarly for the other deletion rules. A cell can be deleted if it contains symbols in the lefthand side of the cell deletion rule, only.

More precisely, since an arbitrarily large number of copies of each 2D picture is available in every node, after an evolutionary step, in each node one gets again an arbitrarily large number of copies of any 2D picture which can be obtained by using any rule associated with that node as defined above. By definition, if L_i is empty for some $1 \le i \le n$, then L_i' is empty as well.

We say that the configuration $C_1 = (L_1, L_2, \ldots, L_n)$ directly changes for the configuration $C_2 = (L_1', L_2', \ldots, L_n')$ by a communication step, written as $C_1 \vdash C_2$ if for every $1 \le i \le n$,

$$L_i' = L_i \setminus \{w \in L_i \mid \tau_i(w) = \underline{true}\} \cup \bigcup_{\{N_i, N_j\} \in E} \{x \in L_j \mid (\tau_j(x) \wedge \rho_i(x)) = \underline{true}\}.$$

Note that the 2D pictures which can pass the output filter of a node are sent out irrespective of they being received by any other node.

Let $\Gamma = (V, N_1, N_2, \ldots, N_n, G, N_{i_0})$ be an CPNEP. By a computation in Γ we mean a sequence of configurations $C_0, C_1, C_2, \ldots$, where C_0 is the initial configuration, $C_{2i} \to C_{2i+1}$ and $C_{2i+1} \vdash C_{2i+2}$ for all $i \ge 0$.

If the sequence is finite, we have a finite computation. The result of any finite or infinite computation is a 2D picture language which is collected in a designated node called the output (master) node of the network. If $C_0, C_1, \ldots$ is a computation, then all 2D pictures existing in the node N_{i_0} at some step t –

the i_0-th component of C_t – belong to the 2D picture language generated by the network. Let us denote this language by $L(\Gamma)$.

Example 3. Consider the CPNEP generating 2D pictures of staircases

$$\Gamma = (\{a, A, B\}, N_1, N_2, N_3, N_4, N_5, K_5, N_5)$$

$$N_1 = (\{a\}, (a, \varepsilon) \to (a, A)(r), (A, \varepsilon) \to (A, A)(r),$$
$$\quad (A, \varepsilon) \to (A, B)(r), (C, \varepsilon) \to (C, A)(r), \{C\}, \{A, B\}, \{B\}, \phi.$$

$$N_2 = (\phi, A \to a, \{A, B\}, \phi, \{a, B\}, \{A\})$$
$$N_3 = (\phi, \begin{pmatrix} \varepsilon \\ B \end{pmatrix} \to \begin{pmatrix} B \\ B \end{pmatrix} (u), \begin{pmatrix} \varepsilon \\ B \end{pmatrix} \to \begin{pmatrix} C \\ B \end{pmatrix} (u), \{B\}, \{A\}, \{C\}, \phi)$$
$$N_4 = (\phi, B \to a, \{B, C\}, \{A\}, \{a, C\}, \{B\})$$

$$N_5 = (\phi, C \to a, \{a, C\}, \{A, B\}, \phi, \{a\})$$

Example 4. Consider the CPNEP generating pictures of token T

$$\Gamma = (\{a, A, B\}, N_1, N_2, K_2, N_2)$$

$$N_1 = (\{aAa\}, (a, \varepsilon) \to (a, a)(r), (\varepsilon, a) \to (a, a)(l),$$
$$\quad \begin{pmatrix} A \\ \varepsilon \end{pmatrix} \to \begin{pmatrix} A \\ A \end{pmatrix} (d), \begin{pmatrix} A \\ \varepsilon \end{pmatrix} \to \begin{pmatrix} A \\ B \end{pmatrix} (d), \phi, \phi, \{B\}, \phi)$$
$$N_2 = (A \to a, B \to a, \{B\}, \phi, \phi, \{a\})$$

Theorem 1. *(i) The families $\mathcal{L}$(2D-CPNEP) and $\mathcal{L}$(BPG) intersect. (ii) The family $\mathcal{L}$(2D-CPNEP) also intersects $\mathcal{L}$(CFPG).*

The statements are clear from Examples 1,2,3,4.

4   3D-Pictural Network of Evolutionary Processors

4.1   Three Dimensional Picture Languages

For a given alphabet V, a 3D picture p of size $l \times m \times n$ over V is a 3D array of the form $p = (a_{ijk})_{i \in \overline{1,l}, \, j \in \overline{1,m}, \, k \in \overline{1,n}}$ with $a_{ijk} \in V$ for $i \in \overline{1,l}, j \in \overline{1,m}, k \in \overline{1,n}$. We denote V^{***} the set of all 3D pictures over V (including the empty picture denoted by λ). A 3D picture (or rectangular)language over V is a subset of V^{***}. A 3D subpicture of a 3D picture p is a 3D sub array of p. A $(2 \times 2 \times 2)$ subpicture of p is called a cube of p. The set of all cubes of p is denoted by $B_{2,2,2}(p)$. In the sequel, we will identify the boundaries of a picture by surrounding it with the marker #.A picture p bounded by markers # is denoted by $\hat{p}$.

4.2   Recognizability of 3D-Rectangles

Here we consider local and recognizable 3D-rectangular languages [3].
A 3D-rectangle over the alphabet $\{a, b, c, d, e, f, g, h\}$ is given below:

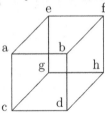

Given a 3D-rectangle p, $B_{i,j,k}(p)$ denotes the set of all sub 3D-rectangles p of
size (i, j, k), cube is a 3D-rectangle of size $(2, 2, 2)$. We denote by $\Sigma^{l \times m \times n}$ the set
of 3D-rectangles of size (l, m, n) over the alphabet Σ. $B_{2,2,2}$ is in fact a set of
cubes.

Definition 3. *Let Σ be a finite alphabet. The 3D-rectangular language $L \subset \Sigma^{***}$
is local if there exists a set of cubes $\triangle \subseteq (\Sigma \cup \{\#\})^{2 \times 2 \times 2}$ such that $L = \{p \in
\Sigma^{***} | B_{2,2,2}(\hat{p}) \subseteq \triangle\}$.*

The family of local picture languages, denoted by 3D-LOC, is a generalization of
the two dimensional case of local languages defined in [5]. Given a set of cubes
$\triangle$, the 3D-local picture language L generated by $\triangle$ is denoted by $L(\triangle)$.

Definition 4. *Let Σ be a finite alphabet. A 3D-rectangular language $L \subseteq \Sigma^{***}$ is
called recognizable if there exists a local 3D-rectangular language L' (given by a
set $\triangle$ of cubes) over an alphabet of Γ and a mapping $\Pi : \Gamma \rightarrow \Sigma$ such that
$L = \Pi(L')$.*

Example 5. The language L of 3D-rectangular pictures over single alphabet of
any size (l, m, n) surrounded by $\#$ symbol on all 6 faces is a local 3D rectangular
language [3]. Note that the 3D-rectangular languages over one letter alphabet
with all sides of equal length is not local but it is a recognizable language.

4.3   The Family 3D-PNEP

We now extend the notion of PNEP [7] to 3D rectangular pictures
    A *3D- pictural network of evolutionary processors* (3D-PNEP for short) of
size n is a construct

$$\Gamma = (V, N_1, N_2, \ldots, N_n, G, N_{i_0}),$$

where the components are as in [7] except that the objects are 3D rectangular
pictures and evolution rules are as follows:

$a \rightarrow b$, $a, b \in V$ (substitution rules),
$a \rightarrow \varepsilon(x)(b)$, $a \in V$ (back face deletion rules),

$a \to \varepsilon(x)(f)$, $a \in V$ (front face deletion rules),
$a \to \varepsilon(y)(r)$, $a \in V$ (right face deletion rules),
$a \to \varepsilon(y)(l)$, $a \in V$ (left face deletion rules),
$a \to \varepsilon(z)(t)$, $a \in V$ (top face deletion rules),
$a \to \varepsilon(z)(bo)$, $a \in V$ (bottom face deletion rules),

$\varepsilon \to a(x)(b)$, $a \in V$ (back face insertion rules),
$\varepsilon \to a(x)(f)$, $a \in V$ (front face insertion rules),
$\varepsilon \to a(y)(r)$, $a \in V$ (right face insertion rules),
$\varepsilon \to a(y)(l)$, $a \in V$ (left face insertion rules),
$\varepsilon \to a(z)(t)$, $a \in V$ (top face insertion rules),
$\varepsilon \to a(z)(bo)$, $a \in V$ (bottom face insertion rules),

$a \to \varepsilon(x)(/)$, $a \in V$ (front or back face deletion rules),
$a \to \varepsilon(y)(-)$, $a \in V$ (left or right face deletion rules),
$a \to \varepsilon(z)(|)$, $a \in V$ (top or bottom face deletion rules),

$\varepsilon \to a(x)(/)$, $a \in V$ (front or back face insertion rules),
$\varepsilon \to a(z)(|)$, $a \in V$ (top and bottom face insertion rules),
$\varepsilon \to a(y)(-)$, $a \in V$ (left or right face insertion rules),

More clearly, the set of evolution rules of any processor contains either substitution, or deletion or insertion rules.

Let $\Gamma = (V, N_1, N_2, \ldots, N_n, G, N_{i_0})$ be an 3D-PNEP. By a computation in Γ we mean a sequence of configurations $C_0, C_1, C_2, \ldots$, where C_0 is the initial configuration, $C_{2i} \to C_{2i+1}$ and $C_{2i+1} \vdash C_{2i+2}$ for all $i \geq 0$ where $\to$ denotes evolution and $\vdash$ denotes communication.

If the sequence is finite, we have a finite computation. The result of any finite or infinite computation is a 3D rectangular picture language which is collected in a designated node called the output (master) node of the network. If $C_0, C_1, \ldots$ is a computation, then all 3D rectangular pictures existing in the node N_{i_0} at some step t – the i_0-th component of C_t – belong to the 3D rectangular picture language generated by the network. Let us denote this language by $L(\Gamma)$.

4.4   Comparison

We start with some examples which constitute our basis for comparing the class of 3D rectangular picture languages generated by 3D-PNEP with other 3D rectangular picture generating devices.

Example 6. Let L_1 denote the set of all 3D rectangular pictures p over the alphabet $\{a\}$. The 3D rectangular language L_1 is described as $L_1 = \{p \in \{a\}^{***} | x(p), y(p), z(p) \geq 1\}$. The language L_1 can be generated in the fourth node N_4 by the following 3D-PNEP of size 4.

$\Gamma = (\{a, W\}, N_1, N_2, N_3, N_4, K_4, N_4)$, where
$N_1 = (\{\varepsilon\}, \{\varepsilon \to a(x)(/), \varepsilon \to W(x)(/)\}, \phi, \{A, T\}, \{W\}, \phi)$

$N_2 = (\phi, \{\varepsilon \to a(y)(-), \varepsilon \to T(y)(-)\}, \{W\}, \phi, \{T\}, \phi)$
$N_3 = (\phi, \{T \to a, W \to a\}, \{W, T\}, \phi, \{a\}, \{W, T\})$
$N_4 = (\phi, \{\varepsilon \to a(z)(|)\}, \{a\}, \{T\}, \phi, \{a\})$
Here $L(\Gamma) = L_1$.

Example 7. Let X be the 3D-rectangular picture of size $(2, 2, 2)$ over the alphabet $\{a\}$. Let L_2 be the set of all 3D-rectangular pictures p over the alphabet $\{a\}$ with all sides equal. This 3D-rectangular language L_2 can be formally described as

$$L_2 = \{p \in \{a\}^{***} \mid x(p) = y(p) = z(p) \geq 1\}.$$

L_2 can be generated in the output node N_5 by the following complete 3D-PNEP of size five:

$$\Gamma = (\{a, A, B, C\}, N_1, N_2, N_3, N_4, N_5, K_5, N_5),$$

where $N_1 = (\{X\}, \{\varepsilon \to A(x)(b)\}, \{a\}, \{A, B, C\}, \{A\}, \emptyset)$
$N_2 = (\emptyset, \{\varepsilon \to B(z)(bo)\}, \{a, A\}, \{B, C\}, \{B\}, \emptyset)$
$N_3 = (\emptyset, \{\varepsilon \to C(y)(l)\}, \{a, A, B\}, \{C\}, \{C\}, \emptyset)$
$N_4 = (\emptyset, \{A \to a, B \to a, C \to a\}, \{A, B, C\}, \emptyset, \{a\}, \{A, B, C\})$
$N_5 = (\emptyset, \{a \to a\}, \{a\}, \{A, B, C\}, \emptyset, \{a\})$.
Here $L(\Gamma) = L_2$

Now we set:

  3D-LOC is the class of local 3D picture languages [3].
  3D-REC is the class of recognizable picture languages
  $\mathcal{L}$(3D-PNEP) is the class of 3D-rectangular picture languages generated by
  3D-PNEP's.

Now we are ready to give the result:

Theorem 2. *(i) The families $\mathcal{L}$(3D-PNEP) and 3D-LOC are incomparable but not disjoint. (ii) The family $\mathcal{L}$(3D-PNEP) intersects the family 3D-REC.*

Proof. (i) The language of 3D-rectangular pictures over $\{a\}$ from Example 6 is in 3D-LOC ([3]) and $\mathcal{L}$(3D-PNEP). On the other hand, the language from Example 7 is not in 3D-LOC ([3]) but is in $\mathcal{L}$(3D-PNEP). The language of 3D-arrays of equal size in which all the diagonal elements are 1 and the remaining elements are 0 and it is known to be in 3D-LOC [3] but it cannot be generated by any 3D-PNEP as, informally speaking, 3D-PNEP's have no ability to fix the position of symbols 1 in the diagonal when using face insertion rule. Hence 3D-LOC and $\mathcal{L}$(3D-PNEP) are incomparable.
(ii)The language generated by the 3D-PNEP in Example 7 is in 3D-REC[3]

5   Conclusion

Here we have extended the study of generation of pictures of rectangular arrays to 2D pictures of arrays not necessarily rectangular by Networks of Evolutionary Processors. Extension of [7] to pictures of 3D rectangular arrays are also considered.

References

[1] J. Castellanos, C. Martín-Vide, V. Mitrana, J. Sempere. *Solving NP-complete problems with networks of evolutionary processors.* IWANN 2001 (J. Mira, A. Prieto, eds.), *LNCS 2084*, Springer-Verlag, 621–628, 2001.

[2] J. Castellanos, C. Martín-Vide, V. Mitrana, J. Sempere. Networks of evolutionary processors, *Acta Informatica*, 39: 517-529, 2003.

[3] K.S. Dersanambika and K. Krithivasan. *Recognizability of 3D Rectangular Pictures*, Paper presented at the 12th International Conference of Jangeon Mathematical Society, University of Mysore, Dec. 2003.

[4] R. Freund. Control Mechanisms on contextfree array gramars, In: Gh. Paun (ed) *Mathematical Aspects of Formal and Natural Languages*, World Scientific, 97-137, 1994.

[5] D. Giammarresi and A. Restivo, Two-dimensional Languages. in *Hand book of Formal languages*, eds. A.Salomaa et al., volume 3, Springer-Verlag, pages215-269, 1997.

[6] C. Martin-Vide, V. Mitrana, M. J. Perez-Jimenez, F. Sancho-Caparrini. Hybrid networks of evolutionary processors,   Genetic and Evolutionary Computation Conference (GECCO 2003), *Lecture Notes in Computer Science 2723*, Springer Verlag, Berlin, 401-412, 2003.

[7] V. Mitrana, K.G. Subramanian, M. Tataram. Pictural networks of evolutionary processors,*RomJIST*,6(1-2): 189-199, 2003.

[8] M. Nivat, A. Saoudi, K.G. Subramanian, R. Siromoney and V.R. Dare. Puzzle Grammars and Context-free Array Grammars, *IJPRAI*,5: 663-675, 1992.

[9] K.G. Subramanian, R. Siromoney, V.R. Dare, and A. Saoudi, Basic Puzzle Languages, *IJPRAI*, 9: 763-775,1995.

[10] V. Mitrana. contextual Insertion and Deletion. *Mathematical Linguistre and related topic*(Gh.Pann ed), The publishing of house of the Romanian Academic 271-278, 1994.

[11] P.S.P. Wang, 3D Sequential/Parallel Universal array grammars for polyhedral Object Pattern analysis, *IJPRAI*, 8: 563-576,1994.

Analysing Sentences with Networks of Evolutionary Processors[*]

Gemma Bel Enguix[1,2] and M. Dolores Jimenez Lopez[2]

[1] Department of Computer Science,
University of Milan-Bicocca,
Via Bicocca degli Arcimboldi, 8, 20126 Milan, Italy
gemma.bel@urv.net
[2] Research Group on Mathematical Linguistics,
Universitat Rovira i Virgili
Pl. Imperial Tarraco, 1, 43005 Tarragona, Spain
mariadolores.jimenez@urv.net

Abstract. A very simple implementation of NEPs is introduced to accept and analyze linguistic structures with the shape $NP\ V\ NP$. The formalization takes advantage of NEPs' features -modularity, specialization and parallelism- to develop a syntactic recognizer that is able to distinguish correct sentences working with lineal strings as input and lineal labeled structures as output.

1 Introduction

Networks of evolutionary processors (NEP) are a new computing mechanism directly inspired in the behavior of cell populations. Every cell is described by a set of words (DNA) evolving by mutations, which are represented by operations on these words. At the end of the process, only the cells with correct strings will survive. In spite of the biological inspiration, the architecture of the system is directly related to the Connection Machine [20] and the Logic Flow paradigm [14]. Moreover, the global framework for the development of NEPs has to be completed with the biological background of DNA computing [27], membrane computing [26] – that focalizes also in the behavior of cells –, and specially with the theory of grammar systems [8], which share with NEPs the idea of several devices working together and exchanging results.

First precedents of NEPs as generating devices can be found in [11] and [10]. The topic was introduced in [3] and [25], and further developed in [2], [4], [9]. A new approach to networks of evolutionary processors as accepting devices has started in [24].

---

[*] This research has been supported by a Marie Curie Fellowship of the European Community programme *Human Potential (IHP)* under contract number HPMF-CT-2002-01582 and by a Marie Curie European Reintegration Grant (ERG) under contract number MERG-CT-2004-510644.

J. Mira and J.R. Álvarez (Eds.): IWINAC 2005, LNCS 3562, pp. 102–111, 2005.

NEPs can be defined as mechanisms which evolve by means of processors that act according some predefined rules, whose outcome travels to the other nodes if they accept it after passing a filtering process. This functioning allows the specialization of each processor, what is a quite suggestive feature for linguistics.

The present work aims to be a preliminary implementation of NEPs for Natural Language Processing (NLP), especially for the recognition and analysis of simple syntactic structures. The main idea is to model NEPs that can accept simple linguistic strings, adapting and simplifying some ideas introduced in [24]. To do so, some important concepts that can relate NEPs and natural language are given in Section 2. In Section 3 we introduce the main features of NEPs for NLP and the formal definition of the system. An example is given in Section 4. Finally, Section 5 is devoted to summarize the results and remark some future lines of research in the area.

2   NEPs, Modularity and Syntax

The idea of several autonomous and specialized processors collaborating in the same generation or recognition process, with a constant exchange of information, can suggest the concept of *modularity*. Modularity has shown to be a very important idea in a wide range of fields. Cognitive science, natural language processing, computer science and, of course, linguistics are examples of fields where modular models have been proposed.

It is a commonplace belief in cognitive science that complex computational systems are at least weakly decomposable into components. In general, modular theories in cognitive science propose a number of independent but interacting cognitive 'modules' that are responsible for each cognitive domain. Specific arrangement of those modules usually varies in each theory, but generally each mental module encapsulates a definable higher mental function. There may be, for example, separate structures for spatial reasoning, mathematical ability, musical talent, phonological skill, oral lexicon, written lexicon, nonverbal thought, and verbal thought to name a few. Even though the idea of modularity has been implicit in cognitive science for a long time, it is with the publication of *The Modularity of Mind* [17] when those implicit ideas that had been current over the previous two decades crystallized into a single recognizable hypothesis: the mind is not a seamless, unitary whole whose functions merge continously into one another; rather, it comprises a number of distinct, specialized, structurally idiosyncratic modules that communicate with other cognitive structures in only very limited ways.

Fodor's theory is not by far the only one about modularity of mind. In fact, in the 1980s there starts a new trend represented by authors such as Chomsky [6], Garfield [18], Jackendoff [21] and, of course, Fodor himself, who defend the non-homogeneous character of mind. The theory of modularity is also present in linguistic approaches. In fact, the modular approach to grammar has been shown to have important consequences for the study of language (cf. [28]). This has led many grammatical theories to use modular models. The idea of having a system made up of several independent components (syntax, semantics, phonology, morphology, etc.) seems to be a good choice to account for linguistic issues. We may cite several modular approaches in Linguistics, from Chomsky's Generative Grammar [5] to Autolexical Syntax [28] or Jackendoff's view of the Architecture of Language Faculty [21] just to name a few. One of the main advantages of modular grammar is that they can reduce delays imposed by monolithic or non-modular grammars. This reduction of delays is due to the fact that, in modular grammars, subsystems of grammatical principles can be applied independently of each other and, sometimes, in a parallel fashion. It seems therefore that this idea of linguistic production is quite related to the working of a NEP.

Several authors have defended as well internal modularity in the different dimensions of grammar [29, 15, 19]. In [7], for example, it is suggested a highly modular organization of *syntax* where modules are determined by representations they recover. Also in [1, 16] different modularizations of the syntactic module are proposed. In those approaches syntax can be divided into small components with a high degree of specialization, explaining independently some aspects of linguistic structures. From this point of view, we think that NEPs provide a quite suitable theoretical framework for the formalization of modularity in syntax for both processes, generation and recognition sentences.

Here we introduce a formalization for implementing the recognition of correct sentences of natural language. The idea is not original, because a preliminary approach to accepting HNEPs has been already introduced in [24]. Nevertheless, this is the first attempt to deal with linguistic issues from this perspective. Our paper tries operating by means of specializing each one of the nodes of a NEP for the labeling of different phrases. In the present paper, we introduce the idea that the system that can recognize strings can also analyze sentences of the natural language. Our goal is to construct a NEP able to decide whether or not a string of natural language is correct, and analyze its structure. We want to do that using only lineal inputs and outputs, this is, our goal is not to generate trees, but recognize and label the internal pattern structure of sentences.

The linguistic NEP for analyzing simple sentences has to recognize every word, make a map of its linguistic features, establish the beginning and the end to the complex units, gather the non-terminal units in phrases and finally give a lineal structural version of the whole sentence. We think the NEPs are a good way for designing processors being allowed to account for only one given syntactical function. In this way, and thanks to the filters, it may be easy to construct a semi-universal automaton for comprehension of almost every well-formed sentence in a language. In the meantime, the device we are to describe now is just modeled for the processing of simple sentences with subcategorized complements NP.

3   Formalizing a NEP for the Analysis of Simple Sentences

In order to implement a NEP for analyzing simple sentences, we propose the nodes of the system to be specialized for accepting and labeling different types of structures inside a sentence: nominal phrases (NP), verbs (V), prepositional phrases (PP), or adverbial complements (AdvP). In the present paper, in order to test if it is possible such a functioning, we are going to construct a device devoted to the recognition of correct syntactic structures with the shape $[S \ V \ O]$, where $S \to [NP]$, $O \to [NP]$, this is, sentences with the form $[[NP] \ V \ [NP]]$, being this a common arrangement in syntax. Speakers use very frequently sentences of this type, such as "*I have a book*". We think that, if it is possible to define such a device, then it will be quite easy to formalize other mechanisms able to deal with more complex linguistic strings.

The system has two main goals: a) to recognize correct strings, like an automaton could do, and b) to give an lineal output explaining the syntactical structure of the input sentence. For the system to be able to achieve the second objective, we introduce a set of labels and rules for labeling sub-strings of non-terminal symbols.

When assigning to the processors their specialized functions, we make some theoretical options. First, a node will be defined whose only mission is to be the input node, reading the terminal symbols of the input string. On the other hand, a node will be specialized only in the labeling and recognition of phrases in order to produce the final output string. Moreover, since the structure we are working with has just two types of sub-structures, namely NP and V, at least two specialized nodes are needed, one for the recognition and labeling of NP and the other one for the recognition and analysis of V. In this way, at least a graph of four nodes has to be designed. Finally, as a methodological option, we design a support node, which will process some of the terminal elements that are part of a NP structure. The reason for having a processor like this in the system is to decrease the complexity in the working of the element that deals with NP, even if it can be thought than the general complexity increases. This support node cannot use labeling rules.

Therefore, for formalizing such a device, the following nodes are needed: a) input node, b) output node, c) a node for labeling NP, d) a node for analyzing V, e) a transition node. In the input filter of specialized nodes, the only elements accepted will be those that can be part of phrases they can recognize. In the output filter of these nodes, only labeled phrases will be allowed to pass and be sent to the other filters. However, in the support node, no labeled element will belong to the input filter, because it can accept just terminal elements, sending non-terminal structural symbols. The graph structure of the NEP (Figure 1) is not complete. Exchange of elements is possible only between nodes that are connected by an edge. Hence, no communication is possible between N_0 and N_4. We think this is an important feature for the decreasing of complexity in the transition, because it avoids the elements of N_0 to be sent directly to N_4, which cannot accept them, because the filter discards terminal symbols in the output node. In the same way, neither N_1 and N_3 are communicated, being N_1 a node for the processing of non-nuclear elements of NP, which cannot be part of V'.

The recognition of the symbols performed by the NEP will be done in two steps: a) classification of the terminal symbols and rewriting by non-terminal

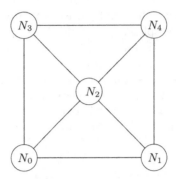

Fig. 1. Graph structure

items corresponding to grammatical categories, and b) recognition and labeling of nominal phrases and sentences. To do that, two types of alphabets are necessary: V, the alphabet of the input symbols, which are terminal strings, this is, lexical items, and Σ the alphabet of categorial symbols, which correspond to grammatical categories together with a matrix of morphological traits, if needed. For simple sentences we are dealing with, the symbols belonging to Σ will be $[N]$, $[V]$, $[ART]$, $[ADJ]$.

For accomplishing grammatical rules of agreement between subject and verb, some of these symbols have to be completed with the introduction of morphological traits. First of all, two different marks will be introduced in the category $[V]$ in order to distinguish between the two different forms of the English verb in present: 1 stands for the general form, and 2 for the third person. In this way, when the node receiving the lexical item analyzes it, it performs the rewriting as $[V]^1$ or $[V]^2$. On the other hand, the verb should bear an explicit mark of the type of complement it needs at the right. In the structures we are working with, it is always a NP, but it could be a PP or $AdvP$. This syntactical constraint is enclosed in brackets after V. Therefore, the final form of the categorial symbol V is $[V(\#)]^{\{1,2\}}$, where the super index mean the agreement with the subject and the symbol $\#$ stands for the syntactical structure the verb asks for. In our case, $(\#)$ will be always NP.

Moreover, in order to fulfill the agreement with the verb, $[N]$ has to be recognized with the same parameters than the verb, $\{1, 2\}$. On the other hand, for being accomplished the agreement between the forms inside the phrase, it has to be marked with s (singular) or p (plural). Therefore, the final form for the categorial symbol $[NP]$ is $[NP]^{\{1,2\}}_{\{s,p\}}$. For distinguishing the article "a" from the article "the", the feature $[ART]$ will be $[ART]_s$ for "a", and $[ART]$ for "the", where the absence of any symbol means it works for both singular and plural. If the agreement is not accomplished inside NP or in the NP at the left of the verb and V, then the sentence will not be recognized. No additional information is required for $[ADJ]$, which has just one form in English.

For delimiting the phrases as the gathering of several elements belonging to Σ and sentences as the gathering of phrases, we introduce a set of labelers, B, which are able to identify different linguistic structures, isolate and classify them.

Finally, in order to perform both types of operations two different sets of rules are considered: N for rewriting and β for labeling. N is an application of Σ in V. β is a packer of non-terminal symbols into phrases. When a string over Σ is labeled by β, then the symbols of internal agreement do not appear any more, and the marks of agreement with other phrases take an external position. For example when $[ART]_s[N]_s^2$ becomes $[NP]$, then s, which is important for the coherence of the phrase, is deleted, and 2 goes to the external label $[\,[ART]\,[N]\,]_{NP}^2$.

With the elements we have just explained, a NEP for sentence analysis can be defined as follows:

Definition 1

$$\Gamma = (V, \Sigma, B, N_1, .., N_i, G)$$

where V is the input vocabulary, Σ is the output vocabulary, B is a set of labelers, $N_i = (M_i, \beta_i, A_i, PI_i, PO_i)$ are the processors of the network, with M_i the set of the evolution rewriting rules of every node, β_i the set of evolution labeling rules, A_i the set of strings over V in the initial configuration, PI_i the input filter over $V \cup \Sigma$ and PO_i the output filter over $V \cup \Sigma$. G is the graph.
$i = 5$ for the sentences we are dealing with, $NP\ V\ NP$, and the methodological options explained above.

The computation works like in a regular NEP, combining evolutionary steps and communication steps. In an evolutionary step both a rewriting and labeling rule can be applied to the same string. While the system is in an evolutionary step, the initial node is reading the new string, and when the system is a communicating step, the node N_0 sends the symbol read to the other nodes, in a way that, two or more operations can be made in the system at the same time. We remark that, in communication steps, a string that has been already sent cannot be sent again if an evolutionary rule is not applied in it.

The system stops when no string is N_0 to be read, no rule of evolution can be applied and no evolutionary step can be performed. The string which has been processed will be correct if, when the system stops, a string with the label $[\,]_O$ is placed in the output node N_4.

4   An Example

In this section, a NEP will be implemented for the recognition of sentences $NP\ V\ NP$, in this case, it will be the string **Your words have a horrible resonance**. As explained in the last section, the number of nodes of the NEP is 5, being the structure as follows:

$$\Gamma = (V, \Sigma, B, N_1, N_2, N_3, N_4, N_5, G)$$

where:

$V = \{$Your, words, have, a, horrible, resonance$\}$

$\Sigma = \{ [ADJ], [ART], [N], [V(NP)] \}$

$B = \{ [\,]_{NP}, [\,]_O \}$

$N_0 = (\emptyset, \emptyset, \{your\}, V^+, V^+)$

$N_1 = (M_1, \emptyset, \emptyset, \{$a, your, horrible$\}, \{[ART], [ADJ]\})$

$\quad M_1 = \{$a $\rightarrow [ART]_s$, $\{$your, horrible$\} \rightarrow [ADJ]\}$

$N_2 = (M_2, \beta_2, \emptyset, \{$words, resonance, $[ART], [ADJ]\}, \{[\,]_{NP}\})$

$\quad M_2 = \{$words $\rightarrow [N]^1_p$, resonance $\rightarrow [N]^2_s\}$

$\quad \beta_2 = \{[ADJ][N]^1_p \rightarrow [\, [ADJ][N] \,]^1_{NP}, [ART]_s[ADJ][N]_s \rightarrow [\, [ART][ADJ][N] \,]^2_{NP}\}$

$N_3 = (M_3, \emptyset, \emptyset, \{$have$\}, \{[V(NP)]\})$

$\quad M_3 = \{$have $\rightarrow [V(NP)]^1\}$

$N_4 = (\emptyset, \beta_4, \emptyset, \{[\,]_{NP}, [V(NP)]\}, \emptyset)$

$\quad \beta_4 = \{[\,]^1_{NP} [V(NP)]^1 [\,]_{NP} \rightarrow [\, [\,]^1_{NP} [V(NP)]^1 [\,]_{NP} \,]_O\}$

The development of the computation is the following:

Initial state C_0

– In C_0, N_0 reads the word *"your"* and the computation starts.

Communication step $C_0 \vdash C_1$

– The reading node N_0 sends the first string, *"your"*, to every node. But just N_1 has it in the input filter, so it is the only node receiving it.

Evolutionary step $C_1 \Longrightarrow C_2$

– The second lexical item *"words"* is read by N_0.
– N_1 applies the rule your $\rightarrow [ADJ]$.

Communication step $C_2 \vdash C_3$

– N_0 sends *"words"* and N_2 accepts it.
– N_1 sends $[ADJ]$ and only N_2 accepts it.

Evolutionary step $C_3 \Longrightarrow C_4$

- The item *"have"* is read by N_0.
- N_2 applies the rule words $\rightarrow [N]^1_p$
- N_2 can apply then the rule $[ADJ][N]^1_p \rightarrow [\ [ADJ][N]\ ]^1_{NP}$

Communication step $C_4 \vdash C_5$

- The item *"have"* is send by N_0 and accepted by N_3.
- N_2 sends $[\ [ADJ][N]\ ]^1_{NP}$, which is accepted by N_4, the only one able to accept labeled strings.

Evolutionary step $C_5 \Longrightarrow C_6$

- N_0 reads *"a"*.
- N_3 applies the rule have $\rightarrow [V(NP)]^1$.

Communication step $C_6 \vdash C_7$

- N_0 sends *"a"*. N_1 accepts it.
- N_3 sends $[V(NP)]^1$ to every node. N_4 is the only one that accepts it.

Evolutionary step $C_7 \Longrightarrow C_8$

- N_0 reads the symbol *"horrible"*.
- N_1 applies the rule a $\rightarrow [ART]_s$.

Communication step $C_8 \vdash C_9$

- N_0 sends the lexical item *"horrible"*, which gos to N_1.
- N_1 sends $[ART]_s$ and N_2 accepts it.

Evolutionary step $C_9 \Longrightarrow C_{10}$

- N_0 reads the item *"resonance"*.
- N_1 applies the rule horrible $\rightarrow [ADJ]$.

Communication step $C_{10} \vdash C_{11}$

- N_0 sends the terminal symbol *"resonance"*, and N_2 accepts it.
- N_1 sends $[ADJ]$ and N_2 accepts it.

Evolutionary step $C_{11} \Longrightarrow C_{12}$

- N_0 has nothing to read. It stops
- N_2 applies the rule ressonance $\rightarrow [N]^2_s$.
- N_2 applies the rule $[ART]_s[ADJ][N]^2_s \rightarrow [\ [ART][ADJ][N]\ ]^2_{NP}$.

Communication step $C_{12} \vdash C_{13}$

- N_2 sends the phrase $[\ [ART][ADJ][N]\ ]_{NP}$ and N_4 accepts it.

Evolutionary step $C_{13} \Longrightarrow C_{14}$

- N_4 applies the rule $[\ ]^1_{NP}\ [V(NP)]^1\ [\ ]_{NP} \rightarrow [\ [\ ]^1_{NP}\ [V(NP)]^1\ [\ ]_{NP}\ ]_O$.

After this step, there is nothing to read, and neither communication nor evolutionary transition can be applied. Therefore, the system stops. The node N_4 has a structure labeled with $[\ ]_O$, what means that a string has been recognized and it is correct.

5   Final Remarks

The paper presents an implementation of NEPs for the recognition of sentences of natural language. The construct is interesting because it presents a system with several nodes, each one of them is specialized in the analysis of different syntactic patterns. We think the model fits with the quite spread cognitive theory of modularity and communication between modules.

The idea of a computational approach to modular syntax is quite new. Several parallelisms have been highlighted between Grammar Systems and NEPs. However, researchers in the field of grammar systems theory have not yet implemented any concrete parser to analyze natural language strings, even though general and promising approaches relating grammar systems to natural language processing and linguistics have been proposed in [12, 23, 22].

An important feature of the system is that it provides a complete sentence analysis in a linear way, avoiding trees, what is more consistent with linguistic intuition of speakers.

The work for the future should be oriented to the implementation of NEPs for carrying out more complex tasks, like parsing of different sentence structures as well as complex syntactic pieces. That work will require more complex architectures and the improvement of the parallelism of the system.

References

1. Altman, G., Modularity and Interaction in Sentence Processing, in Garfield, J.L. ed.), *Modularity in Knowledge Representation and Natural-Language Understanding*, Cambridge, MIT Press (1987): 249-257.

2. Castellanos, J., C. Martín-Vide, V. Mitrana, J.M. Sempere, Solving NP-complet problems with networks of evolutionary processors, in Mira, J. & A. Prieto (eds.), *IWANN 2001*, LNCS 2084, Springer-Verlag (2001): 621-628.

3. Castellanos, J., C. Martín-Vide, V. Mitrana & J.M. Sempere, Networks of Evolutionary processors, *Acta Informatica*. 39 (2003): 517-529.

4. Castellanos, J., Leupold, P. & Mitrana. V., Descriptional and Computational Complexity Aspects of Hybrid Networks of Evolutionary Processors. *Theoretical Computer Science* (in press) (2004).

5. Chomsky, N., *Lectures on Government and Binding. The Pisa Lectures*, Dordrech, Foris Publications (1981).

6. Chomsky, N., *Modular Approaches to the Study of Mind*, San Diego State University Press (1984).

7. Crocker, M.W., Multiple Meta-Interpreters in a Logical Model of Sentence Processing, in Brown, Ch. & Koch, G. (eds.), *Natural Language Understanding and Logic Programming, III*, North-Holland, Elsevier Science Publishers B.V. (1991): 127-145.

8. Csuhaj-Varjú, E., Dassow, J., Kelemen, J. & Păun, G., *Grammar Systems*, London, Gordon and Breach (1993).

9. Csuhaj-Varjú, E., Martín-Vide, C., & Mitrana, V., Hybrid Networks of Evolutionary Processors: Completeness Results. (submitted)

10. Csuhaj-Varjú, E., & Mitrana, V., Evolutionary Systems: A Language Generating Device Inspired by Evolving Communities of Cells, *Acta Informatica* 36 (2000): 913926.

11. Csuhaj-Varjú, E. & Salomaa, A., Networks of Parallel Language Processors, in Păun, Gh. & A. Salomaa, *New Trends in Formal Languages*, LNCS 1218, Berlin, Springer (1997): 299318.

12. Csuhaj-Varjú, E. & Abo Alez, R., Multi-Agent Systems in Natural Language Processing, in Sikkel, K. & Nijholt, A. (eds.), *Parsing Natural Language*, TWLT6, Twente Workshop on Language Technology, University of Twente, (1993): 129-137.

13. Csuhaj-Varjú, E., & Salomaa, A., Networks of Watson-Crick D0L systems, in Ito, M., & Imaoka, T. (eds.), *Proc. International Conference Words, Languages & Combinatorics III*, Singapore, World Scientific (2003): 134 - 150.

14. Errico, L. & Jesshope, C., Towards a New Architecture for Symbolic Processing, in I. Plander (ed.), *Artificial Intelligence and Information-Control Systems of Robots 94*, Singapore, World Sci. Publ. (1994): 3140.

15. Everaert, M., Evers, A., Hybreqts, R. & Trommelent, M. (eds.), *Morphology and Modularity: In Honour of Henk Schultink*, Publications in Language Sciences 29, Foris (1988).

16. Farmer, A.K., *Modularity in Syntax: A Study of Japanese and English*, Cambridge, MIT Press (1984).

17. Fodor, J., *The Modularity of Mind*, Cambridge (MA), The MIT Press (1983).

18. Garfield, J.L. (ed.), *Modularity in Knowledge Representation and Natural-Language Understanding*, Cambridge, MIT Press (1987).

19. Harnish, R.M. & Farmer, A.K., Pragmatics and the Modularity of the Linguisitic System. *Lingua*, 63 (1984): 255-277.

20. Hillis, W.D., *The Connection Machine*, Cambridge, MIT Press (1985).

21. Jackendoff, R., *The Architecture of Language Faculty*, Cambridge, MIT Press (1997).

22. Jiménez-López, M.D., Using Grammar Systems, GRLMC Report 16, Rovira i Virgili University (2002).

23. Jiménez-López, M.D. & Martín-Vide, C., Grammar Systems for the Description of Certain Natural Language Facts, in Păun, Gh. & Salomaa, A. (eds.), *New Trends in Formal Languages. Control, Cooperation, and Combinatorics*, LNCS 1218, Springer (1997): 288-298.

24. Margenstern, M., Mitrana, V., & Perez-Jimenez, M., Accepting Hybrid Networks of Evolutionary Processors, in Ferreti, C., Mauri, G. & Zandron, C., *DNA 10. Preliminary Proceedings*, Milan, University of Milano-Biccoca (2004): 107-117.

25. Martín-Vide, C., Mitrana, V., Perez-Jimenez, M. & Sancho-Caparrini, F., Hybrid Networks of Evolutionary Processors, in *Proceedings of GECCO 2003* (2003): 401412.

26. Păun, Gh.: Computing with Membranes. *Journal of Computer and System Sciences*, 61 (2000): 108-143.

27. Păun, Gh., Rozenberg, G., & Salomaa, A., *DNA Computing. New Computing Paradigms*, Berlin, Springer (1998).

28. Sadock, J.M., *Autolexical Syntax. A Theory of Parallel Grammatical Representations*, University of Chicago Press (1991).

29. Weinberg, A., Modularity in the Syntactic Parser, in Garfield, J.L. (ed.), *Modularity in Knowledge Representation and Natural-Language Understanding*, Cambridge, MIT Press (1987): 259-276.

Simulating Evolutionary Algorithms with Eco-grammar Systems

Adrian Horia Dediu and María Adela Grando*

Research Group on Mathematical Linguistics,
Rovira i Virgili University,
Pl. Imperial Tárraco 1, 43005 Tarragona, Spain
{adrianhoria.dediu, mariaadela.grando}@estudiants.urv.es

Abstract. Starting from an arbitrary evolutionary algorithm, we construct an eco-grammar system that simulates the EA's behavior. There are only several practical applications of eco-grammar systems and our approach tries to bring a new light in this field. We believe that our research opens a new perspective also for evolutionary algorithms that can benefit from the theoretical results obtained in the framework of eco-grammar systems.

Keywords: Grammar systems, Eco-grammar systems, Evolutionary algorithms.

1 Introduction

The history of evolutionary computing goes back to the 1960's, with the introduction of ideas and techniques such as genetic algorithms, evolutionary strategies and evolutionary programming [5].

An *Evolutionary Algorithm* (EA) [2] is a computational model inspired by the Darwinian evolutionist theory. In nature the most adapted organisms have better chances to survive, to reproduce and to have offspring. Computational evolutionary algorithms maintain a population of structures that evolve according to the rules of recombination, mutation and selection. Although simplistic from a biologist's point of view, these algorithms are sufficiently complex to provide robust and powerful adaptive search mechanisms.

The notion of an *Eco-grammar System* (EG System) was introduced in 1993 [2]. Eco-grammars model the interactions between individuals and the environment in "living" systems. Both the environment and the agents evolve. The environment evolves independently of the agents but the agents depends on the environment, they are able to sense and to make changes to the environment. This model captures a lot of life-like features as birth and death of agents, change

---

* This work was possible thanks to the research grant "Programa Nacional para la formación del profesorado universitario", from the Ministery of Education, Culture and Sports of Spain.

J. Mira and J.R. Álvarez (Eds.): IWINAC 2005, LNCS 3562, pp. 112–121, 2005.

of seasons, overpopulation, pollution, stagnation, cyclic evolution, immigration, hibernation, carnivorous animals, and so on and so forth (see [3]).

The aim of this paper is to try to connect the two models mentioned above, eco-grammar systems from formal language theory, and evolutionary algorithms from their practical perspective. Starting from an arbitrary evolutionary algorithm, we construct an eco-grammar system that simulates the EA's behavior. Our work was inspired by an application presented in [8] where eco-grammar systems are used in solving and modelling control strategies problems.

Our motivation for connecting these two models is that on one hand for evolutionary algorithms theoretical basis can explain only partially the empirical results from numerous applications and on the other hand much of the research about eco-grammar systems is theoretical and there are few practical applications (see [4], [11], [7] and [8]). We believe that our research opens a new perspective for evolutionary algorithms that can benefit by the theoretical results obtained in the framework of eco-grammar systems. Also eco-grammar systems can take inspiration from applications using evolutionary algorithm for their theoretical research.

2   Eco-grammar Systems

Before introducing the formal definition of an EG system, we present some basic notations:

An *alphabet* is a finite and nonempty set of symbols. Any sequence of symbols from an alphabet V is called *word* over V. The set of all words over V is denoted by V^* and the empty word is denoted by λ. Further, $V^+ = V \setminus \{\lambda\}$. The number of occurrences of a symbol $a \in V$ in a word $w \in V^*$ is denoted as $(w)_{\#a}$ and the length of w is denoted as $\mid w \mid$. The cardinality of a set S is denoted as $card(S)$.

A Chomsky grammar is a quadruple $G = (N, T, S, P)$, where N is the symbol alphabet, T is the terminal alphabet, $S \in N$ is the axiom, and P is the (finite) set of rewriting rules. The rules are presented in the form $u \rightarrow v$ and used in derivations as follows: we write $x \Longrightarrow y$ if $x = x_1 u x_2$, $y = x_1 v x_2$ and $u \rightarrow v$ is a rule in P (one occurrence of u in x is replaced by v and the obtained string is y). Denoting by $\Longrightarrow^*$ the reflexive and transitive closure of $\Longrightarrow$, the language generated by G is defined by:

$$L(G) = \{x \in T^* \mid S \Longrightarrow^* x\}.$$

The families of languages generated by rules of the form $u \rightarrow v$ where $u \in N$ and $v \in (N \cup T)^*$ are called context-free grammars and denoted by CF.

Similarly a $0L$ system (an interactionless Lindenmayer system) is a triple $G = (V, \omega, P)$ as above, with context-free rules in P, and with complete P (for each $a \in V$ there is a rule $a \rightarrow x \in P$). The derivation is defined in a parallel way: $x \Longrightarrow y$ if $x = a_1 a_2 ... a_n$, $y = z_1 z_2 ... z_n$, for $a_i \in V$ and $a_i \rightarrow z_i \in P$, $1 \leq i \leq n$.

For more information regarding the Formal Languages definitions, notations and properties, the reader is referred to [10].

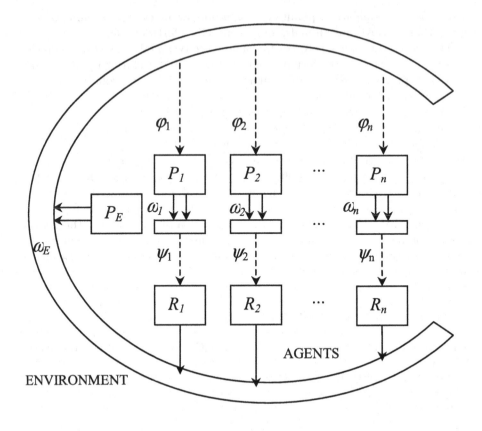

Fig. 1. Description of an eco-grammar system

The interested reader can find an introductory article about eco-grammar systems (EGs) in [2]. For an easier understanding of the formal definitions we present an intuitive image of the system in (Figure 1).

We observe that double arrows stand for parallel rewriting $0L-$ rules while simple arrows are used for $CF-$rules.

Definition 1. *(eco-grammar system) An eco-grammar system of degree n is a $(n+1)$-tuple $\Sigma = (E, A_1, \ldots, A_n)$, where $E = (V_E, P_E)$ is the environment that uses V_E as a finite alphabet, and P_E as a finite set of $0L$ rewriting rules over V_E; A_i, $1 \leq i \leq n$, are agents defined by $A_i = (V_i, P_i, R_i, \varphi_i, \psi_i)$ where V_i is a finite alphabet, P_i a finite set of $0L$ rewriting rules over V_i, R_i is a finite set of simple $CF-$rewriting rules over the environment, $R_i \in V_E^+ \times V_E^*$, φ_i, ψ_i are computable functions that select production sets according to the environment state, respectively the agent state: $\varphi_i : V_E^* \longrightarrow 2^{P_i}$, $\psi_i : V_i^+ \longrightarrow 2^{R_i}$.*

Until now, the definition provides only the description of the eco-grammar system's components. In order to describe the dynamic evolution of EGs, we give the definitions of configuration, derivation and language generated.

Definition 2. *(configuration) A configuration of an eco-grammar system is a tuple* $\sigma = (\omega_E, \omega_1, \ldots, \omega_n)$, $\omega_i \in V_i^*, i \in E \cup \{1, \ldots, n\}$ ω_E *being a string that represents the environment state and* $\omega_1, \ldots, \omega_n$ *are strings that represent the agents' state.*

According to the configuration evolution we define the direct derivation of a configuration in an eco-grammar system.

Definition 3. *(direct derivation) Considering an eco-grammar system* Σ *and two configurations of* c *denoted by* $\sigma = (\omega_E, \omega_1, \ldots, \omega_n)$ *and* $\sigma' = (\omega_E', \omega_1', \ldots, \omega_n')$, *we say that* σ *directly derives* σ' *written as* $\sigma \Longrightarrow_\Sigma \sigma'$ *iff*

- $\omega_i \Longrightarrow \omega_i'$ *according to the selected set of rules for the i-th agent by the* φ_i *mapping,*
- $\omega_E = z_1 x_1 z_2 x_2, \ldots, z_m x_m z_{m+1}$ *and* $\omega_E' = z_1' y_1 z_2' y_2, \ldots, z_m' y_m z_{m+1}'$, *such that* $z_1 x_1 z_2 x_2, \ldots, z_m x_m z_{m+1} \Longrightarrow z_1 y_1 z_2 y_2, \ldots, z_m y_m z_{m+1}$ *as the result of parallel applying the rewriting rules selected by the* ψ_i *mappings for all agents and* $z_1 z_2, \ldots, z_{m+1} \Longrightarrow z_1' z_2', \ldots, z_{m+1}'$ *according to the environment's rules* P_E.

The transitive and reflexive closure of $\Longrightarrow_\Sigma$ is denoted by $\Longrightarrow_\Sigma^+, \Longrightarrow_\Sigma^*$ respectively.

The whole "life" of the system is governed by a universal clock, dividing the time in unit intervals: in every time unit the agents act on the environment then the evolution rules rewrite, in parallel, all the remained symbols in the strings describing the environment and the agents. Thus, the action has priority over evolution.

Definition 4. *(the generated language) The language generated by the environment of an eco-grammar system* Σ, *starting from the initial configuration* σ_0 *is defined as*

$$L_E(\Sigma, \sigma_0) = \left\{ \begin{array}{c} \omega_E \in V_E^* \mid \sigma_j = (\omega_E, \omega_1, \ldots, \omega_n), \\ \sigma_0 \Longrightarrow_\Sigma \sigma_1 \Longrightarrow_\Sigma \ldots \Longrightarrow_\Sigma \sigma_j, j \geq 0 \end{array} \right\}$$

3   Evolutionary Algorithms

EAs, mainly probabilistic searching techniques, represent several convergent research areas. Despite the fact that different sub-domains of EAs, Evolutionary Strategies, Genetic Algorithms, Evolutionary Programming, Genetic Programming, etc. appeared as separate research domains they all have basically a common structure and common components. A searching space for a given problem, a coding scheme representing solutions for a given problem, a fitness function, operators to produce offspring and to select a new generation are the main components of EAs. We can find a very good theory overview for Evolutionary Algorithms in [1]. Briefly, an Evolutionary Algorithm may be defined as a 7-tuple

$$EA = (I, \Phi, \mu, \lambda, \Omega, s, StopCondition) \tag{1}$$

where:

- I represents the set of the searching space instances usually called individuals. Sometimes associated with individuals we can keep useful information for genetic operators. It is usual to associate to each individual its fitness value, using the following notation: $\langle \overrightarrow{i}_q(t), \Phi(\overrightarrow{i}_q(t)) \rangle$, where $\overrightarrow{i}_q(t)$ denotes the vectorial representation of the chromosome of q-th individual in the generation t and $\Phi(\overrightarrow{i}_q(t))$ corresponds to the fitness value associated to that individual. So, we consider $I = \{ \langle \overrightarrow{i}_q(t), \Phi(\overrightarrow{i}_q(t)) \rangle \mid \overrightarrow{i}_q(t) \in Vector \wedge \Phi(\overrightarrow{i}_q(t)) \in F \wedge q \in \mathbf{N} \wedge 1 \leq q \leq \mu + \lambda \wedge t \in \mathbf{N} \wedge t \geq 0 \}$.
- $\Phi : I \to F$ is a fitness function associated to individuals, where F is a finite ordered set corresponding to the fitness function values.
- μ denotes the number of parents.
- λ is the number of offspring inside the population.
- Ω is a set of genetic operators which applied to the individuals of one generation, called parents, produce new λ individuals called offspring.
- s is the selection operator that changes the number of individuals from parents and offspring producing the next generation of parents $(s : I^\mu \times I^\lambda \to I^\mu)$. There are variants of EAs where after one generation the parents are not needed anymore and in this case s selects only from the children population $(s : I^\lambda \to I^\mu)$.
- $StopCondition : F^\mu \times \mathbf{N} \to Boolean$. represents the stop criteria which may be "Stop when a good enough value was reached by an individual fitness function", "Stop after a certain number of generations", "Stop after a maximum time available for computations", etc.

The structure of an **Evolutionary Algorithm** is:

```
1    gen:=0;
2    Initialization process (n, μ, L);
3    Evaluate P(0):=
        {⟨i⃗₁(0), Φ(i⃗₁(0))⟩, . . . , ⟨i⃗_μ(0), Φ(i⃗_μ(0))⟩};
4    do while not(StopCondition(Φ(i⃗₁(gen)), . . . , Φ(i⃗_μ(gen)), gen))
5        Apply genetic operators;
6        Evaluate (P(gen))→ P'(gen)=
            {⟨i⃗'₁(gen), Φ(i⃗'₁(gen))⟩, . . . , ⟨i⃗'_λ(gen), Φ(i⃗'_λ(gen))⟩};
7        Select the next generation P(gen+1):=s(P(gen),P'(gen));
8        gen := gen+1;
     end do;
```

Fig. 2. Description of an Evolutionary Algorithm

In Figure 2 we present a general description of EAs, where *gen* is a numerical variable that indicates the current number of generation, $P(gen)$ represents the set of μ individuals that are potential parents in generation *gen* and $P'(gen)$ is the set of λ offspring we get from the application of genetic operators to individuals from $P(gen)$. Depending on the coding chosen for the EA each gene belongs to a certain data type. Let $L = (C_1, C_2, \ldots, C_n)$ being the list of the sets of genes values. Without loss of generality we consider each C_i, $1 \leq i \leq n$, a finite and discrete set of values of a given type, such that all values of genes in the position i of a chromosome of an individual are of type C_i.

In step 2 of the algorithm the *initialization process*(n, μ, L), defined on $\mathbf{N} \times \mathbf{N} \times C_1 \times C_2 \times \ldots \times C_n$ with values on I^μ, initializes with random values the chromosomes. Recall that $n \in \mathbf{N}$ is the number of genes of the chromosomes and μ is the number of offspring.

Because of the lack of theoretical results able to explain which are the conditions in what EAs perform better, many experimental results try to give us some hints about best practice while implementing a particular EA.

De Jong proposed according to [6] a method to compare the performances of two different strategies. He defined the on-line and off-line performances of a strategy as the average of all fitness functions, respectively the average of maximum fitness function until a given moment. The only conclusion is that we need a large enough number of runs to compute the average results and to compare these average results for different algorithms. Also, De Jong proposed five real valued test functions called $F1, F2, \ldots, F5$ defined on intervals included in $\mathbf{R}^5$ that can be used to compare the performances of evolutionary algorithms. Every function is considered representative for a certain type of optimization problem. Soon, the number of test functions increased and today we can choose from a very extended set of test functions.

4   Simulating an Evolutionary Algorithm with an Eco-grammar System

In Table 1 we show the associations of concepts from EAs and EG systems that we use for the simulation.

Given an arbitrary $EA = (I, \Phi, \mu, \lambda, \Omega, s, StopCondition)$ defined as in (1) and with a description like the one presented in Figure 2, we define an EG system with $\mu + \lambda$ agents, where each agent represents one of the $\mu + \lambda$ individuals of a generation. Recall that n represents the number of genes, C_i, $1 \leq i \leq n$, are sets of genes values, and F is the set of fitness function values.

The string representing the environment´s sate of the EG system that we define has the following structure:

$$\omega_E = \langle h_{1_1}, h_{2_1}, \ldots, h_{n_1} \rangle_1 f_1, Status_1 \ldots \langle h_{1_i}, h_{2_i}, \ldots, h_{n_i} \rangle_i f_i, Status_i \ldots$$
$$\ldots \langle h_{1_{(\mu+\lambda)}}, h_{2_{(\mu+\lambda)}}, \ldots, h_{n_{(\mu+\lambda)}} \rangle_{(\mu+\lambda)} f_{(\mu+\lambda)}, Status_{(\mu+\lambda)},$$
$$ControlSymbolGeneration^{gen}$$

Table 1. Associations between entities in EAs and EGs

Concept	Entity of EA	Entity of EG system
Individual	Vector $\overrightarrow{i}_x(gen)$	String ω_x, state of agent A_x
Evolutive Process	Algorithm of Figure 2	Production set P_E and a symbol in ω_E
Initial Population	Initialization process	Initialization process embedded in φ_x
Evaluation	Function ϕ	Function ϕ, embedded in ψ_x
Termination	Function $StopCondition$	Function $StopCondition$ embedded in φ_x
Number of generation	Variable gen	Number of symbols $Generation$ in ω_E
Genetic operators	Set Ω of operators	Set Ω of operators embedded in φ_x
Selection operator	Function s	Function s embedded in φ_x
Fitness function	Function ϕ	Function ϕ embedded in ψ_x

where:

- $\langle h_{1_i}, h_{2_i}, ..., h_{n_i} \rangle_i$ is the string representing the chromosome of the i-th individual. Because the chromosome of all individuals are kept in ω_E, subindex i is needed to address a particular individual.
- $h_{j_s} \in C_s$ represents the gene value that is in the j-th position of the chromosome of the s-th individual, $1 \leq s \leq n$.
- $f_i, Status_i$ represents the fitness value $f_i \in F$ of the i-th individual with status $Status_i$. Because the fitness values of all individuals are kept in ω_E, subindex i is needed to address the fitness value of a particular individual.
- $Status_i$ is one character in the environment that codes the status of an individual; if it is equal to P then the i-th individual has been selected as potential parent, if $Status_i = O$ it is an offspring and if $Status_i = I$ it is inactive.
- $ControlSymbol \in V_E$, together with the environment productions rules P_E simulate the control sequence described by Figure 2. The V_E alphabet's symbols like $GeneticProcess$, $Selection$, $NewIteration$, $CheckTermination$, $Termination$, etc. are used in ω_E to indicate the step of the algorithm. The productions in P_E control the sequence in which these operations have to take place.
  For example, according to Figure ref fig:algorithm1 after applying genetic operators and getting offspring (step 5), evaluation of their fitness value has to be performed (step 6), so $GeneticProcess \rightarrow OffSpringEvaluation \in P_E$. The end of the iteration process is controlled in the following way: when the symbol $CheckTermination$ is present in the environment the stop condition

is checked (step 4) and if it is satisfied then the symbol is rewritten for *Termination*, if not it is rewritten to *NewIteration* and simulation goes on. The environment and the agents are left in states for which no production from P_E can be applied and mappings φ_i and ψ_i assign the empty set of productions. In this way the EG system does not evolve any more.

− The number of occurrences of the symbol *Generation* $\in V_E$ in ω_E has to be equal to the numeric value of the variable *gen* considered in the algorithm. In this way each time the number of generation has to be increased (step 8) a new symbol *Generation* is introduced in ω_E. And the initialization in the algorithm of the variable *gen* with the value zero (step 1) is simulated by the non occurrence of the variable *Generation* in the *initial configuration* of the EG system, which is defined in the following way: $\sigma_0 = (\omega_E, \omega_1, ..., \omega_{\mu+\lambda})$, where:

  • $\omega_E = Ind_1 \min_{1,I} ...Ind_i \min_{i,I} ...Ind_{(\mu+\lambda)} \min_{(\mu+\lambda),I} Initialization$,
  • $Ind_i \in V_E, 1 \le i \le \mu + \lambda$,
  • $\min_{i,I} \in V_E$, $\min \in F, 1 \le i \le \mu + \lambda$, means that i-th agent starts with an inactive status and with an initial fitness value equal to the minimum value that fitness function can assign to individuals,
  • $\omega_i = Inactive$, for $1 \le i \le \mu + \lambda$, because all the agents are initialized in an inactive state.

With respect to the strings that represent the agents' states have the same coding used for the chromosome of individuals of the EA given. One string describes a vector of genes using brackets $\langle \rangle$ as delimiter symbols and separating genes by commas. And even more complicate coding, like trees, can be model with strings using appropriate balanced brackets. So, the state of the i-th agent can be:

− $\omega_i = \langle h_1, h_2, ..., h_n \rangle StatusSymbol$ where:
  • $\langle h_1, h_2, ..., h_n \rangle$ represents the string corresponding to the chromosome of the i-th individual.
  • $h_j \in C_j$ is the gene value that is in the j-th position of the chromosome of an individual.
  • $StatusSymbol \in \{O, P, S, R\}$ means respectively that the i-th individual is an offspring, a potential parent, has been selected for next generation or has been rejected for next generation.
− $\omega_1 = \alpha Stop$ with $(\alpha = \lambda \lor \alpha = \langle h_1, h_2, ..., h_n \rangle)$ what means that the algorithm´s stop condition is satisfied.
− $\omega_i = Inactive$ what means that the i-th individual is in inactive state.

With respect to the mappings φ_i and ψ_i, $1 \le i \le \mu + \lambda$, the first one embeds the process of random initialization of individuals (step 1), the checking of the stop condition (step 4) and the application of functions from Ω (step 5) and function s (step 7). For mappings φ_i being able to perform step 4, 5 and 7 they need to know the fitness values of the individuals in the population, so these values are kept in ω_E during the whole evolutive process. For the simulation of the step 5 the mappings φ_i also need to know the chromosomes of all the individuals with the status of potential parents. And for step 7 besides the chromosomes of the parents, the chromosomes of the offspring are needed. So before performing this operations, the corresponding chromosomes are copied in ω_E by the mappings ψ_i. Mappings ψ_i are also in charge of embedding the fitness function Φ performing the evaluations corresponding to the steps 3 and 6 of the algorithm.

For the definition of EG system we give we are not interested in the language generated by the environment but just in the string ω_i corresponding to the state of the i-th agent, $1 \leq i \leq \mu + \lambda$, with the best fitness value at the end of the evolution process when the stop condition is satisfied, that we call solution:

$$(solution = \omega_i) \Longleftrightarrow (\omega_i \in V_i^*) \wedge$$
$$(\sigma_0 \Longrightarrow_\Sigma \sigma_1 \Longrightarrow_\Sigma .... \Longrightarrow_\Sigma \sigma_j) \wedge (j \geq 0) \wedge$$
$$\wedge \sigma_j = (ind_1 f_1, Status_1 ... ind_i f_i, Status_i ... ind_{(\mu+\lambda)} f_{(\mu+\lambda)}, Status_{(\mu+\lambda)}$$
$$TerminationGeneration^{gen}, \omega_1, ..., \omega_x, ..., \omega_{\mu+\lambda}) \wedge$$
$$StopCondition \left(f_{s_1}, ..., f_i, ..., f_{s_\mu}, gen \right) \equiv true \wedge$$
$$\wedge (\text{for all } k : 1 \leq k \leq \mu \wedge f_{s_{k_{s_k},P}} \text{ is a substring of } \omega_E \wedge f_x \geq f_{s_k})$$

5   Conclusions and Future Work

In this paper we connect two models, eco-grammar systems and evolutionary algorithms. Starting from an arbitrary evolutionary algorithm we can define an eco-grammar system that simulates the EA's behavior. In this way we open the possibility to use eco-grammar systems not just as language generators but also as searching problem solvers. The range of possible applications is huge, we can mention only several including Location Allocation Problem (an NP-complete problem), game playing, face recognition, financial time-series prediction, etc. We propose to use eco-grammar systems for the mentioned class of problems and for future work we will focus on theoretical aspects regarding the alphabet, the control parameters and appropriate operators for evolutionary algorithms.

References

1. Th. Bäck, *Evolutionary Algorithms in Theory and Practice - Evolution Strategies, Evolutionary Programming, Genetic Algorithms*, Oxford University Press (1996).
2. E. Csuhaj-Varjú, J. Kelemen, A. Kelemenova, G. Păun, Eco-grammar systems: A language theoretic model for AL (Artificial Life), manuscript (1993).
3. E. Csuhaj-Varjú, J. Kelemen, A. Kelemenová, G. Păun, Eco-grammar systems: a grammatical framework for studying lifelike interactions, Artificial Life 3(1) (1997) 1-28.
4. J. Dassow, An example of an eco-grammar system: a can collecting robot, in [9], 240-244.

5. L. Fogel, A. Owens, M. Walsh, *Artificial Intelligence Through Simulated Evolution*, Wiley, New York (1966).
6. D.E. Goldberg, *Genetic Algorithms in Search, Optimization and Machine Learning*, Addison-Wesley (1989).
7. M. Malita, Gh. Stefan, Eco-grammar systems as computing devices, in [9], 260-275.
8. V. Mihalache, General Artificial Intelligence systems as eco-grammar systems, in [9], 245-259.
9. G. Păun (ed.), *Artificial Life. Grammatical Models*, Proc. of the Workshop on Artificial Life, Grammatical Models, Mangalia, 1994, The Black Sea Univ. Press, Bucharest (1995).
10. G. Rozenberg, A. Salomaa (eds.), *Handbook of Formal Languages*, Springer-Verlag, Berlin, (1997).
11. P. Sebestyén, P.Sosik: Multiple robots in space: An adaptive ecogrammar moder, Preproceeding of Grammar System Week 2004, Hungray, Budapest (2004) 284-298.

Timed Accepting Hybrid Networks of Evolutionary Processors

Florin Manea

Faculty of Mathematics and Computer Science, University of Bucharest,
Str. Academiei 14, 70109, Bucharest, Romania
`flmanea@funinf.cs.unibuc.ro`

Abstract. Accepting Hybrid Networks of Evolutionary Processors are
bio-inspired, massively parallel computing models that have been used
succesfully in characterizing several usual complexity classes and also in
solving efficiently decision problems. However, this model does not seem
close to the usual algorithms, used in practice, since, in general, it lacks
the property of stopping on every input. We add new features in order to
construct a model that has this property, and also, is able to characterize
uniformly **CoNP**, issue that was not solved in the framework of regular
AHNEPs. This new model is called Timed AHNEPs (TAHNEP). We
continue by adressing the topic of problem solving by means of this new
defined model. Finally, we propose a tehnique that can be used in the
design of algorithms as efficient as possible for a given problem; this
tehnique consists in defining the notion of Problem Solver, a model that
extends the previously defined TAHNEP.

1 Introduction

In this paper we propose a modified version of accepting hybrid networks of
evolutionary processors, designed in order to provide an accepting method that
seems closer to the regular algorithmic point of view.

As stated in [3], informally, an algorithm is any well-defined computational
procedure that takes some value, or set of values, as *input* and produces, in a finite
time, some value, or set of values, as *output*. An algorithm is thus a *finite sequence
of computational steps* that transform the input into the output. It can also be
seen as a tool for solving a well-specified computational problem. The statement
of the problem specifies in general terms the desired input/output relationship.
The algorithm describes a specific computational procedure for achieving that
input/output relationship. As in usual complexity theory, we assume that we
use algorithms to solve language membership decison problems: given as input
a mechanism defining a language L and a string w we have to decide, using an
algorithm, whether or not $w \in L$.

Networks of evolutionary processors are, as they were defined in [1], a computing model that has its source of inspiration in two points. First they are
inspired by a basic architecture for parallel and distributed symbolic processing,

J. Mira and J.R. Álvarez (Eds.): IWINAC 2005, LNCS 3562, pp. 122–132, 2005.

related to the Connection Machine as well as the Logic Flow paradigm, consisting of several processors, each of them being placed in a node of a virtual complete graph, which are able to handle data associated with the respective node. Each node processor acts on the local data in accordance with some predefined rules, and then local data becomes a mobile agent which can navigate in the network following a given protocol. Only such data can be communicated which can pass a filtering process. This filtering process may require to satisfy some conditions imposed by the sending processor, by the receiving processor or by both of them. All the nodes send simultaneously their data and the receiving nodes handle also simultaneously all the arriving messages. The second point of inspiration the cells' evolution. These are represented by words which describe their DNA sequences. Informally, at any moment of time, the evolutionary system is described by a collection of words, where each word represents one cell. Cells belong to species and their community evolves according to mutations and division which are defined by operations on words. Only those cells are accepted as surviving (correct) ones which are represented by a word in a given set of words, called the genotype space of the species. This feature parallels with the natural process of evolution.

These concepts were put together in the definition of the networks of evolutionary processors: the predefined rules of the nodes in the complete graph are evolutionary rules, namely point mutations in the DNA sequence (insertion, substitution, deletion). Each node has exactly one type of rule associated with it, but, in designing a network one can use different types of nodes; such a network is called hybrid. Although the process defined in our architecture is not an evolutionary process in a strict Darwinian sense, the appliance of an evolutionary rule can be seen as a mutation, and the filltering process can be seen as selection. We can state, consequently, that networks of evolutionary processors are bio-inspired models.

We focus on the way networks of evolutionary processors are used in solving membership decision problems. For this reason there were defined accepting networks of evolutionary processors (AHNEPs, see [8]). Such a network contains two distinguished nodes: the input and the output node. At the begining of the computation the only node that contains local data is the input node, and this local data consists of the string that we want to decide wether it is a member of a given language or not. Then, the application of evolutionary rules and communication are performed alternatively, until the output node receives some data, or until all the nodes should process in a computational step exactly the same data that was processed in the precedent step. In the first case the computation is succesfull, and the input string is accepted, but in the second one, and also in the case of an infinite compuation, the input string is rejected. AHNEPs, defined in this way, were used in problem-solving ([1, 5]), in providing accepting tools for classes in the Chomsky hierarchy[6], and in characterizing the usual complexity classes as **P,NP, PSPACE**, but not **CoNP** (see [7]).

However, the way AHNEPs are defined does not capture the algorithmic point of view, since the rejection answer is not necessarily provided after a finite number of steps. In order to eliminate such drawbacks, we propose a new version of AHNEPs: Timed Accepting Hybrid Networks of Evolutionary Processors (TAHNEP), which consist in a classical AHNEP, a *clock*, and an *accepting-mode bit*. The clock is used for keeping track of the number of steps performed by an AHNEP, and stopping it when it reaches a certain value, while the accepting-mode bit is used in order to permit elegant characterizations of larger classes of languages.

The paper is structured in the following way: first we provide the basic definition for TAHNEPs as an extention of the definitions already known for AHNEPs. Then we present several time complexity results for the newly defined models, and, finally we propose a formalization for the way TAHNEPs can be used more efficiently in problem solving.

2   Basic Definitions

We start by briefly reminding the basic and most important definition for AHNEPs. For a more accurate presentation we refer to [2]. First, we recall that for two disjoint and nonempty subsets P and F of an alphabet V and a word w over V, we define the predicates:

- $\varphi^{(1)}(w; P, F) \equiv P \subseteq alph(w) \ \wedge \ F \cap alph(w) = \emptyset$
- $\varphi^{(2)}(w; P, F) \equiv alph(w) \subseteq P$
- $\varphi^{(3)}(w; P, F) \equiv P \subseteq alph(w) \ \wedge \ F \not\subseteq alph(w)$
- $\varphi^{(4)}(w; P, F) \equiv alph(w) \cap P \neq \emptyset \ \wedge \ F \cap alph(w) = \emptyset.$

An *evolutionary processor over* V is a tuple (M, PI, FI, PO, FO), where:

– Either M is a set of substitution or deletion or insertion rules over the alphabet V. Formally: $(M \subseteq Sub_V)$ or $(M \subseteq Del_V)$ or $(M \subseteq Ins_V)$. The set M represents the set of evolutionary rules of the processor. As one can see, a processor is "specialized" in one evolutionary operation, only.

– $PI, FI \subseteq V$ are the *input* permitting/forbidding contexts of the processor, while $PO, FO \subseteq V$ are the *output* permitting/forbidding contexts of the processor. Informally, the premitting input/output contexts are the set of symbols that should be present in a string, when it enters/leaves the processor, while the forbidding contexts are the set of symbols that should not be present in a string in order to enter/leave the processor.

We denote the set of evolutionary processors over V by EP_V. An *accepting hybrid network of evolutionary processors* (AHNEP for short) is a 7-tuple $\Gamma = (V, U, G, N, \alpha, \beta, x_I, x_O)$, where:

- V and U are the input and network alphabets, respectively, $V \subseteq U$.
- item $G = (X_G, E_G)$ is an undirected graph with the set of vertices X_G and the set of edges E_G. G is called the *underlying graph* of the network. In this paper, we consider *complete* AHNEPs, i.e. AHNEPs having a complete underlying graph denoted by K_n, where n is the number of vertices.

- $N : X_G \longrightarrow EP_U$ is a mapping which associates with each node $x \in X_G$ the evolutionary processor $N(x) = (M_x, PI_x, FI_x, PO_x, FO_x)$.
- $\alpha : X_G \longrightarrow \{*, l, r\}$; $\alpha(x)$ gives the action mode of the rules of node x on the words existing in that node. Informally, this indicates if the evolutionary rules of the processor are to be applied at the leftmost end of the string, for $\alpha = l$, at the rightmost end of the string, for $\alpha = r$, or at any of its position, for $\alpha = *$.
- $\beta : X_G \longrightarrow \{(1), (2), (3), (4)\}$ defines the type of the *input/output filters* of a node. More precisely, for every node, $x \in X_G$, the following filters are defined:

$$\text{input filter: } \rho_x(\cdot) = \varphi^{\beta(x)}(\cdot; PI_x, FI_x),$$
$$\text{output filter: } \tau_x(\cdot) = \varphi^{\beta(x)}(\cdot; PO_x, FO_x).$$

That is, $\rho_x(w)$ (resp. τ_x) indicates whether or not the word w can pass the input (resp. output) filter of x. More generally, $\rho_x(L)$ (resp. $\tau_x(L)$) is the set of words of L that can pass the input (resp. output) filter of x.

- x_I and $x_O \in X_G$ is the *input node*, and the *output node*, respectively, of the AHNEP.

A *configuration* of an AHNEP Γ as above is a mapping $C : X_G \longrightarrow 2^{V^*}$ which associates a set of words with every node of the graph. A configuration may be understood as the sets of words which are present in any node at a given moment. A configuration can change either by an *evolutionary step* or by a *communication step*. When changing by an evolutionary step, the configuration C' is obtained from the configuration C, written as $C \Longrightarrow C'$, iff $C'(x) = M_x^{\alpha(x)}(C(x))$ for all $x \in X_G$. The configuration C' is obtained in *one communication step* from configuration C, written as $C \vdash C'$, iff $C'(x) = (C(x) - \tau_x(C(x))) \cup \bigcup_{\{x,y\} \in E_G}(\tau_y(C(y)) \cap \rho_x(C(y)))$ for all $x \in X_G$.

Let Γ be an AHNEP, the computation of Γ on the input word $w \in V^*$ is a sequence of configurations $C_0^{(w)}, C_1^{(w)}, C_2^{(w)}, \ldots$, where $C_0^{(w)}$ is the initial configuration of Γ defined by $C_0^{(w)}(x_I) = w$ and $C_0^{(w)}(x) = \emptyset$ for all $x \in X_G$, $x \neq x_I$, $C_{2i}^{(w)} \Longrightarrow C_{2i+1}^{(w)}$ and $C_{2i+1}^{(w)} \vdash C_{2i+2}^{(w)}$, for all $i \geq 0$. Note that the two steps, evolutionary and communication, are synchronized and they happen alternatively one after another. By the previous definitions, each configuration $C_i^{(w)}$ is uniquely determined by the configuration $C_{i-1}^{(w)}$. Otherwise stated, each computation in an AHNEP is deterministic. A computation as above immediately halts if one of the following two conditions holds:

(i) After k steps, there exists a configuration in which the set of words existing in the output node x_O is non-empty. In this case, the computation is said to be an *accepting computation*. Since evolution and communication steps are no longer applied, we assume that $C_n^{(w)} = C_k^{(w)}, \forall n > k$.
(ii) There exist two consecutive identical configurations, with the property that the set of words existing in the output node x_O is empty.

In the aforementioned cases the computation is said to be finite. The language accepted by Γ is:

$$L(\Gamma) = \{w \in V^* \mid \text{the computation of } \Gamma \text{ on } w \text{ is an accepting one}\}.$$

Remark 1. *Since every computation in which a string enters the output node is a succesfull one, and the further evolution of the network is not of interest, one may assume, without loss of generality, that the output forbidding filter of the output node contains the whole network alphabet. In the rest of the paper, all the AHNEPs used verify this property.*

We continue by defining the *Timed Accepting Hybrid Networks of Evolutionary Processors*. A TAHNEP is a triple $\mathcal{T} = (\Gamma, f, b)$, where $\Gamma = (V, U, G, N, \alpha, \beta, x_I, x_O)$ is an AHNEP, $f : V^* \to \mathbf{N}$ is a Turing computable function, called *clock*, and $b \in 0, 1$ is a bit called the *accepting-mode bit*. The computation of a TAHNEP $\mathcal{T} = (\Gamma, f, b)$ on the input word w is the (finite) sequence of configurations of the AHNEP Γ: $C_0^{(w)}, C_1^{(w)}, \ldots, C_{f(w)}^{(w)}$. The language accepted by $\mathcal{T}$ is defined as:

- if $b = 1$ then: $L(\mathcal{T}) = \{w \in V^* \mid C_{f(w)}^{(w)}(x_O) \neq \emptyset\}$
- if $b = 0$ then: $L(\mathcal{T}) = \{w \in V^* \mid C_{f(w)}^{(w)}(x_O) = \emptyset\}$

Remark 2. *If $\mathcal{T} = (\Gamma, f, b)$ is a TAHNEP, then Γ is a AHNEP that verifies the property stated in Remark 1. In this conditions, it is immediate that for $b = 1$ we have $L(\mathcal{T}) = \{w \in V^* \mid$ there exists $k < f(w), C_k^{(w)}(x_O) \neq \emptyset\}$, and, for $b = 0$ we have $L(\mathcal{T}) = \{w \in V^* \mid \forall k \leq f(w), C_k^{(w)}(x_O) = \emptyset\}$*

Intuitively we may think that a TAHNEP $\mathcal{T} = (\Gamma, f, b)$ is a triple: an AHNEP, a Turing Machine and a bit. For an input string w we first compute $f(w)$ on the tape of the Turing Machine (by this we mean that on the tape there will exist $f(w)$ elements of 1, and the rest are blanks). Then we begin to use the AHNEP, and at each step we delete an 1 from the tape of the Turing Machine. We stop when no 1 is found on the tape. Finally, we check the accepting-mode bit, and according to its value and $C_{f(w)}^{(w)}(x_O)$ we decide wether w is accepted or not.

We make a few a priori considerations on this model: the usage of the function f, and the way it is implemented, makes this a hybrid definition from another point of view- it uses both classical tools as the Turing Machines, and non-conventional models like AHNEPs; moreover, the fact that every computation is finite brings the model closer to the notion of algorithm. Also, since we presented a simple way of implementing the clock and its interaction with the AHNEP, and since this doesn't affect the computation of the AHNEP, in the following we will neglect the implementation of the clock, and we suppose that it exists, works and it will signal corectly the AHNEP when to stop.

3   Complexity Aspects

In [8] there were given the defintions of some complexity measures for AHNEPs. and the main result obtained was the following theorem; the reader is also referred to [4] for the classical time and space complexity classes defined on the standard computing model of Turing machine.

Theorem 1. [8] **NP** = **PTime**$_{AHNEP}$.

Remark 3. *Recall that for an AHNEP Γ, the fact that $L(\Gamma) \in$ **PTIME**$_{AHNEP}$, expresses the existance of a polynomial g such that a succesfull computation of Γ on a string x takes less than $g(|x|)$ steps. Also, note that the strategy used in constructing a Turing Machine that has the same behaviour as an AH-NEP recognizing a language from **PTIME**$_{AHNEP}$, as it is stated in [8], was to choose non-deterministicaly a path through the network, and then simulate, step by step, the evolution of the input string along that path. The computation of this Turing Machine accepts the input string when the output node contains a string, and rejects it when the simulation lasted for more than $g(|x|)$ evolutionary steps.*

In the case of a TAHNEP $\mathcal{T} = (\Gamma, f, b)$ the time complexity definitions are the following: for the word $x \in V^*$ we define the time complexity of the computation on x as the number of steps that the TAHNEP makes having the word x as input, $Time_{\mathcal{T}}(x) = f(x)$. Consequently, we define the time complexity of $\mathcal{T}$ as a partial function from **N** to **N**, that verifies: $Time_{\mathcal{T}}(n) = \max\{f(x) \mid x \in L(\mathcal{T}), |x| = n\}$. For a function $g : \mathbf{N} \longrightarrow \mathbf{N}$ we define:

$$\mathbf{Time}_{TAHNEP}(g(n)) = \{L \mid L = L(\mathcal{T}) \text{ for a TAHNEP } \mathcal{T} = (\Gamma, f, 1) \text{ with } Time_{\mathcal{T}}(n) \leq g(n) \text{ for some } n \geq n_0\}.$$

Moreover, we write $\mathbf{PTime}_{TAHNEP} = \bigcup_{k \geq 0} \mathbf{Time}_{TAHNEP}(n^k)$.

Note that the above definitions were given for TAHNEPs with the accepting-mode bit set to 1. Similar definitions are given for the case when the accepting-mode bit set to 0. For a function $f : \mathbf{N} \longrightarrow \mathbf{N}$ we define, as in the former case:

$$\mathbf{CoTime}_{TAHNEP}(g(n)) = \{L \mid L = L(\mathcal{T}) \text{ for a TAHNEP } \mathcal{T} = (\Gamma, f, 1) \text{ with } Time_{\mathcal{T}}(n) \leq g(n) \text{ for some } n \geq n_0\}.$$

We define $\mathbf{CoPTime}_{TAHNEP} = \bigcup_{k \geq 0} \mathbf{CoTime}_{TAHNEP}(n^k)$.

We see that, if $\mathcal{T} = (\Gamma, f, 1)$ recognizes a language from $\mathbf{Time}_{TAHNEP}(g(n))$, we can deduce that $f(x) \leq g(n), \forall x \in V^*, |x| = n$ and $\forall n \geq n_0$. If $g(n) = n^k$, than it follows that $f(x) \leq n^k, \forall x \in V^*, |x| = n, \forall n \geq n_0$. Concluding: all the TAHNEPs generating languages contained in $\mathbf{PTIME}_{TAHNEP}$ have the following property: $f(x) \leq g(|x|), \forall x \in V^*$, where g is a polynomial.

It is not hard to see that every language recognized by an TAHNEP that verifies the above property, is in $\mathbf{PTIME}_{TAHNEP}$. For simplicity, we will call a TAHNEP that verifies such a property polynomial.

We can prove a similar result for $\mathbf{CoPTime}_{TAHNEP}$, i.e. a language $L = L(\mathcal{T})$ is contained in $\mathbf{CoPTime}_{TAHNEP}$ if and only if the following are verified: $\mathcal{T} = (\Gamma, f, 0)$ with $f(x) \leq g(|x|), \forall x \in V^*$ and g is a polynomial.

We can now state the most important results of this paper.

Theorem 2. PTime$_{TAHNEP}$ = NP.

Proof. Let $L = L(\mathcal{T})$ be a language from **PTIME**$_{TAHNEP}$, and $\mathcal{T} = (\Gamma, f, b)$. As was explained in Remark 3, we can choose and simulate non-deterministicaly each one of the possible evolutions of the input string x through the underlying network of the AHNEP Γ. Since in this case we are interested only in the first $f(x)$ steps, and there exist a polynomial g such that $f(x) \leq g(|x|)$, we can state that the Turing Machine must simulate, at most, the first $g(|x|)$ steps from the evolution of the input string. As it was proven in Theorem 1, the simulation of such a step can be carried out in polynomial time. Consequently, the Turing Machine will simulate all the $f(x) \leq g(|x|)$ steps in a polynomial time, and then will stop. The computation will be succesfull if at the end of the simulation the output node will contain one or more strings, and unsuccesfull otherwise. From these it follows that L is the language accepted by a non-deterministic Turing Machine working in polynomial time, and **PTime**$_{TAHNEP} \subseteq$ **NP**.

To prove that **NP** $\subseteq$ **PTime**$_{TAHNEP}$ we also make use of Theorem 1. From this theorem it follows that for $L \in$ **NP** there exists an AHNEP Γ and a polynomial g such that: $x \in L$ if and only if $x \in L(\Gamma)$ and $Time_\Gamma(x) \leq g(|x|)$. From this it follows that the TAHNEP $\mathcal{T} = (\Gamma, f, 1)$, where $f(x) = g(|x|)$, accepts L.

From the above we have proven the theorem: **PTime**$_{TAHNEP}$ = **NP**.   □

We can consider that **PTime**$_{TAHNEP}$ is the class of of languages that can be efficiently recognized by means of TAHNEPs, with the accepting-mode bit set to 1. In this setting, the above theorem proves that these languages are exactly the ones contained in **NP**. Also, since every abstract problem can be related to a decision problem (a language-membership problem), and shown to be at least as hard as this one (see [4]), we have proven that the problems that cannot be solved efficiently by non-deterministic Turing Machines are the same with those that cannot be solved efficiently by TAHNEPs.

A similar theorem is stated for the case when the accepting-mode bit is set to 0.

Theorem 3. CoPTime$_{TAHNEP}$ = CoNP.

Proof. Let L be a language from **CoPTime**$_{TAHNEP}$, and L is accepted by a polynomial TAHNEP $\mathcal{T} = (\Gamma, f, 0)$. The complementary of this language **Co**L is the set $V^* - \{w \in V^* \mid C^{(w)}_{f(w)}(x_O) = \emptyset\}$ ($C^{(w)}_{f(w)}$ is a configuration of Γ, and x_O is its output node). From this, and Remark 2, its easy to see that **Co**$L = \{w \in V^* \mid C^{(w)}_{f(w)}(x_O) \neq \emptyset\}$, and, consequently: **Co**$L = L(\mathcal{T}')$, where $\mathcal{T}' = (\Gamma, f, 1)$ is a polynomial TAHNEP. It follows that **Co**$L \in$ **PTime**$_{TAHNEP}$, and, by Theorem 2, we have **Co**$L \in$ **NP**. Therefore, $L \in$ **CoNP**.

The reversal holds as well: let $L \in$ **CoNP**. By Theorem 2, it follows that **Co**L is in **NP**, and, consequently, it is in **PTime**$_{TAHNEP}$. So there exists a polynomial TAHNEP $\mathcal{T} = (\Gamma, f, 1)$ such that $L(\mathcal{T}) =$ **Co**L. We obtain that **Co**$L = \{w \in V^* \mid C^{(w)}_{f(w)}(x_O) \neq \emptyset\}$ (where the configuration in the formula is with respect to Γ, and x_O is its output node). Finally, by Remark 2, we get:

$L = \{w \in V^* \mid C_{f(w)}^{(w)}(x_O) = \emptyset\}$, and as a consequence: $L = L(\mathcal{T}')$, where $\mathcal{T}' = (\Gamma, f, 0)$. Note that $\mathcal{T}'$ is polynomial, since $\mathcal{T}$ is polynomial, and they have the same clock function.

Hence: $L \in \mathbf{CoPTime}_{TAHNEP}$, which concludes our proof.   □

Theorems 2 and 3 and their proofs are interesting in the following context: they provide a common framework for solving both problems from **NP** and from **CoNP**. For example, suppose that we want to solve the membership problem regarding the language L.

– If $L \in$ **NP**, using the proofs of Theorems 1 and 3, and Remark 3, we can construct a polynomial TAHNEP $\mathcal{T} = (\Gamma, f, 1)$ that accepts L.
– If $L \in$ **CoNP**, it results that $\mathbf{Co}L \in$ **NP**, and using the proofs of Theorems 1 and 3, and Remark 3, we can construct a polynomial TAHNEP $\mathcal{T} = (\Gamma, f, 1)$ that accepts $\mathbf{Co}L$. By Theorem 3 we obtain that $(\Gamma, f, 0)$ accepts L.

As a conclusion of the above, note that both TAHNEPs for L and $\mathbf{Co}L$ are using the same AHNEP and clock components, but the accepting mode is different. In other words, we use the same algorithm for the both problems, the only difference being the acceptance criterion. Consequently, it suffies to solve only the membership problem for the languages in **NP**, and, then, by simply changing the acceptance mode we have solved also the problems for the complement of these languages. Following these considerations, we observe that $\mathbf{CoPTime}_{TAHNEP}$ can be seen as the languages that can be recognized efficiently by TAHNEPs with the accepting-mode bit set to 0. Moreover, for every language in this class, there exists another language in $\mathbf{PTime}_{TAHNEP}$ whose recgnition requires the same computational effort, and conversely; this corespondence is defined by associating each language with its complement.

Thus, Theorems 2 and 3 prove that the languages (the decision problems) that are efficiently recognized (solved) by the TAHNEPs (with both 0 and 1 as possible values for the accepting-mode bit) are those from **NP** ∪ **CoNP**.

4   Problem Solving

As in the classical algorithms theory, we focus now on finding the most efficient solution for a problem, not only a polynomial one. But such a thing may require an even more hybrid aproach: by looking at the input we decide -using a deterministic, classic, algorithm- the architecture and features of the TAHNEP that we will use in our solution; finally we run the TAHNEP on that input, and get the acceptance/rejection answer. In this section we propose a formalization for this approach.

A *Problem Solver* (PS) is defined as the following construction: $\mathcal{P} = (F, (\mathcal{T}_n)_{n \in \mathbf{N}})$, where $F : V^* \to \mathbf{N}$ is a Turing computable function, and $\mathcal{T}_n, n \in \mathbf{N}$ is a family of TAHNEPs over the same input alphabet V, the features of every such TAHNEP being related to n. Since the PS uses TAHNEPs, it will be called TAHNEP based; a similar definition could have been given taking $\mathcal{T}_n$ as a family of AHNEPs, and, consequently, the problem solver would be AHNEP based.

Moreover, a TAHNEP based problem solver $\mathcal{P} = (F, (\mathcal{T}_n)_{n \in \mathbf{N}})$ is called uniform if all the TAHNEPs $\mathcal{T}_n$ have the same accepting-mode bit. In the following we deal only with such problem solvers.

The language recognized by a problem solver as above is the set:

$$L(\mathcal{P}) = \{w \in V^* \mid w \in L(\mathcal{T}_{F(w)})\}.$$

A problem solver $\mathcal{P} = (F, (\mathcal{T}_n)_{n \in \mathbf{N}})$ is called polynomial if and only if the following three condition hold:

– F is computable in polynomial time by a deterministic Turing Machine.

– For all $n \in \mathbf{N}$, the number of nodes, the number of evolutionary rules and the cardinality of forbidding and permiting input/output filters of $\mathcal{T}_n$ are bounded by the value $P(n)$ (where P is a polynomial), and there exists a deterministic polynomial Turing Machine algorithm that outputs the complete description of each node of $\mathcal{T}_n$, having as input n.

– For all $n \in \mathbf{N}$, $Time_{\mathcal{T}_n}(x) \leq Q(m), \forall x \in V^*, |x| = m$, and $\forall m \in \mathbf{N}$ (Q being a polynomial also).

The class of TAHNEP based polynomial problem solvers is denoted **PPS**. Informaly, one may see this class as the class of polynomial TAHNEP based algorithms. We also denote **PPS**$_b$ the class of uniform problem solvers $\mathcal{P} = (F, (\mathcal{T}_n)_{n \in \mathbf{N}})$ with the property that $\mathcal{T}_n$ has the accepting-mode bit equal to b, for all $n \in \mathbf{N}$.

We denote by $\mathcal{L}_{PPS}$ the class of languages that are recognized by the problem solvers in **PPS**. Similar, we denote by $\mathcal{L}_{PPS_b}$ the class of languages recognized by the problem solvers in **PPS**$_b$.

The following results describe the computational power of polynomial problem solvers.

Theorem 4. $\mathcal{L}_{PPS_1} = \mathbf{NP}$.

Proof. Throughout this proof we assume that $L \in V^*$.

We first prove that if $L \in \mathbf{NP}$ we can construct a polynomial problem solver that accepts L. From Theorem 2 we obtain that there exists a TAHNEP $\mathcal{T} = (\Gamma, f, 1)$, that accepts L. $\mathcal{T}$ also verifies the following properties: $Time_{\mathcal{T}}(x) \leq Q(n), \forall x \in V^*, |x| = n$, and $\forall n \in \mathbf{N}$ where Q is a polynomial, and there exists $K \in \mathbf{N}$, such that the number of nodes, the number of evolutionary rules and the number of permitting and forbidding input/output filters of Γ is less than K. We define the following TAHNEP based problem solver: $\mathcal{P} = (F, (\mathcal{T}_n)_{n \in \mathbf{N}})$, where $F(w) = 0, \forall w \in V^*$, and $\mathcal{T}_0 = \mathcal{T}_n, \forall n \in \mathbf{N}$. It is not hard to prove that $\mathcal{P}$ is polynomial and $L = L(\mathcal{P})$. Thus L is recognized by $\mathcal{P} \in \mathbf{PPS}_1$.

To prove the reversal, suppose that $L = L(\mathcal{P})$, $\mathcal{P} \in \mathbf{PPS}_1$. We design a non-deterministic Turing Machine with 4 tapes, that simulates the behaviour of $\mathcal{P}$. The input of the Turing Machine is the input word w, placed on the tape 1. First the Turing Machine computes $F(w)$, and places the result on the tape 2. This computation is carried out deterministicaly and in polynomial time. Then, the Turing Machine can compute by a deterministic and polynomial algorithm, starting from $F(w)$ the features (filters and evolutionary rules) of every node of the network $\mathcal{T}_{F(w)}$. The results are placed on the tape 3. Since the number of these features of every node of $\mathcal{T}_{F(w)}$ is polynomial in $F(w)$, the numebr of nodes of $\mathcal{T}_{F(w)}$ is polynomial in $F(w)$, and $F(w)$ is polynomial in $|w|$, it follows that the number of nodes and the number of all the features of $\mathcal{T}_{F(w)}$ is polynomial in $|w|$. Finally, on the tape 4 we simulate the evolution of w in the network $\mathcal{T}_{F(w)}$ by choosing non-deterministicaly the path in the network that it follows. Initially, on the tape 4 is found w. This simultation is carried out in the following way: on the tape 3 we choose a node non-deterministicaly, we see if the string on the tape 4 can enter that node, we choos an evolutionary rule, apply it to the string on tape 4 (and, thus, this string is transformed), and see if the string newly obtained can leave that node; the process is then iterated. Every such step takes a polynomial number of steps, and the process has at most a polynomial number of iterations (it is bounded by $Time_{\mathcal{T}_{F(w)}}(|w|)$). Finally, the string w is accepted if at the end of the iteration the string on the tape 4 entered the output node of $\mathcal{T}_{F(w)}$, otherwise being rejected. From the construction it follows that the non-deterministic Turing Machine we have designed gives the correct output in polynomial time, for every input. Consequently $L \in$ **NP**.

From the above it follows that $\mathcal{L}_{PPS_1} = $ **NP**.   □

Using Theorem 3, and a technique similar with the above, we can prove:

Theorem 5. $\mathcal{L}_{PPS_0} = $ **CoNP**

Mainly, the above definitions formalize the technique already used in solving NP-complete problems via networks of evolutionary processors (for example in [1, 5]). The two results presented prove that, although the strategy taken in solving problems requires a pre-processing phase (in which the architecture of the network is choosen from a set of possibilities), the computational power remains unchanged. However, as was proven in [5], by using problem solvers we can obtain very efficient soutions for hard problems (as NP-complete problems), aspect that is not guaranteed by the constructions in Theorems 2,3.

The results in this section were presented in the framework of TAHNEPs, in order to keep them close to the regular algorithmic point of view, and, also, in order to provide a characterization for the class **CoNP** via polynomial problem solvers. Another reason for this strategy was that the solutions for NP-complete problems presented in [5], the first dealing with problem solving by means of *Accepting* HNEPs, were based on networks that stopped on every input, model that coincides with TAHNEPs having the accepting-mode bit 1. A similar result with Theorem 4 can be proven for AHNEP based problem solvers. However, this is not the case of Theorem 5, since in order to characterize the class **CoNP** we needed the property that the computations stops.

References

1. J. Castellanos, C. Martin-Vide, V. Mitrana, J. Sempere, Solving NP-complete problems with networks of evolutionary processors, *IWANN 2001* (J. Mira, A. Prieto, eds.), LNCS 2084 (2001), 621–628.

2. J. Castellanos, C. Martin-Vide, V. Mitrana, J. Sempere, Networks of evolutionary processors, *Acta Informatica* 39(2003), 517-529..
3. T.H. Cormen, C.E. Leiserson, R.R. Rivest, *Introduction to Algorithms*, MIT Press, 1990.
4. M.R. Garey, D.S. Johnson, *Computers and Intractability. A Guide to the Theory of NP-Completness*, W.H. Freeman and Company, 1979.
5. F. Manea, C. Martin-Vide, V. Mitrana, Solving 3CNF-SAT and HPP in Linear Time Using WWW, *MCU 2004*, LNCS 3354 (2005), 269–280.
6. F. Manea, Using AHNEPs in the Recognition of Context-Free Languages, *Grammars*, in press (also: *Proc. of the Workshop on Symbolic Networks, ECAI 2004*).
7. F. Manea, M. Margenstern, V. Mitrana, M. Perez-Jimenez, A new characterization for NP, submitted.
8. M. Margenstern, V. Mitrana, M. Perez-Jimenez, Accepting hybrid networks of evolutionary processors, *Pre-proc. DNA 10*, 2004, 107–117.
9. C. Martin-Vide, V. Mitrana, M. Perez-Jimenez, F. Sancho-Caparrini, Hybrid networks of evolutionary processors, *Proc. of GECCO 2003*, LNCS 2723 (2003), 401 - 412.

A New Immunotronic Approach to Hardware Fault Detection Using Symbiotic Evolution

Sanghyung Lee[1], Euntai Kim[1], Eunjoo Song[2], and Mignon Park[1]

[1] Dept. of Electrical and Electronic Engr., Yonsei Univ.,
Shinchon-dong 134, Seodaemun-gu, Seoul, Korea
`etkim@yonsei.ac.kr`
[2] Bioanalysis and Biotransformation Research Center,
Korea Institute of Science and Technology,
Hawolgok-dong 39-1, Sungbook-gu Seoul, Korea

Abstract. A novel immunotronic approach to fault detection in hardware based on symbiotic evolution is proposed in this paper. In the immunotronic system, the generation of tolerance conditions corresponds to the generation of antibodies in the biological immune system. In this paper, the principle of antibody diversity, one of the most important concepts in the biological immune system, is employed and it is realized through symbiotic evolution. Symbiotic evolution imitates the generation of antibodies in the biological immune system more than the standard genetic algorithm(SGA) does. It is demonstrated that the suggested method outperforms the previous immunotronic methods with less running time. The suggested method is applied to fault detection in a decade counter (typical example of finite state machines) and MCNC finite state machines and its effectiveness is demonstrated by the computer simulation.

Keywords: immunotronic system, hardware fault detection, tolerance conditions, antibody diversity, symbiotic evolution.

1 Introduction

In the twentieth century, various problems in theoretical immunology were solved and the mathematical models and the analysis methods of immune phenomena were developed. The theoretic developments of the immunology have given rise to the alternative approaches to various fields such as pattern recognition [1], fault and anomaly detection [2], data analysis [3], machine-learning [4]. Among these researches, the development of the immune-inspired hardware fault detection technique or immunotronics (immunological electronics) has been suggested in [5]. Fault detection is one of the key issues in the fault tolerant hardware and there have been many researches on the theme such as n-modular redundancy (NMR) [6], error-detecting and correcting code [7], and self-checking logic circuit [8]. But these methods have shortcomings compared with immune-inspired fault detection technique. In other words, in the classical methods, imperfect matching

J. Mira and J.R. Álvarez (Eds.): IWINAC 2005, LNCS 3562, pp. 133–142, 2005.

between a priori knowledge on the hardware and the fault is not fully utilized. Imperfect matching enables the system to detect faults (nonselfs) though they are not known to the system and faults can be detected by tolerance conditions which are generated from the information of proper states (selfs) by negative selection algorithm [9]. Tolerance conditions in the immunotronic system correspond to antibodies in the biological immune system and the most important problem in designing the immunotronic system is how to generate tolerance conditions effectively and how to distinguish faults (nonselfs) from proper states (selfs). Bradley et al. adopted negative selection algorithm to select the mature tolerance conditions from immature tolerance conditions generated using greedy detetor generation algorithm [5]. Recently, Lee et al. suggested the GA-based design method of the immunotronic system to improve the fault detection rates [10]. But the previous immunotronic system imitated the biological immune system in a superficial manner. That is, the principle of antibody diversity, one of the most important concepts in the biological immune system, has not been taken into consideration in the previous method. This principle enables B cells to generate the diverse antibodies with limited number of DNA and to detect more nonselfs with lesser antibodies [11]. In this paper, tolerance conditions are generated from the perspective of the antibody diversity. Mature tolerance conditions are selected automatically from immature tolerance conditions through symbiotic evolution algorithm equipped with selection, crossover and mutation. It is very similar to the way that the mature antibodies are chosen in a biological immune system. The rest of the paper is organized as follows: In Section 2, the biological immune system and the immunotronic system for hardware fault detection are compared. In Section 3, the symbiotic evolutionary algorithm of generating tolerance conditions are proposed. In Section 4, the algorithm is applied to some typical benchmark FSMs and its performance is demonstrated by the computer simulation. Section 5 concludes the paper with some discussions.

2   The Biological Immune System VS. The Immunotronic System

2.1   The Biological Immune System

The biological immune system protects body from the attack of invaders or antigens such as virus and bacteria. The biological immune system consists of two types of lymphocytes: B cells and T cells. B cells generate antibodies which destroy the antigens. T cells are divided into T-helper cells which help B cells to generate antibodies and T-cytotoxic cells which kill the antigens directly. Interaction of the biological immune system is shown in Fig. 1.

2.2   The Immunotronic System for Fault Detection in Hardware

In general, FSM has transitions between states. The relationship between the biological immune system and the immunotronic system for hardware fault detection is listed in Table 1.

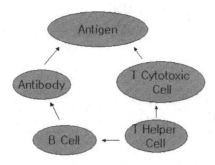

Fig. 1. Interaction of biological immune system

Table 1. The biological immune system vs. the immunotronic system

Biological immune system	Immunotronic system for fault detection
Self	Valid state
Nonself	Invalid state
Antibody	Set of tolerance conditions (Detectors)
To create antibody	Generating of tolerance conditions
Antibody/antigen binding	Pattern matching

In the immunotronic system for hardware fault detection, a set of tolerance conditions is generated using known selfs through negative selection algorithm so that the set of the tolerance conditions detects the nonselfs. How to generate the set of tolerance conditions is the most crucial problem in the design of immunotronic system. In this paper, a new method to generate the tolerance conditions using symbiotic evolution is proposed and this algorithm takes into account the principle of antibody diversity.

3   Design of Fault Detection System Through Symbiotic Evolution Based on the Principle of Antibody Diversity

3.1   The Principle of Antibody Diversity

Scientists long wondered how the biological immune system generates the proper antibodies for infinitely many antigens with a limited number of genes. The answer is that antibody genes are pieced together from widely scattered bits of DNA when they are generated by recombination and mutation. For this reason, antibodies can have the extreme diversity and the immune system has the capacity to recognize and response to about different antigens. This is called the principle of antibody diversity [11]. The process of antibody generation considering the principle of antibody diversity is shown in Fig. 2.

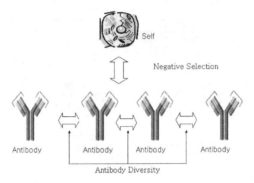

Fig. 2. The process of antibody generation

To detect more nonselfs, we need more tolerance conditions. But the resource of hardware fault detection system is limited, so the smaller the number of tolerance conditions is, the more effective the system is. The implementation of the principle of antibody diversity could be the answer to the problem.

3.2   Symbiotic Evolution

The symbiotic evolution is not new but was introduced in several literatures [12] [13]. Juang et al. applied the symbiotic evolutionary algorithm to the design of fuzzy controllers and demonstrated that the symbiotic evolution is more efficient than the SGA [12]. In this paper, symbiotic evolutionary algorithm is employed to generate the tolerance conditions. Originally, the symbiotic evolution was motivated by the fitness sharing algorithm in an immune model [14] and its application to the immunotronic problem looks very reasonable.

In the SGA, an individual (or a chromosome) is composed of a series of unknown parameters and represents the full solution. The fitness value is assigned to each individual according to the cooperative performance of the constituent unknown parameters. In the symbiotic evolution, however, an individual does not represent the full solution but represents only an unknown parameter, which is only a part of the full solution. A collection of the unknown parameters (or individuals) represents a full solution. Usually, the performance of the collection of the individuals is better than the sum of the performances of the constituent individuals and that is the reason why we call the algorithm as symbiotic evolution. As the concept of symbiotic evolution comes from the model of immune system, the symbiotic evolution is very similar to the generation of antibodies in the biological immune system. So its application to the generation of tolerance conditions in the immunotronic (artificial immune) system makes sense. There is another important characteristic we should note in the biological immune system. The individual antibodies not only cooperate with one another to find antigens, but also compete with one another to survive. That's, they cooperate and compete with one another at the same time. In this paper, we employ the

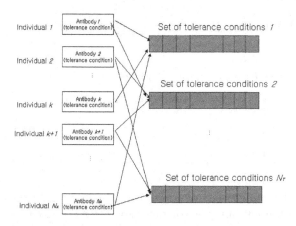

Fig. 3. The basic scheme of symbiotic evolution

symbiotic evolution to generate the tolerance conditions. Tolerance conditions are generated from the perspective of the negative selection and the principle of the antibody diversity. That is, the tolerance conditions are generated in such a way that they are as far away as possible from the selfs in term of Hamming distance (negative selection) and are as different as possible from other tolerance conditions to accomplish the diversity of the antibodies (antibody diversity). As in the biological immune systems, the tolerance conditions not only cooperate with but also compete with one another to accomplish the most efficient antigen finding. In the standard GA, since an individual represents the full solution, a fitness value is assigned to an individual. In contrast, in the symbiotic evolution, a collection of individuals represents a full solution and a fitness value is assigned to the collection of individuals. So, we have to break up the single fitness value and assign the divided values to the constituent individuals. In this paper, we employ the following fitness sharing strategy. The strategy is a bit motivated by Lin's work [12] but it is tailored to this problem. The main scheme of algorithm is shown in Fig. 3.

Here, it is assumed that a population is composed of individuals.

Step 1) Select N_T tolerance conditions from a population of N_A individuals in a random manner, where N_T is the size of the set of the tolerance conditions and $N_T \ll N_A$.

Step 2) Compute the fitness value for the set of the selected N_T tolerance conditions, break up the value and distribute each divided value to the corresponding constituent tolerance condition. The fitness value for the set of the selected N_T tolerance conditions is defined as

$$Fitness = \sum_{k=1}^{N_T}(min_{j=1}^{|S|}(H(\sigma_j, \tau_k))) + \sum_{i=1}^{N_T}\sum_{j=1}^{N_T} H(\tau_i, \tau_j) \tag{1}$$

and the fitness value for each constituent tolerance condition is computed by

$$f_k = min_{j=1}^{|S|}(H(\sigma_j, \tau_k)) + \sum_{i=1}^{N_T}\sum_{j=1}^{N_T} H(\tau_i, \tau_j)/N_T \qquad (2)$$

N_T : the size of the set of tolerance conditions

σ_j : jth self string$(1 \leq j \leq |S|)$

τ_k : kth tolerance condition$(1 \leq k \leq N_T)$

where,

$$H(X,Y) = \sum_{i=1}^{N}(x_i \oplus y_j), X, Y \in \{0,1\} : Hamming\ Distance \qquad (3)$$

This fitness value (1) consists of two terms: The first term prohibits the undesired binding of the selfs and the tolerance conditions (negative selection) and the second term is aimed at the proper allocation of the tolerance conditions to achieve the antibody diversity. This value is broken up and is distributed to each constituent individual as in (2). The fitness value (2) for each tolerance condition is also composed of two terms. The first term is related to the performance of a single tolerance condition τ_k while the second term is related to the symbiotic (or cooperative) performance of the set of the selected tolerance conditions.

Step 3) Repeat the above steps 1 and 2 N_R times until every tolerance condition is selected a sufficient number of times. In each trial, we accumulate the fitness value f_k for each constituent tolerance condition and count the times when each individual is selected.

Step 4) Divide the accumulated fitness value of each individual by the number of times it was selected. In this way, we break up the fitness value assigned to the collection of the tolerance conditions and distribute it to the constituent tolerance conditions (individuals).

The symbiotic evolution employs two basic reproduction operators (crossover and mutation) as in the SGA. For a fair comparison, we use the same design parameters for the SGA and the proposed scheme. The design parameters are shown in Table 2.

Table 2. Genetic parameters

Parameter	SGA	Symbiotic Evolution
The Number of Individual	1000	N_A
The Number of Generation	1000	1000
The Size of Chromosome	p x N_T	p
Mutation Probability	0.7/Length of Individual	0.7/Length of Individual
Crossover Probability	0.7	0.7

4   Simulation and Result

In this section, the proposed algorithm is applied to a decade counter, a typical example of FSM [5] and MCNC benchmark FSMs [15]. An individual string is composed of a user input, previous state and a current state (eg. 01/0000/0001, 00/0010/0010, etc).

4.1   Decade Counter

In this subsection, a decade counter, a typical FSM, taken from the recent paper [5], is considered. There are forty valid transitions in this FSM and the number of selfs is forty. For comparison, we generate tolerance conditions by the proposed method and the SGA-based method [10]. We repeat the simulation fifty times for both methods and one run is composed of one thousand generations. We measure the CPU times and compute the nonself detection rates for various design parameters. The results are shown in Fig. 4. From the figure, it should be noted that the suggested method demonstrates almost the same or a slightly better performance of the previous method [10] but requires much less time than the previous method. The reason for the time-saving might be that, in the SGA-based immunotronic method, an individual is a collection of the tolerance conditions and the collection is evaluated while, in the suggested method, an individual is a tolerance condition and its value is evaluated directly. Thus, in the SGA-based method, it would take more time to find an individual which is all composed of the good tolerance conditions. Instead, in the symbiotic evolution-based method, a tolerance conditions are evaluated directly and if we find N_T good tolerance conditions, we can finish the search. Thus, the suggested method saves much time compared with the previous SGA-based immunotronic method. In addition, it can be seen that the increase in either N_A or N_R leads to not only the increase in the nonself detection rates but also the increase in the CPU time. So it is important to determine the optimal N_A and N_R. Next, the suggested scheme is compared with another previous immunotronic method [5] based on the greedy detector. The performances of the two methods are compared in

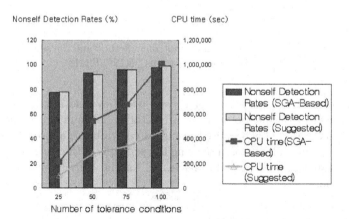

Fig. 4. Comparison of the SGA-based immunotronic method and the suggested method (CPU times and Nonself detection rates)

Nonself Detection Rates (%)

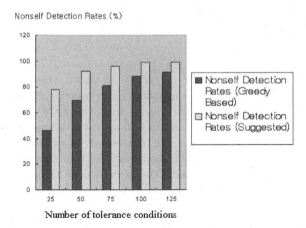

Fig. 5. Comparison of the Greedy-based immunotronic method and the suggested method (Nonself detection rates)

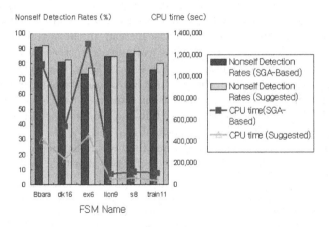

Fig. 6. Comparison of the SGA-based immunotronic method and the suggested method (CPU times and Nonself detection rates) (Various MCNC FSMs)

Fig. 5. The nonself detection rate of the proposed symbiotic scheme is much higher than that of the previous greedy detector-based immunotronic method. This result looks reasonable because the suggested symbiotic approach generates the tolerance condition by accommodating the principle of the antibody diversity and the generated tolerance conditions search for the nonselfs in a wide range of the state space.

4.2   MSNC Benchmark FSMs

In this subsection, we apply the suggested method to MCNC benchmark FSMs. The MCNC FSMs are typical benchmark circuits and widely used in numerous research literatures regarding fault detection [15]. In this example, we set the number of tolerance conditions as half of the number of selfs and run the simulation fifty times. The average running time and the average nonself detection rates are shown in Fig. 6 respectively for various FSMs. As in the previous example, the suggested method demonstrated a slightly better performance (nonself detection rates) within much less time than the SGA-based immunotronic method. In addition, the method suggested herein demonstrates the much better performance than that of the greedy method.

5   Conclusion

In this paper, a novel immunotronic approach to hardware fault detection has been proposed. In the proposed immunotronic algorithm, antibodies (tolerance conditions) were generated using the negative selection and the principle of antibody diversity, two of the most important concepts in the biological immune system. In the implementation of the tolerance condition generation, symbiotic evolution is employed. Two benchmark problems were considered in this paper: decade counter and MCNC FSMs. Through many simulation results, it was seen that the hardware system using the tolerance conditions generated by the proposed algorithm shows the improved nonself detection rates than the greedy method and the consumed CPU time for evolution is considerably reduced compared to the standard GA.

Acknowledgement

This work was supported by the Ministry of Commerce, Industry and Energy of Korea.

References

1. S. Forrest, B. Javornik, R.E. Smith and A.S. Perelson , "Using Genetic Algorithms to Explore Pattern Recognition in the Immune System," Evolutionary Computation, 1(3), pp.191-211, 1993
2. D. Dasgupta and S. Forrest, "An anomaly detection algorithm inspired by the immune system," in Artificial Immune System and Their Applications, D. Dasgupta, Ed. Berlin Germany : Spinger-Verlag, pp.262-277, 1998
3. J. Timmis, M. Neal, J. Hunt, "Data analysis using artificial immune systems, cluster analysis and Kohonen networks: some comparisons," Proc. of IEEE SMC '99 Conference, Vol. 3 , 12-15 pp.922 - 927, 1999
4. R. Xiao, L. Wang, Y. Liu "A framework of AIS based pattern classification and matching for engineering creative design," Proc. of International Conference on Machine Learning and Cybernetics, Vol. 3 , 4-5, pp.1554 - 1558, Nov. 2002
5. D.W. Bradley and A.M. Tyrrell, "Immunotronics-Novel Finite-State-Machine Architectures With Built-In Self-Test Using Self-Nonself Differentiation," IEEE Trans. on Evolutionary Computation, Vol.6, No. 3, pp. 227-238, Jun. 2002

6. Y. Chen and T. Chen, "Implementing fault-tolerance via modular redundancy with comparison," IEEE Trans. on Reliability, Volume: 39 Issue: 2 , pp. 217 -225, Jun 1990.
7. S. Dutt and N.R Mahapatra, "Node-covering, error-correcting codes and multi-processors with very high average fault tolerance," IEEE Trans. Comput., Vol. 46, pp.997-1914, Sep.1997
8. P. K. Lala, Digital Circuit Testing and Testablilty, New York: Academic, 1997
9. S. Forrest, L. Allen, A.S. Perelson, and R. Cherukuri, "Self-Nonself Discrimination In A Computer," Proc. of IEEE Symposium on Research in Security and Privacy, pp.202-212, 1994.
10. S. Lee, E. Kim, M. Park, "A Biologically Inspired New Hardware Fault Detection : immunotronic and Genetic Algorithm-Based Approach," International Journal of Fuzzy Logic and Intelligent Systems, vol. 4 , no. 1, pp7-11, June, 2004.
11. R.A. Goldsby, T.J. Kindt, and B.A Osborne, Kuby Immunology, 4th ed. W.H Freeman and Company: New York, 2000.
12. C. Juang, J. Lin, and C. Lin, "Genetic Reinforcement Learning through Symbiotic Evolution for Fuzzy Controller Design," IEEE Trans. on Systems, Man And Cybernetics-Part B Cybernetics, Vol.30, No. 2 April 2000.
13. D.E. Moriarty and R. Miikkulanien, "Efficient reinforcement learning through symbiotic evolution," Mach. Learn, vol.22, pp.11-32, 1996.
14. R.E. Smith, S. Forrest and A.S. Perelson, "Searching for diverse, cooperative populations with genetic algorithms," Evol. Comput,, vol.1, no.2 pp 127-149 1993.
15. S. Yang "Logic Synthesis and Optimization Benchmarks User Guide Version 3.0," Technical Report, Microelectronics Center of North Carolina, 1991.

A Basic Approach to Reduce the Complexity of a Self-generated Fuzzy Rule-Table for Function Approximation by Use of Symbolic Regression in 1D and 2D Cases

G. Rubio, H. Pomares, I. Rojas, and A. Guillen

Department of Computer Architecture and Computer Technology,
University of Granada (18071). Spain

Abstract. There are many papers in the literature that deal with the problem of the design of a fuzzy system from a set of given training examples. Those who get the best approximation accuracy are based on TSK fuzzy rules, which have the problem of not being as interpretable as Mamdany-type Fuzzy Systems. A question now is posed: How can the interpretability of the generated fuzzy rule-table base be increased? A possible response is to try to reduce the rule-base size by generalizing fuzzy-rules consequents which are symbolic functions instead of fixed scalar values or polynomials, and apply symbolic regressions technics in fuzzy system generation. A first approximation to this idea is presented in this paper for 1-D and 2D functions.

1   Introduction

The problem of estimating an unknown function f from samples of the form $(\overrightarrow{x}^k, z^k)$; k=1,2,..,K; with $z^k = f(\overrightarrow{x}^k) \in \mathbb{R}$ and $\overrightarrow{x} \in \mathbb{R}^N$ (i.e. function approximation from a finite number of data points), has been and is still a fundamental issue in a variety of scientific and engineering fields. Inputs and outputs can be continuous and/or categorical variables. This paper is concerned with continuous output variables, thus considering regression or function approximation problems [5].

Generally, there are three ways to solve the function approximation problem from a set of numerical data:

1. by building a mathematical model for the function to be learned
2. by building a model-free system
3. by seeking human experts' advice

One limitation of the first method is that accurate mathematical models for complex non-linear systems either do not exist or can only be derived with great difficulty. Therefore, the theory of traditional equation-based approaches is well developed and successful in practice only for linear and simple cases [5].

J. Mira and J.R. Álvarez (Eds.): IWINAC 2005, LNCS 3562, pp. 143–152, 2005.

Recently, model-free systems, such as artificial neural networks or fuzzy systems, have been proposed to avoid the knowledge-acquisition bottleneck [1], [2]. Fuzzy systems provide an attractive alternative to the "black boxes" characteristic of neural network models, because their behavior can be easily explained by a human being.

Many fuzzy systems that automatically derive fuzzy IF-THEN rules from numerical data have been proposed in the bibliography to overcome the problem of knowledge acquisition [2], [7]. An important study in this context was carried out in [9].

The approaches presented in [3],[8],[7] need a fixed structure for the rules. However, the distribution of the membership functions (shape and location) has a strong influence on the performance of the systems. Its usually difficult to define and tune the membership functions and rules. These limitations have justified and encouraged the creation of intelligent hybrid systems that overcomes the limitations of individual techniques. Genetic algorithms (GA's) and artificial neural networks (ANN's) offer a possibility to solve this problem [2], [1], [6].

This paper proposes a generalization based on the learning method proposed in [5] to automatically obtain the optimum structure of a fuzzy system and derive fuzzy rules and membership functions from a given set of training data, using a hybridization between fuzzy systems and traditional equation-based approaches using symbolic regression [4]. We will propose and study a basic application of the idea for one dimensional (1-D) continuous functions. Our aim is to obtain an analytical partitioned description of a 1-D function domain, using symbolic regression to determine some possible analytical equations for each partition while, at the same time, the number and definition of each partition is optimized. Each partition of the input domain is associated with a trapezoidal membership function that can be intuitively interpreted without effort.

2   Statement of the Problem

We consider the problem of approximating a continuous single-input single-output function to clarify the basic ideas of our approach, since the extension of the method to a multiple-input is straightforward. Let us consider a set D of desired input-output data pairs, derived from an unknown 1D or 2D function or system F. Each vector datum $(\overrightarrow{x}^k, y^k)$ can be expressed as (x^k, y^k) or (x_1^k, x_2^k, y^k) and k=1,2,...,K. Our fuzzy system comprises a set of n IF-THEN fuzzy rules having the following form for 1D case:

$$IF\ x\ is\ X_i\ THEN\ y = f_i(x) \tag{1}$$

where i=1..n with n being the number of membership functions of the input variable and $f_i(x)$, j=1..n, is a analytical equation associated to rule (partition of x domain). For the 2D case the fuzzy rules have a similar form:

$$IF\ x_1\ is\ X_1^i\ AND\ x_2\ is\ X_2^j\ THEN\ y = f_{i,j}(x) \tag{2}$$

where i=1..n, j=1..m with n,m being the number of membership functions of the input variables x_1 and x_2 respectively and $f_{i,j}(x)$ is a analytical equation associated to the rule (partition of 2D space of x_1 and x_2 domains). Using the above notation, the output of our fuzzy system can be expressed as (for the 1D case):

$$\widetilde{F}(x^k; R, C) = \frac{\sum_{i=1}^{n} f_i(x) \cdot U_i(x)}{\sum_{i=1}^{n} U_i(x)} \tag{3}$$

and in 2D case as:

$$\widetilde{F}(\overrightarrow{x}^k; R, C) = \frac{\sum_{i=1}^{n} \sum_{j=1}^{m} f_{i,j}(x_1, x_2) \cdot U_{i,j}(x_1, x_2)}{\sum_{i=1}^{n} \sum_{j=1}^{m} U_{i,j}(x_1, x_2)} \tag{4}$$

where an explicit statement is made of the dependency of the fuzzy output, not only on the input vector, but also on the matrix of rules R and on all the parameters that describe the membership functions C. The problem considered in this paper may be stated in a precise way as that of finding a configuration C and generating a set of fuzzy rules from a data set D of K input-output pairs, such that the fuzzy system correctly approximates the unknown function F. The function to be minimized is the sum of squared errors:

$$J(R, C) = \sum_{k \in D} (F(x^k) - \widetilde{F}(x^k; R, C))^2. \tag{5}$$

The index selected to determine the degree of accuracy of the obtained fuzzy approximation is the Normalized Root-Mean-Square Error (NRMSE) defined as:

$$NRMSE = \sqrt{\frac{\overline{e^2}}{\sigma_y^2}} \tag{6}$$

where σ_y^2 is the mean-square value of the output data, and $\overline{e^2}$ is the mean-square error between the obtained and the desired output. This index is independent of scale factors or number of data.

3   Proposed Approach

In this section we present the basics of the algorithm we have implemented and studied. Figure 1 shows a flowchart describing the structure of the algorithm.

Before starting the algorithm, we must find an upper bound to the number of partitions for the input variable. For this purpose, we compute the number of local maxima and minima of the underlying function, with the idea that in the worst case, linear (or planar) functions can be use to approximate the segments in between. Throughout the algorithm, the number of partitions will be optimized starting from this maximal value. As can be seen from Fig. 1, the algorithm performs a search for the best system in a top-down fashion, i.e. starting

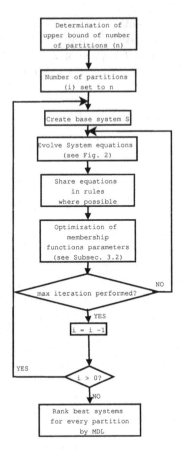

Fig. 1. Algorithm's flowchart

from the biggest number of partitions, it tries to optimize both the location of each partition and the function used in the consequent of each fuzzy rule. After generating every system, it is analyzed to try to share equations between rules in order to reduce the number of rules. The procedure is to try to apply for each rule, the consequents (equations) of rules of adjacent partitions and use the best one. The procedure iterates until no new change can be performed. Finally, we optimize the membership functions parameters (see subsection 3.2) using the Levenberg-Marquardt algorithm and calculate the NRMSE for the final system.

For a given number of partitions, the algorithm in charge of making such optimization is presented in figure 2.

Figure 2 represents the sub-algorithm that is the core of our method. It considers each one of the i partitions at a time. For each partition, an equation F is generated and its constants optimized as explained in subsection 3.1; then iteratively we generate mutated forms of F, optimizing its constants and we accept or reject it if it is better than the actual F. The algorithm iterates until a minimum value of NRMSE or a maximum number of iterations performed are not reached.

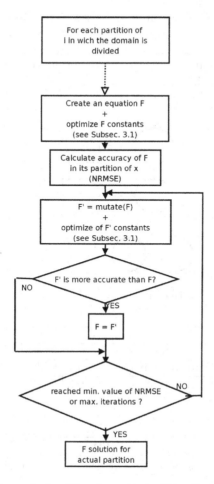

Fig. 2. Local search algorithm for optimization of equation structure

3.1 Representation of Rules Consequents: Symbolic Equations

In order to have the less limited possible set of usable functions as consequents
and use symbolic integration techniques, we have represented them as a vector
of elements that represent a equation in Reverse Polish Notation (RPN) with a
set of possible nodes:

1. Variable: x.
2. Constant: a real value.
3. Binary operators: +, -, *, 'safe /' . We have defined 'a safe b' as $\frac{a}{1+b^2}$, the
   reason is to prevent divisions by zero.
4. Unary operators: - (negation), cos, sin, tanh, atan, 'safe ln(a)', 'limited e^a'.
   We have defined 'safe ln (a)' as $ln(1 + a^2)$, the reason is to prevent ∞
   values; and 'limited e^a' as e^{-a^2}, the reason is to prevent strong function
   value variations.

Nodes that are 'variable' or 'operator' have associated 2 implicit constants: 1 multiplicative and another additive. Note that this constants are distinct than the 'constant' node type. This large number of constants by equation gives a lot of flexibility to the procedure and helps in the optimization stage (for example, the input and output domains can be implicitly normalized in an automatic way).

For example, the function encoded as:

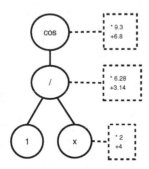

Fig. 3. Tree view of example equation

Nodes	x	const	/	cos
constants				
Multiplicative	3	1	3.14	9.3
Additive	4		6.28	0

is

$$f(x) = 9.3 \cdot cos \left(3.14 \cdot \frac{1}{1 + (3x + 4)^2} + 6.28 \right) + 0 \qquad (7)$$

To limit the complexity of the results we have quoted the size of the equations to 10 elements in the executions.

With this representation, we can apply "mutation" operators that make possible to use local search algorithms to equations in order to optimize their structure itself. A mutation consist in 75% of probability of choosing one of the nodes in the equation tree and substitute all the subtree beyond by another one of the same length; and 25% of probability of substituting the equation by a new one. See figure 4.

3.2   Membership Functions of the System

Intuitively, the results of the algorithm should be an indication of the kind of functions that seem to follow the output of the function to model, in some intervals of the input variable; but we should not forget the "transitions" between them. For these, trapezoidal membership functions can be of much help to implement that idea.

As is well known, we can define a trapezoidal membership function over the interval [0,1] using 4 parameters, $0 \leq z1 \leq c1 \leq c2 \leq z2 \leq 1$. In our

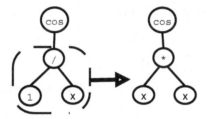

Fig. 4. Tree view of a mutation example

approach, we have also restricted the membership functions making them share some parameters in order to insure a soft union of functions parts, in the form determined by their indexes. For example, for $t+1$ partitions, we have:

$$U_0(x) = \begin{cases} 1 & 0 \leq x < c_{0,2} \\ \frac{c_{1,1}-x}{c_{1,1}-c_{0,2}} & c_{0,2} \leq x < c_{1,1} \\ 0 & otherwise \end{cases}$$

$$\cdots$$

$$U_i(x) = \begin{cases} \frac{x-c_{i-1,2}}{c_{i,1}-c_{i-1,2}} & c_{i-1,2} \leq x < c_{i,1} \\ 1 & c_{i,1} \leq x < c_{i,2} \\ \frac{c_{i+1,1}-x}{c_{i+1,1}-c_{i,2}} & c_{i,2} \leq x < c_{i+1,1} \\ 0 & otherwise \end{cases} \tag{8}$$

$$\cdots$$

$$U_t(x) = \begin{cases} \frac{x-c_{t-1,2}}{c_{t,1}-c_{t-1,2}} & c_{t-1,2} \leq x < c_{t,1} \\ 1 & c_{t,1} \leq x < 1 \\ 0 & otherwise \end{cases}$$

Extension to 2D case is simply:

$$U_{i,j}(x_1, x_2) = U_i(x_1) * U_2(x_2) \tag{9}$$

4 Simulation Results

To see how the proposed approach works and to facilitate its understanding, in this Section we use an 2 artificial examples.

4.1 1D Case

For 1D case, we have generated a function by combining the following three functions:

$$\begin{aligned} f_1(x) &= e^{-5x} \cdot sin(2\pi x) \\ f_2(x) &= \ln(1 + x^2) + 0.2 \\ f_3(x) &= 0.25 * sin(5\pi x + \pi/2) + 0.6 \end{aligned} \tag{10}$$

using trapezoidal membership functions as

$$f(x) = f_1(x)U(x, 0, 0, 0.13, 0.33) +$$
$$f_2(x)U(x, 0.13, 0.33, 0.47, 0.67) + f_3(x)U(x, 0.47, 0.57, 1, 1). \quad (11)$$

We decided to parse the function with 1000 homogenously distributed points, without noise.

The results of apply the proposed algorithm to the 1D example are given by Tables 2:

Table 2. Statistical results obtained by the proposed methodology for the 1D example

N of Partitions	Best NRMSE	MEAN NRMSE	VAR NRMSE	AVG of final N of rules
5	0.060	0.066	0.000	5.000
4	0.064	0.078	0.000	3.000
3	0.094	0.134	0.000	3.000
2	0.129	0.178	0.034	2.000
1	0.034	0.113	0.008	1.000

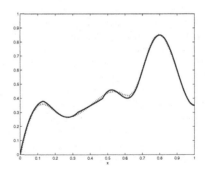

Fig. 5. Output (dotted line) of the best solution found by the algorithm

Table 3. Number of fuzzy rules needed to get an NRMSE similar to the one obtained by the proposed algorithm for 1D example

N of Rules	NRMSE
40	0.061
35	0.062
30	0.063
25	0.065
20	0.071
15	0.088
10	0.149

It is interesting to note that the solution with just one partition (therefore with a single expression) has been the one with the best performance result.

In order to compare with other works, we have applied an adaptation of the method presented in [5] to obtain fuzzy systems with NRMSE similar to the one generated by our method. Table 3 shows these results.

4.2   2D Case

For 2D example we use the following expression:

$$f(x_1, x_2) = \sin(3\pi \cdot x_1) + \cos(3\pi \cdot x_2) \tag{12}$$

We decided to parse the function with 10000 random points, without noise.

The results of apply the proposed algorithm to 2D example are given by Table 4. We have also applied the same method [5] that for 1D case, with results given by Table 5:

Table 4. Statistical results obtained by the proposed methodology for the 2D example

N of Partitions (x1,x2)	Best NRMSE	MEAN NRMSE	VAR NRMES	AVG of final N of Rule
(4, 4)	0.201	0.482	0.197	11.400
(4, 3)	0.083	0.248	0.023	8.600
(4, 2)	0.160	0.377	0.057	7.000
(4, 1)	0.610	0.665	0.002	3.500
(3, 4)	0.000	0.064	0.011	4.300
(3, 3)	0.346	0.881	0.231	6.700
(3, 2)	0.114	0.264	0.005	5.800
(3, 1)	0.236	0.348	0.019	3.000
(2, 4)	0.267	0.531	0.110	5.000
(2, 3)	0.318	0.614	0.141	3.200
(2, 2)	0.046	0.556	0.122	3.000
(2, 1)	0.819	0.857	0.004	1.000
(1, 4)	0.151	0.295	0.024	4.000
(1, 3)	0.000	0.125	0.070	1.300
(1, 2)	0.080	0.249	0.053	2.000
(1, 1)	0.038	0.593	0.092	1.000

Table 5. Number of fuzzy rules needed to get an NRMSE similar to the one obtained by the proposed algorithm for 2D example

x_1/x_2	5	4	3	2	1
5	0.244	0.270	0.734	0.729	0.761
4	0.353	0.369	0.788	0.785	0.812
3	0.250	0.275	0.738	0.732	0.764
2	0.691	0.700	0.978	0.970	0.995
1	0.693	0.704	0.980	0.976	1.000

5   Conclusions

This paper has dealt with the problem of reducing the size of the rule base in the design of a fuzzy system from a set of given training examples. For that purpose, we have generalized the fuzzy-rules consequents allowing them to be symbolic functions instead of fixed scalar values or polynomials, and applying symbolic regressions technics in the fuzzy system generation. The results indicate that this is a possible good application of symbolic regression to reduce the fuzzy rules base complexity, but it's a little far from being effective to increase the interpretability of the whole system. In future works, the method will be extended to an arbitrary number of input, single-input functions, using Gaussian membership functions, more practical for n-D problems. We will also try to define a theory about symbolic equation creation and operation for human understandability, with the idea to endow the equations generated with as much relevant meaning as possible.

References

1. M.Funabashi, A.Maeda, "Fuzzy and neural hybrid expert systems: synergetic AI", IEEE Expert, pp.32-40,1995.
2. J.S.R.Jang, C.T.Sun, E.Mizutani, "Neuro-Fuzzy and soft computing", Prentice Hall, ISBN 0-13-261066-3, 1997.
3. B.Kosko, Neural Networks and Fuzzy Systems. Englewood Cliffs, NJ; Prentice-Hall, 1992.
4. J.R.Koza, "Genetic programming: on the programming of computers by means of natural selection". MIT Press, Cambridge, MA, USA, 1992.
5. H.Pomares, I.Rojas, J.Ortega, J.Gonzalez, A. Prieto "A Systematic Approach to a Self-Generating Fuzzy Rule-Table for Function Approximation", IEEE Trans. on Syst. Man. and Cyber., vol.30, no.3, June 2000.
6. I.Rojas, J.J.Merelo, J.L.Bernier, A.Prieto, "A new approach to fuzzy controller designing and coding via genetic algorithms", IEEE International Conference on Fuzzy Systems, Barcelona, pp.1505-1510, July 1997.
7. R.Rovatti, R.Guerrieri, "Fuzzy sets of rules for system identification", IEEE Trans.on Fuzzy Systems, vol.4, no.2, pp.89-102, 1996.
8. T.Sudkamp, R.J.Hammell, "Interpolation, Completion, and Learning Fuzzy Rules", IEEE Trans. on Syst. Man and Cyber., vol.24, no.2, February, 1994, pp.332-342.
9. L.X.Wang, J.M.Mendel, "Generating fuzzy rules by learning from examples", IEEE Trans. On Syst. Man and Cyber, vol.22, no.6, November/December, pp.1414-1427, 1992.

Parallel Evolutionary Computation: Application of an EA to Controller Design

M. Parrilla, J. Aranda, and S. Dormido-Canto

Dpto. de Informática y Automática,
E.T.S. de Ingeniería Informática, UNED
{mparrilla, jaranda, sebas}@dia.uned.es

Abstract. The evolutionary algorithms can be considered as a powerful and interesting technique for solving large kinds of control problems. However, the great disadvantage of the evolutionary algorithms is the great computational cost. So, the objective of this work is the parallel processing of evolutionary algorithms on a general-purpose architecture (cluster of workstations), programmed with a simple and very well-know technique such as message passing.

1 Introduction

Efficiency of evolutionary algorithms in the optimization problem solution lead to consider them as an alternative method to solve control systems problems. The Evolutionary Algorithms (EA) present a series of advantages with respect to other methods that are most effective in some situations but present applicability limitations. Some of this methods are:

- **Linear Programming**. Only applicable to problems with linear functions.
- **Non-linear Optimization Methods Based on the Gradient**. Applicable to problems with non-linear functions. The functions must be continuous and differentiable, at least at the neighborhood of the optimum. The methods based on the gradient also work with linear problems, but in this case, the linear programming is preferable.
- **Exhaustive Search**. Applicable on those cases where there is a limited number of solutions to problem.

However, the evolutionary algorithms are independent of the function to optimize, and can be applied when the number of possible solutions is unlimited.

The application of evolutionary algorithms to control can be classified in two main groups: first, the *off-line* applications, the most cases are included in this group; in these applications the EA can be employed as a search and optimization engine to select suitable control laws for a plant to satisfy given performance criteria or to search for optimal parameter setting for a particular controller structure. And second, the *on-line* applications, where the EA may be used as a learning mechanism to identify characteristics of unknown or

J. Mira and J.R. Álvarez (Eds.): IWINAC 2005, LNCS 3562, pp. 153–162, 2005.

non-stationary systems or for adaptive controller tuning for known or unknown plants. The *on-line* methods present two essential problems: on the one hand, the high computational cost, making difficult that the parameters are available when they are needed, and on the other hand, and even a more important thing, is the fact that the stochastic nature of this kind of algorithms could lead to that the best solution obtained doesn't comply with the minimum requirements. These disadvantages have made to focus almost all works in the off-line methods, leaving the on-line methods for pure research works.

In certain circumstances, the algorithm execution time comes a very important factor. In the case of *on-line* methods is necessary to have the parameters in the appropriate moment, a higher power in computation is required. Other case is when the algorithm become more complex, as in [10], here the controller is unknown and the algorithm assumes responsibility for determining the number of zeros and poles in the controller and its tuning.

When computational complexity is increased, the use of multiple processors to solve the problem, acquire significance. Available options, can be grouped basically in two categories: to use a supercomputer with multiple processors, or to use a *cluster* of PC's.

The advantages of *clusters* of PC's over supercomputers are: a much lower cost, and a higher capacity to be upgraded. In this work, a *cluster* of 14 PC's was used. Regarding the *software*, Matlab with the PVMTB functions library were used.

2   Controllers Design by Evolutionary Algorithms

In the early 1990s, evolutionary algorithms were first investigated as an alternative method of tuning PID controllers.

Oliveira *et al* [8] used a standard genetic algorithm to get initial estimates for the values of PID parameters. They applied their methodology to a variety of linear time-invariant systems.

Wang and Kwok [14] tailored a genetic algorithm to PID controller tuning. They stressed the benefit of flexibility with regard to cost function, and alluded to the concept of Pareto-optimality to simultaneously address multiple objectives.

More recently, Vlachos *et al* [13] applied a genetic algorithm to the tuning of decentralized PI controllers for multivariable processes. Controller performance was defined in terms of time-domain.

Onnen *et al* [9] applied genetic algorithms to the determination of an optimal control sequence in model-based predictive control. Particular attention was paid to non-linear systems with input constraints.

Genetic algorithms have also been successfully applied in the field of H-infinity control. Chen and Cheng [3] proposed a structure specified H-infinity controller. The genetic algorithm was used to search for good solutions within the admissible domain of controller parameters.

They have also been extended to simultaneously address multiple design objectives, achieved via the incorporation of multiobjective genetic algorithm

(MOGA). Multiple design objectives may be defined, in both the time and frequency domain, resulting in a vector objective function. In one such study, Fonseca and Fleming [4] applied a MOGA to the optimization of the low-pressure spool speed governor of a Rolls-Royce Pegasus gas turbine engine.

Research has also been directed toward the so-called intelligent control systems. The two most popular techniques are *fuzzy control* and *neural control*.

Ichikawa and Sawa [5] used a neural network as a direct replacement for a conventional controller. The weights were obtained using a genetic algorithm. Each individual in the population represented a weight distribution for the network.

Tzes *et al* [11] applied a genetic algorithm to the *off-line* tuning of Gaussian membership functions, developing a fuzzy model that described the friction in a dc-motor system.

Evolutionary methods have also been applied to the generation of control rules, in situations where a reasonable set of rules is not immediately apparent. Matsuura *et al* [7] used a genetic algorithm to obtain optimal control of sensory evaluation of the sake mashing process. The genetic algorithm learned rules for a fuzzy inference mechanism, which subsequently generated the reference trajectory for a PI controller based on the sensory evaluation. Varsek *et al* [12] also used genetic algorithms to develop rule bases, applied to the classic inverted pendulum control problem.

3   Problem Description

The problem formulation corresponds to the RCAM design problem proposed for the GARTEUR Action Group FM(AG08), [6]. The non-linear model proposed was used to generate a linear model around the following conditions: airspeed $V_A = 80$ m/s, altitude h $= 1000$ m, mass $= 120000$ kg, center of gravity cgx $= 0.23$, cgz $= 0.1$ and transport delay $\delta = 0$. From the linearized model obtained, only the longitudinal mode was used, because is trivial to extend the algorithm to the lateral mode, once designed.

The linearized model, in the state-space representation, is:

$$
\begin{pmatrix} \dot{q} \\ \dot{\theta} \\ \dot{u}_B \\ \dot{w}_B \\ \dot{X}_T \\ \dot{\chi}_{TH} \end{pmatrix} = \begin{pmatrix} -0.9825 & 0 & -0.0007 & -0.0161 & -2.4379 & 0.5825 \\ 1 & 0 & 0 & 0 & 0 & 0 \\ -2.1927 & -9.7758 & -0.0325 & 0.0743 & 0.1836 & 19.6200 \\ 77.3571 & -0.7674 & -0.2265 & -0.6683 & -6.4785 & 0 \\ 0 & 0 & 0 & 0 & -6.6667 & 0 \\ 0 & 0 & 0 & 0 & 0 & -0.6667 \end{pmatrix} \begin{pmatrix} q \\ \theta \\ u_B \\ w_B \\ X_T \\ \chi_{TH} \end{pmatrix} + \begin{pmatrix} 0 & 0 \\ 0 & 0 \\ 0 & 0 \\ 0 & 0 \\ 6.6667 & 0 \\ 0 & 0.6667 \end{pmatrix} \begin{pmatrix} d_T \\ d_{TH} \end{pmatrix}
$$

$$
\begin{pmatrix} q \\ n_x \\ n_z \\ w_V \\ V_A \end{pmatrix} = \begin{pmatrix} 1 & 0 & 0 & 0 & 0 & 0 \\ 0.0075 & 0 & -0.0033 & 0.0076 & 0.0187 & 2 \\ -0.2661 & 0 & -0.0231 & -0.0681 & -0.6604 & 0 \\ 0 & -79.8667 & -0.0283 & 0.9996 & 0 & 0 \\ 0 & 0 & 0.9996 & 0.0290 & 0 & 0 \end{pmatrix} \begin{pmatrix} q \\ \theta \\ u_B \\ w_B \\ X_T \\ \chi_{TH} \end{pmatrix} \tag{1}
$$

where the states are: pitch rate (q), pitch angle (θ), x component of the inertial velocity in body-fixed reference frame (u_B), z component of the inertial velocity in body-fixed reference frame (w_B), state corresponding to the tailplane (X_T) and state corresponding to the engines throttles (χ_{TH}).

The outputs are: pitch rate (q), horizontal load factor (n_x), vertical load factor (n_z), z component of vertical velocity in the vehicle-carried reference frame (w_V) and air speed (V_A).

The objectives considered are in the design specifications, in the document [6]. For longitudinal mode, the design specifications are summarized as follows:

- Closed-loop stability: it's the most basic objective to be satisfied.
- The control system should be able to track step reference signals: in V_A with a rise time $t_r < 12s$, a setting time $ts < 45s$ and overshoot $M_p < 5\%$, and in flight path angle (γ) with $t_r < 5s$, $t_s < 20s$ and $M_p < 5\%$. But γ isn't available in the model. To cope with such a problem, the relation $sin(\gamma) = \frac{-w_V}{V}$ can be used, where V is the total inertial velocity.
- Ride quality criteria: vertical accelerations would be minimized.
- Saturations limits in control signals, would be observed.
- Robustness criteria: the gain margin is required to be at least 10 dB and the phase margin is required to be at least 50.

In a previous work [1], the authors of this document solved the problem by means of an sequential evolutionary algorithm. They took a fixed controller structure for this, as showed in figure 1, where a static gain matrix (K_p) was directly applied on the 5 model outputs, and another gain matrix (K_i) was applied on the integral of the errors in V_A y w_V, to eliminate steady state errors. The algorithm got the gains, K_p y K_i, meeting the design specifications.

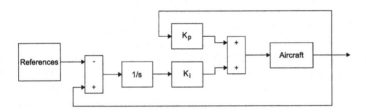

Fig. 1. Structure of longitudinal controller

The sequential evolutionary algorithm used, is shown in figure 2. A initial population of chromosomes is randomly generated, whose members would be possible solutions to the problem, properly coded. Chromosomes would be evaluated and sorted according to fitness. The chromosome with the best fitness would be established as the problem solution. After that a loop would start, where new generations of chromosomes would be obtained from previous generation, by applying evolutionary operators over the parents selected in a previous step. The new generation would be again evaluated and sorted according to fitness. If the best chromosome in the current population was more suitable than the previously established as solution to problem, it would replace it. Finally if end conditions are satisfied, the program would finish, otherwise a new loop iteration would start.

In a more ambitious project, it's possible to let the algorithm to find the controller structure, or in a multivariable system, ask the algorithm for the matrix of transfer functions representing the controller, taking the algorithm charge

```
k ← 0
get random initial population P_k(x)
for each x_i in P_k(x)
    decode x_i
    evaluate x_i ← get fitness
endfor
sort P_k(x) according to fitness
solution ← best x_i in P_k(x)
while not end conditions
    k ← k + 1
    select parents by tournament selection
    get new population P_{k+1}(x)
    for each x_i in P_k(x)
        decode x_i
        evaluate x_i ← get fitness
    endfor
    sort P_k(x) according to fitness
    if the best x_i in P_k(x) is better than solution
        solution ← best x_i in P_k(x)
    endif
endwhile
```

Fig. 2. Sequential Evolutionary Algorithm

in that case of determining the number of zeros and poles for each transfer functions. To solve this kind of problems, the algorithm would have to tune a very high number of controllers, which meant that the time of execution increase dramatically. In that cases, the necessity to speedup the tune process arise.

If the sequential algorithm in figure 2 is parallelized, and a good speedup is obtained, the parallelization result could be used as a component of a more complex program: the parallelized algorithm would be called every time a controller has to be tuned. Now the parallelization of the sequential algorithm is described and also the speedup obtained are shown.

4   Parallelizing the Algorithm

Matlab is a standard in control, thanks to its specialized *toolboxes*. However, it lacked ability to carry out parallel programing. To cover this gap, Baldomero [2], designed PVMTB (Parallel Virtual Machine ToolBox), a *toolbox* including almost all functionalities in PVM, the known parallelization library by message passing.

Thanks to PVMTB, all the control specialized Matlab functions can be used to design a parallel evolutionary algorithm. The parallelizing process will be described below.

Matlab with PVMTB, and a Master/Slave strategy were used.

In order to parallelize effectively the sequential algorithm previously described, a study to determine the most computing intensive parts was carried out, resulting that generating new chromosomes from parents and determining fitness were the most time consuming tasks. Therefore, this were the stages where parallelization would be focused.

An inherent feature of evolutionary algorithms was also taken into account: for each generation, all chromosomes will have to be created and evaluated before continuing with a new generation. If the workload isn't uniformly distributed between the different processes, those that firstly finish their work will have

```
start up PVM: pvm_start_pvmd();
start up slave tasks: pvm_spawn();
k ← 0
get random initial population P_k(x)
for each x_i in P_k(x)
    decode x_i
    evaluate x_i ← get fitness
endfor
sort P_k(x) according to fitness
solution ← best x_i in P_k(x)
while not end conditions
    k ← k + 1
    select parents by tournament selection
    for each slave
        send a pair of parents: pvm_send();
    endfor
    while there is a pair of parents not used
        receive 2 evaluated chromosomes from one slave: pvm_recv();
        send a pair of parents to this slave: pvm_send();
    endwhile
    while num chromosomes asked for < size of P_k(x)
        receive 2 evaluated chromosomes from one slave: pvm_recv();
        ask for 2 inmigrants to this slave: pvm_send();
    endwhile
    while num chromosomes received < size of P_k(x)
        receive 2 evaluated chromosomes from one slave: pvm_recv();
    endwhile
    sort P_k(x) according to fitness
    if the best x_i in P_k(x) is better than solution
        solution ← best x_i in P_k(x)
    endif
endwhile
for each slave
    send signal to quit
endfor
halt PVM: pvm_halt;
```

Fig. 3. Master Process Algorithm

to wait a lot, because they can't start with a new generation until the others processes finish with the current one. That obviously imply a decrease of the speedup obtained with parallelization.

An added difficulty is the fact that times to get the fitness of different chromosomes, can be very different. A chromosome implying a unstable system, would be quickly evaluated, but those that give rise to a stable system would have to be studied more slowly, to determine the system features, increasing the evaluation time. To minimize this problem, the master divided the work in small tasks, and quickly assign a task to slaves waiting for a work.

The master also had to receive the results from slaves, and to organize them as they arrived. It wasn't necessary to parallelize the evaluation of the initial population, because first chromosomes generally give rise to unstable systems, and are quickly evaluated. The algorithm corresponding to the master process is shown in figure 3.

Slave processes for its part, were concerned with generating a pair of offsprings from the pair of parents passed by the master, evaluating them, and sending the new chromosomes obtained and its fitness to master. They also generated and evaluated pairs of immigrants, randomly obtained, when master requested them. Slaves would generate and evaluate chromosomes until the master send them the end signal. The algorithm used by slaves is shown in figure 4.

```
while not signal to quit
    get master message: pvm_recv();
    if parents provided
       generate 2 offsprings
    else
       generate 2 inmigrants
    endif
    for each generated chromosome
       decode chromosome
       evaluate chromosome
    endfor
    send chromosomes and fitness to master: pvm_send();
endwhile
quit
```

Fig. 4. Slave Process Algorithm

5   Hardware and Software Description

The *cluster* used in this work had 14 PC's with AMD K7 500 MHz processors, 384 MB of RAM memory and a hard drive of 7GB each of them. The nodes (1 Master + 13 Slaves) were connected by Fast-Ethernet switch. The operating system was Linux (Red-Hat 6.1).

The algorithm was implemented using a *toolbox* of parallel processing, developed in Matlab by Baldomero, J.F., [2]: PVMTB (Parallel Virtual Machine

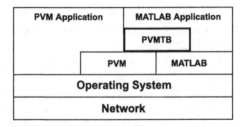

Fig. 5. High level overview of PVM

ToolBox), based on the standard library PVMTB. With PVMTB, Matlab users can quickly build parallel programs, using a message passing system like PVM.

Figure 5 shows a high level overview diagram of PVMTB. The *toolbox* makes PVM and Matlab-API (Application Program Interface) calls to enable messages between Matlab processes.

6   Performance Results

After 4 executions of the parallel algorithm, to tune the controller in figure 1, an average of the times of execution and the speedups obtained, was calculated. Each execution was repeated for each possible number of processors. Results are shown in figure 6, and its numeric values are grouped in table 1.

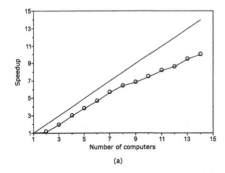
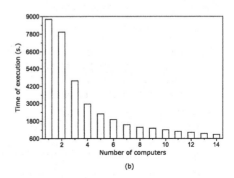

Fig. 6. (a) Speedup vs. number of computers. (b) Time of execution vs. number of computers

7   Conclusions

Features of evolutionary algorithms make them appropriated to deal with problems, difficult to be solved by other methods. A drawback is its high computational cost, making them impossible to be applied to solve complex problems, in some cases. But evolutionary algorithms are easily parallelized by nature, and *clusters* of PC's provide a low-cost alternative to supercomputers.

Table 1. Speedup and Time of execution vs. number of computers

Number of computers	Time of execution (seg.)	Speedup	Standard deviation (speedup)
1	8905.9259	1	0
2	7953.4766	1.1197555	0.0079589
3	4597.6172	1.9388577	0.0611588
4	2991.8349	2.9776865	0.0541746
5	2310.3739	3.8557793	0.0610961
6	1904.5179	4.6804295	0.1380201
7	1566.8359	5.6894018	0.1702718
8	1392.8489	6.3950083	0.0752763
9	1312.2312	6.7910607	0.1919798
10	1194.6286	7.4551005	0.0350726
11	1087.4028	8.1928836	0.1823867
12	1034.5336	8.6091675	0.0984071
13	941.04137	9.4676986	0.2028066
14	890.80885	10.023033	0.5184515

Acknowledgment

Part of this work was supported by MCyT of Spain under contract DPI2003-09745-C04-01.

References

1. Aranda, J.; De la Cruz, J.M.; Parrilla, M. and Ruipérez, P.: *Evolutionary Algorithms for the Design of a Multivariable Control for an Aircraft Flight Control*, AIAA Guidance, Navigation, and Control Conference and Exhibit, Denver, CO. (August 2000)
2. Baldomero, J.F.: *PVMTB: Parallel Virtual Machine ToolBox*, II Congreso de Usuarios Matlab'99, Dpto. Informática y Automática. UNED. Madrid. (1999) pp. 523-532
3. Chen, B. S., and Cheng, Y. M.: *A Structure-Specified H-Infinity Optimal Control Design for Practical Applications: A Genetic Approach*, IEEE Transactions on Control Systems Technology, Vol. 6. (November 1998) pp707-718
4. Fonseca, Carlos M., and Fleming, Peter J.: *Multiobjective Optimization and Multiple Constraint Handling with Evolutionary Algorithms-Part I: A Unified Formulation and Part II: Application Example*, IEEE Transactions on Systems. Man and Cybernetics. Part A: Systems and Humans. Vol. 28. No. 1. (January 1998) pp26-37 and pp38-47
5. Ichikawa, Y., and Sawa, T.: *Neural Network Application for Direct Feedback Controllers*, IEEE Transactions on Neural Networks, Vol. 3, No. 2. (March 1992) pp224-231
6. Lambrechts, P.F. et al.: *Robust flight control design challenge problem formulation and manual: the research civil aircraft model (RCAM)*. Technical publication TP-088-3, Group for Aeronautical Research and technology in EURope GARTEUR-FM(AG-08). (1997)

7. Matsuura, K.; Shiba, H.; Hirotsune, M., and Nunokawa, Y.: *Optimal control of sensory evaluation of the sake mashing process*, Journal of Process Control, Vol. 6, No. 5. (1996) pp323-326

8. Oliveira, P.; Sequeira, J., and Sentieiro, J.: *Selection of Controller Parameters using Genetic Algorithms*, Engineering Systems with Intelligence. Concepts, Tools, and Applications, Kluwer Academic Publishers, Dordrecht, Netherlands. (1991) pp431-438

9. Onnen, C.; Babuska, R.; Kaymak, U.; Sousa, J. M.; Verbruggen, H. B., and Isermann, R.: *Genetic Algorithms for optimization in predictive control*, Control Engineering Practice, Vol. 5, Iss. 10. (1997) pp1363-1372

10. Parrilla, M.; Aranda, J. and Díaz J.M. *Selection and Tuning of Controllers, by Evolutionary Algorithms: Application to Fast Ferries Control*. CAMS2004, IFAC. (2004)

11. Tzes, A.; Peng, P. Y., and Guthy, J.: *Genetic-Based Fuzzy Clustering for DC-Motor Friction Identification and Compensation*, IEEE Transactions on Control Systems Technology, Vol. 6, No. 4. (July 1998) pp462-472

12. Varsek, A.; Urbancic, T., and Fillipic, B.: *Genetic Algorithms in Controller Design and Tuning*, IEEE Transactions on Systems, Man, and Cybernetics, Vol. 23, No. 5. (September/October 1993) pp1330-1339

13. Vlachos, C.; Williams, D., and Gomm, J. B.: *Genetic approach to decentralized PI controller tuning for multivariable processes*, IEEE Proceedings - Control Theory and Applications, Vol. 146, No. 1, (January 1999) pp58-64

14. Wang, P., and Kwok, D. P.: *Autotuning of Classical PID Controllers Using an Advanced Genetic Algorithm*, International Conference on Industrial Electronics, Control, Instrumentation and Automation (IECON 92), Vol. 3. (1992) pp1224-1229

MEPIDS: Multi-Expression Programming for Intrusion Detection System

Crina Groşan[1,2], Ajith Abraham[2], and Sang Yong Han[2]

[1] Department of Computer Science,
Babeş-Bolyai University, Kogălniceanu 1,
Cluj-Napoca, 3400, Romania
[2] School of Computer Science and Engineering,
Chung-Ang University, Seoul, South Korea,
cgrosan@cs.ubbcluj.ro, ajith.abraham@ieee.org, hansy@cau.ac.kr

Abstract. An Intrusion Detection System (IDS) is a program that analyzes what happens or has happened during an execution and tries to find indications that the computer has been misused. An IDS does not eliminate the use of preventive mechanism but it works as the last defensive mechanism in securing the system. This paper evaluates the performances of Multi-Expression Programming (MEP) to detect intrusions in a network. Results are then compared with Linear Genetic Programming (LGP) approach. Empirical results clearly show that genetic programming could play an important role in designing light weight, real time intrusion detection systems.

1   Introduction

An intrusion is defined as any set of actions that attempt to compromise the integrity, confidentiality or availability of a resource. Intrusion detection is classified into two types: misuse intrusion detection and anomaly intrusion detection [13]. Misuse intrusion detection uses well-defined patterns of the attack that exploit weaknesses in system and application software to identify the intrusions. These patterns are encoded in advance and used to match against the user behavior to detect intrusion. Anomaly intrusion detection uses the normal usage behavior patterns to identify the intrusion. The normal usage patterns are constructed from the statistical measures of the system features. The behavior of the user is observed and any deviation from the constructed normal behavior is detected as intrusion [7], [15]. Data mining approaches for intrusion detection were first implemented in mining audit data for automated models for intrusion detection [2], [6], [9]. Several data mining algorithms are applied to audit data to compute models that accurately capture the actual behavior of intrusions as well as normal activities. Audit data analysis and mining combine the association rules and classification algorithm to discover attacks in audit data. Soft Computing (SC) is an innovative approach to construct computationally intelligent systems consisting of artificial neural networks, fuzzy inference systems,

J. Mira and J.R. Álvarez (Eds.): IWINAC 2005, LNCS 3562, pp. 163–172, 2005.

approximate reasoning and derivative free optimization methods such as evolutionary computation etc. [14]. This paper compares a Genetic Programming (GP) technique performance – Multi-Expression Programming (MEP) – with Linear Genetic Programming (LGP) [3], Support Vector Machines (SVM) [16] and Decision Trees (DT) [5]. Rest of the paper is organized as follows. Section 2 provides the technical details of MEP. In Section 3, a description of the intelligent paradigms used in experiments is given. Experiment results are presented in Section 4 and some conclusions are also provided towards the end.

2   Multi Expression Programming (MEP)

A GP chromosome generally encodes a single expression (computer program). By contrast, Multi Expression Programming (MEP)[11], [12] chromosome encodes several expressions. The best of the encoded solution is chosen to represent the chromosome (by supplying the fitness of the individual).

The MEP chromosome has some advantages over the single-expression chromosome especially when the complexity of the target expression is not known. This feature also acts as a provider of variable-length expressions. Other techniques (such as Gramatical Evolution (GE) [14] or Linear Genetic Programming (LGP) [4]) employ special genetic operators (which insert or remove chromosome parts) to achieve such a complex functionality.

2.1   Solution Representation

MEP genes are (represented by) substrings of a variable length. The number of genes per chromosome is constant. This number defines the length of the chromosome. Each gene encodes a terminal or a function symbol. A gene that encodes a function includes pointers towards the function arguments. Function arguments always have indices of lower values than the position of the function itself in the chromosome.

The proposed representation ensures that no cycle arises while the chromosome is decoded (phenotypically transcripted). According to the proposed representation scheme, the first symbol of the chromosome must be a terminal symbol. In this way, only syntactically correct programs (MEP individuals) are obtained.

An example of chromosome using the sets $F= \{+, *\}$ and $T= \{a, b, c, d\}$ is given below:

1: a
2: b
3: $+$ 1, 2
4: c
5: d
6: $+$ 4, 5
7: $*$ 3, 6

The maximum number of symbols in MEP chromosome is given by the formula:

$Number\_of\_Symbols = (n{+}1) * (Number\_of\_Genes - 1) + 1,$

where n is the number of arguments of the function with the greatest number of arguments.

The maximum number of effective symbols is achieved when each gene (excepting the first one) encodes a function symbol with the highest number of arguments. The minimum number of effective symbols is equal to the number of genes and it is achieved when all genes encode terminal symbols only.

The translation of a MEP chromosome into a computer program represents the phenotypic transcription of the MEP chromosomes. Phenotypic translation is obtained by parsing the chromosome top-down. A terminal symbol specifies a simple expression. A function symbol specifies a complex expression obtained by connecting the operands specified by the argument positions with the current function symbol.

For instance, genes 1, 2, 4 and 5 in the previous example encode simple expressions formed by a single terminal symbol. These expressions are:

$E_1 = a,$
$E_2 = b,$
$E_4 = c,$
$E_5 = d,$

Gene 3 indicates the operation + on the operands located at positions 1 and 2 of the chromosome. Therefore gene 3 encodes the expression: $E_3 = a + b$. Gene 6 indicates the operation + on the operands located at positions 4 and 5. Therefore gene 6 encodes the expression: $E_6 = c + d$. Gene 7 indicates the operation * on the operands located at position 3 and 6. Therefore gene 7 encodes the expression: $E_7 = (a + b) * (c + d)$. E_7 is the expression encoded by the whole chromosome.

There is neither practical nor theoretical evidence that one of these expressions is better than the others. This is why each MEP chromosome is allowed to encode a number of expressions equal to the chromosome length (number of genes). The chromosome described above encodes the following expressions:

$E_1 = a,$
$E_2 = b,$
$E_3 = a + b,$
$E_4 = c,$
$E_5 = d,$
$E_6 = c + d,$
$E_7 = (a + b) * (c + d).$

The value of these expressions may be computed by reading the chromosome top down. Partial results are computed by dynamic programming and are stored in a conventional manner.

Due to its multi expression representation, each MEP chromosome may be viewed as a forest of trees rather than as a single tree, which is the case of Genetic Programming.

2.2   Fitness Assignment

As MEP chromosome encodes more than one problem solution, it is interesting to see how the fitness is assigned.

The chromosome fitness is usually defined as the fitness of the best expression encoded by that chromosome.

For instance, if we want to solve symbolic regression problems, the fitness of each sub-expression E_i may be computed using the formula:

$$f(E_i) = \sum_{k=1}^{n} |o_{k,i} - w_k|,$$

where $o_{k,i}$ is the result obtained by the expression E_i for the fitness case k and w_k is the targeted result for the fitness case k. In this case the fitness needs to be minimized.

The fitness of an individual is set to be equal to the lowest fitness of the expressions encoded in the chromosome:

When we have to deal with other problems, we compute the fitness of each sub-expression encoded in the MEP chromosome. Thus, the fitness of the entire individual is supplied by the fitness of the best expression encoded in that chromosome.

3   Intelligent Paradigms

3.1   Linear Genetic Programming (LGP)

Linear genetic programming is a variant of the GP technique that acts on linear genomes [4]. Its main characteristics in comparison to tree-based GP lies in that the evolvable units are not the expressions of a functional programming language (like LISP), but the programs of an imperative language (like c/c ++). An alternate approach is to evolve a computer program at the machine code level, using lower level representations for the individuals. This can tremendously hasten the evolution process as, no matter how an individual is initially represented, finally it always has to be represented as a piece of machine code, as fitness evaluation requires physical execution of the individuals.

3.2   Support Vector Machines

Support Vector Machines [16] have been proposed as a novel technique for intrusion detection. A Support Vector Machine (SVM) maps input (real-valued) feature vectors into a higher dimensional feature space through some nonlinear mapping. These are developed on the principle of structural risk minimization.

SVM uses a feature called kernel to solve this problem. Kernel transforms linear algorithms into nonlinear ones via a map into feature spaces.

3.3   Decision Trees

Decision tree induction is one of the classification algorithms in data mining [5]. Each data item is defined by values of the attributes. The Decision tree classifies the given data item using the values of its attributes. The main approach is to select the attributes, which best divides the data items into their classes. According to the values of these attributes the data items are partitioned. This process is recursively applied to each partitioned subset of the data items. The process terminates when all the data items in current subset belongs to the same class.

4   Experiment Setup and Results

The data for our experiments was prepared by the 1998 DARPA intrusion detection evaluation program by MIT Lincoln Labs [10]. The data set has 41 attributes for each connection record plus one class label as given in Table 1. The data set contains 24 attack types that could be classified into four main categories Attack types fall into four main categories:

DoS: Denial of Service
Denial of Service (DoS) is a class of attack where an attacker makes a computing or memory resource too busy or too full to handle legitimate requests, thus denying legitimate users access to a machine.

R2L: Unauthorized Access from a Remote Machine
A remote to user (R2L) attack is a class of attack where an attacker sends packets to a machine over a network, then exploits the machine's vulnerability to illegally gain local access as a user.

U2Su: Unauthorized Access to Local Super User (root)
User to root (U2Su) exploits are a class of attacks where an attacker starts out with access to a normal user account on the system and is able to exploit vulnerability to gain root access to the system.

Probing: Surveillance and Other Probing
Probing is a class of attack where an attacker scans a network to gather information or find known vulnerabilities. An attacker with a map of machines and services that are available on a network can use the information to look for exploits.

Our experiments have two phases namely training and testing phases. In the training phase, MEP models were constructed using the training data to give maximum generalization accuracy on the unseen data. The test data is then passed through the saved trained model to detect intrusions in the testing phase. The 41 features are labeled as shown in Table 1 and the class label is named as *AP*.

Table 1. Variables for intrusion detection data set

Variable No.	Variable name	Variable type	Variable label
1	duration	continuous	A
2	protocol_type	discrete	B
3	service	discrete	C
4	flag	discrete	D
5	src_bytes	continuous	E
6	dst_bytes	continuous	F
7	land	discrete	G
8	wrong_fragment	continuous	H
9	urgent	continuous	I
10	hot	continuous	J
11	num_failed_logins	continuous	K
12	logged_in	discrete	L
13	num_compromised	continuous	M
14	root_shell	continuous	N
15	su_attempted	continuous	O
16	num_root	continuous	P
17	num_file_creations	continuous	Q
18	num_shells	continuous	R
19	num_access_files	continuous	S
20	num_outbound_cmds	continuous	T
21	is_host_login	discrete	U
22	is_guest_login	discrete	V
23	count	continuous	W
24	srv_count	continuous	X
25	serror_rate	continuous	Y
26	srv_serror_rate	continuous	X
27	rerror_rate	continuous	AA
28	srv_rerror_rate	continuous	AB
29	same_srv_rate	continuous	AC
30	diff_srv_rate	continuous	AD
31	srv_diff_host_rate	continuous	AE
32	dst_host_count	continuous	AF
33	dst_host_srv_count	continuous	AG
34	dst_host_same_srv_rate	continuous	AH
35	dst_host_diff_srv_rate	continuous	AI
36	dst_host_same_src_port_rate	continuous	AJ
37	dst_host_srv_diff_host_rate	continuous	AK
38	dst_host_serror_rate	continuous	AL
39	dst_host_srv_serror_rate	continuous	AM
40	dst_host_rerror_rate	continuous	AN
41	dst_host_srv_rerror_rate	continuous	AO

This data set has five different attack types (classes) namely *Normal, DoS, R2L, U2R* and *Probes*. The training and test data comprises of 5,092 and 6,890 records respectively [8]. All the training data were scaled to (0-1). Using the data set, we performed a 5-class classification.

The settings of various linear genetic programming system parameters are of utmost importance for successful performance of the system [1]. The population size was set at 120,000 and a tournament size of 8 is used for all the 5 classes. Crossover and mutation probability is set at 65-75% and 75-86% respectively for the different classes. Our trial experiments with SVM revealed that the polynomial kernel option often performs well on most of the datasets. We also constructed decision trees using the training data and then testing data was passed through the constructed classifier to classify the attacks [13]. Parameters

Table 2. Parameters used by MEP

Attack type	Parameter value				
	Pop. Size	Generations	Crossover (%)	No. of mutations	Chromosome length
Normal	100	100	0.9	3	30
Probe	200	200	0.9	4	40
DOS	500	200	0.8	5	40
U2R	20	100	0.9	3	30
R2L	800	200	0.9	4	40

Table 3. Performance comparison

Attack type	Classification accuracy on test data set (%)			
	MEP	DT	SVM	LGP
Normal	**99.82**	99.64	99.64	99.73
Probe	95.52	99.86	98.57	**99.89**
DOS	98.91	96.83	99.92	**99.95**
U2R	**99.75**	68.00	40.00	64.00
R2L	**99.72**	84.19	33.92	99.47

Table 4. MEP evolved functions for the attack classes

Attack type	Evolved Function
Normal	var12 * log_2(var10 + var3)
Probe	$(log_2$(var2) < (fabs((var36 * var27) > (var27 + var35 − var34) ? (var35 * var27) : (var27 + var35 − var34))) ? $(log_2$(var2)) : (fabs((var36 * var27) > (var27 + var35 − var34) ? (var36 * var27) : (var27 + var35 − var34)));
DOS	$0.457*(var8+(ln(var6))*(lg(var41))−−var40+var23+var8)$
U2R	$sin(var14) − −var33$
R2L	0.36 + (var11 < 0.086 ? var11 : 0.086 + 0.086) > (var6 > $(log_2(log_2$(var12 * var3))) ? var6 : $(log_2(log_2$(var12 * var3)))) ? (var11 < 0.086 ? var11 : 0.086 + 0.086) : (var6 > $(log_2(log_2$(var12 * var3))) ? var6 : $(log_2(log_2$(var12 * var3)))) + var6

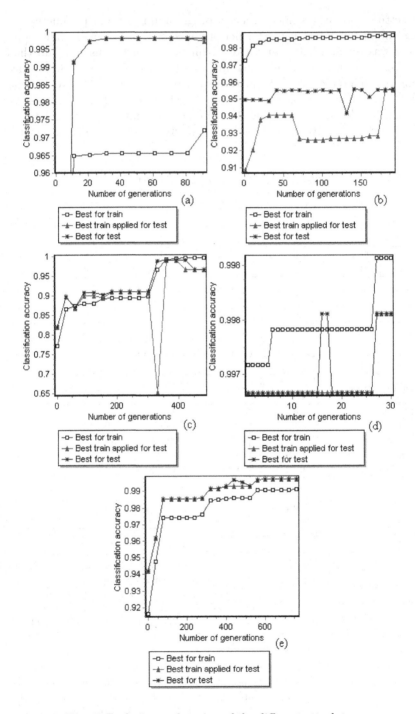

Fig. 1. Evolutionary learning of the different attack types

used by MEP are presented in Table 2. We made use of +, - , *, /, sin, cos, sqrt, ln, lg, $\log_2$, min, max, and abs as function sets.

In Table 4 the variable combinations evolved by MEP are presented. Results presented in Table 3 are based on these evolved functions.

As evident from Table 3, MEP gives the best results for detecting Normal patterns, U2R and R2L attacks. While DT, SVM and LGP did not perform well U2R attacks, MEP obtained the best results for this class (99.76% accuracy). Results for MEP presented in this table are obtained by applying the test data using the training function which gave the best results during training.

In Figure 1, the classification accuracy for the best results obtained for training data, results obtained for the test data using the best training function and the best results obtained for the test data are depicted. Figure 1 (a) corresponds to Normal patterns, Figure 1 (b) corresponds to Probe, Figure 1 (c) corresponds to DOS Figure 1 (d) corresponds to U2R and Figure 1 (e) corresponds to R2L respectively.

In some classes the accuracy figures tend to be very small and may not be statistically significant, especially in view of the fact that the 5 classes of patterns differ in their sizes tremendously. For example only 27 data sets were available for training the U2R class. More definitive conclusions can only be made after analyzing more comprehensive sets of network traffic.

5   Conclusions

In this paper, we have illustrated the importance of genetic programming based techniques for modeling intrusion detection systems. MEP outperforms LGP for three of the considered classes and LGP outperform MEP for two of the classes. MEP classification accuracy is grater than 95% for all considered classes and for four of them is greater than 99.65%. It is to be noted that for real time intrusion detection systems MEP and LGP would be the ideal candidates because of its simplified implementation.

Acknowledgements

This research was supported by the MIC (Ministry of Information and Communication), Korea, under the Chung-Ang University HNRC-ITRC (Home Network Research Center) support program supervised by the IITA (Institute of Information Technology Assessment).

References

1. Abraham A., Evolutionary Computation in Intelligent Web Management, Evolutionary Computing in Data Mining, Ghosh A. and Jain L.C. (Eds.), Studies in Fuzziness and Soft Computing, Springer Verlag Germany, Chapter 8, pp. 189-210, 2004.

2. Barbara D., Couto J., Jajodia S. and Wu N., ADAM: A Testbed for Exploring the Use of Data Mining in Intrusion Detection. SIGMOD Record, 30(4), pp. 15-24, 2001.

3. Brameier. M. and Banzhaf. W., A comparison of linear genetic programming and neural networks in medical data mining,Evolutionary Computation," IEEE Transactions on, Volume: 5(1), pp. 17-26, 2001.

4. Brameier M. and Banzhaf W, Explicit control of diversity and effective variation distance in Linear Genetic Programming. In Proceedings of the fourth European Conference on Genetic Programming, Springer-Verlag Berlin, 2001.

5. Brieman L., Friedman J., Olshen R., and Stone C., Classification of Regression Trees. Wadsworth Inc., 1984.

6. Cohen W., Learning Trees and Rules with Set-Valued Features, American Association for Artificial Intelligence (AAAI), 1996.

7. Denning D., An Intrusion-Detection Model, IEEE Transactions on Software Engineering, Vol. SE-13, No. 2, pp.222-232, 1987.

8. KDD Cup 1999 Intrusion detection data set: http://kdd.ics.uci.edu/databases/ kddcup99/kddcup.data_10_percent.gz

9. Lee W. and Stolfo S. and Mok K., A Data Mining Framework for Building Intrusion Detection Models. In Proceedings of the IEEE Symposium on Security and Privacy, 1999.

10. MIT Lincoln Laboratory. http://www.ll.mit.edu/IST/ideval/

11. Oltean M. and Grosan C., A Comparison of Several Linear GP Techniques, Complex Systems, Vol. 14, No. 4, pp. 285-313, 2004.

12. Oltean M. and Grosan C., Evolving Evolutionary Algorithms using Multi Expression Programming. Proceedings of The 7th European Conference on Artificial Life, Dortmund, Germany, pp. 651-658, 2003.

13. Peddabachigari S., Abraham A., Thomas J., Intrusion Detection Systems Using Decision Trees and Support Vector Machines, International Journal of Applied Science and Computations, USA, Vol.11, No.3, pp.118-134, 2004.

14. Ryan C. et al, 1998. Gramatical Evolution:Evolving programs for an arbitrary language. In Proceedings of the first European Workshop on Genetic Programming, Springer-Verlag, Berlin

15. Summers R.C., Secure Computing: Threats and Safeguards. McGraw Hill, New York, 1997.

16. Vapnik V.N., The Nature of Statistical Learning Theory. Springer, 1995.

A Study of Heuristic Techniques Inspired in Natural Process for the Solution of the Container Fill Problem

M. Delgado Pineda, J.M. De Pedro Salinas, and J. Aranda Almansa

Universidad de Educación a Distancia,
E.T.S.I. Informática

Abstract. This article shows some techniques based on simulated annealing and genetic alghoritms for the resolution of a filling problem of a container of two dimensions using rectangular pieces of sizes not congruent. This problem is quite related to problems like bin-packing or strip-packing. The comparative was made using several type problems and having into account parameters like run time, number of solutions that converge to the optimum one and quality of the found non-optimum solutions.

1 Introduction

Most problems found in industry and government are either computationally intractable by their nature, or sufficiently large so as to preclude the use of exact algorithms. In such cases, heuristic methods are usually employed to find good, but not necessarily guaranteed optimum solutions.

The effectiveness of these methods depends upon their ability to adapt to a particular realization, avoid to go into local optima, and exploit the basic structure of the problem, such as a network or a natural ordering among its components.

Furthermore, restart procedures, controlled randomization, efficient data structures, and preprocessing are also beneficial. Building on these notions, various heuristic search techniques have been developed that have demonstrably improved our ability to obtain good solutions to difficult combinatorial optimization problems. The most promising of such techniques include simulated annealing [10], tabu search [8], ant colony optimization [11], genetic algorithms [9] and GRASP (Greedy Randomized Adaptive Search Procedures) [7].

In this article is set out the problem of filling a container in its bi-dimensional version, which consists of locating on a rectangular surface of fixed length and width, also rectangular pieces of a certain set, so that it does not find overlapping nor spillage and that the occupied space is the maximum (the totality of the surface, if possible).

There are two potential approaches to the problem depending on the possibility or not to be able to apply turns to the pieces, only the problem without turns

J. Mira and J.R. Álvarez (Eds.): IWINAC 2005, LNCS 3562, pp. 173–181, 2005.

was treated. This problem is included within the packing and cutting problems, problems widely studied due to its importance in certain industrial and transport sectors [1] [3]. This kind of problems belongs to the class of non-polynomial problems (NP), therefore an exact solution in a reasonable time cannot be given, in spite of having algorithms like the comprehensive search that could find this solution.

The heuristic techniques acquire a great importance in the solution of these other similar problems since they provide relatively good solutions in an acceptable time. The problems are much more complex in practice, such as the problem of cutting nonrectangular pieces of different forms, but this characteristic increases the complexity, therefore the comparison is restricted to the so-called rectangular problems, that is to say, those in which the objects to pack or to cut have a rectangular form. This paper showns two simulated annealing algorithm and one genetic algorithm for the solution of this kind of problem.

2   Simulated Annealing

Simulated annealing is a generalization of a Monte Carlo method for examining the equations of state and frozen states of n-body systems [6]. The concept is based on the manner in which liquids freeze or metals re-crystallize in the process of annealing. In an annealing process a melt, initially at high temperature and disordered, is slowly cooled so that the system at any time is approximately in thermodynamic equilibrium. As cooling proceeds, the system becomes more ordered and approaches a "frozen" ground state at $T = 0$. Hence the process can be thought of as an adiabatic approach to the lowest energy state. If the initial temperature of the system is too low or cooling is done insufficiently slowly the system may become quenched forming defects or freezing out in meta-stable states (ie. trapped in a local minimum energy state).

The original Metropolis scheme was that an initial state of a thermodynamic system was chosen at energy E and temperature T, holding T constant the initial configuration is perturbed and the change in energy dE is computed.

If the change in energy is negative the new configuration is accepted. If the change in energy is positive it is accepted with a probability given by the Boltzmann factor $e^{-\frac{dE}{T}}$. This processes is then repeated sufficient times to give good sampling statistics for the current temperature, and then the temperature is decremented and the entire process repeated until a frozen state is achieved at $T = 0$.

The connection between this algorithm and mathematical minimization was first noted by Pincus, but it was Kirkpatrick [10] who proposed that it form the basis of an optimization technique for combinatorial (and other) problems. The current state of the thermodynamic system is analogous to the current solution to the combinatorial problem, the energy equation for the thermodynamic system is analogous to at the objective function, and ground state is analogous to the global minimum. The major difficulty (art) in implementation of the algorithm is that there is no obvious analogy for the temperature T with respect to a free parameter in the combinatorial problem. Furthermore, avoidance of entrainment in local minima (quenching) is dependent on the "annealing schedule", the choice of initial temperature, how many iterations are performed at each temperature, and how much the temperature is decremented at each step as cooling proceeds.

Two alternatives for the solution by means of techniques based on simulated annealing appear next: The first is based on the original model of Metropolis, the second supposes that a search for each temperature is done.

2.1   Algorithm of Metropolis

The laws of the thermodynamics say that to a temperature T the probability of a power increase of magnitude dE can be approximated by $P[dE] = e^{-\frac{dE}{kT}}$, where k is a physical constant denominated Boltzmann

In the model of Metropolis a random disturbance in the system is generated and the resulting changes of energy are calculated: if there is a power fall, the change is accepted automatically; on the contrary, if a power increase takes place, the change was accepted with a probability indicated by the previous expression. The idea consists of beginning with a high temperature and to be decreasing it very slowly until a base state is reached, in our case the temperature is only a parameter of control of the probability of accepting a worsening in the solution, nevertheless we used this nomenclature to state the similarity with the physical process that serves as base to our algorithm. The code of the algorithm of Metropolis would be.

$S = Initial\_solution : T = Initialize : Gamma = Initialize$
REPEAT
$S^* = M(S)$
IF $Eval(S^*) < Eval(S)$ THEN $S = S^*$ ELSE
      IF $Random(0, 1) < exp((Eval(S) - Eval(S^*))/(kT))$ THEN $S = S^*$
      ELSE $S = S$
      $T = T * Gamma$
UNTIL SHUTDOWN CRITERION.

where T is the temperature, k the constant of Boltzman, this constant appears in the natural processes of crystallization but in this case it is only another parameter of control which will take the value $k = 1$, $Gamma$ is a constant that takes a value between 0 and 1 and it is used to diminish T, imitating in this form the cooling of the system and $M()$ a function that generates a new solution by means of a small disturbance, for example, it could be the function that was in the previous section.

This type of algorithms, allows with a certain probability, that tends to zero when the number of iterations tends to infinite, the worsening in the evaluation of the solutions, which will allow us to leave possible local minimums.

Three groups of tests was made, changing the different parameters of the algorithm to try to see which ones are better adapted to the treated problems. In this first experiment the algorithm was applied 100 times with 500 iterations each, the initial temperature, 100, and a $Gamma = 0.99$. In the second ex-

Gamma	Initial T	Final T	Finished	Average aptitude	Average time
0,99	100	0,66	9	23,06	8,60
0,95	100	7,27,E-10	12	21,56	7,39
0,99	10	0,07	14	23,34	6,74

Fig. 1. Comparison of different parameters in Algorithm of Metropolis

periment the algorithm was applied 100 times with 500 iterations each, thesame initial temperature and a *Gamma* = 0.95. In the third experiment the algorithm was applied 100 times with 500 iterations each, the initial temperature, 10, and a *Gamma* = 0.99. If a comparative of the results of the three experiments is made, several conclusions can be drawn: The best results were obtained in the third experiment, nevertheless, the best average aptitude was obtained in the second, this is due to the formula of acceptance of solutions

IF $Eval(S^*) < Eval(S)$ THEN $S = S^*$
ELSE
    IF $Random(0,1) < exp((Eval(S) - Eval(S*))/(kT))$ THEN $S = S^*$
    ELSE $S = S$

As the temperature faster decreases and lower is, also the probability of allowing a worsening of the solution is smaller, for this reason, in experiment two the temperature descends quickly then exists a convergence towards local minimums and therefore the average aptitude is smaller, in experiment one the temperature descends very slowly and at any moment there is possibility of accepting a solution of worse aptitude, for that reason, the average aptitude is worse. In case three the initial temperature smaller, the stabilization of the solutions takes place before, but as the temperature descends very slowly it does not lose the possibility of leaving global minimums until the end. With the second problem several groups of tests was made.

In the first experiment the algorithm was applied 100 times with 500 iterations each, the initial temperature, 10 and a *Gamma* = 0.99. In the second experiment the algorithm was applied 50 times with 1000 iterations each, the initial temperature, 10000, and a *Gamma* = 0.99.As we can observe the results in the first experiment have been better than in the second, as much in time as in average aptitude, for that reason we make a third experiment with 1000 iterations and the parameters of the first experiment adapted to the number of iterations, that is to say, the initial temperature must be of 1515 so that the final one agrees with the one of the first experiment. In the third experiment the algorithm was applied 100 times with 1000 iterations each, then initial temperature, 1515, and *Gamma* = 0.99.

As it is observed there exists a light improvement of the average aptitude as well as of the best found aptitude. Taking as a base the algorithms of metropolis, in the early 80s Kirkpatrick and Cerny developed algorithms for the design of circuits VLSI and the resolution of TSP (Traveling Salesman Problem) and showed how they could be applied to combinatorial optimization problems, like

Number of solutions	100
Solutions that found the optimal	-
Better aptitude	11,00
Average aptitude	30,54
Average time	18,29
Total time	1.829,00

Fig. 2. Experiment 1: Results for case 2

Number of solutions	50
Solutions that found the optimal	-
Better aptitude	13,00
Average aptitude	31,32
Average time	45,44
Total time	2.272,00

Fig. 3. Experiment 2: Results for case 2

Number of solutions	100
Solutions that found the optimal	-
Better aptitude	7,00
Average aptitude	28,90
Average time	40,38
Total time	4.038,00

Fig. 4. Experiment 3: Results for case 2

the one which occupies us, the algorithm that they developed was termed Simulated Annealing [2], which is the new algorithm that was treated.

A Simulated Annealing Algorithm Alternative. The basic difference between the former algorithm and the following one is that it search for each temperature [1] [2], which is translated in the following code.

$S = Initial\_solution : T = Initialize : Gamma = Initialize$
$Niter = Number$ of iterations that we wished to be produced
REPEAT
    REPEAT
    $S^* = M(S)$
    IF $Eval(S^*) < Eval(S)$ THEN $S = S^*$
    ELSE
        IF $Random(0,1) < exp(Eval(S)-Eval(S^*))/T)$ THEN $S = S^*$
        ELSE $S = S$
    UNTIL NUMBER OF ITERATIONS $= Niter$
    T= T * Gamma
UNTIL SHUTDOWN CRITERION.

Experiment	1	2
Number of solutions	100	100
Explorations by temperature	10	20
Solutions that found the optimal	20	17
Better aptitude	1,00	1,00
Average aptitude	16,98	22,21
Average time	48,67	38,48
Total time	4.867,00	3.848,00

Fig. 5. Results SA for case 1

Experiment	1	2	3
Number of solutions	15	15	25
Explorations by temperature	20	30	30
Solutions that found the optimal	-	-	2
Better aptitude	25,00	15,00	1,00
Average aptitude	34,40	21,93	22,44
Average time	353,76	263,55	480,30
Total time	5.306,00	3.953,00	12.008,00

Fig. 6. Results SA for case 2

This type of algorithms, allows with a certain probability, that tends to zero when the number of iterations tends to infinite, the worsening in the evaluation of the solutions, which will allow us to leave possible local minimums. Unlike the previous algorithm, this algorithm makes an exploration for each temperature, which increases the time necessary for its execution but also the probability of finding the global optimum. Two groups of tests was made, changing the different parameters of the algorithm to try to see which ones are better adapted to our problems: In the first experiment the algorithm was applied with (Times, Iterations, Initial Temperature, $Gamma$)=(100, 500 , 100, 0, 99). These parameters were the best ones in the previous section. $Niter =$ 10, that is that ten iterations was madee for each power state (for each temperature).

In the second experiment the algorithm was applied with (100, 200, 100, 0, 99). These parameters were the best ones in the Algorithm of Metropolis, with. $Niter = 20$.

Three groups of tests was made with the second problem changing the different parameters of the algorithm to try to see which ones are better adapted to this problem.

- First experiment. The algorithm was applied with (15, 500, 1515, 0.99). $Niter = 20$.
- Second experiment. The algorithm was applied with (15, 200 ,, 10000, 0.95). $Niter = 30$.
- Third experiment. The algorithm was applied with (25, 500 , 10,0.99). $Niter = 30$.

Fitness	Case 1
Fitness = 1	74
2 <= Fitness <= 10	5
11 <= Fitness < =20	16
Fitness > 20	5

Fig. 7. Results GA for case 1

It is observed that the results of the third experiment are substantially better than those of first and second experiment, nevertheless, the necessary time was greater also. It seems logical to think that the parameters used in the third case are more appropriate that the two firsts, in the first case the final temperature was of 9.95, a too high temperature that allows worsening with too much probability and therefore exists a excessively random behavior, in the second experiment this final temperature was of 0.35, therefore, the algorithm tends to allow worsening less and less and therefore the search can be considered less random. An improvement in the average aptitude exists and the time was even smaller, nevertheless, the global optimum was not found. The results of third experiment were quite good, the global optimum was found, the 13, 33% of the solutions had a satisfactory convergence, however, the necessary time was greater than in the other experiments. Another interesting point is the increase in the number of explorations for each power.

3   Genetic Algorithms

The Genetic Algorithms (AGs) introduced by John Holland in 1970 [4] are techniques of optimization that are based on concepts like the natural and genetic selection. In this method the variables are represented like genes in a chromosome [4].

In this case the representation of the solutions is the already presented, the initial possible solutions population was generated of randomly, the evaluation function was the free space, the crossing operator was PMX that along with BLF is the one that presents better results [5], the selection was made by means of the roulette method and a mutation was a transposition of two elements of the solution.. A series of tests of 100 executions was made with the first problem and the following parameters:

(Recombination, Mutation, Initial solution, Number of solution, [Number of iterations with solution, Number of iterations without solution])

- Recombination: The recombination was by pairs using operator PMX and passing to the following generation the children.
- Mutations: The mutations was by change, and its probability was of 3%.
- Initial solutions: Random.

Solutions that finalized	Average time finalized	Average number of iterations for finalized
74	14,47	135
Total average fitness	Total average time	Average number of iterations total
4,89	30,77	332

Fig. 8. Results GA for case 1

– Number of solutions: 20.
– Number of iterations: 1000 or to find the optimum one.

(PMX & passing the childen, 3%, Random, 20, [1000, _])

The total time of execution was of $3,077$ seconds, that is , 51 minutes, to speed up the tests we will include a new criterion of shutdown, if in 100 iterations the best fitness found does not experience improvement we finalized the experiment.

Five experiments with 100 executions was made with the following parameters:

1. (PMX & passing the childen, 3%, Random, 20, [1000, 1000]).
2. (PMX & passing the childen, 5%, Random, 20, [1000, 100]).
3. (PMX & passing the childen, 3%, Random, 20, [1000, 100]).
   We calculate the recombination probability by adding 20 to each one of the fitness of the solutions, to equal the probabilities and avoid a premature convergence to local minimums.

$$SF = \sum_i (Fit(S_i) + 20); \quad S_iA = \frac{SF}{Fit(S_i)};$$
$$SFI = \sum_i S_iA; \quad P(S_i) = \frac{Fit(S_i)+20}{SFI}.$$

4. (PMX & passing the childen, 3%, Random, 20, [1000, 100]).
   We calculate the recombination probability by elevating to the fourth power the original probability, this favors the propagation of the solutions with better fitness, but also the danger to fall in local minimums.

$$SF = \sum_i Fit(S_i); \quad S_iA = \left(\frac{SF}{Fit(S_i)}\right)^4;$$
$$SFI = \sum_i S_iA; \quad P(S_i) = \frac{Fit(S_i)+20}{SFI}.$$

5. (PMX & passing the best one, 3%, Random, 20, [1000, 100]).

The best experiment was the first one for the time used and for the results obtained. We did not put the criterion of shutdown of 100 iterations without improvement, obtained many more convergences, 74% as opposed to 49%, which represents a 51% of convergences, nevertheless, the used time went as 3077 seconds opposed to 636, which supposes a 384% but time.

4   Conclusions

This kind of problems is very treatable with techniques based on simulated annealing. The genetic algorithms present initially the problem of the suitable election of the representation of the solutions.

The genetic algorithms present good results because a certain intermediate solution contains a sub-chain that comprises the optimal solution, and when crossing several of these solutions those sub-chains are prospering and joining themselves until obtaining the optimum one, nevertheless, in this problem this is not fulfilled since what is evaluated is the space occupied within the container, however, in the crossing of the genetic algorithm the arrangements of pieces are crossed.

Remark: All experiments were made in a computer with Pentium IV, CPU 2,4 GHz, 522,228 KB of RAM, and. the code of programs were Visual BASIC code.

Acknowledgements

This research was supported by the international joint research grant of the IITA (Institute of Information Technology Assessment) foreign professor invitation program of the MIC (Ministry of Information and Communication), Korea.

References

1. E. Hopper and B. C. H. Turton, An empirical investigation of Mete-heuristic and Heuristic algorithms for a 2D packing problem, E.J.O.R 128/1,pp 34-57, (2000)..
2. K. A. Dowsland, Heuristic desirn and fundamentals of the Simulated Annealing, Revista Iberoamericana de Inteligencia Artificial. N 19 (2003), pp 93-102..
3. A. Gómez, Resolución de problemas de packing en una empresa de calcamonias mediante algoritmos geneticos, XII congreso nacional y VII congreso Hispano-Francs de AEDEM, (1998), pp 999-1008.
4. Z. Michalewicz, Genetic Algorithms + Data Structures = Evolution Programs, Springer-Velag, (1994).
5. A. Gómez, Resolución del problema de Strip-Packing mediante la metaheurstica algoritmos genticos, Boletn de la SEIO (2000), pp. 12-16.
6. N. Metrópolis A.W. Rosenbluth, M.N. Rosenbluth, and A.H. Teller. Equation of state calculations by fast computing machines. (1953). The Journal of Chemical Physics, vol. 21, no. 6, pp. 1087-1091
7. T.A. Feo and M.G.C. Resende. Greedy randomized adaptive search procedures. Journal of Global Optimization, 6:109-133. 1995.
8. F. Glover and M. Laguna. Taby search. Kluwer Academic Publisher, 1997.
9. D.E. Goldber. Genetic algorithm in search, optimization and machine learning. Addison-Wesley, 1989.
10. S. Kirkpatrick. Optimization by simulate annealing: quantitative studies. Journal of Statistical Physics, 34:975-986. 1984.
11. M. Dorigo and T. Sttzle. The ant colony optimization metaheuristic: Algorithms, applications, and advances, 2001. in Handbook of Metaheuristics, F. Glover and G. Kochenberger.

Attribute Grammar Evolution

Marina de la Cruz Echeandía, Alfonso Ortega de la Puente,
and Manuel Alfonseca

Universidad Autónoma de Madrid, Departamento de Ingeniería Informática
{marina.cruz, alfonso.ortega, manuel.alfonseca}@uam.es

Abstract. This paper describes Attribute Grammar Evolution (AGE),
a new Automatic Evolutionary Programming algorithm that extends
standard Grammar Evolution (GE) by replacing context-free grammars
by attribute grammars. GE only takes into account syntactic restrictions
to generate valid individuals. AGE adds semantics to ensure that both
semantically and syntactically valid individuals are generated. Attribute
grammars make it possible to semantically describe the solution. The
paper shows empirically that AGE is as good as GE for a classical problem, and proves that including semantics in the grammar can improve
GE performance. An important conclusion is that adding too much semantics can make the search difficult.

1 Introduction

1.1 Syntax and Semantics of High Level Programming Languages

The differences between syntax and semantics in high level programming languages are rather artificial.

Turing Machines are associated to Chomsky 0 grammars [1], while the syntax
of high level programming languages is usually expressed by means of context
free grammars. Context free grammars are associated to pushdown automata,
which have less expressive power than Turing Machines. The expressive power
gap between Chomsky 0 and context free grammars is usually called the "semantics" of high level programming languages. This gap mainly deals with context
dependent constructions, such as the mandatory declaration of the variables before their use or, the constrains about number and type of the arguments in
functions calls, which must agree with their declaration.

1.2 Attribute Grammars

Attribute grammars [2] are one of the tools used to describe high level programming languages completely (their syntax and their semantics). Attribute
grammars extend context free grammars by adding these components to them:

- Each non terminal symbol has a set of attributes. Attributes are similar to
  the variables in programming languages; they have a name and their values
  belong to a given domain.
- Each rule contains expressions to compute the value of the attributes.

J. Mira and J.R. Álvarez (Eds.): IWINAC 2005, LNCS 3562, pp. 182–191, 2005.

A detailed description of attribute grammars and some examples of their use can be found in references [3, 4]

1.3   Grammar Evolution (GE)

GE [5] is an automatic programming evolutionary algorithm independent of the target programming language, which includes a standard representation of genotypes as strings of integers (codons), and a context free grammar, as inputs for the deterministic mapping of a genotype into a phenotype. This mapping minimizes the generation of syntactically invalid phenotypes. Genetic operators act at the genotype level, while the fitness function is evaluated on the phenotypes.

The *genotype to phenotype mapping* is an algorithm that iterates on the string of codons and derives words by applying the context free grammar. It starts with the first codon and the axiom of the grammar, and finishes when the genotype is exhausted or when there are no more non-terminal symbols in the current derived word. This last condition means that the mapping has derived a word belonging to the language of the grammar, i. e. a syntactically correct program.

To process each codon, the next non terminal symbol is selected from the current string (usually the leftmost one), the (n+1) rules applicable to the non terminal are enumerated (from 0 to n), and the current codon is mapped into one of them (usually by computing *codon mod n*).

1.4   Practical Considerations

In the first step, this paper solves a symbolic regression problem by means of GE. The algorithm has been implemented in Java, we have tried to keep as close as possible to the original description in reference[5].

The symbolic regression problem tries to find a symbolic expression fitting a given function on a set of control points.

In Genetic Programming [6], most of the problems can be reduced to the symbolic regression problem. Therefore, this paper will be restricted to that problem. The target function chosen is $f(x) = x^4 + x^3 + x^2 + x$

The solution of this problem in [5] is done by means of the following features and parameters:

Several characteristics of our experiments are the same as in [5]:

- The context free grammar

```
<expr>::=<expr> + <expr>
        |<expr> - <expr>
        |<expr> * <expr>
        |(<expr>)
        |<pre_op>(<expr>)
        |<var>
```

```
<pre_op>::=sin
          |cos
          |exp
          |log
<var>::=x
```

- The set of control points: 21 values uniformly taken from $[-1, 1]$.
- The fitness function: the sum of the absolute error over the set of the 21 control points.
- The population size: 500.
- The length of the genotypes: variable length, initially between 1 and 10.
- Codons: they belong to the $[0, 256]$ interval.
- Probability of crossover: 0.9.
- Bit mutation with probability 0.1.

We have tried to reproduce the same example, with the following differences:

- The above grammar is ambiguous, as it does not define any operator precedence. In the Java application, we have removed the ambiguity and designed the associated pushdown automata to make the evaluation of the individuals easier. For the shake of simplicity, the grammar appearing in the remainder of this paper is still the ambiguous version.
- The high cumulative success frequency described in reference[5] is not reached unless the parents are chosen with a fitness proportional strategy, and the next populations are generated by means of a generational scheme, rather than a steady state scheme.
- The size of the genotypes increase with the number of generations. It has been empirically observed that the number of unused codons also increases. The crossover operator described in reference [5] uses one single random crossover point. Thus, the number of descendants that map to the same phenotype as their parents, correspondingly increases. To solve this situation, we have restricted the crossover point choice to the used portion of each genotype, rather than to the whole genotype.
- In our work, mutation changes a single codon by a random value in the $[0, 256]$ interval. The best probability rate of mutation was fond empirically as 0.5, although the performance is very similar for mutations in $\{0.5, 0.7, 0.8\}$.

2   Attribute Grammar Evolution (AGE)

2.1   Previous Similar Works

This is not the first attempt to extend genetic programming by adding the complete description of a programming language: references [7, 8, 9, 10] describe some Prolog based approaches. These algorithms are criticized by some authors [11, 12] because the logic engine makes it difficult to control some parameters of the search algorithm, and because the backtracking tends to worsen the final performance and does not ensure that the computation finishes in all the possible cases.

AGE mainly differs from the previous references in the following:

- It does not depend on any programming paradigm.
- Ross and Hussain's works represent the genotypes by means of trees.
- Man Leung Wong and Kwong Sak Leung's works are mainly interested in data mining and machine learning.

2.2 An Algorithm to Evaluate the Attributes While Building the Derivation Tree

Attribute grammars are exhaustively used in the design of parsers for program translators. AGE uses them to derive the phenotypes. The attributes are evaluated by means of the derivation tree. Each time that a node of the tree is expanded, the values of the attributes that can be evaluated are computed in the following way:

- Attributes inherited from the parent symbol are evaluated directly.
- If the node symbol is prefixed by other symbols to the right of where it appears, attributes inherited from the left siblings are also evaluated.
- After expanding the last child of a node, the parent synthesized attributes are evaluated.

The axiom of the grammar has only synthesized attributes. The leaves of the tree are associated to terminal symbols wihtout any attributes of their own, but which may be used to input data into the derivation tree.

AGE uses the attributes to describe the conditions that a phenotype must comply with to be considered semantically valid. As soon as one of these constraints is violated, the derivation process is aborted.

2.3 Modifying the Genotype to Phenotype Mapping

AGE adds the evaluation of the attributes to the previously described GE mapping. The following steps outline the algorithm applied to every codon in the genotype:

1. Choose the leftmost non-terminal symbol in the current word.
2. Select in the tree the node associated with the symbol. This is the current node.
3. Update the attributes in the derivation tree.
4. Number in zero origin the right hand sides of all the rules for this non-terminal symbol.
5. Select the right hand side of the rule whose number equals *codon mod number of right hand sides* for this non-terminal.

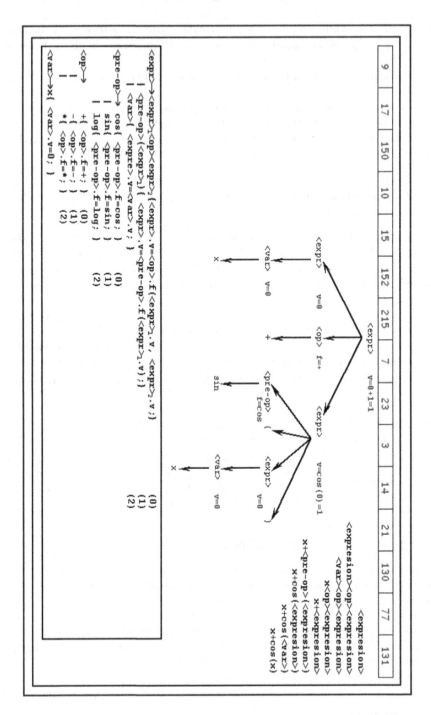

Fig. 1. AGE genotype to phenotype mapping of $x + cos(x)$. The attribute grammar computes its value in $x = 0$

6. Derive the next word by replacing the non-terminal by the selected right hand side.
7. Insert a new node for each symbol to the right hand side of the rule and make all the new nodes the children of the current one.

Figure 1 graphically shows the derivation of the expression $x + cos(x)$ from a genotype and the following attribute grammar:

```
<expr>::=<expr>₁ <op> <expr>₂{<expr>.v=<op>.f(<expr>₁.v, <expr>₂.v);}
        |<pre_op>(<expr>₁){<expr>.v=<pre_op>.f(<expr>₁.v);}
        |<var>{<expr>.v=<var>.v;}
<pre_op>::=sin{<pre_op>.f=cos;}
          |cos{<pre_op>.f=sin;}
          |log{<pre_op>.f=log;}
<op>::=+{<op>.f=+;}
      |-{<op>.f=-;}
      |*{<op>.f=*;}
<var>::=x{<var>.v=0;}
```

Where:

- Non-terminal symbols `<expr>` and `<var>` have an attribute `v`, that stands for the value of the expression. Notice that the last rule inputs 0 as the value of the variable `x`, the value of `<var>` attribute `v`.
- Non-terminal `<pre_op>` and `<op>` have an attribute `f` that represents the function that will be applied when computing the value of the expression.
- The association of an attribute to a symbol is represented by means of a dot as in the C language.

2.4   GE vs. AGE Performance

Our first experiment avoids the generation of a phenotype that may be undefined on any control point. This can only happen when the expression contains log (subexpression) and the value of the subexpression is less or equal to 0.

The goal of the first experiment is estimating the possible loss of performance due to the steps added to the mapping algorithm. We are not really improving GE, because no new semantics is actually added to the algorithm. There are, however, some differences between AGE and GE in this case: GE generates semantically invalid phenotypes that will probably be punished with the worst fitness value, but AGE prefentsthe generation of such expressions as soon as possible.

This experiment uses the same parameters as those described for GE, except for the following:

- The attribute grammar used is the following:

```
<expr>::=<expr>₁ + <expr>₂{<expr>.vⱼ=<expr>₁.vⱼ+<expr>₂.vⱼ ∀i∈[0,20]}
        |<expr>₁ - <expr>₂{<expr>.vⱼ=<expr>₁.vⱼ-<expr>₂.vⱼ∀i∈[0,20]}
        |<expr>₁ * <expr>₂{<expr>.vⱼ=<expr>₁.vⱼ*<expr>₂.vⱼ∀i∈[0,20]}
```

$$|(\texttt{<expr>}_1)\{\texttt{<expr>}.v_i=\texttt{<expr>}_1.v_i \ \forall i \in [0,20]\}$$
$$|\texttt{<pre\_op>}(\texttt{<expr>}_1)\{\texttt{<expr>}.v_i=\texttt{<pre\_op>}.f(\texttt{<expr>}_1.v_i) \ \forall i \in [0,20]\}$$
$$|\texttt{<var>}\{\texttt{<expr>}.v_i=\texttt{<var>}.v_i \ \forall i \in [0,20]\}$$

```
<pre_op>::=sin{<pre_op>.f=sin}
          |cos{<pre_op>.f=cos}
          |exp{<pre_op>.f=exp}
          |log{<pre_op>.f=log}
<var>::=x{<var>.v₀=-1
          <var>.v₁=-0.9
          <var>.v₁=-0.8
          ...
          <var>.v₂₀=1}
```

This grammar is very similar to the one used in the genotype to pheno-type mapping example. The main difference is the existence of 21 attributes (v_i $\forall i \in [0,20]$) to record the value of the expression on each control point.

– The probability of mutation has been empirically optimized to 0.7 as shown in figure 2.

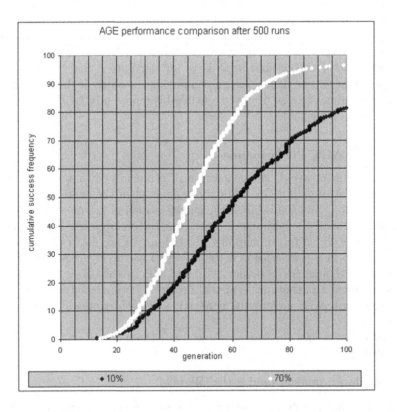

Fig. 2. AGE performance comparison after 500 runs. The worst performance corre-sponds to a probability of mutation equals 0.1 and the best to 0.7

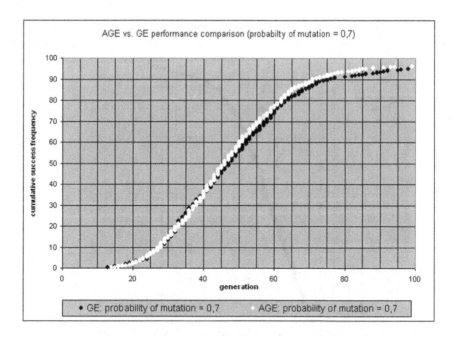

Fig. 3. GE vs. AGE performance comparison after 500 runs with the best probability of mutation (0.7). As with the rest of probabilities of mutation, there is no remarkable performance change

Figure 3 shows the performance comparison between GE and AGE for the best probability of mutation after 500 runs of the algorithm. As expected, there is no remarkable improvement in performance. However, there neither is any remarkable loss. So, we can deduce that AGE is as good as GE to solve this problem. As many of the problems solved by GP are reducible to the symbolic regression domain, we are optimistic about the generalization of thse results to other kinds of problems.

The goal of our second experiment is to check if, after adding some semantics to the grammar, the algorithm improves its performance. We shall add constrains to consider invalid any phenotype that does not exactly fit the target function in any of the three control points: -1, 0 or 1.

For this experiment, the probability of mutation is set to 0.7, one of the better values both in GE and in AGE.

Figure 4 shows a significant improvement in performance after 500 runs of the algorithm: the cumulative frequency of success grows faster and reaches higher values.

It is worth noticing that, even although the semantic constrains are rather loose, the improvement is significant. On the other hand, other tests have shown that increasing too much the semantics causes bad performance. A possible reason is that the expressive power of attribute grammars allows a full description of the solution to the problem. But, in this case, the only valid phenotypes is the

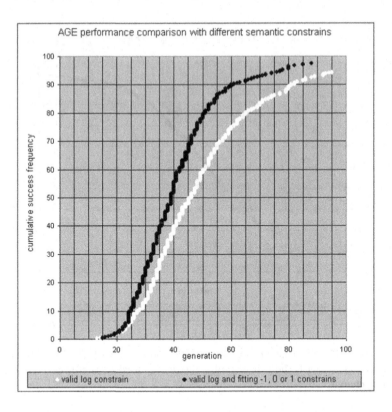

Fig. 4. AGE performance comparison with different semantic constrains. The improvement in the performance after adding a rather loose constrain is significant

solutions ,and the search is not directed by the genetic engine: it really becomes a random search.

This is a very important topic for further research: determining how many semantics must be added to get the optimal performance.

3   Conclusions and Future Research

This work describes AGE (attribute grammar evolution), a new automatic evolutionary programming algorithm that extends GE (grammar evolution) by adding semantic constrains which make it possible the removal from the population of, syntactically and semantically invalid phenotypes.

The differences with other approaches to genetic programming that use syntactic and semantic descriptions are discussed in the paper.

It is shown that AGE is as good as GE for a standard problem in GP. It is also proved that adding rather loose semantic constrains improves significantly the performance of AGE.

In the future, we plan to extend GE with different formal descriptions of the programming language and to apply this approach to automatically solve problems in other domains. We also plan to study in depth the determination of the adequate semantics needed to get the optimal performance.

References

1. A. N. Chomsky, Formal properties of grammars, in Handbook of Math. Psych., vol. 2, John Wiley and Sons, New York. 1963, pp. 323-418
2. D. E. Knuth, Semantics of Context-Free Languages, in Mathematical Systems Theory, vol. 2, n° 2, 1968, pp. 127-145
3. A.V. Aho, R. Sethi, J.D.Ullman (1986) Compilers: Principles, Techniques and Tools. Addison-Wesley
4. D.Grune, et al. (2000) Modern Compiler Design. John Wiley & sons Ltd.
5. M. ONeill & R. Conor, Grammatical Evolution, evolutionary automatic programming in an arbitrary language, Kluwer Academic Phblishers, 2003.
6. W.Banzaf, P.Nordic, R.E.Keller, F.D.Francone. Genetic Programming. An introduction Morgan and Kaufmann Publishers, Inc.1998
7. B.J. Ross, Logic-based Genetic Programming with Definite Clause Translation Grammars 1999
8. M.L.Wong and K.S. Leung. Evolutionary Program Induction Directed by Logic Grammars. Evolutionary Computation, 5 (2): 143-180, 1997
9. T.S.Hussain and R.A.Browse. Attribute Grammars for Genetic Representations of Neural Networks and Syntactic Constraints of Genetic Programming. In AIVIGI'98: Workshop on Evolutionary Computatino, 1998
10. T.S.Hussain. (2003) Attribute Grammar Encoding of the Structure and Behaviour of Artificial Neural Networks. Ph.D.thesis. Queens University. Kingston, Ontario, Canada
11. J. N. Shutt, Recursive Adaptable Grammars. A thesis submitted to the Faculty of the Worcester Politechnic Institute in partial fulfillment of the requirements for the degree of Master of Science in Computer Science, Agoust 10, 1993 (enmended December 16, 2003)
12. H. Christiansen, A Survey of Adaptable Grammars, in ACM SIGPLAN Notices, vol. 25, n°11. November 1990, pp. 3544.

Evolution and Evaluation in Knowledge Fusion System

Jin Gou, Jiangang Yang, and Qian Chen

Ningbo Institute of Technology,
Zhejiang University, 315100 Ningbo, China
{goujin, yangjg, lovehome}@zju.edu.cn

Abstract. The paper presents a method to control evolution of pattern in a knowledge fusion system. A self-adapt evaluation mechanism to assign proper value dynamically to weight parameters is also described. Some rules are defined with aid of the matrix theory to promise the controllablity and describability to the evolution process. A new knowledge object, called LKS (local knowledge state), that can redirect path in knowledge fusion system and evolve to other knowledge object(s) is formed in that model. Experimental results of a case study show that it can improve the efficiency and reduce computational complexity of a knowledge fusion system.

1 Introduction

1.1 Knowledge Fusion and Knowledge Engineering

Knowledge fusion is an important component of knowledge science and engineering, which can transform and integrate diversiform knowledge resources to generate new knowledge objects[1]. So information and knowledge from distributed knowledge resources can be shared and cooperating. Presently, Multi-agent technology and grid computation method are used to integrate specifically knowledge in a way, which pay more attention to implement the fusion process[2, 3, 4].

1.2 Knowledge Fusion Requires a Reduction of Computational Complexity

Knowledge fusion will result in an enormous amount of knowledge base resources on the web. Since ontology technologies carry the promise to enable widespread sharing and coordinated use of networked knowledge resources with only one logical base, we adopt a kind of meta-knowledge model to build a knowledge fusion platform.

However, there are two limitations of those knowledge fusion systems. First, distributed knowledge objects in those systems are massively. That complicates allocation and management of knowledge resources. It will make the system working inefficiently and a useless knowledge object may lead to a terrible waste

J. Mira and J.R. Álvarez (Eds.): IWINAC 2005, LNCS 3562, pp. 192–201, 2005.

of work. Second, structure of knowledge representation is relatively diversiform, which leads to slather unify process[5, 6].

The purpose of this paper is to setup a controllable evolution process model and present an evaluation method that can revises knowledge fusion process with dynamic self-adjusted parameters. We start in the next section with a description of the overall structure of the frameset. Section 3 introduces the evolution module with a control matrix. The evaluation algorithm and the methods to revise fusion parameters are showed in section 4. Section 5 describes a referential application on a case study in a knowledge-based grid environment. This paper finishes with a brief conclusion.

2   Architecture

The knowledge fusion framework focuses the evolution process and evaluation method. Fig 1 shows the overall architecture.

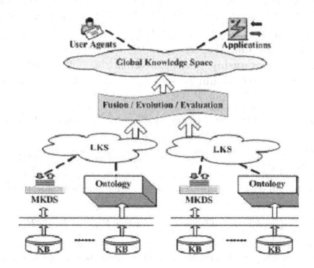

Fig. 1. Overall architecture

Connotational formalized representation of distributed knowledge resources should be described underlying certain rules and added into the ontology before a knowledge fusion process can go on, while extracted meta-knowledge sets should be registered into network path module of MKDS (meta-knowledge directory service). Then the fusing algorithm runs on those knowledge resources' new structure, which will be called LKS (local knowledge state) in the paper. Through evaluation and evolution, LKSs form global knowledge space that can be used in applications. To setup the ontology base is an initial work of the frameset, so we elaborate on this issue in the following.

2.1   Ontology Base and Meta-knowledge Sets

Ontology base is a complex description sets. Let O and O_i denote the ontology base and the ith object instance in it. The O_i will be used in the paper, which is defined by

$$O_i = \{(P_j^i, T_j^i, D_j^i)\} \tag{1}$$

Where P_j^i denotes the jth attribute of the object O_i, T_j^i denotes the type of P_j^i, D_j^i denotes its value, expression or behavior. The domain for variant j is decided by its scope knowledge ontology[7, 8].

Let S_k denotes meta-knowledge sets of the kth knowledge base node that can be defined by

$$S_k = \{((C_1^k, E_1^k), (C_2^k, E_2^k), \ldots, (C_n^k, E_n^k))\} \tag{2}$$

Where C_i^k denotes the ith character object of S_k, E_i^k denotes the description content of C_i^k .

Relationships among character objects are not defined here because those will be described in the ontology base.

In order to implement the interchanged process between meta-knowledge sets and knowledge space, the definiendum (C_i^k, E_i^k) stands for not embodied characters but denotative objects of knowledge ontology. That operation must be synchronous with the initialization of meta-knowledge sets.

To incarnate the relationships among diversiform meta-knowledge sets, following rules must be obeyed during the ontology base's constructed procedure.

Rule 1. If not-non relationship between C_i^k and C_j^l exists, CREAT $(P_m^k, T_m^k, D_m^k) \in O_k$.

If $(k \neq l)$, CREAT $(P_n^l, T_n^l, D_n^l) \in O_l$.. Where $P_m^k = P_n^l = $ "$R(C_i^k, C_j^l)$" and $D_m^k = D_n^l = R(C_i^k, C_j^l)$

Rule 2. If none relationship between C_i^k and C_j^l exists, CREAT $(P_m^k, T_m^k, D_m^k) \in O_k$.

If $(k \neq l)$, CREAT $(P_n^l, T_n^l, D_n^l) \in O_l$.. Where $P_m^k = P_n^l = $ "$R(C_i^k, C_j^l)$" and $D_m^k = D_n^l = $ NULL

The reliability parameters can be regarded as a sub set of the ontology. So they meet the requirement of all other elements in ontology.

2.2   Fusion Algorithm

Describe the supposition as constructing meta-knowledge sets, function as fusing and generating new knowledge space.

$$FA(Fitness, Fitness\_threshold, p, r, m, \mu, \varsigma) \tag{3}$$

A detailed explanation of such a knowledge fusion function can be found in another paper[9] written by us. It is an important module in a whole knowledge fusion system, so we should run it over in short. And here two new parameters are added to FA for evaluation module in this paper. The difference between[9] and this paper: the former introduced how to setup a knowledge fusion framework and the latter is emphasis on how to optimize though evolution and evaluation mechanism. What's more, overall architecture here is in higher level compared to the former.

Where Fitness denotes assessing function for fusion matching grade, which can endow the given supposition sets with a matching grade. Fitness_threshold denotes the threshold value beyond which fusion process can't persist[10]. p is a supposed number which should be adjusted according to the result of unifying among diversiform meta-knowledge sets. If dimension of some sets are too small, NULL can be used to instead of it. r and m are intercross and aberrance percent. μ and ζ are reliability parameters.

Step 1. Select p meta-knowledge sets randomly to setup initially supposition colony marked as H. Calculate $Fitness(S)$ for each S belonged to H.

Step 2. If $max\{Fitness(S)\} < fitness\_threshold$, do the following operation circularly:

Select $(1-r)p$ members of H into H_s

Intercross, aberrance, update and assess.

Where μ and ζ can be given value according to rules in section 4.

Step 3. Find the supposition whose fit grade is the max among those returned from H. Append it to the solution knowledge space.

Select process in step 2. Lies on whether a relational attribute value is not equal to NULL exists in ontology base that belongs to meta-knowledge set of the supposition. Intercross requires uniform relation description in the ontology base. Aberrance goes along with ontology description corresponded with supposition meta-knowledge sets. That progress of the three steps can be summarized into four sub-steps:

Select, Intercross

Then adjust ω_i in μ.

Aberrance.

3   Evolution of Knowledge Object

There are two ways of implementing knowledge space. One is to integrate knowledge objects into a physical warehouse with central management. The other is to store only character information of meta-knowledge's directory as a path pointer. For the massive information in practice, we choose the latter.

Mapping knowledge objects into formalized meta-knowledge set and regarding each element of the set as a knowledge point, we construct a LKS K_i consisted of those points whose relationship grade is great than a threshold. Knowledge filed in K_i is expressed by a finite question ontology that is knowledge states are correlated or not.

3.1   Evolution Types

Details of the evolution rules to LKSs can be summarized into three types based on the type theory in software engineering. They are simple coalition, simple migration and cooperating evolution.

Simple coalition means that certain LKSs can be integrated into a new one according to some restriction and control rules. Ontology should be reconstructed because the connotation of former knowledge states is changed. Simple migration means that a LKS can be redefined into another new one according to some restriction and control rules. Ontology need not to be reconstructed but to adjusted its elements. Cooperating evolution means that certain LKSs according to some restriction can cooperate and be reciprocity to create new one(s). Probably some would turn into dead state during that process. Cooperating evolution includes two subtypes. The first one is that certain LKS should be "divided" into several other new ones if the frequency of being used to resolve diverse problems is too large, while most of those process need only a small part of the whole knowledge set in that LKS. The other one is a migration from several LKSs to another several ones.

3.2   Evolution Process

Define a state matrix to quantitate relationship among knowledge objects. Let its dimension equal to the amount of them, while each element's value lies on whether those two objects belong to a same LKS.

According to this method we can define an initial state matrix A^0 and a result matrix R. Actually any evolution of knowledge system state is corresponding to a control matrix B^i, and the whole evolution process can be expressed as B where $b_{ij} \in \{0,1\}$ and $B = B^0 B^1 \dots$. Initial value of B is given by experts' experience and usually it is a thin matrix. They obey the rule:

$$A^0 B = R \tag{4}$$

So in a learning procedure we can get B^T's value with R^T's linear equation group through iterative solution.

For each $r_{ij} \in R$,

```
IF (j > i) AND (r_ij > threshold_coalition)
{IF (r_ii > threshold_LKS) OR (r_jj > threshold_LKS)
UNITE Object_i AND Object_j ;
ELSE
{LKS_New.Create ; //Create a new LKS
LKS_New.ADD(element_i) ;
LKS_New.ADD(element_j) ;
UPDATE R ;
}
}
```

If any diagonal element's value is less than threshold_annihilate, annihilation takes place. $Object_i \xrightarrow{\subseteq} Object_j$, $Object_i.Destory$. Where j meets the requirement as followed:

$$r_{ij} = Max_{k=i+1}^m \{r_{ik}\} \tag{5}$$

4 Evaluation Method

Rationality of knowledge objects' structure must be feedback by applications. So the fusion process should be evaluated by the degree of match between fusion result and applications to adjust weight parameters in FA.

4.1 Data Set of Reliability Parameters

Let $O_0 = \{(P^0 T^0 D^0)\}$ denotes the data set of reliability parameters of the knowledge fusion system. Where P^0 at least includes two elements, the physical reliability μ and the logical reliability ζ.

μ descripts the proportion of a knowledge object in its father knowledge entity. We can trust a knowledge object at prime tense, so a holistic knowledge set has a μ whose value is 1.0. An exit knowledge node set (ens) is created which contains the lowest level nodes in the fusion system. For those nodes S_i not at the lowest level, an immediate successor set (iss) is created which contains the nodes immediately below S_i and have ontology-dependent on it. And if we regard the first knowledge entity as a bottom layer node then elements on higher layers should have a less value of μ. During the knowledge fusion process we only use some sub sets or elements of it, so the corresponding physical reliability must be just a part of 1.0, which we can calculate as following expression:

$$\mu(S_i) = 1.0, (S_i \in ens) \tag{6}$$

$$\mu(S_i) = \omega_i \times \sum i, NOT(S_i \in ens) \tag{7}$$

Where ω_i is the weight of the node S_i and its default value is 1.0. μ incarnates the similar degree between two ontology elements while being fused.

ζ is a dualistic data which descripts the quantity that knowledge object has been used and the frequency it has been right used. So the parameter can be defined as $\zeta = (\alpha,\beta)$. When we create a whole new knowledge entity, it means we have used and right used it this time, so the initial value of ζ must be (100%, 1). According to those mentioned in last paragragh, in the fusion system any knowledge node can inherit the value of ζ from its father node on the lower neighbour layer when it is firstly created. Because the frequency it has been used is changing during applications and such changes must be feed back to the corresponding dualistic data, ζ may keep changing. The process of changing β's value is implemented by control module of the whole knowledge fusion system. When a correct result from applications is coming, ζ's value can be refreshed according to followed:

$$\zeta^{T+1} = (\beta^T + 1)/(\alpha^T \beta^T + 1), if(Result = TRUE) \tag{8}$$

$$\zeta^{T+1} = (\beta^T + 1)/\alpha^T \beta^T, if(Result = FALSE) \tag{9}$$

ζ incarnates the adjustment of μ and it descripts the reliability according to what is actually happening.

4.2   Revising μ and ζ

ω_i in μ has been adjusted automatically in the algorithm introduced above. Then let us see how to revise the other one. If we describe knowledge as meta-knowledge sets after fusion process and each element in those sets as knowledge point, the sets consisted of whole knowledge states generated by FA can be regarded as solution knowledge space which marked as K. Knowledge filed in K is expressed by a finite question ontology that is knowledge states are correlated or not.

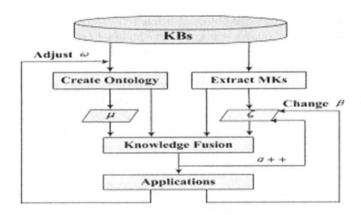

Fig. 2. Flow chart of processing the parameters

Solution knowledge can be generated as follows:

Step 1. Create ontology corresponding with question (O_p) and meta-knowledge (restriction) sets (S_p). The idiographic method can be found in section 2

Step 2. Search all knowledge states in K for S_a whose relationship grade to question state is the max. For each knowledge state related to question ontology, seek out its relationship grade. It is also the percent of related knowledge points in knowledge state and those in question state.

Knowledge state S_k relates to question ontology O_p must meet requirement as follows: Exists $(P_j^p, T_j^p, D_j^p) \in O_p$, $(P_j^k, T_j^k, D_j^k) \in O_k$ and $P_j^p = P_j^k$ and $D_j^p \neq$ NULL, $D_j^k \neq$ NULL.

5   A Referential Application

In this section, we shift our focus beyond the abstract architecture and point to a case study. Table 1 shows the attributes of every node in a knowledge-based grid environment.

Result of qualitative analysis on partly data is in Table 2, where we can see the model in the paper can reduce size of the application questions especially when the information is massive.

Quantitive analysis of the case focuses on efficiency and controllability. Emulate on control matrix B, we can see that the state pattern of fusion result goes to stable when B's density locates in a certain range. The trendline is shown in fig 3.

Table 1. Physical characters of KB-Grid nodes

IP address	OS	Strict	Knowledge type
192.168.0.2-192.168.0.8	Windows	Not	Information table
192.168.0.9-192.168.0.15	Linux	Not	Frame
192.168.0.16-192.168.0.22	Linux	Yes	Procedure present
192.168.0.23-192.168.0.29	Windows	Yes	Conceptual graph
192.168.0.30-192.168.0.36	Windows	Not	Vector

Table 2. Result of qualitative analysis on partly data

-	Transform times	Losing connotation	Find solution	Reused	Uniform
Based on interchange	Prod	Yes	No	Can	Yes
Method in the paper	Sum	No	Yes	Can	Yes

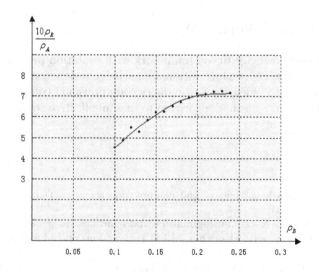

Fig. 3. Trendline of knowledge state

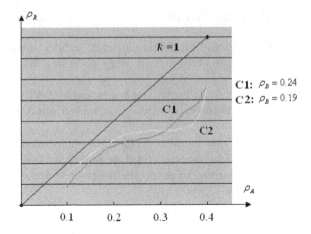

Fig. 4. State of interzone's ports

What's more, we can also find in that figure that change of expression ρ_R /ρ_A is not prominent any longer when ρ_B ∈ [0.19, 0.24]. To get a clearer view on effect of control matrix and feedback mechanism, let ρ_B equal to the interzone's ports and we can get a result as shown in fig 4.

According to those emulates result, the trend of knowledge state when ρ_B equal to the interzone's ports can be formalized as followed:

$$\overline{\rho_B(C_1)} \approx \overline{\rho_B(C_2)} \tag{10}$$

$$\frac{1}{Max_{\rho_A}|\rho_B(C_1) - \rho_B(C_2)|} \to \infty \tag{11}$$

6   Summary and Future Work

We present a new knowledge fusion framework with evolution process and evaluation method. Compared with traditional knowledge fusion system, method in the paper improves the efficiency and reduces computational complexity. In the future, we will use some optimal algorithm to find an effective search method in global knowledge space.

References

1. Ru-qian, L.: Knowledge science and computation science. Tsinghua University Press, Beijing (2003)
2. Tetsuo S, Jun U, O.K.e.a.: Fusing multiple data and knowledge sources for signal understanding by genetic algorithm. IEEE Transactions on Industrial Electronics **43** (1996) 411–421
3. Preece A D, Hui K Y, G.W.A.e.a.: The kraft architecture for knowledge fusion and transformation. Knowledge Based Systems **13** (2000) 113–120

4. M., J.: Structured knowledge source integration and its applications to information fusion. In: Proceedings of the 5th International Conference on Information Fusion, Maryland (2002) 1340–1346
5. Slembek., I.: An evaluation framework for improving knowledge-intensive business processes. In: Proceedings of 14th International Workshop on Database and Expert Systems Applications (DEXA'03), Prague, Czech Republic (2003) 813–817
6. J., G.J.: Optimization of control parameters for genetic algorithms. IEEE Trans on SMC **16** (1986) 122–128
7. S. Castano, A.F.: Knowledge representation and transformation in ontology-based data integration. In: Workshop W7 at the 15th European Conference on Artificial Intelligence, France, Lyon (2002) 107–121
8. WN., B.: Construction of Engineering Ontologies. PhD thesis, University of Twenty (1997)
9. Gou J, Yang JG, Q.H.: A knowledge fusion framework in the grid environment. In: Proceedings of 4th International Conference Computational Science, Poland (2004) 503–506
10. M., M.T.: Machine learning. China machine press, Beijing (2003)

The Allele Meta-model – Developing a Common Language for Genetic Algorithms

Stefan Wagner and Michael Affenzeller

Institute for Formal Models and Verification,
Johannes Kepler University,
Altenbergerstrasse 69, A-4040 Linz - Austria
{stefan, michael}@heuristiclab.com

Abstract. Due to the lot of different Genetic Algorithm variants, encodings, and attacked problems, very little general theory is available to explain the internal functioning of Genetic Algorithms. Consequently it is very difficult for researchers to find a common language to document quality improvements of newly developed algorithms.

In this paper the authors present a new Allele Meta-Model enabling a problem-independent description of the search process inside Genetic Algorithms. Based upon this meta-model new measurement values are introduced that can be used to measure genetic diversity, genetic flexibility, or optimization potential of an algorithm's population. On the one hand these values help Genetic Algorithm researchers to understand algorithms better and to illustrate newly developed techniques more clearly. On the other hand they are also meaningful for any GA user e.g. to tune parameters or to identify performance problems.

1 Introduction

Genetic Algorithms (GAs) developed in 1975 by J. H. Holland [11] are a heuristic optimization technique based on the natural evolution process. Although they represent a very strong simplification of the complex processes observed in nature we commonly subsume with the term evolution, GAs are very successful in different fields of industrial and scientific application. In spite of their success there is still very little mathematical theory available that might comprehensively explain the different processes and their interactions inside a GA. One of the main reasons for this drawback is the tremendous lot of different GA variants, encodings, operators, and attacked problems. So most theoretical considerations like the Schema Theorem introduced by J. H. Holland [11] or the Building Block Hypothesis presented by D. E. Goldberg [9] concentrate mainly on one specific form of individual encoding (in most cases binary encoding) and can therefore hardly be generalized. Other approaches aiming to develop a common theory for Evolutionary Computation in general (cf. [14], [18]) are faced with severe difficulties due to the huge variety in the field of GAs.

As a consequence of this lack of theoretical background most scientists working in the area of GAs have a very profound intuitive knowledge about GAs

J. Mira and J.R. Álvarez (Eds.): IWINAC 2005, LNCS 3562, pp. 202–211, 2005.

and GA behavior but it is very difficult for them to show some kind of hard facts documenting their results. Most GA researchers will agree on the fact that effects like selection, selection pressure, genetic diversity, (unwanted) mutations, or premature convergence strongly interact with each other. These interactions play an important role concerning achievable solution quality. However, very few papers can be found that suggest ways to measure and visualize these forces affecting GA populations.

In this contribution the authors present some new measurement values that might help to overcome this situation. First of all a new allele oriented meta-model for GAs is presented making it possible to understand and to discuss processes inside the algorithm in a problem-independent way. Based upon this allele model new measurement values are presented measuring genetic diversity, genetic variety, goal-orientedness, and optimization potential during a GA run. So these values help any GA developer e.g. to precisely show improvements when introducing new algorithm variants and to discuss and compare the behavior of new algorithms with already existing ones. Furthermore, the values can also be used by any GA user to get some more meaningful feedback from the algorithm helping to tune parameters, to identify performance problems, or to develop a deeper understanding for GAs in general.

2   An Allele Oriented Meta-model for GAs

GAs mainly use the three genetic operators selection, crossover, and mutation to manipulate solution candidates in order to achieve better results. Only one of these three operators, namely selection, is independent of the chosen problem encoding, as selection depends only on the individuals' fitness values. So when trying to make any general propositions about GAs this high amount of problem (i.e. encoding) dependency is a severe difficulty. Therefore, most of the existing theoretical considerations like the Schema Theorem or the Building Block Hypothesis only concentrate on binary encoding, as it was the first encoding variant suggested by J. H. Holland and is still used in numerous applications.

However, in the last years various other forms of individual encoding (e.g. permutation-based, real-valued, etc.) have been introduced. These codifications have shown high potential in lots of applications like e.g. the Traveling Salesman Problem, Scheduling Problems, or Genetic Programming. In those cases using binary encoding was not at all intuitive and required the development of sophisticated and rather inappropriate crossover and mutation operators. As different solution encodings also led to the development of new crossover and mutation operators, it is very difficult to generalize the theoretical statements of Holland and Goldberg in order to be applicable for these new GA variants.

So before we can think of new measurement values describing the internal functioning of GAs, it is necessary to develop a new problem-independent view. Otherwise the newly proposed insights would also only be applicable for a specific kind of GA applications. This conclusion was the cornerstone for the development of the *Allele Meta-Model* of GAs. The basic question that needs to be answered is, what the atomic entities, GAs work with, are and if and how these entities can be stated in a problem-independent way. In fact, in the case of binary encoding the Building Block Hypothesis already highlighted the importance of small parts of genetic material (low-order schemata with short defining length and above average fitness) that are assembled by the algorithm to generate better solutions. When abstracting this consideration a GA in general can be seen as a process that combines different parts of genetic code.

In biology the concrete realization of a gene is called an allele and represents the logical entity on top of the molecular level. So it seems quite reasonable to use the term allele also in the context of GAs describing the basic entity that represents genetic information, forms chromosomes, and describes traits. In the case of binary encoding an allele can e.g. be a bit at a specific position of the individual's bit string. As the concept of alleles is problem-independent and not restricted to binary encoding, it can be defined for other GA encodings as well. In the case of permutation encoding e.g. the sequence of two numbers of the permutation or a number at a specific position of the permutation can be considered as an allele depending on the optimization problem and the interpretation of its encoding. Obviously the identification and interpretation of alleles, i.e. the layer below the Allele Meta-Model, is highly problem and encoding dependent. Though on top of the Allele Meta-Model GAs can be discussed in a completely problem-independent way including crossover and mutation concepts.

Based upon the Allele Meta-Model the Standard Genetic Algorithm (SGA) (as described in e.g. [7], [8], [16], [17], [21]) can be reformulated in the following way: Each individual is represented by a randomly initialized set of alleles. Each set is analyzed by an evaluation function returning a fitness value of the individual. In the selection step individuals, i.e. allele sets, with high fitness are selected for reproduction. Then the crossover operator is used to merge two parental allele sets to one new set in order to generate a new solution, i.e. child. Thereby it might happen, that not all alleles contained in the new solution are also elements of at least one of the parental allele sets. This situation occurs when using some more complex encoding requiring crossover operators with some kind of repair strategy and is referred to as unwanted mutations. Finally the mutation operator might be used to exchange some alleles of the child's allele set by some other ones.

3   Allele Frequency Analysis

As population genetics focus on the changing of allele frequencies in natural populations the Allele Meta-Model of GAs builds the bridge between GAs and population genetics. So it should be a very fruitful approach to consider various aspects and terms of population genetics for GA analysis (cf. [2], [3]). Population genetics define various forces influencing allele frequencies in natural populations: the Hardy-Weinberg Law, Genetic Drift, Selection, Mutation, etc. (a good overview can be found in [10]). However, as the basic population model in population genetics assumes diploid individuals these insights have to be adapted accordingly in order to be valid for GAs which are haploid per design. All these different forces lead to one of the following four different results concerning the frequency of alleles (p denotes the probability that a specific allele is contained in the genetic code of an individual, or in other words is element of an individual's allele set):

- $\mathbf{p} \rightarrow \mathbf{1}$: The allele is fixed in the entire population.
- $\mathbf{p} \rightarrow \mathbf{0}$: The allele is lost in the entire population.
- $\mathbf{p} \rightarrow \hat{\mathbf{p}}$: The allele frequency converges to an equilibrium state.
- $\mathbf{p} \rightarrow \mathbf{p}$: The allele frequency remains unchanged.

So in this context the global goal of GAs can be reformulated in the following way: Use selection, crossover and mutation to modify the allele frequencies of the population in such a way that all alleles of a global optimal solution are fixed in the entire population. All other alleles belonging to suboptimal solutions should be lost.

As a consequence it should be very insightful to monitor the distribution of alleles in a GA population during the execution of the algorithm in order to observe the success of a GA concerning the success criterion stated above. In fact this is the main idea of the Allele Frequency Analysis (AFA) and its new measurement values. Consequently it is necessary to distinguish between two different types of alleles: On the one hand there are alleles belonging to a global optimal solution and on the other hand all other alleles being definitively not optimal. As a matter of course such a distinction can only be made when using benchmark problems with known optimal solutions. For better differentiation the first kind of alleles are referred to as *relevant alleles* (cf. building blocks) in the following.

Based on the Allele Meta-Model of GAs some measurement values can be introduced that represent the basis of the AFA:

- **Total Number of Different Alleles (A):**
  The total number of different alleles contained in the entire population is a precise measurement for genetic diversity. The more different alleles are available in the population the more diverse is the genetic code of the individuals. In the case of combinatorial optimization problems it's important to bear in mind that the total number of different alleles in the whole solution space is usually not that large as the complexity of combinatorial optimization problems is caused by the millions of possible combinations of alleles. E.g. a 100 cities Traveling Salesman Problem has only $99 + 98 + 97 + \ldots + 1 = \frac{100 \cdot 99}{2} = 4950$ different edges, i.e. alleles.

- **Number of Fixed Alleles (FA):**
  A second measurement value of interest is the number of fixed alleles in the entire population, i.e. the number of alleles contained in the allele set of every individual. It indicates the genetic flexibility of the population as any fixed allele cannot be altered by crossover anymore (apart from unwanted mutations). Consequently especially the fixing of suboptimal alleles is very harmful for GA performance because it benefits premature convergence.

- **Total Number of Relevant Alleles (RA):**
  If benchmark problems with known optimal solutions are used for analyzing GAs, it is possible to identify alleles of global optimal solutions and to count their total number in the entire population. This value provides information about how goal-oriented the evolutionary search process is. Ideally the total number of relevant alleles should be steadily increasing until all relevant alleles are fixed in the entire population.

- **Number of Fixed Relevant Alleles (FRA):**
  The number of fixed relevant alleles is a fraction of all fixed alleles and estimates the success chances of the GA. Especially the deviation between the number of fixed relevant and fixed alleles is very informative as it points out how many suboptimal alleles are already fixed in the population which indicates the severity of premature convergence.

- **Number of Lost Relevant Alleles (LRA):**
  Contrary to the number of fixed relevant alleles the number of lost relevant alleles shows how many relevant alleles are not included in the population's gene pool anymore. In an ideal (hyperplane sampling) GA this value should always be 0 as such lost relevant alleles can only be regained by mutation which is contradictory to the idea of hyperplane sampling. Again this measurement value helps to appraise premature convergence.

- **Distribution of Relevant Alleles (DRA):**
  Concerning relevant alleles there is finally the opportunity to monitor the distribution of all relevant alleles in the entire population during the GA execution. In fact this is not a single measurement value but a very good technique to visualize the dynamics of a GA and in particular the interplay between hyperplane sampling (crossover) and neighborhood search (mutation).

- **Selection Pressure (SP):**
  Last but not least there is another measurement value not directly motivated by the Allele Meta-Model. The concept of selection pressure was first introduced by Charles Darwin as a result of birth surplus [6]: A population is producing more offspring than the actual environmental resources can keep alive. Consequently some of the not so fit children die before they reach the age of sexual maturity. This fact causes a so-called selection pressure among the offspring requiring a minimum fitness to survive in order to pass on their own genetic information. In the context of Evolutionary Computation selection pressure has been defined for some algorithms that also produce a birth surplus like Evolution Strategies (ES) [19], the Breeder GA [15], SEGA [1], or SASEGASA [4]. E.g. selection pressure is defined as $\frac{\lambda}{\mu}$ for the (μ, λ)-ES where μ denotes the population size and λ stands for the number of procreated offspring. A large value of $\frac{\lambda}{\mu}$ indicates a high selection pressure (small population size, lots of children) and vice versa. In the above mentioned algorithms selection pressure turned out to have a great influence on the algorithms' performance.

However, in the general case of GAs selection pressure cannot be defined so easily as a GA is normally procreating exactly as many children as needed. So how can selection pressure be measured, if there is no birth surplus? One possible suggestion is to define selection pressure as the ratio of the selection probability of the fittest individual to the average selection probability of all individuals (see e.g. [5]). However, this definition has two weaknesses: First, it is an a priori definition of selection pressure, which doesn't take stochastic effects into account that might have a great influence especially when using rather small populations. Second, the average selection probability depends on the chosen selection strategy and cannot be calculated that easily in some cases (e.g. when using Tournament Selection concepts). So the authors decided to calculate selection pressure in an a posteriori way independent of the used selection operator.

Selection pressure can be abstracted somehow as a measurement value indicating how hard it is for an individual to pass on its genetic characteristics from one generation to the next. So it seems to be reasonable to define selection pressure in a classical GA as the ratio between the population size and the number of individuals selected as parents of the next generation. If the individuals of the next generation are procreated by a few parent individuals only, selection pressure is very high and vice versa. So selection pressure can be calculated according to the following formula:

$$SP = 1 - \frac{|PAR|}{|POP|} \tag{1}$$

where $|PAR|$ stands for the number of different selected parents and $|POP|$ represents the population size. So a minimum selection pressure of 0 indicates that all individuals of the parental generation got the chance of mating and a maximum selection pressure of $1 - \frac{1}{|POP|}$ represents the situation that all offspring are mutated clones of a single super-individual.

4   Allele Frequency Analysis in Practice

In this section an example run of the Standard Genetic Algorithm (SGA) is performed with the HeuristicLab[1] optimization environment [22] to outline the potential of the AFA. To highlight the encoding independency of the AFA a test problem not restricted to binary encoding was chosen. The authors decided to use the Traveling Salesman Problem (TSP) (e.g. described in [13]) as the TSP is a very well-known combinatorial optimization problem with exactly one optimal solution in a majority of cases. Furthermore, a lot of different encodings, crossover operators and mutation concepts for GAs are available (cf. [12]). As test problem instance the ch130 TSP benchmark problem taken from TSPLIB [20], a comprehensive collection of TSP benchmark problems, is used. For this problem

---

[1] More details can be found on the HeuristicLab homepage http://www.heuristiclab.com.

Table 1. Parameter Settings

Generations	2'500	Selection Operator	Tournament Selection
Population Size	250	Crossover Operator	Order Crossover (OX)
Mutation Rate	5%	Mutation Operator	Simple Inversion Mutation (SIM)
Tourn. Group Size	3		

Table 2. Solution Quality and AFA Results

Optimal Solution Fitness	6'110	A	150
Best Found Solution Fitness	7'099	RA	19'738
Average Solution Fitness	7'118.06	FA	113
Evaluated Solutions	625'000	FRA	69
Average SP	0.361	LRA	51

not only the quality, i.e. the tour length, of the optimal solution is known (6'110) but also the optimal tour itself, which makes it well-suited for the AFA. Furthermore, path representation is used for solution encoding, whereby the alleles are represented by the edge information contained within the permutation in that case.

However, it has to be mentioned once again that the AFA is not restricted to a specific form of TSP encoding or to the TSP in general. The AFA can be performed for any kind of optimization problem, if the definition of alleles is reasonable.

The parameter values used for the example run are listed in Table 1. Moreover, the achieved results concerning solution quality and the AFA values of the last generation are presented in Table 2.

However, the development of solution quality and AFA values during the GA run is far more interesting than the final results as it helps to gain some insight about what's going on inside the algorithm. In Figure 1 the fitness value of the best found solution and the average solution fitness, as well as the A and RA value (second y-axis) are shown. Additionally, Figure 2 shows the progress of A as well as FA, FRA, and LRA (second y-axis). It can be seen that the diversity of the population (A) is decreasing proportional with the decrease of the fitness value.[2] Contrariwise the number of relevant alleles (RA) is increasing as the solution quality increases. These aspects are not really surprising as the GA uses evolutionary forces to improve the genetic material of its population. Consequently, disadvantageous alleles will be eliminated due to selection. The population size is not modified during the run and so the total number of alleles is staying constant leading to an increase of advantageous alleles reflected in the increasing RA.

A more interesting aspect revealed by the charts is the drastic loss of genetic diversity in the first few generations. The A value is dropping from almost 8.000

---

[2] Note that in the context of the TSP a decreasing fitness value is equivalent to an increasing solution quality as the total tour length is used as fitness value which should be minimized.

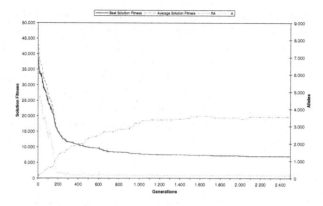

Fig. 1. Best Solution Fitness, Average Solution Fitness, A, and RA

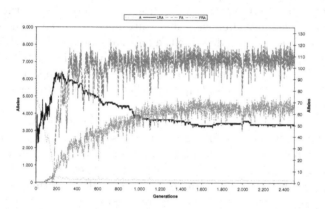

Fig. 2. A, FA, FRA, and LRA

at the beginning to approximately 4.500 within the first 20 generations. This dramatic diversity reduction comes along with a significant increase of solution quality. However, also a lot of relevant alleles are lost indicated by the LRA value jumping to almost 60 during this period. Although mutation and also unwanted mutations are able to regain some of the lost relevant alleles the algorithm doesn't fully recover from this initial diversity loss during the whole run, leading to premature convergence in the end.

After this initial phase genetic diversity is further strongly decreased and on the opposite the number of lost relevant alleles increases, indicating a rather high selection pressure (as expected when using Tournament Selection with a group size of 3). Then shortly before generation 200 is reached genetic diversity is reduced that much that first alleles are fixed in the entire population (FA). The FA value increases very quickly and from that moment on genetic flexibility of the population is very low. Crossover is not able to combine alleles in order to generate better solutions anymore and the algorithm needs mutation and also unwanted mutations to induce new alleles into the gene pool of its population. Obviously, mutation is able to find some of the missing relevant alleles in the last phase of the GA as the number of relevant alleles and consequently also the solution quality further increases slowly. These newly found relevant alleles are then propagated via crossover among the allele sets of all individuals leading to an increasing FRA value. However, not all relevant alleles are regained by mutation and so the algorithm prematurely converges at a best found solution fitness value of 7'099 which is 16.19% worse than the optimal solution.

5   Conclusion

In this paper the authors present the Allele Meta-Model for GAs. By introducing alleles as the atomic entities GAs work with, it gets possible to consider the whole process of a GA in a problem-independent way. Furthermore, inspired by the area of population genetics the Allele Meta-Model builds the basis for introducing some new measurement values subsumed with the term Allele Frequency Analysis. These values describe the internal state inside an algorithm by measuring genetic diversity, genetic flexibility, goal-orientedness, or optimization potential. Furthermore, in a short experimental part the paper also illustrates how the measurement values also help to predict premature convergence and to identify its reasons.

Finally it can be stated, that the Allele Meta-Model and the Allele Frequency Analysis are not only meaningful for any GA researcher helping to document improvements of newly developed algorithms and providing a common language, but also for GA users, as the calculated values provide essential feedback about the algorithm and help to tune parameters, to identify performance problems, and to gain deeper understanding for GAs in general.

References

1. M. Affenzeller. Segregative genetic algorithms (SEGA): A hybrid superstructure upwards compatible to genetic algorithms for retarding premature convergence. *International Journal of Computers, Systems and Signals (IJCSS)*, 2(1):18–32, 2001.
2. M. Affenzeller. Population genetics and evolutionary computation: Theoretical and practical aspects. *Accepted to be published in: Journal of Systems Analysis Modelling Simulation*, 2004.
3. M. Affenzeller and S. Wagner. The influence of population genetics for the redesign of genetic algorithms. In Z. Bubnicki and A. Grzech, editors, *Proceedings of the 15th International Conference on Systems Science*, volume 3, pages 53–60. Oficyna Wydawnicza Politechniki Wroclawskiej, 2004.
4. M. Affenzeller and S. Wagner. SASEGASA: A new generic parallel evolutionary algorithm for achieving highest quality results. *Journal of Heuristics - Special Issue on New Advances on Parallel Meta-Heuristics for Complex Problems*, 10:239–263, 2004.

5. T. Bäck. Selective pressure in evolutionary algorithms: A characterization of selection mechanisms. In *Proceedings of the First IEEE Conference on Evolutionary Computation*, pages 57–62. IEEE Press, 1994.
6. C. Darwin. *The Origin of Species*. Wordsworth Classics of World Literature. Wordsworth Editions Limited, 1998.
7. L. Davis. *Handbook of Genetic Algorithms*. Van Nostrand Reinhold, 1991.
8. D. Dumitrescu, B. Lazzerini, L. C. Jain, and A. Dumitrescu. *Evolutionary Computation*. The CRC Press International Series on Computational Intelligence. CRC Press, 2000.
9. D. E. Goldberg. *Genetic Algorithms in Search, Optimization and Machine Learning*. Addison Wesley Longman, 1989.
10. D. L. Hartl and A. G. Clark. *Principles of Population Genetics*. Sinauer Associates Inc., 2nd edition, 1989.
11. J. H. Holland. *Adaption in Natural and Artifical Systems*. University of Michigan Press, 1975.
12. P. Larranaga, C. M. H. Kuijpers, R. H. Murga, I. Inza, and D. Dizdarevic. Genetic algorithms for the travelling salesman problem: A review of representations and operators. *Artificial Intelligence Review*, 13:129–170, 1999.
13. E. L. Lawler, J. K. Lenstra, A. Rinnooy-Kan, and D. B. Shmoys. *The Travelling Salesman Problem*. Wiley, 1985.
14. H. Mühlenbein. Towards a theory of organisms and evolving automata. In A. Menon, editor, *Frontiers of Evolutionary Computation*, volume 11 of *Genetic Algorithms and Evolutionary Computation*, chapter 1. Kluwer Academic Publishers, 2004.
15. H. Mühlenbein and D. Schlierkamp-Voosen. The science of breeding and its application to the breeder genetic algorithm BGA. *Evolutionary Computation*, 1(4):335–360, 1993.
16. Z. Michalewicz. *Genetic Algorithms + Data Structures = Evolution Programs*. Springer-Verlag, 1992.
17. M. Mitchell. *An Introduction to Genetic Algorithms*. The MIT Press, 1996.
18. N. J. Radcliffe. Forma analysis and random respectful recombination. In R. K. Belew and L. B. Booker, editors, *Proceedings of the 4th International Conference on Genetic Algorithms*, pages 222–229. Morgan Kaufmann Publishers, 1991.
19. I. Rechenberg. *Evolutionsstrategie*. Friedrich Frommann Verlag, 1973.
20. G. Reinelt. TSPLIB - A traveling salesman problem library. *ORSA Journal on Computing*, 3:376–384, 1991.
21. M. Tomassini. A survey of genetic algorithms. *Annual Reviews of Computational Physics*, 3:87–118, 1995.
22. S. Wagner and M. Affenzeller. Heuristiclab: A generic and extensible optimization environment. In *Accepted to be published in: Proceedings of ICANNGA 2005*, 2005.

Using Bees to Solve a Data-Mining Problem Expressed as a Max-Sat One

Karima Benatchba, Lotfi Admane, and Mouloud Koudil

Institut National de formation en Informatique,
BP 68M, 16270, Oued Smar, Algeria
{k_benatchba, l_admane, m_koudil}@ini.dz
http://www.ini.dz

Abstract. The "NP-Complete" class gathers very significant practical problems such as Sat, Max-Sat, partitioning There is not polynomial algorithm for the resolution of these problems. As a result, the interest in heuristics and meta-heuristics is still growing. In this paper, we present a very recent metaheuristic introduced to solve a 3-sat problem. This metaheuristic can be classified as an evolutionary algorithm. It is based on the process of bees' reproduction. We adapted it for the resolution of the Max-Sat problem. We tested it on a medical benchmark obtained from a data-mining problem that we translated into a Max-Sat problem.

Keywords: Data-Mining, Satisfiability, Maximum-Satisfiability, Optimization, Optimisation using bees, MBO algorithm.

1 Introduction

Most of the NP-Complete problems studied in Combinatorial Optimization are simple to express (the salesman problem, SAT, partitionning...) and can be easily solved in theory, just by enumerating all the possible solutions and choosing the best. This is how the exact methods work. In practice, this approach is used, and even recommended, for small sizes problems. However, using exact methods quickly becomes impossible as the size of the problem increases. This is due to the computing time which grows exponentially with the number of variables. Indeed, to set the truth table of a Boolean formula of 64 variables (264 possible combinations) would take several years on a current PC.

Because of the practical importance and applications of most NP-hard problems, it is often essential to find approximate solutions in a reasonable time. In most of the practical cases, having an optimal solution is not of primary importance and a solution of rather good quality found in a reduced time can be more interesting than an optimal solution found in a longer time. The approaches based on heuristics, such as the metaheuristics, meet this requirement. They find acceptable solutions, even optimal ones, in a relatively short time.

Optimization by Bees' Colony is a new evolutionary metaheuristic inspired by the social organization of bees and their process of reproduction. The characteristic of this method is that it mixes the principle of general metaheuristics

J. Mira and J.R. Álvarez (Eds.): IWINAC 2005, LNCS 3562, pp. 212–220, 2005.

and dedicated heuristics in the same algorithm. We use the algorithm of optimization by Bees', in this paper, to solve a data mining problem expressed as a Max-Sat One. The problem considered is a supervised learning one that we translate into a Boolean formula in its Conjunctive Normal Form (CNF). The obtained Max-Sat problem is solved using the metaheuristic. The best solutions found can easily be back-translated into rules that can be applied to the data sets in order to verify that they really satisfy a maximum number of instances in the original learning problem.

Section 2 gives brief definitions on the Satisfiability, Maximum Satisfiability and Data-Mining problems. The third one introduces some characteristics of the real bees colonies. Sections 4 and 5 present the artificial modeling of the reproduction for bees and the algorithm associated. Section 6 introduces the medical benchmark used to test the algorithm. Section 7 gives some results obtained on the benchmark and section 8 concludes this paper.

2 Sat, Max-Sat and Data Mining

The SAT problem is central in the class of NP-Complete problems. It is the first problem to have been proved to be NP-Complete [1]. Several problems having practical applications can be reduced to a SAT one in a polynomial time [2]. We define in the following the problem of Satisfiability of a Boolean formula (SAT problem). Let X= x1, x2 xn and C = C1, C2 Ck be respectively a set of n variables and a set of k clauses. F is a Boolean formula in its conjunctive normal form (called system SAT) where:

$$F = \wedge Ci(1 \leq i \leq k) \quad and each \quad Ci = \vee xj(1 \leq j \leq n); \tag{1}$$

xj being a literal (a propositional variable or its negation). F is said to be satisfiable if and only if there exists a truth assignment I such as I(F) is true, I being a function which associates to each variable a truth value (Yes or No).

The non satisfiability of a SAT system leads us to ask the following question: "How many clauses of F can be satisfied simultaneously?" This problem is called the maximum satisfiability problem (Max-Sat). It is an optimization problem that has been classified as a NP-Complete problem [3].

Data-mining [4], [5] is the process of exploration and analysis, by automatic or semi-automatic means, a large quantities of data in order de discover meaningful patterns and rules [6]. The main tasks of Data-Mining are classification, estimation, prediction, affinity grouping, association rules, clustering, and visualization. The knowledge discovery can be of two types: directed (classification, estimation, prediction) or undirected (affinity grouping, association rules, clustering, and visualization). In the first case, the "directed knowledge discovery", tries to explain or classify a particular field of data, using the remainder fields to classify new individuals. The second case relates to the discovery of forms or similarities between groups of records, without using a specific target field, or a collection of preset classes.

The problem, we deal with in this paper, is a classic prediction problem that is often expressed as follows: Knowing a certain number of characteristics on a given individual (called explanatory variables) noted (X1, X2 ... X N), with which certainty can one deduce the value of an unknown characteristic called variable to explain and noted Y. As we want to treat this problem as a Max-Sat one, we translate it into a CNF Boolean formula using a specific algorithm developed in [7], [8].

3   Real Bees' Colonies

The bees are social insects living in hives in very organized colonies. A colony of bees is made up of three categories of adults: the queen, drones and workers. The bees appear among the most studied social insects, undoubtedly because they are among the most organized animals and as a result the most attractive ones. Indeed, the bees live in community divided into several social layers. They communicate between them by several means (dance of the bees, pheromone) and cooperate in the realization of various tasks such as the construction and the maintenance of the hive, the harvest of pollen, etc.

The process of bees' reproduction is very singular. Each social layer has a role to play in the birth of a new bees'. In what follows, we will look more closely to the process of reproduction, as the algorithm of optimisation by Colony of Bees (MBO) is based on that process. It was introduced in 2001 by Abbas and Al. [9], [10], [11] and used by the authors to solve the 3-Sat problems.

In nature, a colony of bees is made up of one or more queens, drones (male bees), workers and broods. The queens are the principal reproductive individuals of the colony. Only they can lay fertilized eggs. The drones are in charge of fertilizing the new queen and are, thus, the "fathers" of the colony. The workers have, as a main task, to take care of broods and can, for certain species, lay eggs too. Broods are made up of fertilized eggs and unfertilized eggs. The first ones will give birth to queens or workers; whereas, the second ones generate drones.

Each queen carries out once in her life a mating-flight. During her flight, the queen will mate the drones that followed her. A queen mats herself with seven to twenty drones per flight. After each mating, the sperm of the drone is added to the spermatheca of the queen (reserve of spermatozoids). The queen will use, all her life, this reserve which constitutes the genetic potential of the colony. Each time that a queen has to lay fertilized eggs, it withdraws a mixture of sperm of its spermatheca to fertilize them.

4   Optimisation by Bees' Colonies

Optimisation using colonies of bees is a new intelligent technique inspired from the biological process of bees' reproduction. It is an evolutionary method which includes, nevertheless, strategies based on the neighbourhood approach [9].

The reproduction of bees' has been modeled to solve optimization problems. It gave birth to the Marriage in Honey-Bees Optimization algorithm (MBO). In this algorithm, artificial queen, workers, drones and broods are used. An artificial queen has a genotype, set of genes which constitutes its heredity. The queen's genotype can be considered as a complete solution to the problem. Energy and speed are associated to a queen and used during her mating-flight. In addition, each queen will have a spermatheca that will collect the sperms of the droves encountered. The queen lays eggs that will become new queens (new solutions).

The artificial workers are specific heuristics to the problem which role is to improve the genotypes of broods (future queens), making it possible to obtain better solutions. Thus, in the artificial model, taking care of eggs corresponds to improving their genotype. In other words artificial workers are algorithms that improve the current solution locally.

Artificial Drones have only half of a genotype, they are haploid. In the artificial model, drones have a complete genotype and a mask being used to mask half of the genes (selected randomly). The unmasked half of the genotype constitutes the sperm of the drone. For example, in the Max-Sat, half of a solution is a solution where only half of the variables of the formula appear. At the time of fecundation, the male genes (elements of its genotype) are crossed-over with the genotype of the queen to form a new genotype (complete solution).

The mating-flight can be considered as a series of transitions among a set of states. Each queen goes from a state to another according to her speed and mats herself with the drones met at each state according to a probability rule. A state is in fact a drone; as a result, both terms will have the same significance in what follows. At the beginning of each mating-flight, each queen has an initial energy. It decreases during the flight and the queen returns to her nest when its energy reaches a critical point or when its spermatheca is full. The probability that a queen R mats herself with a drone B is given by the following formula :

$$Prob(R, B) = Exp(-difference/Speed(t)) \qquad (2)$$

Where Prob (R, B) represents the probability of a successful mating between drone B and queen R. Difference represents the absolute difference between the "fitness" of the queen and that of the drone. Speed (t) is the speed of the queen at the time t. A quick look at this equation shows that the probability of a successful mating is higher either at the beginning of the flight because the energy of the queen is the highest (and thus its speed also), or when the fitness of the drone is close to that of the queen.

After each transition, the speed and the energy of the queen are decreased by using the following functions: Speed $(t +1) = a *$ Speed (t) where $a \in]0 \ 1[$ Energy $(t+1) =$ Energy $(t) -$ step.

Where step represents for a given queen the quantity of energy spent at each stage. It is initialized as follows:

$$step = \frac{0.5 * Initial\_Energy}{Size\_of\_the\_spermatheca} \qquad (3)$$

The process of reproduction tends to improve the genotype of the queen through the generations and thus improve the initial solution.

5   The MBO Algorithm for the Max-Sat Problem

The MBO algorithm starts by initializing a set of workers. It consists of choosing a certain number of heuristics and their parameters. Then, the genotype of each queen is initialized to a random value. A series of mating-flights is then programmed where energy, speed and the initial position of each queen are randomly generated. We can consider that the MBO algorithm is divided into two main parts. The first one corresponds to the mating-flight of the queens, while the second one consists in generating new eggs for the colony. During the mating-flight, queens will collect sperms that will be added to her spermatheca. It will consist of a set of partial solutions that will be used to produce new ones that will guide the search in different areas of the search space (diversification). During this process, each queen starts to move among the states according to her speed and mats herself with the drones met according to the probability rule (1). To avoid consanguineous mating (convergence), the drones must be independent of the queens and, for this purpose, are generated randomly. When a drone mats itself with a queen, its sperm is added to the spermatheca of the queen. At the end of their flight, each queen joins its hive and start laying eggs (creation of broods). It is the second part of the algorithm. During this operation, each queen withdraws randomly sperm from her spermatheca in order to crossover it with her genome (genes carried by the chromosomes), thus complementing the withdrawn sperm (incomplete solutions). The result of this crossover will give a fertilized egg. A mutation is then applied to the eggs (diversification). Then, the workers (heuristics) are used to improve the quality of each egg (solution). The final stage is the replacement where the queens of less quality are replaced by eggs of better quality. The unused eggs are eliminated and another mating-flight can start.

The steps of the MBO algorithm adapted to the Max-Sat problem are given in fig. 1.

A genotype (solution) for the Max-Sat problem is an array of size n, where n is the number of variables in the formula. Each entry is set to 0 if the corresponding variable is false, 1 if the corresponding variable is true. The fitness of a genotype used is the quotient between the number of satisfied clauses by the genotype and the total number of clauses in the formula. As we have seen, a worker is a heuristic that will improve the genotype of the queen and the broods. In our implementation of the MBO, the user can chose among the following methods: GSAT [12], GWSAT [13], GTSAT [14], HSAT [15] and local search method (LS).

The four first methods are dedicated ones for the Sat problem.

To simulate the displacement of the queen towards another drone in the next step of the mating flight, one flips the bits of the current drone according to a probability represented by the current speed of the queen.

Procedure MBO$_{Max-Sat}$()
  $Initialis NbQueen, NB_Workers, Nb_Eggs /*numberof queens, WorkersandEggs$
  $InitialiseMsizeof thespermatheca$
  $Initialiseworkerswiththesameheuristic$
  $Initialiseeachqueenwithar andomlygeneratedgenotype$
  $Calltheworkerstoimprovethef itnessof eachqueen$
  $WhileStoppingcriterianotmetdo$
   $ForEachQueen$
    $Initialiseenergyandspeedof eachQueen$
    $step = 0, 5 * energy/M$
    $GenerateaDroneGnrerunf aux - bourdonenutilisantposition$
    $Whileenergy > 0Do$
     $If theDroneisselectedaccordingtotheprobabilityrulethen$
      $If thespermathecaof theQueenisnot fullthen$
       $Addthespermof theDronetotheQueen'sspermatheca$
      $EndIf$
     $EndIf$
     $energy = Energy - step/*DecreasethevalueoftheQueen'senergy$
     $vitesse = 0, 9 * vitesse/*DecreasethevalueoftheQueen'sSpeed$
     $Flipthegenes(bits)of theDroneusingSpeedasaprobability.$
    $EndWhile$
   $EndFor$
   $ForEachegg$
    $SelectaQueenwithagoodf itness$
    $Selectrandomlyaspermf romthisQueen'sspermatheca$
    $GenerateanEggbycrossingovertheQueen'sgenotypewiththeselectedsperm$
    $Improvethegenotypeof theeggusingtheworkers$
   $EndFor$
   $Whilethebestegghasabetterf itnessthantheworstQueenDo$
    $ReplacetheworstQueenwiththebestEgg$
    $DeletetheEggf romthelist$
   $EndWhile$
  $EndWhile$
$End.$

Fig. 1. The MBO for the Max-Sat problem

6   Presentation of the Bench Mark Used

The training set used as benchmark in this study is extracted from a medical
one, aiming at analyzing the most revealing symptoms of the presence or not
of a laparotomy of the principal bile duct (LVBP). A sample of 150 individuals
was selected. Each individual being characterized by a set of seven variables.
Six explanatory ones: 1) DDT indicates the existence or not of gastric disorders,
2) ICTERE indicates the existence or not of jaundice, 3) STONE indicates the
existence or not of stones in the vesicle, 4) CYST indicates the existence or not
of cystic bile duct, 5) VBP indicates the existence or not of a dilation of the bile
duct, 6) BILE indicates the state of cleanliness of the bile; and finally a variable

to explain 7) LVBP which indicate the existence or not of laparotomy of the principal bile duct. The translation of the benchmark, using [7], [8] gave a CNF formula with 68 variables and 346 clauses.

7   Tests and Results

In the following, some of the tests obtained using the MBO algorithm on the Max-Sat problem are presented. It consists of two tables presented below (Table 1 and Table 2). In the first table we use GSAT as the worker for improving found solutions through different iterations. By using the same worker for the different simulations, we wanted to test different values for the parameters of the method to find the best ones. The parameters of the MBO method are the following: NbR: Number of queens; Sp:Size of the spermatheca of each queen; NbF: number of mating-flights for each queen, V: initial speed of each queen, En: initial energy of each queen, a : decreasing speed factor; Q0: Mating probability; Work: type of worker. The last three columns of the table give the results which consist of cost: the cost of the function, the total number of satisfied clauses by the solution found. Gen the generation where it was found and the ET the execution time.

The best result obtained with the MBO was the solution satisfying 332 clauses out of 346. It was obtained using only one queen, the size of the spermatheca is 10, the number of mating-flight is 2 or 3, the initial speed is 0.7, the initial energy is 0.6, a is 0.9 and Q0 is 0.4. One notices that when we increase the number of queens to 2 (MBO 11), the execution time increases a lot. With One queen, one has to set the size of the spermatheca to 7 to reach the solution satisfying 332 clauses (MBO 14).

Table 1. Results of the application of MBO on the medical benchmark

MBO	NbR	Sp	NbVol	V	En	Alpha	Q0	Ouv	Cost	Gen	Time
MBO1	1	10	3	0.7	0.6	0.9	0.4	GSAT	332	2	2:55:94
MBO2	1	10	3	0.9	0.6	0.9	0.4	GSAT	332	1	2:05:29
MBO3	1	10	3	0.5	0.6	0.9	0.4	GSAT	327	2	2:51:99
MBO4	1	10	3	0.7	0.9	0.9	0.4	GSAT	332	1	2:10:98
MBO5	1	10	3	0.7	0.7	0.9	0.4	GSAT	332	2	2:49:34
MBO6	1	10	3	0.7	0.6	0.4	0.4	GSAT	330	1	2:51:98
MBO7	1	10	3	0.7	0.6	0.5	0.4	GSAT	330	1	2:50:29
MBO8	1	10	3	0.7	0.6	0.7	0.4	GSAT	331	1	1:06:45
MBO9	1	10	2	0.7	0.6	0.9	0.4	GSAT	331	1	1:20:99
MBO10	1	10	1	0.7	0.6	0.9	0.4	GSAT	330	1	1:21:95
MBO11	2	10	2	0.7	0.6	0.9	0.4	GSAT	332	1	4:03:93
MBO12	1	1	2	0.7	0.6	0.9	0.4	GSAT	330	2	0:33:03
MBO13	1	4	2	0.7	0.6	0.9	0.4	GSAT	331	1	0:55:09
MBO14	1	7	2	0.7	0.6	0.9	0.4	GSAT	332	2	2:01:47

Table 2. Influence of the workers on the performance of the MBO

MBO	NbR	Sp	NbVol	V	En	Alpha	Q0	Ouv	Cost	Gen	Time(s)
MBO1	1	10	2	0.7	0.6	0.9	0.4	LS	331	2	1:05:93
MBO2	1	10	2	0.7	0.6	0.9	0.4	GSAT	332	2	2:55:94
MBO3	1	10	5	0.7	0.6	0.9	0.4	GWSAT	328	1	2:02:97
MBO4	1	10	5	0.7	0.6	0.9	0.4	HSAT	330	2	3:52:85

In the tests presented in table 2, we wanted to show the impact of the workers on the quality of the solution found. As mentioned before several methods where used as workers: a local search algorithm LS, GSAT, HSAT, GWSAT. Each worker was used with its best parameters. One can notice that the best solution found was the MBO using GSAT as a worker. It was the only one that could reach the solution satisfying 332 clauses out of 346. Non of the other workers could give this solutions.

8   Conclusion

Optimization by Bees Colony (MBO) is a metaheuristic that one can classify in the family of evolutionary methods because it shares certain concepts with the Genetic Algorithms (genes, population, crossover, replacement...). Nevertheless, MBO presents a great number of particular characteristics in the way it operates. Indeed, this new method combines the advantages of metaheuristics (general methods, adaptable to a lot of problems) and advantages of dedicated heuristics (designed to a particular problem, therefore potentially more effective). The reason is that, on the one hand, the queens and the drones used by MBO can be adapted to optimisation's problems and, on the other hand, the workers constitute the dedicated heuristics to the problem, thus allow the MBO algorithm to benefit from specific information to each problem.

Because it is a very recent metaheuristic (in 2001), the MBO method was applied, to our knowledge, only to the 3-Sat problem [10]. In this paper, we adapted it to the Max-Sat problem and tested on a real case benchmark. It comes from a supervised learning problem that was translated into a Max-Sat problem. Different simulations and tests performed show the importance of the different parameters of the method especially the workers. One main drawback of this method is the time it takes for the resolution when the number of queens increases. For this reason, a parallel approach using this metaheuristic seems to be very interesting

Ackowledgements

We would like to thank our students Sofiane MADJOUR and Yahia ALILECHE for their collaboration in this work.

References

1. Cook S.A., "The complexity of theorem-proving procedures". Proc. 3rd ACP symp. On theory of computing Association of Computing Machinery, New York, p151-1158, 1971.
2. Garey R.M. and Johnson, "Computers and Intractability, A guide to the Theory of NP-Completeness". W.H. Freeman and CO., San Francisco, 1979.
3. Andr P., "Aspects probabilistes du problme de la satisfaction d'une formule boolenne. Etude des problmes SAT, #SAT et Max-SAT", Thse de doctorat de l'Universit Paris VI, 1993.
4. Dhar V. and Tuzhilin A, "Abstract-Driven Pattern Discovery in Databases " IEEE Trans. on Knowledge and Data Eng., 1993.
5. Fayyad U.M., Piatesky-Shapiro G., Smyth P. and Uthurusamy R., "Advances in Knowledge Discovery and Data-mining" AAAI Press/ the MIT Press, 1996.
6. Berry M.J., Linoff G.S., "Mastering Data Mining", Wiley Computer Publishing Ed., 2000.
7. Admane L., Benatchba K., Koudil M. and Drias H., "Evolutionary methods for solving data-mining problems", IEEE International Conference on Systems, Man & Cybernetics, Netherlands, October 2004.
8. Benatchba K., Admane L., Koudil M. and Drias H. "Application of ant colonies to data-mining expressed as Max-Sat problems", International Conference on Mathematical Methods for Learning, MML'2004, Italy, June 2004.
9. Abbas H.A., "MBO : Marriage in Honey Bees Optimization, A Haplometrosis Polygynous Swarming Approach", The Congress on Evolutionary Computation, CEC2001, Seoul, Korea, May 2001.
10. Abbas H.A., "A single queen single worker honey bees approach to 3-sat", The Genetic and Evolutionary Computation Conference, GECCO2001, San Francisco, USA July 2001.
11. Abbas H.A., J. Teo, "A True Annealing Approach to the Marriage in Honey-Bees Optimization Algorithm", The Inaugural Workshop on Artificial Life (AL'01), pp.1-14, ISBN 0-7317-0508-4, Adelaide, Australia, December 2001.
12. Selman B., Levesque H., Mitchell D., "A New Method for Solving Hard Satisfiability Problems", Proceedings of the Tenth National Conference on Artificial Intelligence, pp.440-446, San Jose, CA, July 1992.
13. Selman B., Kautz H.A., Cohen B., "Noise Strategies for Improving Local Search", Proceedings of the Twelfth National Conference on Artificial Intelligence (AAAI'94), pp.337-343, Seattle Washington, 1994.
14. Stutzle T., Hoos H., "A review of the literature on local search algorithms for MAX-SAT", Technical Report AIDA-01-02, Darmstadt University of Technology, Computer Science Department, Intellectics Group, 2001.
15. Gent I.P. and Walsh T., "Towards an Understanding of Hill-climbing Procedures for SAT", Proceeding of National Conference on Artificial Intelligence (AAAI), pp.28-33, 1993.

A Comparison of GA and PSO for Excess Return Evaluation in Stock Markets

Ju-sang Lee, Sangook Lee, Seokcheol Chang, and Byung-Ha Ahn

Mechatronics Dept., Gwangju Institute of Science and Technology,
1 Oryong-dong, Buk-gu, Gwangju 500-712 Republic of Korea
{jusang, yashin96, stniron, bayhay}@gist.ac.kr

Abstract. One of the important problems in financial markets is making the profitable stocks trading rules using historical stocks market data. This paper implemented Particle Swarm Optimization (PSO) which is a new robust stochastic evolutionary computation Algorithm based on the movement and intelligence of swarms, and compared it to a Genetic Algorithm (GA) for generating trading rules. The results showed that PSO shares the ability of genetic algorithm to handle arbitrary nonlinear functions, but with a much simpler implementation clearly demonstrates good possibilities for use in Finance.

1 Introduction

Particle Swarm Optimization (PSO) is an evolutionary computation technique developed by Kennedy and Eberhart in 1995 [1]. The basic algorithm is very easy to understand and implement. It is similar in some ways to genetic algorithms, but requires less computational bookkeeping and generally fewer coding lines [2]. Recently, this new stochastic evolutionary computation technique, based on the movement and intelligence of swarms has been successfully applied to artificial neural network [3], assignment problem [4], electromagnetics [5], size and shape optimization [6], power system [7], chemical process [8] and so on.

There is a growing interest in their application in financial economics but so far there has been little formal study, whereas other evolutionary algorithms such as genetic algorithm (GA) [9], [10], genetic programming [11], [12] have studied for over a decade. There have been a large number of literatures on the effectiveness of various technical trading rules[1] over the years.

The majority of them have found that such rules do not provide effective output. For instance, Alexander [13], Fama and Blume [14] concluded that the

---

[1] Technical trading rules uses only historical data, usually consisting of only past prices but sometimes also includes volume, to determine future movements in financial asset prices. This rules are devised to generate appropriate buying and selling signals over short term periods (commonly some times per year) while fundamental trading rules (buy-and-hold) doing it during relatively long term periods (commonly one trade from a few years to over decades).

J. Mira and J.R. Álvarez (Eds.): IWINAC 2005, LNCS 3562, pp. 221–230, 2005.

filter rules are not profitable and needs dynamic adjustment at the expense of transaction cost. Moreover Fama [15] dismissed technical analysis as a futile undertaking as it deserved *movement* characterization. In a more recent study, Brock *et al.* [16] considered the performance of various simple moving average rules in the absence of transaction costs. Their finding also reported that the rules can identify periods to be in the market (long the Dow Jones index) when returns are high and volatility is low and vice versa.

This paper shows how PSO can be used to derive trading system with the moving average rules that are not ad hoc but are in a sense optimal. Nowhere is evidence that the result of this experiment makes consistent profits but suggest an unbiased rules chosen by an evolutionary algorithm to technical traders or trading systems. It demonstrates how PSO can be applied to find technical trading rules in comparison to a GA. The results show that PSO is better suited to GA from an efficiency of view.

Section 2 describes PSO algorithms. Section 3 shows how the rules are found and evaluated, and addresses the robustness of the results and Section 4 Computational results obtained with the KOSPI200 are reported. Section 5 contains concluding remarks.

2   Overviews

2.1   GA Overview

GA is an algorithm used to find approximate solutions to difficult-to-solve problems, inspired by and named after biological processes of inheritance, mutation, natural selection, and the genetic crossover that occurs when parents mate to produce offspring. It is a particular class of evolutionary algorithms in which a population of abstract representations of candidate solutions to an optimization problem are stochastically selected, recombined, mutated, and then either eliminated or retained, based on their relative fitness.

John Holland [17] was the pioneering founder of much of today's work in genetic algorithms, which has moved on from a purely theoretical subject (though based on computer modeling), to provide methods which can be used to solve some difficult problems today.

The structure of a general GA is shown in Fig. 1. The problem to be solved is represented by a list of parameters which can be used to drive an evaluation procedure.

Possible solutions to the problem (referred to as chromosomes) are also represented in this parameter list form. These chromosomes are evaluated, and a value of goodness or fitness is returned. The next step of the algorithm is to generate a second generation pool of parameter lists, which is done using the genetic operator selection, crossover, and mutation. These processes result in a second generation pool of chromosomes that is different from the initial generation, which is then evaluated and the fitness values for each list is obtained. Generally the average degree of fitness will have increased by this procedure for

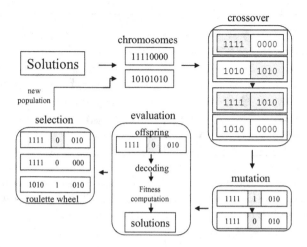

Fig. 1. The general structure of GA

the second generation pool. The process continues by generating third, fourth, fifth,... (and so on) generations, until one of the generations contains solutions which are good enough.

2.2   PSO Overview

PSO has been shown to be effective in optimizing difficult multidimensional discontinuous problems in a variety of fields [3]. Developed in 1995 by Kennedy and Eberhart [1], the PSO can best be understood through an analogy similar to the one that led to the development of the PSO. Robinson [5] explained the PSO with a swarm of bees in a field. Their goal is to find the location in the field with the highest density of flowers. Without any knowledge of a priori, the bees are attracted both to the best location (group best) of highest concentration found by the entire swarm, and the best location (personal). Eventually, after being attracted to areas of high flower concentration, all the bees swarm around the best location (group best) Fig 2.

The following equations are utilized, in computing the positions and velocities:

$$\mathbf{v}_n = w * \mathbf{v}_n + c_1 * rand() * (p_{best,n} - \mathbf{x}_n) + c_2 * rand() * (g_{best,n} - \mathbf{x}_n) \quad (1)$$

$$\mathbf{x}_n = \mathbf{x}_n + \mathbf{v}_n \quad (2)$$

where, $\mathbf{v}_n$ is the velocity of the particle in the nth dimension and $\mathbf{x}_n$ is the particle's coordinate in the nth dimension, $p_{best,n}$ is the best location of individual, $g_{best,n}$ is the best location among the individual (group best). W is the weight factor, c_1, c_2 are the positive constants having values, rand() is the random numbers selected between {0,1}.

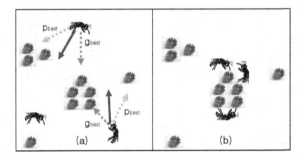

Fig. 2. a) Bees are attracted both to the best location (group best) by the entire swarm and the best location (personal). (b) Eventually, all the bees swarm around the best location (group best)

3   Specification of Trading Rules

3.1   Decision of the Investment Position

Financial market traders use trading rules which can be based on either technical (short term, normally several times trading per year) or fundamental (relatively long term, normally one trading during a few years) analysis to assist them in determining their investments. The trading rules considered in this study is based on a simple market timing strategy, consisting of investing total funds in either the stock market or a risk free security. If stock market prices are expected to increase on the basis of a buy signal from a technical trading rule, then the risk free security is sold and stocks are bought until a sell signal returns.

A technical indicator which includes in moving averages, channels, filters, momentum, is a mathematical formula that transforms historical data on price and /or volume or each other to form trading rules. This paper considered in moving average rules to enhance the performance of PSO.

3.2   Moving Average Rules

Moving averages are used to identify trends in prices. A moving average (MA) is simply an average of current and past prices over a specified period of time (starting from the current time). We define an MA of length θ as follows:

$$MA_t(\theta) = \frac{1}{\theta} \sum_{i=0}^{\theta-1} P_{t-i} \tag{3}$$

$$F(\Theta)_t = \begin{cases} 0 \text{ if } MA_t(\theta_1) - MA_t(\theta_2) > 0 \\ 1 \text{ else.} \end{cases} \tag{4}$$

where $\Theta = \theta_1, \theta_2, \forall \theta_1, \theta_2 \in \{1, 2, 3, ...\}, \theta_1 < \theta_2$

The periodicity of the two MAs have a range defined by $1 < \theta_1 < L$ and $2 < \theta_2 < L$ where L represents the maximum length of the moving average (For this study L $= 250$ days which is a common one year except closed dates).

If $F(\Theta)_1 = 1$ (a shorter term MA crosses a longer term MA), the trading rule returns 'buy' signal at that price and hold until $F(\Theta)_1 = 0$ (a shorter term MA down-crosses a longer term MA). This simple rule can find decision rules that divide days into two distinct categories, either 'in' the market (earning the market rate of return) or 'out' of the market (earning the risk-free rate of return). A trading rule is a function that returns either a 'buy' or a 'sell' signal for any given price history. The trading strategy specifies the position to be taken the following day, given the current position and the trading rule signal. If the current state is 'in' (i.e., holding the market) and the trading rule signals 'sell', switch to 'out' (move out of the market); if the current state is 'out' and the trading rule signals 'buy', switch back to 'in'. In the other two cases, the current state is preserved.

4   Methodology

4.1   Objective Function

Trading rules as defined by the indicator functions given in Equations (4), return either a buy or sell signal. These signals can be used to divide the total number of trading days (N), into days either 'in' the market (earning the market rate of return r_{mt}) or 'out' of the market(earning the risk-free rate of return r_{ft}). Thus the trading rule returns over the entire period of 0 to N can be calculated as

$$r_t = \sum_{i=1}^{N} F(\Theta)_t r_{mt} + \sum_{i=1}^{N}(1 - F(\Theta)_t)r_{ft} - T(tc) \tag{5}$$

where $r_{mt} = \ln(\frac{P_t}{P_{t-1}})$

which includes the summation of the daily market returns for days 'in' the market and the daily returns on the risk free security for days 'out' of the market. An adjustment for transaction cost is given by the last term on the right hand side of Equation (5) which consists of the product of the cost per transaction (tc) and the number of transaction (T). Transaction costs of 0.2 percent per trade are considered for the in-sample optimization of the trading rules.

4.2   Implementation of PSO

Discrete Binary Version
The revision of the particle swarm algorithm from a continuous to a discrete operation may be more fundamental than the simple coding changes would imply. The floating-point algorithm operated through a stochastic change in the rate of change of position. The original $x_n = x_n + v_n$(Equation 2) seems very different from

$$IF(rand() < S(\mathbf{v}_n)), then x_n = 1; else x_n = 0 \tag{6}$$

where the function is a sigmoid limiting transformation and rand() is a quasi-random number selected from a uniform distribution in $\{0, 1\}$ [18].

Parameter Settings

The PSO algorithm has four parameters settings $\{W, c_1, c_2, V_{max}\}$, defined as: V_{max} = Maximum velocities on each dimension, W = Weight factor, c_1 = Constant 1, c_2 = Constant 2.

Without any boundaries or limits on the velocity, particles could essentially search of the physically meaningful solution space. To solve this problem, Eberhart [14] considered a maximum allowed velocity called V_{max}. V_{max} is as a constraint to control the global exploration ability of a particle swarm. If V_{max} is too large, particles might fly past good solutions. On the other hand, if V_{max} is too small, particles may not fly beyond locally good solutions. V_{max} was best set around 10-20% of the dynamic range of the variable on each dimension and we set $V_{max} = 6$ which means that the probabilities is limited to $S((v)_n)$, between 0.9975 and 0.0025.

The concept of a weight factor was developed to better control exploration and exploitation. Eberhart [14] found out that suitable selection of the weight factor provides a balance between global and local exploration and exploitation and encourage form 0 to 1.

The acceleration constants c_1 and c_2 represent the weighting of the stochastic acceleration terms that pull each particle toward p_{best} and g_{best} positions. Thus, adjustment of these constants changes the amount of 'tension' in the system. Low values allow particles to roam far from target regions before being tugged back, while high values result in abrupt movement toward, or past, target regions. Early experience with particle swarm optimization (trial and error, mostly) led us to set the acceleration constants c_1 and c_2 each equal to 2.0 for almost all applications [19].

4.3   Implementation of GA

Parameter Settings

The GA has four parameters settings $\{b, p, c, m\}$, defined as: b = Number of elements in each vector, p = Number of vectors or candidates in the population, c = Probability associated with the occurrence of crossover, m = Probability associated with the occurrence of mutation.

In order to use a GA to search for the optimal parameter values for the rules considered above, potential solutions to this optimization problem are represented using vectors of binary. In order to satisfy the limiting values given in equations (3), (4), the binary representations for θ_1 and θ_2 are each given by a vector consisting of eight elements. Therefore, the binary representation can be defined by a row vector consisting of twenty three elements. Tests were implemented a generational GA Model with a ranking-based procedure developed by [20], one point crossover operator equal to 0.6, a mutation operator equal to 0.05 [12].

5   Computational Results

5.1   Data and Parameters

For the study, we used daily data for the Korea Composite Stock Price Index (KOSPI) 200 from 2 September 1994 to 31 August 2004. Call rate in Bank of KOREA was chosen to risk-free rates during those terms. Transaction Costs (tc) of 0.2 percent per trade were considered for the optimization of the trading rules. Table 1 displays the data and parameter values that are used by the PSO and GA for the trading rules.

Table 1. Data and parameters

a) PSO parameters

P_t	r_f	tc	V_{max}	W	C_1	C_2
KOSPI200	Call rate	0.002	6	0	2	2

b) GA parameters

P_t	r_f	tc	b	p	c	m
KOSPI200	Call rate	0.002	23	10-300	0.6	0.05

5.2   Numerical Results

Both algorithms for the computational tests were implemented in C programming language on an Intel 2.4GHz Pentium 4, 512MB PC. To measure the effectiveness and efficiency of PSO, results were compared with general GA. For each algorithm, the results were averages of ten test trials. Figure 3 shows the chart of returns by changing the population size and iteration number. The fitness of PSO is better as compared with GA. For example, the average of 10 returns via PSO with 50 pop sizes and 50 iterations is 144.0363%, but the same condition of GA is only 122.7074%.

Both algorithms for the computational tests were implemented in C programming language on an Intel 2.4GHz Pentium 4, 512MB PC. To measure the effectiveness and efficiency of PSO, results were compared with general GA. For

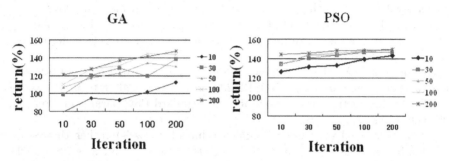

Fig. 3. Average returns of 10 trials by changing the population size and iteration from 10 to 200

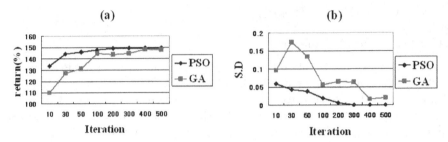

Fig. 4. (a) Returns comparison between PSO and GA by changing the iteration from 10 to 500 with fixed population size at 100. (b) standard deviation comparison between PSO and GA by changing the iteration from 10 to 500 with fixed population size at 100

each algorithm, the results were averages of ten test trials. Figure 3 shows the chart of returns by changing the population size and iteration number. The fitness of PSO is better as compared with GA. For example, the average of 10 returns via PSO with 50 pop sizes and 50 iterations is 144.0363%, but the same condition of GA is only 122.7074%.

Figure 4 shows the average returns and standard deviations when the pop size is fixed to 100 and varying the Iterations from 10 to 500. The standard deviations of PSO with 300, 400, and 500 Iterations are equal to zero (we can implicitly conclude that PSO must be reach the global maximum value). On the other hand, the standard deviations of GA are remained as 0.062, 0.015 and 0.019.

6   Conclusion

This paper is the first study that demonstrates how PSO can be applied to find technical trading rules and shows the effectiveness and efficiency compare with GA. PSO can reach the global optimal value with less iteration and keep equilibrium versus GA.

In this paper, PSO also shows the possibility that it can solve the complicated problems without using the crossover, mutation, and other manipulations as GA but using only basic equations. PSO will be used to explore large numbers of different possibilities and can be easily extended to other areas that need many fundamental variables.

We consider MA rules to enhance the performance of revised PSO (discrete binary version) and used the KOSPI 200 daily data, Call rate in BANK of KOREA to demonstrate the superiority and standard PSO parameters to avoid a potential bias.

We have focused on market timing (technical analysis) rather than stock selection in our tests, now is a good time to think some about how PSO might be used to facilitate economics from a fundamentalist perspective because PSO is able to explore large numbers of different possibilities.

The particle swarm optimizer shares the ability of the genetic algorithm to handle arbitrary nonlinear cost functions, but with a much simpler implementation it clearly demonstrates good possibilities for widespread use in Finance.

References

1. James Kennedy, Russell Eberhart, "Particle Swarm Optimization" *IEEE* (1995)
2. Daniel W. Boeringer, Douglas H., Werner, "Particle Swarm Optimization Versus Genetic Algorithms for Phased Array Synthesis", *IEEE Transaction on Antennas and Propagation*, Vol, 52, No.3, (2004).
3. R. C. Eberhart, Y. Shi, "Evolving artificial neural networks," in Proc. *Int. Conf. Neural Networks and Brain*, Beijing, P.R.C., (1998).
4. Ayed Salman, Imtiaz Ahmad, Sabah Al-Madani, "Particle swarm optimization for task assignment problem", *Microprocessors and Microsystems* 26 (2002) pp.363~371.
5. Jacob Robinson, Yahya Rahmat-Samii, "Particle Swarm Optimization in Electormagnetics", *IEEE Transactions on Antennas and Propagation*, Vol, 52, No. 2, (2004).
6. P.C. Fourie and A.A. Groenwold, "The particle swarm optimization algorithm in size and shape optimization", *Struct Multidisc Optim* 23, (2002), pp.259~267.
7. M.A. abido, "Optimal Design of Power–System Stabilizers Using Particle Swarm Optimization", *IEEE Transactions on Energy Conversion*, Vol. 17, No. 3, (2002).
8. Cla'udia O. Ourique, Evaristo C. Biscaia, Jr, Jose' Carlos Pinto, "The use of particle swarm optimization for dynamical analysis in chemical processes", *Computers and Chemical Engineering* 26 (2002) pp.1783~1793.
9. Franklin Allen, Risto Karjalainen "Using genetic algorithms to find technical trading rules", *Journal of Financial Economics*, 51 (1999), pp. 245~271.
10. Bruce K. Grace, "Black&Scholes option pricing via genetic algorithms", *Applied Economics Letters*, (2000), pp. 129~132.
11. M A H Dempster and C M Jones, "A real-time adaptive trading system using genetic programming", *Quantitative Finance Volume*, 1, (2001), pp. 397–413.
12. Jean-Yves Potvina, Patrick Sorianoa, Maxime Vall, "Generating trading rules on the stock markets with genetic programming", *Computers & Operations Research*, 31, (2004) pp. 1033–1047.
13. Alexander, S.S., "Price movements in speculative markets: trends or random walks". In: Cootner, P. (Ed.), *The Random Character of Stock Market Prices*, vol. 2, MIT Press, Cambridge, (1964) pp. 338~372.
14. Fama, E.F., Blume, M.E., "Filter rules and stock market trading. Security prices: a supplement", *Journal of Business* 39 (1966), 226~241.
15. Fama, E.F., "Efficient capital markets: a review of theory and empirical work". *Journal of Finance* 25, (1970) 383~417.
16. Brock, W., Lakonishok, J., LeBaron, B., "Simple technical trading rules and the stochastic properties of stock returns", *Journal of Finance* 47, (1992) 1731~1764.
17. Holland, J.H., "adaptation in Natural and Artificial Systems", University of Michigan Press (1975).
18. James Kennedy, Russell C. Eberhart, "A discrete binary version of the particle swarm algorithm" , *IEEE*, (1997).

19. Russell C. Eberhart, Yuhui Shi, "Particle Swarm Optimization : Development, Applications and Resources", *IEEE*, (2002).
20. Whitley D. "The GENITOR Algorithm and Selection Pressure: Why Rank-based Allocation of Reproductive Trials is Best". In: Schaffer D.J (Ed.) Proceddings of the Third International Conference on Genetic Algorithms, Morgan Jaufmann, San Mateo, (1989), pp.116-121.

Nonlinear Robust Identification Using Multiobjective Evolutionary Algorithms

J.M. Herrero, X. Blasco, M. Martínez, and C. Ramos

Predictive Control and Heuristic Optimization Group,
Department of Systems Engineering and Control,
Polytechnic University of Valencia
juaherdu@isa.upv.es, http://ctl-predictivo.upv.es

Abstract. In this article, a procedure to estimate a nonlinear models set (Θ_p) in a robust identification context, is presented. The estimated models are Pareto optimal when several identification error norms are considered simultaneously. A new multiobjective evolutionary algorithm $\epsilon-MOEA$ has been designed to converge towards Θ_P^*, a reduced but well distributed representation of Θ_P since the algorithm achieves good convergence and distribution of the Pareto front $J(\Theta)$. Finally, an experimental application of the $\epsilon-MOEA$ algorithm to the nonlinear robust identification of a scale furnace is presented. The model has three unknown parameters and ℓ_∞ and ℓ_1 norms are been taken into account.

1 Introduction

When modelling dynamic systems by first principles, the problem of parametric uncertainty always appears. Model parameters are never exactly known and system identification can be used in order to identify their values from measured input-output data. Uncertainty can be caused mainly by measurement noise and model error (e.g. unmodelled dynamics) and it always appears as an error between model and process outputs (identification error for a specific experiment).

Well established identification techniques exist for linear models [6], [10], but not for nonlinear ones. Most of these techniques rely on cost function optimization where an error norm is considered. In particular, if the cost function is convex, local optimizers offer the solution, but in the nonconvex case, global optimization becomes necessary.

In general, nonlinear models produce nonconvex optimization problems, and in this case, Evolutionary Algorithms [3], [5] offer a good solution. Furthermore, considering several norms of the identification error simultaneously (e.g. ℓ_∞, ℓ_1, ℓ_2, $Fair$, $Huber$, $Turkey$), the quality of the estimated models can be improved [8]. Thus, identification of nonlinear processes is stated as a multiobjective optimization problem. To solve this problem, a Multiobjective Evolutionary Algorithm ($\epsilon-MOEA$), based on ϵ-dominance concept [1], [7], has been developed. The algorithm converges to a well distributed sample of the Pareto Front.

The paper is organized as follows. In section 2, robust identification problem is posed when different error norms are considered simultaneously. Section 3 des-

J. Mira and J.R. Álvarez (Eds.): IWINAC 2005, LNCS 3562, pp. 231–241, 2005.

cribes the $\mathscr{e}$–$MOEA$ algorithms to find the Pareto optimal set. Finally, section 4 presents experimental results when $\mathscr{e}$–$MOEA$ is applied to the identification of a thermal process represented by a nonlinear model with three unknown parameters.

2   Robust Identification Approach

In this work the following structure is assumed for the nonlinear model:

$$\dot{\mathbf{x}}(t) = f(\mathbf{x}(t), \mathbf{u}(t), \theta), \quad \hat{\mathbf{y}}(t) = g(\mathbf{x}(t), \mathbf{u}(t), \theta) \tag{1}$$

where

- $f(.), g(.)$ are the nonlinear functions of the model,
- $\theta \in D \subset R^q$ is the vector of unknown model parameters,
- $\mathbf{x}(t) \in R^n$ is the vector of model states,
- $\mathbf{u}(t) \in R^m$ is the vector of model inputs,
- $\hat{\mathbf{y}}(t) \in R^l$ is the vector of model outputs.

Let $\mathbf{E}(\theta) = \mathbf{Y} - \hat{\mathbf{Y}}(\theta)$, where

- $\mathbf{E}(\theta)$ is the identification error,
- $\mathbf{Y}$ are the process output measurements , $[\mathbf{y}(0), \mathbf{y}(T)...\mathbf{y}(NT)]$, when the inputs $\mathbf{U} = [\mathbf{u}(0), \mathbf{u}(T)...\mathbf{u}(NT)]$ are applied to the process,
- $\hat{\mathbf{Y}}$: are the simulated[1] outputs $[\hat{\mathbf{y}}(0), \hat{\mathbf{y}}(T)...\hat{\mathbf{y}}(NT)]$, when the same inputs $\mathbf{U}$ are applied to the model.

Denote $\|\mathbf{E}(\theta)\|_{p_i}$ as the p_i-norm of the identification error, with $i \in A :=$ $[1, 2 \ldots s]$. When several norms of the identification error $\mathbf{E}(\theta)$ are considered simultaneously, model identification can be posed as a multiobjective optimization problem

$$\min_{\theta \in D} \mathbf{J}(\theta) \tag{2}$$

where

$$\mathbf{J}(\theta) = \{J_1(\theta), J_2(\theta), \ldots, J_s(\theta)\} = \{\|\mathbf{E}(\theta)\|_{p1}, \|\mathbf{E}(\theta)\|_{p2}, \ldots, \|\mathbf{E}(\theta)\|_{ps}\} .$$

Consequently, there is not a unique optimal model and a Pareto optimal set Θ_P (solutions where no-one dominates others) must be found. Pareto dominance is defined as follows.

A model θ_1, with cost function value $\mathbf{J}(\theta_1)$ dominates another model θ_2 with cost function value $\mathbf{J}(\theta_2)$, denoted by $\mathbf{J}(\theta_1) \prec \mathbf{J}(\theta_2)$, iff

$$\forall i \in A, J_i(\theta_1) \leq J_i(\theta_2) \wedge \exists k \in A : J_k(\theta_1) < J_k(\theta_2) .$$

---

[1] Model outputs are calculated integrating equations (1). T is the sample time and N is the number of measurements.

Therefore the Pareto optimal set $\mathbf{\Theta}_P$, is given by

$$\mathbf{\Theta}_P = \{\theta \in D \mid \nexists \; \tilde{\theta} \in D \; : \; \mathbf{J}(\tilde{\theta}) \prec \mathbf{J}(\theta)\} \; . \tag{3}$$

$\mathbf{\Theta}_P$ is unique and normally includes infinite models. Hence a set $\mathbf{\Theta}_P^*$, with a finite number of elements from $\mathbf{\Theta}_P$, will be obtained[2].

3 $\mathscr{E}-MOEA$

The variable ϵ MOEA ($\mathscr{E}-MOEA$) is a multiobjective evolutionary algorithm based on ϵ-dominance concept.

Consider the cost function space splitted up in a fixed number of boxes (for each dimension, n_box_i cells of ϵ_i wide)[3]. That grid preserves diversity of $J(\mathbf{\Theta}_P^*)$ since one box can be occupied by just one solution. This fact avoids the algorithm converging to just one point or area inside the cost function space (Fig. 1).

The concept of ϵ-dominance is defined as follows. For a model θ, $box_i(\theta)$ is defined by[4]

$$box_i(\theta) = ceil\left(\frac{J_i(\theta) - J_i^{min}}{\epsilon_i}\right) \; .$$

Let $\mathbf{box}(\mathbf{J}(\theta)) = \{box_1(\theta), \ldots, box_s(\theta)\}$. A model θ_1 with cost function value $\mathbf{J}(\theta_1)$ ϵ-dominates the model θ_2 with cost function value $\mathbf{J}(\theta_2)$, denoted by $\mathbf{J}(\theta_1) \prec_\epsilon \mathbf{J}(\theta_2)$, iff

$$\mathbf{box}(\theta_1) \prec \mathbf{box}(\theta_2) \vee (\mathbf{box}(\theta_1) = \mathbf{box}(\theta_2) \wedge \mathbf{J}(\theta_1) \prec \mathbf{J}(\theta_2)) \; .$$

Hence, a set $\mathbf{\Theta}_P^*$ is ϵ-Pareto iff

$$\mathbf{\Theta}_P^* \subseteq \mathbf{\Theta}_P \wedge (\mathbf{box}(\theta_1) \neq \mathbf{box}(\mathbf{J}(\theta_2))), \;\; \forall \theta_1, \theta_2 \in \mathbf{\Theta}_P^*, \;\; \theta_1 \neq \theta_2 \; . \tag{4}$$

$\mathscr{E}-MOEA$ algorithm obtains the ϵ-Pareto front $J(\mathbf{\Theta}_P^*)$, a well-distributed sample of the Pareto front $J(\mathbf{\Theta}_P)$. The algorithm, which adjusts the width ϵ dynamically, is composed of three populations (see Fig. 2).

1. Main population $P(t)$ explores the searching space D during the algorithm iterations (t). Population size is $Nind_P$.
2. Archive $A(t)$ stores the solution $(\mathbf{\Theta}_P^*)$. Its size $Nind_A$ can be variable and it will never be higher than $Nind\_max\_A$ which depends on the number of the boxes $(\mathbf{n\_box})$.
3. Auxiliary population $G(t)$. Its size is $Nind_G$, which must be an even number.

The pseudocode of the $\mathscr{E}-MOEA$ algorithm is given by

---

[2] Notice that Θ_P^* is not unique.
[3] $\epsilon_i = (J_i^{max} - J_i^{min})/n\_box_i$.
[4] The function $ceil(x)$ returns the smallest integer greater or equal than x.

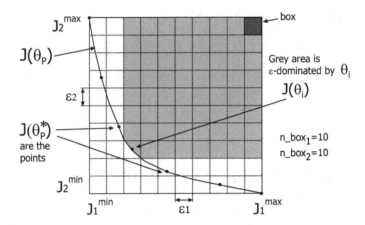

Fig. 1. ϵ-dominance concept. ϵ-Pareto Front $J(\mathbf{\Theta}_P^*)$ in a bidimensional problem

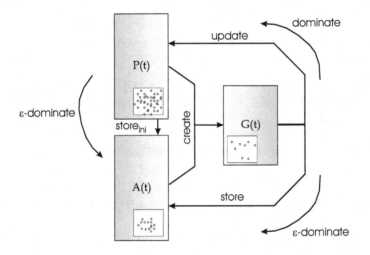

Fig. 2. $\not{\epsilon}-MOEA$ algorithm structure

1. $t := 0$
2. $A(t) := \emptyset$
3. $P(t) := ini\_random(D)$
4. $eval(P(t))$
5. $A(t) := store_{ini}(P(t), A(t))$
6. $while \ t < n\_iterations \ do$
7. $G(t) := create(P(t), A(t))$
8. $eval(G(t))$
9. $A(t+1) := store(G(t), A(t))$
10. $P(t+1) := update(G(t), P(t))$
11. $t := t+1$
12. $end \ while$

Main steps of the algorithm are detailed:

Step 3. $P(0)$ is initialized with $Nind_P$ individuals (models) randomly selected from the searching space D.

Step 4 and 8. Function **eval** calculates cost function value (2) for each individual in $P(t)$ (step 4) and $G(t)$ (step 8).

Step 5. Function **store**$_{ini}$ checks individuals of $P(t)$ that might be included in the archive $A(t)$ as follows:

1. $P(t)$ individuals non-dominated are detected, θ_{ND}.
2. Cost function space limits are calculated from $J(\theta_{ND})$.
3. Individuals of θ_{ND} are analyzed, one by one, and those not ϵ-dominated by individuals of $A(t)$, will be included in $A(t)$.

Step 7. Each iteration, function **create** creates $G(t)$ as follows:

1. Two individuals are randomly selected, one from $P(t)$, θ_1, and another from $A(t)$, θ_2.
2. θ_1 and θ_2 are crossed over by extended lineal recombination technique creating two new individuals θ_1' and θ_2'.
3. θ_1' and θ_2' are included in $G(t)$.

This procedure is repeated $Nind_G/2$ times until $G(t)$ will be filled up.

Step 9. Function **store** checks, one by one, which individuals of $G(t)$ must be included in $A(t)$ based on their location in the cost function space (see Fig. 3). Thus $\forall \theta_G \in G(t)$

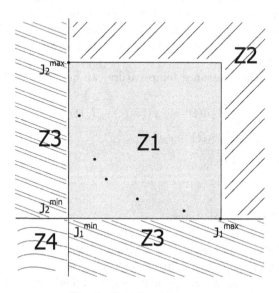

Fig. 3. Cost function space

1. If θ_G lies on the area $Z1$ and is not ϵ-dominated by any individual of $A(t)$, it will be included in $A(t)$[5]. Individuals of $A(t)$ which are ϵ-dominated by θ_G will be eliminated.
2. If θ_G lies on the area $Z2$ then it is not included in the archive, since it is dominated by all individuals of $A(t)$.
3. If θ_G lies on the area $Z3$, new cost function limits and ϵ_i widths are recalculated. One by one the individuals of $A(t)$ are again analyzed using the same procedure as **store**$_{ini}$ function. Finally, θ_G is included in $A(t)$.
4. If θ_G lies on the area $Z4$, all individuals of $A(t)$ are deleted since all of them are ϵ-dominated by θ_G. θ_G is included and cost function space limits are $\mathbf{J}(\theta_G)$.

Step 10. Function **update** updates $P(t)$ with individuals from $G(t)$. Every individual θ_G from $G(t)$ is randomly compared with an individual θ_P from $P(t)$. If $\mathbf{J}(\theta_G) \prec \mathbf{J}(\theta)$ then θ_G replaces θ, on the other hand, θ is maintained.

Finally, individuals from $A(t)$ compound the solution Θ_P^* to the multiobjective minimization problem.

4   Robust Identification Example

Consider a scale furnace with a resistance placed inside. A fan continuously introduces air from outside (air circulation) while energy is supplied by an actuator controlled by voltage. Using a data acquisition system, resistance temperature and air temperature are measured when voltage is applied to the process.

Fig. 4 shows the input signal applied and the output signal measured for an experiment of length $N = 6000$. These signals will be used for the robust identification problem.

The dynamics of the resistance temperature can be modelled by

$$
\begin{aligned}
\dot{x}_1(t) &= \frac{1}{1000}\left(k_1 u(t)^2 - k_2\left((x_1(t) - T_a(t))\right)\right) + O_{ffset} , \\
\dot{x}_2(t) &= (1/k_3)(x_1(t) - x_2(t)) , \\
\hat{y}(t) &= x_2(t) ,
\end{aligned}
\tag{5}
$$

where

 - $\dot{x}_1(t), \dot{x}_2(t)$ are the model states,
 - $u(t)$ is the input voltage with rank 0 - 100 (%),
 - $\hat{y}(t)$ is the resistance temperature ($^\circ C$) (model output),
 - $T_a(t)$ is the temperature inside furnace ($^\circ C$),
 - $\theta = [k_1, k_2, k_3]$ are the model parameters to be identified,

---

[5] When the individual is not ϵ-dominated and its box is occupied, that individual lying farthest from the box centre will be eliminated.

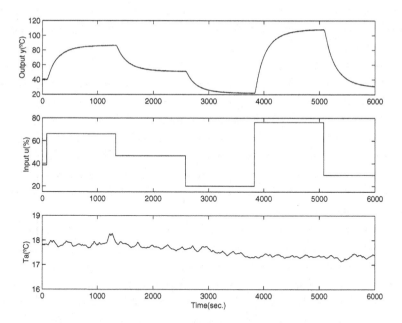

Fig. 4. Output process $y(t)$, input process $u(t)$ and disturbance $T_a(t)$. $T = 1 sec.$ (sample time), $N = 6000$ (experiment length)

- O_{ffset} is the correction to ensure zero steady-state error at a particular operating point [6].

To determine Θ_P^*, two norms $p_1 = \ell_\infty$ and $p_2 = \ell_1$ have been considered, to limit maximum and average identification error respectively. Therefore two cost functions have to be minimized to solve the problem stated at (2)

$$J_1(\theta) = \|\mathbf{E}(\theta)\|_\infty , \tag{6}$$

$$J_2(\theta) = \frac{\|\mathbf{E}(\theta)\|_{\ell_1}}{N} . \tag{7}$$

The parameters of the $\mathcal{E}-MOEA$ algorithm are set to:

- The searching space D is $k_1 \in [0.05 \ldots 0.12]$, $k_2 \in [3.0 \ldots 8.0]$ and $k_3 \in [1.0 \ldots 25.0]$.
- $Nind_P = 8000$, $Nind_G = 8$, and $n\_iterations = 3000$.
- Cost funcion space parameter **n_box** $= [10, 10]$ (number of divisions for each dimension).

At the end of the optimization process, the solution Θ_P^* is $A(3000)$, which contains six optimal models. Fig. 5 shows Θ_P^* with its projections on the planes (k_1, k_3), (k_2, k_3) and (k_2, k_1) respectively.

---

[6] Before applying input (see Fig. 4) the process is in steady-state. Assuming that $\dot{x}_1(0) = \dot{x}_2(0) = 0$ and $O_{ffset} = -\frac{1}{1000}(k_1 u(0)^2 - k_2(x_1(0) - T_a(0)))$, is not necessary to identify $x_1(0)$ and $x_2(0)$ since $x_1(0) = x_2(0) = y(0)$.

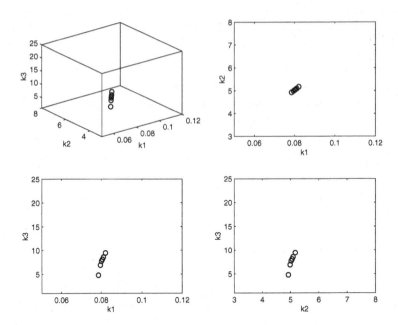

Fig. 5. The Pareto optimal set $\boldsymbol{\Theta}_P^*$ and its projections on (k_1, k_3), (k_2, k_3) and (k_2, k_1) planes respectively

Fig. 6 shows associated Pareto front $J(\boldsymbol{\Theta}_P^*)$. Notice that the algorithm has achieved an appropriate distribution of the Pareto front[7] due to the gridding of the cost function space. The ε–$MOEA$ algorithm succeeds in characterizing the front despite its short length and the differences among models.

The Pareto optimal set $\boldsymbol{\Theta}_P^*$ is a sample of non-dominated solutions well distributed over the Pareto front $J(\boldsymbol{\Theta}_P)$. Since the final objective is to obtain a unique model (θ^*), a way to select it has to be established[8]. In this work, the proposal is based on the minimum distance to an ideal value cost function space.

The ideal value is obtained from the extremes of the sampled Pareto front $J(\boldsymbol{\Theta}_P^*)$:

$$\theta_{\ell_1} = arg \min_{\theta \in D} J_1(\theta) = [0.078, 4.92, 4.77]; \ J_1(\theta_{\ell_1}) = \mathbf{0.42}, \ J_2(\theta_{\ell_1}) = 2.9 \ ,$$

$$\theta_\infty = arg \min_{\theta \in D} J_2(\theta) = [0.082, 5.17, 9.44]; \ J_1(\theta_\infty) = 0.480, \ J_2(\theta_\infty) = \mathbf{1.72} \ .$$

Hence,

$$\mathbf{J}_{ideal} = (J_1^{min}, J_2^{min}) = (0.42, 1.72) \ .$$

---

[7] If a better characterization of Pareto front is required, **n_box** parameter should be increased.

[8] In multiobjective literature this is not a trivial issue and there is several methodologies regarding it.

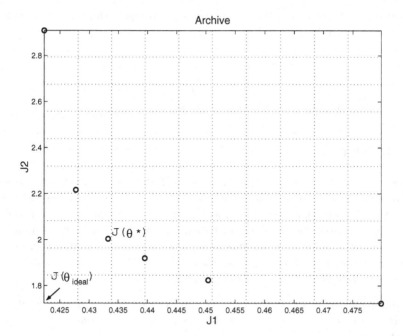

Fig. 6. The Pareto front $J(\Theta_P^*)$

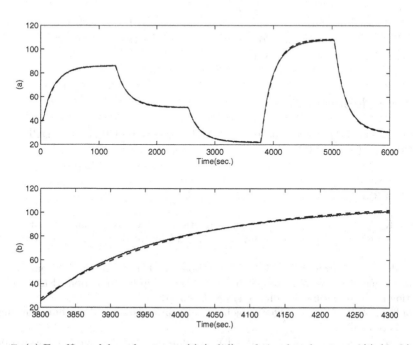

Fig. 7. (a) For θ^* model, real output $y(t)$ (solid) and simulated output $\hat{y}(t)$ (dash). (b) Notice differences between both outputs

The proposed solution θ^* is the nearest model (in the cost function space) to the ideal model (see Fig. 6):

$$\theta^* = [0.08, 5, 7.7]; \; J_1(\theta^*) = 0.43, \; J_2(\theta^*) = 2 \;.$$

The estimated nonlinear model θ^* can be validated by comparing its simulated output against measured output data (see Fig. 7).

5   Conclusions

A multiobjective evolutionary algorithm $\varepsilon\!-\!MOEA$, based on ϵ-dominance concept, has been developed for robust identification of nonlinear processes. Consequently, robust identification is posed as a multiobjective optimization problem and $\varepsilon\!-\!MOEA$ estimates the nonlinear model set Θ_P^* by assuming, simultaneously, the existence of several bounds in identification error. $J(\Theta_P^*)$ results in a well distributed sample of the optimal Pareto front $J(\Theta_P)$.

The algorithm presents the following features:

– Assuming parametric uncertainty, all kind of processes can be identified if their outputs can be calculated by model simulation. Differentiability respect to the unknown parameters is not necessary.
– The algorithm dynamically adjusts the Pareto front precision without increasing the archive size $Nind\_max\_A$.
– The algorithm adapts the extremes of the Pareto front with independence of the parameter **n_box**. Consequently, good distribution and convergence to the Pareto front is achieved.

Acknowledgements

This work has been partially financed by AGL 2002-04108-C02-01 and DPI 2001-3106-C02-02 MCYT (SPAIN)

References

1. Deb, K., Mohan, M., Mishra, S.: A fast multi-objective evolutionary algorithm for finding well-spread pareto-optimal solutions. Technical Report 2003002 KanGAL (2003)
2. Garulli, A., Kacewicz, B., Vicino, A., Zappa, G.: Error Bounds for Conditional Algorithms in Restricted Complexity Set Membership Identification. IEEE Transaction on Automatic Control, **45**(1) (2000) 160-164
3. Goldberg, D.E.: Genetic Algorithms in search, optimization and machine learning. Addison-Wesley (1989)
4. Herrero, J.M., Blasco, X., Salcedo, J.V., Ramos, C.: Membership-Set Estimation with Genetic Algorithms in Nonlinear Models. In Proc. of the XV international Conference on Systems Science, (2004)

5. Blasco, X., Herrero, J.M., Martínez, M., Senent, J.: Nonlinear parametric model identification with Genetic Algorithms. Application to thermal process. Connectionist Models of Neurons, Learning Processes, and Artificial Intelligence (Springer). Lecture notes in computer science 2084 (2001)
6. Johansson, R.; System modeling identification. Prentice Hall (1993)
7. Laumanns, M., Thiele, L., Deb, K., Zitzler, E.: Combining convergence and diversity in evolutionary multi-objective optimization. Evolutionary computation, **10**(3) (2002)
8. Ram, B., Gupta, H. Bandyopadhyay, P., Deb, K., Adimurthy, V.: Robust Identification of Aerospace Propulsion Parameters using Optimization Techniques based on Evolutionary Algorithms. Technical Report 2003005 KanGAL (2003)
9. Reinelt, W., Garulli, A., Ljung, L.: Comparing different approaches to model error modelling in robust identification. Automatica, **38**(5) (2002) 787-803
10. Pronzalo, L., Walter, E.: Identification of parametric models from experimental data. Springer Verlang (1997)

Genetic Algorithms for Multiobjective Controller Design

M.A. Martínez, J. Sanchis, and X. Blasco

Predictive Control and Heuristic Optimization Group,
Department of Systems Engineering and Control,
Polytechnic University of Valencia
xblasco@isa.upv.es
http://ctl-predictivo.upv.es

Abstract. Multiobjective optimization strategy so-called *Physical Programming* allows controller designers a flexible way to express design preferences with a 'physical' sense. For each objective (settling time, overshoot, disturbance rejection, etc.) preferences are established through categories as *desirable, tolerable, unacceptable, etc.* assigned to numerical ranges. The problem is translated into a unique objective optimization but normally as a multimodal problem. This work shows how to convert a robust control design problem into a multiobjective optimization problem and to solve it by Physical Programming and Genetic Algorithms. An application to the American Control Conference (ACC) Robust Control Benchmark is presented and compared with other known solutions.

1 Introduction

Usually, controller design is done to fulfill a set of conflicting specifications. For instance, a robust controller for model variations is not compatible with high performance, the same appends if it is necessary a bounded control effort, etc. Therefore, the design of a controller can be understood as the search of the best tradeoff among all specifications, then a multiobjective optimization (MO) is a reasonable alternative.

The solution to a MO problem is normally not unique, the best solution for all objectives does not exist. There is a set of good solutions and it is said they are non-dominated solutions (none other is better for all objectives). Then it is defined the concept of Pareto Set (set of non-dominated solutions) and Pareto Front objectives values for Pareto Set solutions. Several techniques have been developed to obtain Pareto Set [10], [3]. The next step consists of the selection of a solution from Pareto Set, this is a subjective and non-trivial procedure that depends on designer preferences. Decision Maker (DM) algorithms are focused on helping designers in this task.

A traditional way to solve a MO problem (including DM) is to translate it into a one objective problem by means of a weighted sum of all objectives. Weights are usually ajusted by a trial and error procedure and it is difficult for the designers to translate their knowledge and preferences.

J. Mira and J.R. Álvarez (Eds.): IWINAC 2005, LNCS 3562, pp. 242–251, 2005.

To overcome this disadvantage Physical Programming methodology [7] sets specifications design in an understandable intuitive language for designers. Preferences for each objective (settling time, maximum control effort, etc.) are specified, in a flexible and natural way, by means of so-called Class Functions and range of preferences. All settings of MO problem and DM become more transparent for designer who only needs algorithms to compute objectives and to define ranges of preference for each objective. Notice that this new optimization problem could be multimodal then an adequate optimization technique has to be used, Genetic Algorithms supply good solutions for this situation.

This paper is organized as follows. Section 2 shows Physical Programming (PP) methodology and concepts associated to Class Function that allows designers express their preferences in a understandable 'physical' way. Section 3 describes the nonlinear optimization technique that has been used: Genetic Algorithms (GA). Section 4 describes the application problem selected to illustrate the benefits of the proposed methodology (PP+GA). In particular, the ACC Robust Control Benchmark is used. Finally, section 5 shows the results and compares them with other known results.

2   Physical Programming

Physical Programming (PP) is a methodology to solve a MO problem including information about preferences for each objective and converting the problem into a unique objective problem. To do it, PP introduces innovations on how to include designer knowledge and preference. In a first step, PP converts designer knowledge about the types of objectives and their desired values into previously established Class Functions. With this step all variables are normalized and preferences are included. Next step consists of aggregating all Class Functions in a unique function and using an optimization technique to solve this new problem. It is usual that the new problem was multimodal (several local minima), then optimization technique had to be powerful enough to solve it. GAs have demonstrated good performance in such type of problems and it will be the proposal for this article.

If $\mathbf{x}$ is the vector of parameters, designer has to supply a way to evaluate each specification by means of a function, $g_i(\mathbf{x})$, where subindex i is associated to each specification. For each function a Class Function $\bar{g}_i(g_i(\mathbf{x}))$ is defined. The shape of each Class Function $\bar{g}_i$ is a key point to express designer preferences.

These functions can be *hard* or *soft* depending on the type of objective involved in the problem. To convert all common situations the list of class to define could be:

Soft classes	Hard classes
– Class-1s: Smaller is better.	– Class-1h: Must be smaller.
– Class-2s: Larger is better.	– Class-2h: Must be larger.
– Class-3s: A value is better.	– Class-3h: Must be equal.
– Class-4s: A range is better.	– Class-4h: Must be in a range.

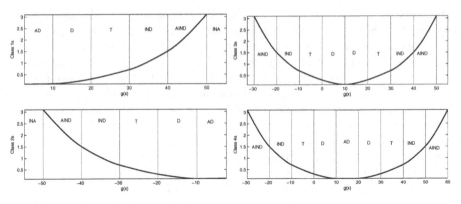

Fig. 1. Classes 1s, 2s, 3s and 4s

Once a Class Function is selected for an objective, the designer has to choose g_i^k values to establish preference ranges. For example, for a class-1s type g_i^k values are $g_i^1 \dots g_i^5$ and the associated ranges:

- *Highly Desirable (AD):* $g_i \leq g_i^1$
- *Desirable (D):* $g_i^1 \leq g_i \leq g_i^2$
- *Tolerable (T):* $g_i^2 \leq g_i \leq g_i^3$

- *Undesirable (IND):* $g_i^3 \leq g_i \leq g_i^4$
- *Highly undesirable (AIND):* $g_i^4 \leq g_i \leq g_i^5$
- *Unacceptable (INA):* $g_i \geq g_i^5$

These ranges are defined in physical units of the specification referred to, that means the designer expresses preferences in a natural and intuitive way.

Fig. 1 shows an example of different soft classes (all other soft classes are defined in a similar way) for different sets of designer range selection. In particular class-1s of the figure could correspond to an overshoot specification in a controller design problem:

- Highly Desirable : $\delta \leq 10\%$
- Desirable : $10\% \leq \delta \leq 20\%$
- Tolerable : $20\% \leq \delta \leq 30\%$

- Undesirable : $30\% \leq \delta \leq 40\%$
- Highly undesirable : $40\% \leq \delta \leq 50\%$
- Unacceptable: $\delta \geq 50\%$

Regardless of the soft class function type, g_i image has an adimensional and strictly positive value ($\overline{g}_i$). Images of g_i^k values, $\overline{g}^k$, have a key role, and are the same value for all soft Class Function (then it is not necessary to distinguish them with a subscript i). This characteristic produces a normalizing effect in $\overline{g}_i$ space for all g_i variables.

For *Hard* classes, only two values are set *possible* or *impossible* and they represent hard restrictions in a classical optimization problem.

Aggregated function combines all normalized objectives in a unique function. Then the problem is translated into a one-objective optimization problem. Physical Programming establishes the following Aggregated Function:

$$J(\mathbf{x}) = log_{10}\frac{1}{n_{sc} \cdot \overline{g}^5}\left[\sum_{i=1}^{n_{sc}}\overline{g}(g_i(\mathbf{x}))\right] \tag{1}$$

(n_{sc}) is the number of soft classes.

Minimization problem to solve is then:

$$\mathbf{x}^* = arg[\min_{\mathbf{x}} J(\mathbf{x})] \tag{2}$$

subject to Hard Class and $\mathbf{x}$ restrictions. Remember that $\mathbf{x}$ is the controller parameter vector and it could be subject to design restrictions.

Problem (2) is, in general, a nonlinear optimization problem that could be solved by several optimization techniques.

Logarithmic formulation of (1) tries to expand search range to obtain a reduction of iteration in search algorithm selected. Function (1) is based on addition of all soft classes but with the important characteristic that all of them weight the same. And, because of class function, all terms are independent of original specification showing a similar shape.

Relationships $\mathbf{x}$ with g_i (problem dependent) and these ones with $\bar{g}$ (class function dependent) do not allow to assure that equation (1) has only one minimum. With several specifications the problem is more complex and multiple minima presence is almost certain, then the quality of the solution is highly sensitive to the optimization technique used.

As it is said, presence of local minima in (1) depends on relationships $g_i(\mathbf{x})$ and $\bar{g}_i(g_i(\mathbf{x}))$. $g_i(\mathbf{x})$ is imposed by the specific problem but $\bar{g}_i(g_i)$ can be defined to satisfy several properties that make easier optimization algorithm work.

With these considerations, soft class functions have to be defined with following properties [7]: strictly positive, first derivative continuity and second derivative strictly positive.

Considering theses properties, it is possible to develop a method to build class function [6] based on spline segments.

3   Nonlinear Optimization with Genetic Algorithm

Minimization of the aggregated function subject to restrictions could be a nonlinear multimodal optimization problem. To solve it, it is necessary to use an adequate optimization technique. Genetic Algorithms (GAs) have demonstrated their capabilities in this type of situation and then it is the proposed option for this article.

A GA is an optimization technique that looks for the solution of the optimization problem, imitating species evolutionary mechanism [4], [5]. In these type of algorithms, a set of individuals (so-called population) changes generation by generation (evolution) adapting better to environment.

In an optimization problem, there is a function to optimize (cost function) and a zone where to look for (search space). Every point of the search space had an associated value of the function.

The different points of the search space are the different individuals of population. Similarly to natural genetic, every different individual is characterized by a chromosome, in the optimization problem, this chromosome is made by the point coordinates in the search space $\mathbf{x} = (x_1, x_2, \ldots, x_n)$. Following the simile, each coordinate corresponds to a gene.

The cost function value for an individual has to be understood as the adaptation level to the environment for such individual. For example, if it is a minimization problem for function $J(\mathbf{x})$, it is understood an individual is better adapted than another if it has a lower cost function value.

Evolutionary mechanism, that is, the rules for changing populations throughout generations is performed by Genetic Operators. A general GA evolution mechanism could be described as follows:

From an initial population (randomly generated), the next generation is obtained as:

1. Some individuals are selected for the next generation. This selection is made depending on adaptation level (cost function value). Such individuals with better $J(\mathbf{x})$ value have more possibilities to be selected (to survive).
2. To explore search space, an exchange of information between individuals is performed by means of crossover. That produces a gene exchange between chromosomes. The rate of individuals to crossover is fixed by P_c, crossover probability.
3. An additional search space exploration is performed by mutation. Some individuals of the new generation are subject to a random variation in their genes. The rate of individuals to be mutated is set by mutation probability P_m.

In this general framework, there are several variation in the GA implementation; different gene codification, different genetic operator implementation, etc. [4], [9].

Implementation for the present work has the following characteristics:

1. Real value codification [9], each gene has a real value.
2. $J(\mathbf{x})$ is not directly used as cost function. A 'ranking' operation is performed [1]. First individuals are sorted in decreasing $J(\mathbf{x})$ value and next, $J(\mathbf{x})$ is substituted by its position in such distribution, then each individual has a new cost function value $J'(\mathbf{x})$. Ranking operation prevents the algorithm from exhausting, it avoids clearly dominant individuals prevailing too soon.
3. Selection is made by the operator known as *Stochastic Universal Sampling (SUS)* [2]. Survival probability of an individual $(P(\mathbf{x}_i))$ is guaranteed to be:

$$P(\mathbf{x}_i) = \frac{J'(\mathbf{x}_i)}{\sum_{j=1}^{Nind} J'(\mathbf{x}_j)} \qquad (3)$$

where $Nind$ is the number of individuals.

4. For crossover it is used *intermediate recombination* operator [11]. Chromosomes sons ($\mathbf{x}_1'$ and $\mathbf{x}_2'$) are obtained through following operation on chromosomes fathers ($\mathbf{x}_1$ and $\mathbf{x}_2$):

$$\mathbf{x}_1' = \alpha_1 \cdot \mathbf{x}_1 + (1 - \alpha_1)\mathbf{x}_2 \; ; \; \alpha_1 \in [-d, 1 + d]$$
$$\mathbf{x}_2' = \alpha_2 \cdot \mathbf{x}_2 + (1 - \alpha_2)\mathbf{x}_1 \; ; \; \alpha_2 \in [-d, 1 + d] \qquad (4)$$

The operation could be done for the chromosome or for each gene separately. In this last case random parameters α_1 and α_2 have to be generated for each gene increasing search capabilities but with a higher computational cost. Implemented GA has been adjusted as follows: $\alpha_1 = \alpha_2$ and generated for each chromosome, $d = 0$ and $P_c = 0.8$.

5. Mutation operation is done with a probability $P_m = 0.1$ and a normal distribution with standard deviation set to 20% of search space range.

4 Robust Control Benchmark Solution

Wie and Bernstein [13] proposed a serie of problems for robust control where controller designer has to achieve a tradeoff between maximizing stability and robust performances of the system, and minimizing control effort.

Fig. 2 shows the process described in the benchmark. It consists of a flexible structure of two masses connected by a spring. State space model is:

$$\begin{bmatrix} \dot{x}_1 \\ \dot{x}_2 \\ \dot{x}_3 \\ \dot{x}_4 \end{bmatrix} = \begin{bmatrix} 0 & 0 & 1 & 0 \\ 0 & 0 & 0 & 1 \\ -k/m_1 & k/m_1 & 0 & 0 \\ k/m_2 & -k/m_2 & 0 & 0 \end{bmatrix} \begin{bmatrix} x_1 \\ x_2 \\ x_3 \\ x_4 \end{bmatrix} + \begin{bmatrix} 0 \\ 0 \\ 1/m_1 \\ 0 \end{bmatrix} u + \begin{bmatrix} 0 \\ 0 \\ 0 \\ 0 \end{bmatrix} w \tag{5}$$

$$y = x_2 + v$$

where x_1 and x_2 are positions, x_3 and x_4 are speeds for mass 1 and 2.

Nominal values for m_1 and m_2 masses and for spring constant k are 1. Control action u is the strength applied to mass 1 and the controlled variable is mass 2 position affected by noise measurement y. Moreover, it exists a disturbance w on mass 2. [14] proposed three control scenario, but in this article only the first one is considered:

– Closed loop system has to be stable for $m_1 = m_2 = 1$ and $k \in [0.5 \dots 2]$.
– Maximum settling time for nominal system $(m_1 = m_2 = k = 1)$ has to be 15 sec. for unity impulse in perturbation w at time $t = 0$.
– Phase and gain margins have to be reasonable for a reasonable bandwidth.
– Closed loop have to be relatively insensitive to high frequency noise in the measurements.

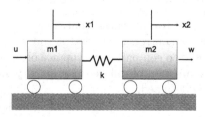

Fig. 2. Two masses and spring system with uncertainties in parameters

Table 1. Preferences for controller design

	g_i^1	g_i^2	g_i^3	g_i^4	g_i^5
$Re(\lambda)_{max}$	-0.01	-0.005	-0.001	-0.0005	-0.0001
u_{max}	0.8	0.85	0.95	1	2
t_{est}^{max}	15	40	80	90	100
$noise_{max}$	1.8	2	2.2	2.5	3
u_{nom}	0.9	1.2	2	2.5	3
t_{est}^{nom}	14	14.2	14.4	14.6	15

- Control effort has to be minimized.
- Controller complexity has to be minimized.

Numerator and denominator coefficients conform the vector parameters to obtain by optimization, $\mathbf{x}$. Specification or design objectives ($g_i(\mathbf{x})$) have to be quantities that designer wants to maximize, minimize, set to a specific value, etc. For the robust control benchmark six functions that supply specification values for a controller will be used:

1. **Nominal settling time** (t_{est}^{nom}): Maintaining interpretation made in [12] and [8]. It is assumed that controlled variable achieves steady state for a unit impulse in perturbation w when it is in ± 0.1 units range.
2. **Worst case settling time** (t_{est}^{max}): Maximum settling time (in ± 0.1 range) for a given controller evaluated in the worst case $k = 0.5$ or $k = 2$.
3. **Robust stability and robust performances** ($Re(\lambda)_{max}$): [12] shows that phase and gain margins for the worst case are poor indicators of robustness. Instead of, closed loop poles for the worst case can be used: $Re(\lambda)_{max} = max_{k \in [0.5...2]} Re(\lambda[A(k)])$ (A is the closed loop system matrix).
4. **Noise sensibility** ($noise_{max}$): For a frequency range, noise sensitivity is measured by the ratio of noise amplification in front of a $-20db/dec$ slope.

$$noise_{max} = max_{k \in [0.5,2]} \left(\left| \frac{u(jw)}{v(jw)} \right| \middle/ \left| \frac{1}{jw} \right| \right) \; ; \; w \in [100 \ldots 10000] \quad (6)$$

5. **Nominal control effort** (u_{nom}): Maximum control action produced for a unit impulse in disturbance at nominal case.
6. **Maximum control effort** (u_{max}): Maximum control action produced for a unit impulse in the disturbance when there are uncertainties.

Then there are six algorithms (or functions) that supply the above described values for every combination of controller numerator and denominator coefficients. These functions can be used to compare performance obtained by other authors. Controllers with transfer function of different complexity and strictly proper (a priori selected) have been designed.

Class 1s functions have been selected for all objectives. Table 1 shows preferences (by means of extremes values for each preference interval) for the six specifications.

5   Results

In this section, results for the robust control benchmark using Physical Programming with Genetic Algorithms are analyzed. There is previous work that collects solutions [12] but it focussed on designing also obtained with an application of Physical Programming [8] but with classical nonlinear optimization techniques. The proposed solution goes beyond previous ones.

Table 2 shows two controllers proposed by other authors and referenced as good solutions (W34 and M34). Controllers obtained with PP+GA are also presented at Table 2 (PPGA34 and PPGA23*). Comparison between them has been done analyzing the six proposed specifications. For every controller, numeric values of each specification and obtained range are shown at Table 3.

Controllers obtained using PP+GA are better for all specifications. Even with controller of lower complexity, the performances are improved, this is the case of PPGA23*. Solution proposed by [8] M34 is not even a local minimum. Minimization of aggregated function with restrictions causes a very complex multimodal problem that GA achieves to solve. Notice that for PPGA controllers

Table 2. Solutions for the robust control benchmark

Proposal	(m, n)	f.d.t.
Wie (W34) [13]	$(3, 4)$	$\dfrac{-2.13s^3 - 5.327s^2 + 6.273s + 1.015}{s^4 + 4.68s^3 + 12.94s^2 + 18.36s + 12.68}$
Messac (M34) [8]	$(3, 4)$	$\dfrac{-0.66s^3 - 4.101s^2 + 4.558s + 0.627}{s^4 + 3.416s^3 + 10.15s^2 + 13.52s + 9.281}$
PPGA34	$(3, 4)$	$\dfrac{-0.3226s^3 - 2.276s^2 + 4.79s + 0.7539}{s^4 + 2.075s^3 + 8.664s^2 + 11.32s + 7.825}$
PPGA23*	$(2,3)$	$\dfrac{-1.5704s^2 + 3.1911s + 0.52}{s^3 + 5.2347s^2 + 7.2333s + 5.2436}$

Table 3. Specifications and obtained ranges. AD-Highly Desirable, D-Desirable, T-Tolerable, IND-Undesirable, AIND-Highly Undesirable, INA-Unacceptable

Controller	$Re(\lambda)_{max}$	u_{max}	t_{est}^{max}	$noise_{max}$	u_{nom}	t_{est}^{nom}
W34	-0.0427	0.6793	22.125	2.1317	0.5595	16.7756
	AD	AD	D	T	AD	IND
M34	-0.0542	0.6681	18.375	0.6620	0.5127	13.3580
	AD	AD	D	AD	AD	AD
PPGA34	-0.0166	0.7194	11.625	0.3241	0.5872	10.8731
	AD	**AD**	**AD**	**AD**	**AD**	**AD**
PPGA23*	-0.0219	0.5259	15.000	1.5698	0.4170	10.8731
	AD	**AD**	**AD**	**AD**	**AD**	**AD**

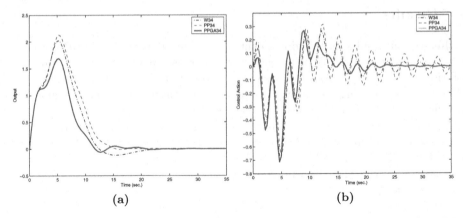

(a)   **(b)**

Fig. 3. (a) Nominal settling time for closed loop with W34, PP34 and PPGA34. (b) Control action in the worst case ($k = 0.5$) for W34, PP34 and PPGA34

all specifications are in AD range, then it is possible (because of GA application) to improve performance selecting a more restrictive performance.

Fig. 3(a) shows process output for a unit impulse in the disturbance. PPGA34 presents better performances than other controllers of the same complexity.

Better performances are obtained without increasing excessively control effort (Fig. 3(b)), because PPGA34 also considers preferences in this specification.

6   Conclusions

A new methodology to design controllers has been presented. It allows to incorporate, in a simple and intuitive way, different types of specification. Then it is possible to design robust controller in a simpler way.

Fundamentals about incorporating specifications and design preferences are based on Physical Programming methodology. The multiobjective problem is translated into a normalized domain by means of Class Functions, then all these Class Functions are aggregated in a single cost function. It is possible to achieve a successful minimization if an adequate optimization technique is selected and tuned. Genetic Algorithms have a fundamental role and have demonstrated very good performances in such multimodal complex problems (ACC Robust Control Benchmark). All solutions have improved previous work, even such obtained with Physical Programming but with other optimizers.

The methodology is applicable to other design domains different from controller design. The only necessary condition consists of the fact that the design has to be based on a several criteria optimization.

Acknowledgements

Partially financed by AGL2002-04108C02-01 and DPI2001-3106C02-02 MCYT.

References

1. T. Back. *Evolutionary Algorithms in Theory and Practice*. Oxford University Press, New York, 1996.
2. J.E. Baker. *Reducing Bias and Inefficiency in the Selection Algorithms*, volume Proceedings of the Second International Conference on Genetic Algorithms, pages 14–21. Lawrence Erlbaum Associates,, Hillsdale, NJ, Grefenstette, J.J. (ed.) edition, 1987.
3. Carlos A. Coello Coello, David A. Van Veldhuizen, and Gary B. Lamont. *Evolutionary algorithms for solving multi-objetive problems*. Kluwer Academic Publishers, 2002.
4. D.E. Goldberg. *Genetic Algorithms in search, optimization and machine learning*. Addison-Wesley, 1989.
5. J.H. Holland. *Adaptation in natural and artificial systems*. Ann Arbor: The University of Michigan Press, 1975.
6. Miguel A. Martínez, Javier Sanchis, and Xavier Blasco. Algoritmos genéticos en physical programming. applicación al benchmark de control robusto de la acc. *Revista Iberoamericana de Automática e Informática Industrial*, Submitted 2005.
7. A. Messac. Physical programming: effective optimitation for computational design. *AIAA Journal*, 34(1):149–158, January 1996.
8. A. Messac and B. H. Wilson. Physical programming for computational control. *AIAA Journal*, 36(1):219–226, February 1999.
9. Z. Michalewicz. *Genetic Algorithms + Data Structures = Evolution Programs*. Springer series Artificial Intelligence. Springer, 3rd edition edition, 1996.
10. Kaisa M. Miettinen. *Nonlinear multiobjective optimization*. Kluwer Academic Publishers, 1998.
11. H. Mühlenbein and D. Schlierkamp-Voosen. Predictive models for the breeder genetic algorithm. i. continuous parameter optimization. *Evolutionary Computation*. The MIT Press, 1(1):25–49, Spring 1993.
12. R. Stengel and C. Marrison. Robustness of solutions to a benchmark control problem. *Journal of Guidance, Control and Dynamics*, 15(5):1060–1067, September 1992.
13. B. Wie and D. Bernstein. A benchmark problem for robust control design. In *Proceedings of the American Control Conference*, pages 961–962, San Diego, CA, 1990.
14. B. Wie and D. Bernstein. Benchmark problems for robust control design. *Journal of Guidance, Control and Dynamics*, 15(5):1057–1059, September 1992.

Grammar Based Crossover Operator in Genetic Programming

Daniel Manrique, Fernando Márquez, Juan Ríos, and Alfonso Rodríguez-Patón

Artificial Intelligence Dpt., Facultad de Informática*,
Universidad Politécnica de Madrid, 28660 Boadilla del Monte, Madrid, Spain
{dmanrique, jrios, arpaton}@fi.upm.es

Abstract. This paper introduces a new crossover operator for the genetic programming (GP)paradigm, the grammar-based crossover (GBX). This operator works with any grammar-guided genetic programming system. GBX has three important features: it prevents the growth of tree-based GP individuals (a phenomenon known as code bloat), it provides a satisfactory trade-off between the search space exploration and the exploitation capabilities by preserving the context in which subtrees appear in the parent trees and, finally, it takes advantage of the main feature of ambiguous grammars, namely, that there is more than one derivation tree for some sentences (solutions). These features give GBX a high convergence speed and low probability of getting trapped in local optima, as shown throughout the comparison of the results achieved by GBX with other relevant crossover operators in two experiments: a laboratory problem and a real-world task: breast cancer prognosis.

1 Introduction

Genetic programming is a means of automatically generating computer programs by employing operations inspired by biological evolution, such as reproduction, crossover and mutation, to breed a population of trial solutions that improves over time [1].

The crossover operator bears most responsibility for the acceptable evolution of the genetic programming algorithm, because it governs most of the search process (evolution). It usually operates on two parents and produces two offspring using parts of each parent. One of the first of these important operators was Standard crossover defined by Koza [2], which randomly swaps subtrees in both parent trees to generate the offspring. The main advantage of such crossovers is that they maintain diversity, because crossing two identical trees can yield different trees. This excessive exploration capability makes the trees large and complex [3], as the search space of a GP problem is potentially unlimited and individuals may grow in size during the evolutionary process. This situation,

---

* Special thanks to Soledad Hurtado and David Pérez for their reviews, technical support and discussions about the topic of this paper.

J. Mira and J.R. Álvarez (Eds.): IWINAC 2005, LNCS 3562, pp. 252–261, 2005.

known as bloat or code bloat, has a very high computational cost, degrading the convergence speed of GP [4].

Size Fair [5] and Fair [6] crossover operators were developed to limit bloat, increasing exploitation capabilities to stop these operators from conveniently progressing to larger areas of the search space where solutions are more plentiful. The result of this overexploitation is a greater probability of getting trapped in local optima [7].

The quest for a satisfactory trade-off between exploration and exploitation to produce a genetic programming system that yields good results led to the so-called closure problem, which consists of avoiding the generation of syntactically illegal programs: individuals that do not represent possible solutions to the problem [8]. This is another of the drawbacks facing GP when executing a crossover operator, because it notably detracts from the convergence process. Context-free grammar genetic programming was introduced to overcome this problem. This approach uses the parse trees of a given context-free grammar as the individuals of the population [9]. Nowadays, one of the main crossover operators most commonly used with this representation is Whigham's crossover (WX) [8],[10],[11]. What this operator does is basically to randomly choose a node from the second parent tree that represents the same non-terminal symbol as the node chosen, also randomly, in the first parent tree. Having chosen the crossover nodes, the operator swaps the subtrees below these non-terminals. The main shortcoming of WX is that there are other possible choices of nodes in the second parent that are not explored and which could lead towards the sought-after solution. This limitation of the exploration capability leads to higher probability of getting trapped in local optima.

This paper introduces a new crossover operator, called grammar-based crossover operator (GBX), designed to work in context-free grammar genetic programming systems. The three main features of this operator address the above-mentioned problems in GP: it prevents the generation of illegal trees, it explores all nodes in the parent trees that can generate new legal individuals that lead to the sought-after solution (a feature that is enhanced by using ambiguous grammars) [12], and it has an efficient bloat control mechanism. Experimental tests have been run to search for a mathematically true arithmetical expression and the solution to a real-world problem: breast cancer prognosis. The results achieved, which have been compared with other important operators like WX, Fair and Standard crossover, show that GBX has well-balanced exploitation and exploration capabilities, which increases the local search speed of the genetic programming system, while avoiding local optima.

2   The Genetic Programming System

The genetic programming system is able to find solutions to a problem starting from the formal definition of the syntactical constraints of the problem using a context-free grammar, whose language represents the search space, and an evaluation function, which provides a fitness value for each of the possible solutions generated to drive the search process and finally choose the optimal solution.

A context-free grammar G is defined as a string-rewiring system comprising a 4-tuple $G = (S, \Sigma_N, \Sigma_T, P)$, where S represents the start symbol or axiom of the grammar, Σ_N is the alphabet of non-terminal symbols, Σ_T is the alphabet of terminal symbols, $\Sigma_N \cap \Sigma_T = \emptyset$, and P is the set of production rules, written in BNF (Backus-Naur Form). Based on this grammar, the individuals that are part of the genetic population are defined as derivation trees, where the root is the axiom S, the internal nodes contain only non-terminal symbols and the external nodes (leaves) contain only terminal symbols. A derivation tree represents the series of derivation steps that generate a sentence, which is a possible solution to the problem. Therefore, an individual codifies a sentence of the language generated by the grammar as a derivation tree.

The input for the evaluation function F is any of the individuals that are members of the genetic programming algorithm population. This evaluation function provides a fitness measure that indicates how good the solution codified by the individual is for the problem at hand.

The structure of the proposed system is shown in Figure 1 and consists of two modules.The input for the evolution engine module is the grammar that defines the search space of the problem used to generate the random initial population of trees with a previously established maximum depth. The evolution engine module employs the genetic operators (selection, crossover, mutation and replacement) to search for the optimal solution. This module implements the GBX. The input for the evaluation module is an evaluation function used to calculate the fitness of the individuals (trees) generated by the evolution engine module, from the initial population or from the genetic operators, as well as the mean fitness of the entire population. To do so, a decodification process is implemented, which consists of concatenating the terminal symbols included in the leaves of the derivation tree to get the original sentence.

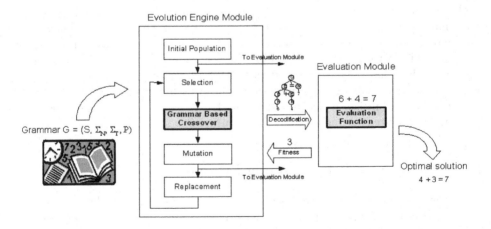

Fig. 1. The genetic programming system schema

3   The Grammar-Based Crossover

GBX is a general-purpose operator for optimization problems with context-free grammar genetic programming. To give a better understanding of the GBX, a simple laboratory example is given, in which a genetic programming system is used to find out a mathematically true expression (equality). The syntax of the equalities used in the population can be defined using the grammar expressed in formula 1. Note that this is an ambiguous grammar, since a sentence can be derived from more than one derivation tree. Specifically, neither the non-terminal symbol F, nor the production rules that include it are needed. However, this example has been used to better illustrate the features of GBX.

$$G = (S, \Sigma_N, \Sigma_T, P) \quad with :$$
$$\Sigma_T = \{0, 1, 2, 3, 4, 5, 6, 7, 8, 9, +, -, =\}$$
$$\Sigma_N = \{S, E, F, N\}$$
$$P = \{S \rightarrow E = N \tag{1}$$
$$E \rightarrow E + E | E - E | F + E | F - E | N$$
$$F \rightarrow N$$
$$N \rightarrow 0|1|2|3|4|5|6|7|8|9\}$$

One possible solution to the problem would be, for example, $4 + 3 = 7$. The expression $6 + 4 = 7$ is correct, as it follows the syntax defined by the grammar, but it is not a solution. Suppose we want to cross the expressions $6 + 4 = 7$ and $3 + 2 + 4 = 8$, Figure 2 shows both codifications as derivation trees, which is how they are managed by GBX. In Figure 2, the root represents the axiom of the grammar, circles are non-terminals and the other symbols are terminals. The GBX consists of eight steps:

1. A node with a non-terminal symbol is randomly chosen from the first parent (except the axiom). This node is called crossover node or crossover place. The crossover node of the stated example, corresponding to the non-terminal E, is highlighted in gray in Figure 3.
2. The parent node of the node chosen in step 1 is searched. As we are working with a context-free grammar, this is a non-terminal symbol, which is the

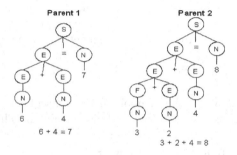

Fig. 2. The two individuals used for explanatory purposes

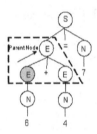

Fig. 3. Crossover node (gray), parent node (arrow) and main derivation (dashed line)

antecedent of one or more production rules. The consequents of all these productions are stored in an array R, which completes this second step.

In the case of the stated example, the parent node of crossover node E is also E, which is illustrated in Figure 3. According to the production rules defined for the grammar we are using (formula 1), E is the antecedent of five productions $E ::= E + E$, $E ::= E - E$, $E ::= F + E$, $E ::= F - E$ and $E ::= N$, hence the array $R = [E + E; E - E; F + E; F - E; N]$.

3. The derivation produced by the parent of the crossover node is called main derivation $(A ::= C)$. Additionally, the derivation length (l) is defined as the number of non-terminals and terminals included in the consequent of a derivation. In this third step of GBX, the 3-tuple $T(l, p, C)$ is calculated and stored. This 3-tuple contains the main derivation length, the position (p) of the crossover node in the main derivation and the consequent of the main derivation (C). The dashed line in Figure 3 shows the main derivation for the stated example: $E ::= E + E$. The main derivation length is 3 $(l = 3)$, the position of the crossover node in its consequent is 1 $(p = 1)$ and its consequent is $E + E$ $(C = E + E)$. Hence, the 3-tuple $T = (3, 1, E + E)$.

4. All the consequents with different lengths from the main derivation are deleted from array R. In the stated example, array R contains five consequents of lengths 3 and 1, and the consequent of production rule $E ::= N$ is, therefore, deleted, leaving $R = [E + E; E - E; F + E; F - E]$.

5. For each element of R, all the symbols, except the one that is located in the position of the crossover node, are compared with the symbols of the consequent of the main derivation. Then, all the consequents with any difference are deleted from R. In the stated example, the crossover node is located in the first position in the consequent of the main derivation and, therefore, $+E$, $-E$, $+E$ and $-E$ are compared with $+E$. As the second and forth consequents differ (- and +), they are deleted from R, leaving $R = [E + E; F + E]$.

6. The set X formed by all the symbols in the consequents of R that are in the same position as the crossover node is calculated. This set establishes the non-terminal symbols that can be chosen as crossover nodes in the second parent. As there are two consequents in the proposed example and the crossover node is in first place, $X = \{E, F\}$.

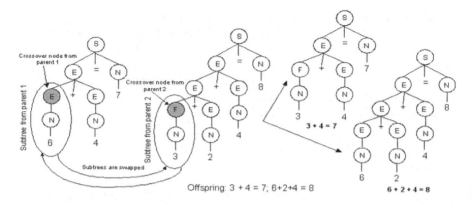

Fig. 4. Once the crossover nodes are chosen, GBX interchanges the two subtrees to generate two new valid ones

7. The crossover node within the second parent is randomly chosen from any of the nodes that have a symbol that is a member of the set X. The crossover node chosen in the second parent is shaded gray in Figure 4.
8. The two new individuals produced as the offspring are calculated by swapping the two subtrees whose roots are the calculated crossover nodes of the two parents. Figure 4 shows this process.

From the algorithm proposed above, we find that GBX has three main attractive features:

1. It is easy to incorporate control bloat in the seventh step by discarding all the nodes that generate offspring trees deeper than a previously established value D. This operation can be done very efficiently before swapping subtrees by adding the length of the first parent from the root to the crossover node, plus the length of the subtree of the second parent to be swapped. If the result of this operation is lower than or equal to D, then the first swap can be made to generate the first offspring. Interchanging the roles of the parents, a similar procedure can be followed to produce the second offspring.

   For the example in Figure 4, the length of the first parent from the root to the crossover node is 2 and the length of the second subtree is also 2. If $4 \leq D$ (as assumed in the example), then we get the individual $3 + 4 = 7$. Interchanging the roles, the length of the first subtree is 3, whereas it is 2 for the second tree, and it is assumed that $5 \leq D$ to produce the new individual $6 + 2 + 4 = 8$.
2. The steps of the algorithm for GBX assure that the offspring produced are always composed of two legal trees. This is very positive for the evolution process, as all the individuals generated are possible solutions to the problem.
3. GBX takes into account all the possible nodes of the second parent that can generate legal individuals starting from the crossover node previously chosen from the first parent. The proposed example has been especially prepared

to illustrate this feature: having chosen the crossover node from parent 1 (non-terminal E), GBX calculates that all the nodes of the second parent with non-terminal symbols E or F are valid for crossover. Possible nodes for crossover in the second parent generated by other operators, like WX, would, in this case, contain only the non-terminal symbol E. The possibility that GBX offers of also being able to choose node F is what leads to the final solution $(3 + 4 = 7)$ in a single iteration.

4   Results

Two different experiments were carried out to test GBX: the laboratory problem syntactically defined in the grammar of formula 1, designed to show what benefits can be gained by using ambiguous grammars, and the real-world task of diagnosing breast cancer. These tests were run using the WX, Fair and Standard crossover operators, apart from GBX, with the aim of comparing the results yielded by each one.

For each experiment, a set of 100 runs was performed. After some tuning runs, we decided the following settings for the operator rates of the genetic programming system: 75% crossover, 20% straight reproduction and 5% mutation. The tournament size was 7 and SSGA was the replacement method.

The objective of the first experiment was to search for a true mathematical expression that follows the syntax defined by the context-free grammar of formula 1. The population size employed was 20, the size ceiling limit was set to 6. The fitness function consisted of calculating the absolute value of the difference between the left- and right-hand sides of the expression. Figure 5 shows the evolution process for the average fitness of the population using each of the four crossover operators. It takes GBX an average of 8.1 generations to get true expressions, WX 10.4, Fair 15.2 and Standard crossover 16.7 It is clear that GBX outperforms the other three crossover operators. This is because GBX is able to take advantage of the ambiguity of the grammar to explore all possible paths that lead to the sought-after solution.

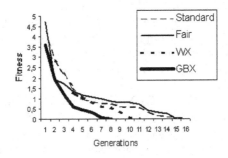

Fig. 5. Fitness evolution for the expressions search problem

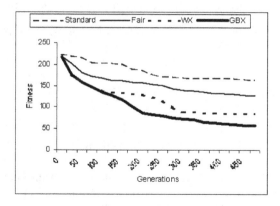

Fig. 6. Fitness evolution for breast cancer prognosis

The second experiment involved searching a knowledge base of fuzzy rules that could give a prognosis of suspicious masses being cancer. This is a complex classification problem where a set of features that describes a breast abnormality has to be assigned to a class: benign or malignant. The data employed for this problem included 315 sets of features from lesions that had been detected in mammograms of real patients. Each lesion was defined by 7 features that described the morphology (size, border, shape), localization (zone, side, depth) and patient's age. The grammar employed was formed by 17 non-terminal symbols, 46 terminal symbols and 47 production rules.

The population size employed was 1000, the size ceiling limit was set to 40. The fitness function consisted of calculating the number of misclassified patterns. Figure 6 shows the evolution process for each of the four crossover operators.

In this experiment, the Standard crossover yields the worst results, because it maintains an excessive diversity in the population, which prevents it from focusing on one of the possible solutions. The effect of Fair crossover is the opposite, leading very quickly to one of the possible solutions with an initially high convergence speed, which slows down if convergence is towards a local optimum. WX produces good results, as is to be expected of a crossover operator designed to work on grammar-guided genetic programming systems, bettered only by the proposed crossover. This evidences the benefits of the possibility of exploring all the nodes that generate legal derivation trees in the second parent.

Table 1 shows the analytical results of this experiment: the average fitness of the best individual in 500 generations, the average number of iterations needed to reach a fitness of 150 (165 correctly classified patterns) and the percentage of times that the above objective is achieved (not trapping in local optima). The stop condition for calculating the values of Table 1 was when there was no observed improvement after 50 consecutive generations. Note the similarity between WX and GBX in terms of the average number of iterations. However,

Table 1. Analytical results for the breast cancer prognosis problem

Crossover Op	Avg. fitness	Avg. number of iterations	Succesful runs
Standard	163.74	715.15	61/100
Fair	127.33	250.28	85/100
WX	84.74	72.34	93/100
GBX	57.72	77.82	97/100

GBX is less likely to get trapped in local optima and has a better final fitness than the other three.

The fuzzy knowledge base built by the genetic programming system with GBX was used to prognosticate the 315 breast lesions and correctly classified 81% of benign and 83% of malignant cases. These results are similar to classifications by expert radiologists in the case of malignant lesions and slightly better for benign cases, because doctors prognosticate doubtful findings as malignant.

5   Conclusions

This paper presented a new crossover operator, the grammar-based crossover, to work on grammar-guided genetic programming systems. The results section demonstrated that this operator strikes a satisfactory balance between exploration and exploitation capabilities, which gives it a high convergence speed, while eluding local optima. To be able to achieve these results, the proposed crossover includes two important features. The first is a mechanism for bloat control, whose computational cost is practically negligible. The second feature, which is novel in grammar-guided genetic programming systems, means that, having chosen a crossover node in the first parent, any nodes likely to generate legal individuals in the second parent rather than just nodes with the same non-terminal symbols can be chosen.

In the proposed experiments, the population was generated randomly. This is too complex, especially for the breast cancer prognosis problem, where 1000 individuals need to be generated. Research efforts now target the use of the context-free grammar to generate the initial population more intelligently and, therefore, more efficiently to satisfy the syntactical constraints and size ceiling limit. This same algorithm will be used to assure that the mutation operator is not confined to the replacement of one sub-derivation tree by another randomly generated subtree rooted in the same non-terminal symbol, but can include other legal non-terminal symbols.

In the two experiments described in this paper, the use of an ambiguous grammar benefits GBX. This points to other promising research work to find the answers to questions such as how and/or to what extent the ambiguity of the grammar benefits the convergence of GBX and how much ambiguity is needed to get the best results in terms of efficiency.

References

1. Koza, J.R.: Genetically Breeding Populations of Computer Programs to Solve Problems in Artificial Intelligence. Stanford University, Department of Computer Science. CS-TR-90-1314 (1990)
2. Koza, J.R.: Genetic Programming: On the Programming of Computers by Means of Natural Selection. MIT Press, Cambridge, MA (1992)
3. Singleton, A.: Greediness. Posting of 8-21-1993 on genetic-programming@cs.standford.edu mailing list (1993)
4. Silva S., Almeida, J.: Dynamic Maximum Tree Depth. A Simple Technique for Avoiding Bloat in Tree-Based GP. In: Cantú-Paz, E. et al. (Eds.): Proceedings of the Genetic and Evolutionary Computation Conference GECCO. Lecture Notes in Computer Science, Vol. 2724. Springer-Verlag, Berlin Heidelberg New York (2003) 1776 - 1787
5. Langdon, W.B.: Size Fair and Homologous Tree Crossovers for Tree Genetic Program-ming. Genetic Programming and Evolvable Machines, Vol. 1, Kluwer Academic Publishers, Boston. (2000) 95-119
6. Crawford-Marks R., Spector, L.: Size control via Size Fair Genetic Operators in the PushGP Genetic Programming System. Proceedings of the Genetic and Evolutionary Computation Conference, New York, USA. (2002) 733-739
7. Barrios, D., Carrascal, A., Manrique, D., Ríos, J.: Optimization with Real-Coded Genetic Algorithms Based on Mathematical Morphology. Intern. J. Computer Math., Vol. 80, No. 3 (2003) 275-293
8. Whigham, P.A.: Grammatically-Based Genetic Programming. Proceedings of the Workshop on Genetic Programming: From Theory to Real-World Applications, California, USA. (1995) 33-41
9. Hussain, T.S.: Attribute Grammar Encoding of the Structure and Behavior of Artificial Neural Networks. PhD Thesis, Queen's University. Kingston, Ontario, Canada. (2003)
10. Rodrigues, E., Pozo, A.: Grammar-Guided Genetic Programming and Automatically De-fined Functions. Proceedings of the 16th Brazilian Symposium on Artificial Intelligence, Recife, Brazil. (2002) 324-333
11. Grosman, B., Lewin, D.R.: Adaptive Genetic Programming for Steady-State Process Mod-eling. Computers and Chemical Engineering, Vol. 28. (2004) 2779-2790
12. Hoai, N.X., McKay, R.I.: Is Ambiguity Useful or Problematic for Grammar Guided Genetic Programming? A case of Study. Proceedings of the 4th Asia-Pacific Conference on Simu-lated Evolution and Learning, Singapore. (2002) 449-453

GA-Selection Revisited
from an ES-Driven Point of View

Michael Affenzeller, Stefan Wagner, and Stephan Winkler

Institute for Formal Models and Verification,
Johannes Kepler University, Altenbergerstrasse 69,
A-4040 Linz - Austria
{michael, stefan, stephan}@heuristiclab.com

Abstract. Whereas the selection concept of Genetic Algorithms (GAs) and Genetic Programming (GP) is basically realized by the selection of above-average parents for reproduction, Evolution Strategies (ES) use the fitness of newly evolved offspring as the basis for selection (survival of the fittest due to birth surplus). This contribution proposes a generic and enhanced selection model for GAs considering selection aspects of population genetics and ES. Some selected aspects of these enhanced techniques are discussed exemplarily on the basis of standardized benchmark problems.

1 Introduction

In contrast to other heuristic optimization techniques Genetic Algorithms and certainly also Genetic Programming (GP) take a fundamentally different approach by considering recombination (crossover) as their main operator. The essential difference to neighborhood-based techniques is given by the fact that recombination is a sexual operator, i.e. properties of individuals from different regions of the search space are combined in new individuals. Therefore, the advantage of applying GAs to hard optimization problems lies in their ability to scan broader regions of the solution space than heuristic methods based upon neighborhood search do. Nevertheless, also GAs are frequently faced with a problem which, at least in its impact, is quite similar to the problem of stagnating in a local but not global optimum. This drawback, called premature convergence in the terminology of GAs, occurs if the population of a GA reaches such a suboptimal state that the genetic operators are no longer able to produce offspring that are able to outperform their parents (e.g. [5], [1]). This happens if the genetic information stored in the individuals of a population does not contain that genetic information which would be necessary to further improve the solution quality. Therefore, in contrast to the present contribution, the topic of premature convergence is considered to be closely related to the loss of genetic variation in the entire population in GA-research [11], [15]. In this contribution we do not identify the reasons for premature convergence in the loss of genetic variation in general but more specifically in the loss

J. Mira and J.R. Álvarez (Eds.): IWINAC 2005, LNCS 3562, pp. 262–271, 2005.
© Springer-Verlag Berlin Heidelberg 2005

of what we call essential genetic information, i.e. in the loss of alleles which are part of a global optimal solution. Therefore, we will denote the genetic information of the global optimal (which may be unknown a priori) solution as essential genetic information in the following. If parts of this essential genetic information are missing, premature convergence is already predetermined in some way.

A very essential question about the general performance of a GA is, whether or not good parents are able to produce children of comparable or even better fitness (the building block hypothesis implicitly relies on this). In natural evolution, this is almost always true. For GAs this property is not so easy to guarantee. The disillusioning fact is that the user has to take care of an appropriate coding in order to make this fundamental property hold.

In order to somehow overcome this strong requirement we try to get to the bottom of reasons for premature convergence from a technical as well as from a population genetics inspired point of view and draw some essential interconnections.

The basic idea of the new selection model is to consider not only the fitness of the parents in order to produce a child for the ongoing evolutionary process. Additionally, the fitness value of the produced child is compared with the fitness values of its own parents. The child is accepted as a candidate for the further evolutionary process if and only if the reproduction operator was able to produce a child that could outperform the fitness of its own parents. This strategy guarantees that evolution is presumed mainly with crossover results that were able to mix the properties of their parents in an advantageous way. I.e. **survival of the fittest alleles is rather supported than survival of the fittest individuals** which is a very essential aspect for the preservation of essential genetic information stored in many individuals (which may not be the fittest in the sense of individual fitness).

2   Some Basic Considerations About GA-Selection

In terms of goal orientedness, selection is the driving force of GAs. In contrast to crossover and mutation, selection is completely generic, i.e. independent of the actually employed problem and its representation. A fitness function assigns a score to each individual in a population that indicates the 'quality' of the solution the individual represents. The fitness function is often given as part of the problem description or based upon the objective function. In the Standard GA the probability that a chromosome in the current population is selected for reproduction is proportional to its fitness. However, there are also many other ways of accomplishing selection. These include linear-rank selection or tournament selection (cf. e.g. [7], [10]).

However, all evenly mentioned GA-selection principles have one thing in common:

They all just consider the aspect of sexual selection, i.e. mechanisms of selection only come into play for the selection of parents for reproduction. The enhanced selection model which will be described in the following section defies this limitation by considering selection in a more general sense.

Selection and Selection Pressure: In the terminology of population genetics the classical GA selection concept is known as sexual selection. In the population genetics view, sexual selection covers only a rather small aspect of selection which appears when individuals have to compete to attract mates for reproduction. The population genetics basic selection model considers the selection process in the following way:

random mating $\rightarrow$ selection $\rightarrow$ random mating $\rightarrow$ selection $\rightarrow$ ......

I.e. selection is considered to depend mainly on the probability of surviving of newborn individuals until they reach pubescence which is called viability in the terminology of population genetics. The essential aspect of offspring selection in the interpretation of selection is rarely considered in conventional GA selection. The classical (μ, λ) Evolution Strategy in contrast does this very well: Reconsidering the basic functioning of a (μ, λ) ES in terms of selection, μ parents produce λ ($\lambda \geq \mu$) offspring from which the best μ are selected as members of the next generation. In contrast to GAs where selection pressure is predetermined by the choice of the mating scheme and the replacement strategy, ES allow an easy steering of selection pressure by the ratio between μ and λ. The selection pressure steering model introduced in Section 3 picks up this basic idea of ES and transforms these concepts for GAs in order to be have an adjustable selection pressure (independent of the mating scheme and replacement strategy) at one's disposal.

Our advanced selection scheme allowing self-adaptive steering of selection pressure aims to transform the basic ideas for improving the performance of GAs. In doing so the survival probability is determined by a comparison of the fitness of the newly generated individual with the fitness values of its parents.

A very important consequence of selection in population genetics as well as in evolutionary computation is its influence on certain alleles. As a matter of principle there are four possibilities for each allele in the population:

- The allele is fixed in the population.
- The allele is lost in the population.
- The allele frequency converges to an equilibrium state.
- The allele frequency remains unchanged.

The basic approaches for retarding premature convergence discussed in GA literature aim at the maintenance of genetic diversity. The most common techniques for this purpose are based upon preselection [3], crowding [4], or fitness-sharing [6]. The main idea of these techniques is to maintain genetic diversity by the preferred replacement of similar individuals [3], [4] or by the fitness-sharing of individuals which are located in densely populated regions [6]. While methods based upon [4] or [6] require some kind of neighborhood measure depending on the problem representation, [6] is additionally quite restricted to proportional selection. Moreover, these techniques have the common goal to maintain genetic diversity which is very important in natural evolution where a rich gene pool is the guarantor in terms of adaptiveness w.r.t. changing environmental conditions.

In case of artificial genetic search as performed by a GA the optimization goal does not change during the run of a GA and the fixing of alleles of high quality solutions is desirable in the same manner as the erasement of alleles which are definitely not part of a good solution in order to reduce the search space and make genetic search more goal-oriented. I.e. we claim that pure diversity maintenance mechanisms as suggested in [3], [4], or [6] do not support goal-oriented genetic search w.r.t the locating of global optimal solutions.

3 An Enhanced Selection Model for Self-adaptive Steering of Selection Pressure

The basic idea to create and evaluate a certain amount (greater or equal population size) of offspring, to be considered for future members of the next generation, is adapted from Evolution Strategies. Self-adaption comes into play when considering the question which amount of offspring is necessary to be created at each round, and which of these candidates are to be selected as members of the next generation, i.e. for the ongoing evolutionary process. In order to keep the concepts generic, no problem specific information about the solution space is allowed to be used for stating the self-adaptive model. Thus, it is desirable to systematically utilize just the fitness information of the individuals of the actual generation for building up the next generation of individuals, in order to keep the new concepts and methods generic. In principle, the new selection strategy acts in the following way:

The first selection step chooses the parents for crossover either randomly or in the well-known way of GAs by roulette-wheel, linear-rank, or some kind of tournament selection strategy. After having performed crossover and mutation with the selected parents we introduce a further selection mechanism that considers the success of the apparently applied reproduction in order to assure the proceeding of genetic search mainly with successful offspring in that way that the used crossover and mutation operators were able to create a child that surpasses its parents' fitness. Therefore, a new parameter, called success ratio ($SuccRatio \in [0,1]$), is introduced. The success ratio gives the quotient of the next population members that have to be generated by successful mating in relation to the total population size. Our adaptation of Rechenberg's success rule [8] for GAs says that a child is successful if its fitness is better than the fitness of its parents, whereby the meaning of 'better' has to be explained in more detail: is a child better than its parents, if it surpasses the fitness of the weaker, the better, or is it in fact some kind of mean value of both?

For this problem we have decided to introduce a cooling strategy similar to Simulated Annealing. Following the basic principle of Simulated Annealing we claim that an offspring only has to surpass the fitness value of the worse parent in order to be considered as 'successful' at the beginning and while evolution proceeds the child has to be better than a fitness value continuously increasing

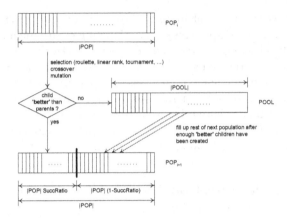

Fig. 1. Flowchart for embedding the new selection principle into a Genetic Algorithm

between the fitness of the weaker and the better parent. As in the case of Simulated Annealing, this strategy effects a broader search at the beginning whereas at the end of the search process this operator acts in a more and more directed way. Having filled up the claimed ratio ($SuccRatio$) of the next generation with successful individuals in the above meaning, the rest of the next generation $((1 - SuccRatio) \cdot |POP|)$ is simply filled up with individuals randomly chosen from the pool of individuals that were also created by crossover but did not reach the success criterion. The actual selection pressure $ActSelPress$ at the end of a single generation is defined by the quotient of individuals that had to be considered until the success ratio was reached and the number of individuals in the population in the following way: $ActSelPress = \frac{|POP|SuccRatio + |POOL|}{|POP|}$.

Figure 1 shows the operating sequence of the above described concepts. With an upper limit of selection pressure ($MaxSelPress$) defining the maximum number of children considered for the next generation (as a multiple of the actual population size) that may be produced in order to fulfill the success ratio, this new model also functions as a precise detector of premature convergence:

If it is no longer possible to find a sufficient number of ($SuccRatio \cdot |POP|$) offspring outperforming their own parents even if ($MaxSelPress \cdot |POP|$) candidates have been generated, premature convergence has occurred.

As a basic principle of this selection model a higher success ratio causes higher selection pressure. Nevertheless, higher settings of success ratio and therefore of selection pressure do not necessarily cause premature convergence as the preservation of fitter alleles is additionally supported and not only the preservation of fitter individuals.

Also it is possible within this model to state selection pressure in a very intuitive way that is quite similar to the notation of selection pressure in Evolution Strategies. Concretely, we define the actual selection pressure as the ratio of individuals that had to be generated in order to fulfill the success ratio to the population size. For example, if we work with a population size of say 100 and it would be necessary to generate 1000 individuals in order to fulfill the success ratio, the actual selection pressure would have a value of 10. Via these means we are in a position to attack several reasons for premature convergence as illustrated in the following sections. Furthermore, this strategy has proven to act as a precise mechanism for self-adaptive selection pressure steering, which is of major importance in the migration phases of parallel evolutionary algorithms. The aspects of offspring selection w.r.t. parallel GAs are combined in the parallel SASEGASA-algorithm [1].

4   Empirical Discussion

The empirical section is subdivided into two parts:

The first subsection aims to highlight the main message of the paper (preservation of essential alleles). As the scope of the present work does not allow a deeper and more sophisticated analysis of different problem situations, the second part of the experimental discussion gives some references to related contributions which include a more detailed and statistically more relevant experimental discussion on the basis of several benchmark but also practical problems on which we have applied the new selection model recently. All empirical work shown and referred in this section have been implemented and performed using the HeuristicLab environment [12].[1]

4.1   Conservation of Essential Genetic Information

This subsection aims to point out the importance of mutation for the recovery of essential genetic information in the case of conventional GAs in order to oppose these results with the results being achieved with the enhanced selection model discussed in this paper. By reasons of compactness, the results are mainly shown on the basis of diagrams and give only a brief description of introduced operators, parameter settings, and test environments. Furthermore, the chosen benchmark instance is of rather small dimension in order to allow the observation of essential alleles during the run of the algorithm.

The results displayed in Figure 2 (left diagram) show the effect of mutation for reintroducing already lost genetic information. The horizontal line of the diagram shows the number of iterations and the vertical line stands for the solution quality. The bottom line indicates the global optimal solution which is known for this benchmark test case. The three curves of the diagram show the performance of a Genetic algorithm with no mutation, with a typical value of 5% mutation as well as a rather high mutation rate of 10%. For each of the three curves the lower line stands for the best solution of the actual population and the upper line shows the average fitness value of the population members. The results with no mutation are extremely weak and the quality curve stagnates very soon and far away from the global optimum. The best and average solution quality are

---

[1] For more detailed information concerning HeuristicLab the interested reader is referred to http://www.heuristiclab.com/

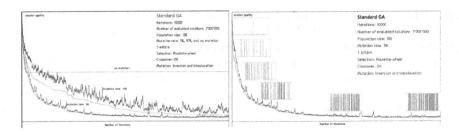

Fig. 2. The effect of mutation for certain mutation rates (left diagram) and the distribution of essential genetic information for a mutation rate of 5% (right diagram) both in case of a standard GA for the ch130 benchmark TSP

the same and no further evolutionary process is possible - premature convergence has occurred. As already stated before, mutation is a very essential feature of standard GAs in order to avoid premature convergence. But also a rather high mutation rate of 10% produces results which are not very satisfying and indeed the best results are achieved with a mutation rate which is very typical for GA applications - namely a mutation rate of 5%. Considering a standard benchmark problem like the ch130 (a 130 city TSP taken from the TSPLib [9]) with one single best solution allows to consider the edges of the shortest path as the essential alleles whose preservation during the run can be observed. The following figures indicate the spreading of essential alleles during the runs of the certain algorithms. This is visualized by inserting bar charts which have to be considered as snapshots after a certain number of iterations approximately corresponding to the position in the figure. The higher a certain bar (130 bars for a 130-city TSP) the higher the relative occurrence of the corresponding essential allele in the population.

The right diagram of Figure 2 shows the distribution of essential alleles over the iterations for a standard GA with a mutation rate of 5%. The interesting thing is that some minor ratio of essential alleles is rapidly fixed in the population and the majority of essential alleles which are still missing have disappeared in the entire population. During the further run of the algorithm it is only mutation which can reintroduce this essential genetic information. As it could be seen in Figure 2, without mutation premature convergence would already have occurred at this early state of evolutionary search. But with an appropriate mutation rate (5% in this example) more and more essential alleles are discovered ending up with quite a good solution. But there is still a gap to the global optimum caused by that alleles which could not be recovered due to mutation. The next figures will show how the new selection concept is able to close this gap and make the algorithm much more independent of mutation.

So let us take a closer look at the distribution of essential genetic information in the population when using the enhanced selection concepts. The left diagram of Figure 3 shows the quality curve and the distribution of essential alleles for

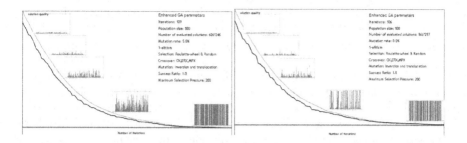

Fig. 3. The distribution of essential genetic information when using the enhanced selection concept considering the ch130 benchmark TSP with 5% mutation (left diagram) and with no mutation (right diagram)

a mutation rate of 5% (which was able to achieve the best results in case of a standard GA).

When applying the GA with the new selection principle to the same benchmark test case one can see that the global optimal solution is detected in only about 100 iterations. Nevertheless, the computational effort is comparable to the standard GA as much more individuals have to be evaluated at each iteration step due to the higher selection pressure. Considering the distribution of essential alleles we see a totally different situation. Almost no essential alleles get lost and the ratio of essential alleles continuously increases in order to end up with a final population that contains almost all pieces of essential genetic information and therefore achieving a very good solution. This shows that the essential alleles are preserved much more effectively and indicates that the influence of mutation should be much less. But is this really the case? In order to answer this question, let us consider the same example with the same settings - only without mutation. And indeed the assumption holds and also without mutation the algorithm finds a solution which is very close to the global optimum (see right diagram of Figure 3). The essential alleles interfuse the population more and more and almost all of them are members of the final population. Reconsidering the standard GA without mutation the algorithm was prematurely converging very soon with a very bad total quality.

4.2   References to Recent Related Works

The basic concepts of the enhanced selection ideas as published in the present paper have already emerged more than one year ago. As the actual focus (like also stated in the present contribution) is to study the properties of the new selection concepts systematically, the potential w.r.t. achievable advancements in global solution quality were obvious immediately. Therefore, the main aim of the first works in this area was to check the generality of the new algorithmic concepts by applying them to various theoretically as well as practically relevant problems. And indeed this worked out very well and it was possible to demonstrate similar effects and achievements in global solution quality in various areas of application under very different problem codifications with exactly that enhanced generic selection techniques as being proposed in this paper.

While the last subsection considered only relatively small TSP instances in order to illustrate some selected aspects, journal article [1] includes a detailed and comprehensive empirical analysis also based on TSP instances of much higher dimension. Furthermore, [1] gives a comprehensive solution analysis based on several real valued n-dimensional test functions (like the n-dimensional Rosenbrock, Rastrigin, Griewangk, Ackley, or Schwefel's sine root function). Also here it is possible to locate the global optimal solution in dimensions up to $n = 2000$ with exactly the same generic extensions of the selection model as being stated here - only the crossover- and mutation-operators have been replaced with standard operators for real-valued encoding.

But also in practical applications like the Optimization of Production Planning in a Real-World Manufacturing Environment based on an extended formulation of the Job-shop Scheduling Problem [2] a significant increase in solution quality could be accomplished with the described methodology. Especially in combination with Genetic Programming self-adaptive selection pressure steering has already proven to be very powerful. In [13] and [14] we report first results achieved in the context of nonlinear structure identification based on time-series data of a diesel combustion engine. Concretely the aim of this project is the development of models for the NOx emission. Already until now it has become possible with a GP-based approach equipped with offspring selection to identify models which are superior to the models achieved with conventional GP-techniques and also superior to machine learning techniques which have also been considered in earlier stages of this project. Very recently we have adapted this GP-approach for the application on symbolic as well as logistic regression problems. First results achieved on benchmark classification problems (taken from the UCI machine learning repository) indicate a high potential also in these areas of application.[2]

5   Conclusion and Future Perspectives

This paper discusses a new generic selection concept and points out its ability to preserve essential genetic information more goal-oriented than standard concepts. Possibly the most important feature of the newly introduced concepts is that the achievable solution quality can be improved in a non-problem specific manner so that it can be applied to all areas of application for which the theory of GAs and GP provides suitable operators. Further aspects worth mentioning concern the robustness and self-adaptiveness of the population genetics and ES inspired measures: Selection pressure can be steered self-adaptively in a way that the amount of selection pressure actually applied is that high that further

---

[2] First results tables for the thyroid and Wiskonsin breast cancer data-sets are shown on http://www.heuristiclab.com/results/regression.

progress of evolutionary search can be achieved. Possible future research topics in that area are certainly to open new areas of application also and especially in GP related applications where the aspect of preservation of essential genetic information is especially important as the disruptive properties of GP-operators tend to add impurities into the genetic information for the ongoing search when using standard parent selection models.

References

1. M. Affenzeller and S. Wagner. SASEGASA: A new generic parallel evolutionary algorithm for achieving highest quality results. *Journal of Heuristics, Special Issue on New Advances on Parallel Meta-Heuristics for Complex Problems*, 10(3):239–263, 2004.
2. R. Braune, S. Wagner, and M. Affenzeller. Applying genetic algorithms to the optimization of production planning in a real-world manufacturing environment. *Proceedings of the European Meeting on Cybernetics and Systems Research - EMCSR 2004*, pages 41–46, 2004.
3. D. Cavicchio. *Adaptive Search using Simulated Evolution*. unpublished doctoral thesis, University of Michigan, Ann Arbor, 1970.
4. K. DeJong. *An Analysis of the Behavior of a Class of Genetic Adaptive Systems*. PhD thesis, Department of Computer Science, University of Michigan, 1975.
5. D. Fogel. An introduction to simulated evolutionary optimization. *IEEE Trans. on Neural Network*, 5(1):3–14, 1994.
6. D. Goldberg. *Genetic Alogorithms in Search, Optimization and Machine Learning*. Addison Wesley Longman, 1989.
7. Z. Michaliwicz. *Genetic Algorithms + Data Structurs = Evolution Programs*. Springer-Verlag Berlin Heidelberg New York, 3. edition, 1996.
8. I. Rechenberg. *Evolutionsstrategie*. Friedrich Frommmann Verlag, 1973.
9. G. Reinelt. TSPLIB - A traveling salesman problem library. *ORSA Journal on Computing 3*, 3:376–384, 1991.
10. E. Schoeneburg, F. Heinzmann, and S. Feddersen. *Genetische Algorithmen und Evolutionsstrategien*. Addison-Wesley, 1994.
11. R. Smith, S. Forrest, and A. Perelson. Population diversity in an immune system model: Implications for genetic search. *Foundations of Genetic Algorithms*, 2:153–166, 1993.
12. S. Wagner and M. Affenzeller. HeuristicLab: A generic and extensible optimization environment. *Accepted to be published in: Proceedings of the International Conference on Adaptive and Natural Computing Algorithms (ICANNGA) 2005*, 2005.
13. S. Winkler, M. Affenzeller, and S. Wagner. Identifying nonlinear model structures using genetic programming techniques. *Proceedings of the European Meeting on Cybernetics and Systems Research - EMCSR 2004*, pages 689–694, 2004.
14. S. Winkler, M. Affenzeller, and S. Wagner. New methods for the identification of nonlinear model structures based upon genetic programming techniques. *Proceedings of the 15th International Conference on Systems Science*, pages 386–393, 2004.
15. Y. Yoshida and N. Adachi. A diploid genetic algorithm for preserving population diversity - pseudo-meiosis GA. *Lecture Notes in Computer Science*, 866:36–45, 1994.

Agent WiSARD in a 3D World

E. Burattini[1], P. Coraggio[1], M. De Gregorio[2], and M. Staffa[1]

[1] Dip. di Scienze Fisiche,
Università degli Studi di Napoli "Federico II", Napoli, Italy
[2] Istituto di Cibernetica "Eduardo Caianiello",
CNR, Pozzuoli (NA), Italy
{ernesto.burattini, paolo.coraggio}@na.infn.it
m.degregorio@cib.na.cnr.it

Abstract. This paper investigates the integration of verbal and visual information for describing (explaining) the content of images formed by three–dimensional geometrical figures, from a hybrid neurosymbolic perspective. The results of visual object classifications involving top–down application of stored knowledge and bottom–up image processing are effectively explained relying on both words and pictures. The latter seems particularly suitable in explanations concerning high–level visual tasks involving both top–down reasoning and bottom–up perceptual processes.

1 Introduction

The system Agent WiSARD proposed in this paper is a neurosymbolic hybrid system whose aim is that of reconstructing and explaining images taken from a three–dimensional simplified block world. The system is the natural evolution of a two–dimensional images reconstruction tool [1][2] in which images formed by triangles, squares and rectangles were analyzed and explained to a user. Agent WiSARD is capable both of recognizing and classifying the single blocks forming the image and of "explaining" it to a user interested in image peculiarities. The blocks can stand alone or have covered parts hidden by other superimposed blocks: in this latter case, the system is able to find and visualize the hidden angles showing which block lays upon the others.

High–level vision, which involves both top–down application of stored knowledge and bottom–up image processing, is a suitable testing domain for the effectiveness of explanations integrating pictures with verbal accounts of problem–solving processes [3]. Indeed, high–level visual tasks may involve problem-solving processes that require the kinds of explanation that are appropriate for knowledge-based systems [4]. Neurosymbolic computational systems can exploit largely complementary advantages of both neural and symbolic information processing [5][6]. In view of their learning and associative capacities, neural networks are known to perform well in visual pattern recognition, whereas symbol manipulation algorithms are tailored for formal reasoning. Combining these features in a neurosymbolic system is clearly appealing for modelling tasks that involve both reasoning and low–level visual processing. An additional motive of interest for

J. Mira and J.R. Álvarez (Eds.): IWINAC 2005, LNCS 3562, pp. 272–280, 2005.

neurosymbolic systems, more specifically related to the explanation problem, is emphasized here: a form of bidirectional associative behavior of neural networks is exploited to produce mental images [7][8].

The neural module of Agent WiSARD is a weightless neural system formed by RAM–discriminators called WiSARD (**W**ilkie, **S**tonham and **A**leksander's **R**ecognition **D**evice) [9]. The standard weightless discriminator model was slightly modified in order to generate a grey–level, non crisp example of the class of simple visual features it was trained to detect [10]. This change does not affect the standard recognition characteristics of the classical WiSARD while improves the description power of the whole hybrid system [11]. On the other hand, the symbolic system is formed by a BDI agent (Belief–Desire–Intention) [12][13]. Agent technology offers a good flexibility both in treating data coming from the external environment (in our case from the neural module), and in interacting with users interested in image features.

Though this approach could remind in part the one proposed by Fukushima "neocognitron" in which the neural network model proposed was based on features search and recognition [14][15], Agent WiSARD is able to explain the picture proposed and to generate mental images (by means of the same neural network) in order to enrich its explanation power about the recognized features.

In the following sections we will introduce the two modules forming the hybrid system and show how the system reconstructs and describes the whole scene.

2   The Neural Module

WiSARD is an adaptive pattern recognition device which is based on neural principles [16]. It is a weightless system whose basic components are RAM–discriminators. In order to generate examples ("mental images") of visual features that the system is able to recognize, RAM–discriminators were modified in what their memory locations may hold and, correspondingly, in their training algorithm [10]. The modified WiSARD can both scan the input image discriminating some peculiar features and generate and show the "mental" images of the classes it is trained to detect [8].

2.1   Agent WiSARD Training Set

One of the neural module's tasks is to identify and locate corners and points of intersection belonging to the whole image (that is formed by blocks). The image to be analyzed is scanned by means of a tool called "spot". The part of the image present into the spot is fed as input to the neural module. The spot is a rectangular "eye" (15–pixel side) that scans only those parts of the image containing black pixels. When a significant number of black pixels are present into the spot, WiSARD starts up its discriminators in order to classify the content of the spot.

The set of classes the system is able to discriminate is shown in Fig. 1.

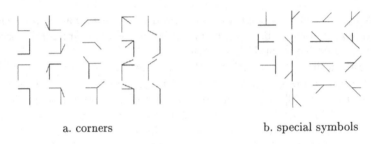

a. corners   b. special symbols

Fig. 1. Classes of corners and special symbols

The corners (Fig. 1.a) are those belonging to blocks, while the special symbols (Fig. 1.b) are typical of block superimposition. The system has been trained with a set of images (examples) for each class of corners and special symbols.

2.2   "Mental Images"

Generating examples of such geometric features involves the construction of grey–level images, obtained on the basis of information held in the modified RAM–discriminator memory locations. Due to the bidirectionality of the modified WiSARD, for each class of discriminated symbols the system is capable of giving grey–level, non crisp example of the class of simple visual features it was trained to detect [8] (see Fig. 2).

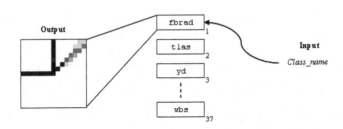

Fig. 2. Mental image of a given class

Summing up, given a bitmap image of a virtual block world as input, the neural module scans the image looking for some peculiar elements. In fact, it is able to discriminate between corners and special symbols integrating these information with their spatial co–ordinates. The neural module is so capable of providing, for each corner and special symbol of the input image:

 – their relative position with respect to a Cartesian axes system originating in the center of the image;
 – the belonging class of the discriminated corner or special symbol;

– the response (i.e. how much the image seems to belong to a particular class) the discriminated corner or special symbol has been detected with respect to the training set.

These information are contained in a list of ground first order logic predicates of the following form: location(name,symbol_type,r,x,y); that is, the neural module has discriminated in (x, y), a symbol_type with a certain response r. name is a label the system assigns to the discriminated symbol. This list of predicates are written in a text file and passed to the symbolic module that assumes them as its starting base beliefs.

3   The Symbolic Module

The symbolic module is formed by a BDI agent designed and implemented in ProAgent [17]. Its task is that of reconstruct (recognize) 3D shapes in a virtual block world scene starting from information collected by the neural module, and eventually that of interact with a user interested in some image peculiarities. The BDI model considers agents as characterized through three mental states: Beliefs (cognitive state) Desires (motivational state) and Intentions (deliberative state). These three states determine agent behavior. Agent knowledge is represented by a set of beliefs (a database of ground first order logic predicates). As the agent acts in its environment the Belief set can dynamically change, asserting (if the agent settles new relationships) or retracting (if it has to drop hypotheses that no longer hold) beliefs. Taking advantages of a hypothesize–and–test cycle [18], the agent transform this set in order to check the right relationship between symbols and thus reconstructing the scene.

In order to reconstruct the scene, the agent task has been structured in a hierarchical manner. The agent has firstly to identify all the possible alignment between the symbols detected by WiSARD, then it has to reconstruct the faces of each block and finally, thanks to information about the faces, to reconstruct free and covered blocks present in the analyzed scene. It is worth notice that it is crucial to keep information that could be useful later in order to have a better description of the scene. For this reason, during its action (and so through its plans), the agent abduces and verifies a set of hypotheses H. Then a subset $H_1 \subseteq H$, satisfying certain geometrical constraints, will be confirmed.

3.1   Image Reconstruction

The first step in reconstructing the scene is that of finding out all the possible alignments between corners. In fact, this kind of alignment suggests the possible existence of lines between certain points.

Not all the discriminated symbols can be coupled in order to establish the right relationship between them. In fact, in order to assert that two symbols are effectively aligned, it is necessary to verify whether some conditions, based on the type of alignment, between the values of the co–ordinates hold. For example, a horizontal alignment between two symbols is established if they have the same ordinate. Once established the alignment between symbols, the agent is able to reconstruct the lines composing the faces of the blocks. This is done activating the set of plans devoted to this goal. The effect is an increasing of the Beliefs set of the agent with predicates of the following type:

```
hor_line(Name,Type,S1,S2,Dist)    ver_line(Name,Type,S1,S2,Dist)
obldx(Name,Type,S1,S2,Dist)
```

The predicates name specify the kind of line (respectively an horizontal, vertical or slanting line), Name is a label assigned by the system, Type identifies the line characteristic (if it is free, covered or that covers another line), S1 and S2 are the line extremes, Dist is the distance between S1 and S2.

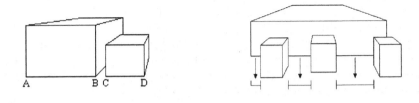

Fig. 3. Free lines (left) and covered line (right)

Referring to Fig. 3 left, the agent is able to find out all the plausible alignments, more precisely, the point A is aligned both with B and D while C is aligned with D. In this case the agent has to set the right alignments dropping the incorrect relationship. In a first stage of its computation, the agent will hold both hypotheses. The wrong one will be dropped later when the agent will try to reconstruct the block faces. In fact it will be looking for all alignment bringing to plausible faces. In the case shown in Fig. 3 left, the agent will state that the correct lines are $\overline{AB}$ and $\overline{CD}$.

In the previous example, it has been shown a case in which only free lines where involved, but a more interesting case is the presence, in the scene, of a block covered by others as shown in Fig. 3 right.

The side of a block face can be formed by more than one line. More precisely, the side can be formed by:

- a free line: if it is just a segment whose vertices are corners;
- a covered line: if it is either a single segment or multiple segments with at least a special symbol as vertex (see Fig. 3 right).

Different predicates will be asserted as agent new beliefs according with different situations. At this point the agent is able to reconstruct the block faces trying to combine the lines it found out. Once that all the possible faces have been asserted, the system can retract those alignment that are no more plausible.

The same kind of reasoning is performed by the system in order to set the blocks in the scene. Having knowledge about all the plausible faces, the system selects the ones that can be put together in order to form blocks. All the remaining information will be, again, retracted by the system.

This approach in building ground beliefs for the input image (finding possible alignments between corners and corners and special symbols, then constructing the possible faces and finally reconstructing the blocks) in a bottom–up manner, helps the image reconstruction process and, as it will be shown in the next section, gives the system much explanation power.

3.2   Scene Explanation

All the information collected into the set of beliefs allow the agent to explain the scene. In fact it is able to describe (in a "pseudo–natural" language way):

- how many blocks are in the scene (free and non free blocks);
- which is the nearest block;
- which blocks cover or are covered by others;
- the whole scene visualizing the elements involved in the explanation.

This system characteristic is firstly due to the fact that agent belief set has been refined through plans application (i.e. its deliberative ability) allowing the agent itself to have an "understanding" of its environment. Moreover, the modularity of the reconstruction process, and in particular of the plan library structure, makes easier the construction of a set of predicates devoted to scene explanation. In fact, the agent begins its analysis in first finding out all the free blocks (if any), then the others. This is easily done if the agent "knows" which faces composing the blocks cover or are covered by other faces.

4   Modules Interaction and System Answers

The interaction between the two modules occurs in two different moments: when the neural module terminates its analysis and feeds the symbolic module with the elements the agent needs to classify the image and when the agent answers to a user question showing the image details by means of "mental images".

The mutual data exchange between the two modules characterizes the system. The neural network is able to detect both the image spatial properties, identifying angles and special symbols coordinates, and the class they belong to, classifying their shape. The symbolic system processes data coming from the neural module comparing the process output with its visual models - in somehow codified in its plans. The agent seeks for relationships between figure components, shifting its attention only upon certain elements of the whole image, creating new information the agent itself needs to reconstruct the image. Information flow in two directions:

- from screen to agent: we have got the "visual" recognition of single figures;
- from agent to screen: when the agent communicates to neural module to visualize, associating the same color, a figure. In this way we obtain the "mental image" of single components forming the overall image.

In both cases, the information flows through the neural module.

Once the image has been reconstructed (i.e. interpreted), the system is able to "talk about" it answering to users questions using both "verbal" and "pictorial" components. In fact, beside the answer written in the agent dialog interface (figure 4), the agent communicates to the neural module the vertices to be shown. In order to give a better description, vertices belonging to the same picture are visualized with the same color. A user can ask, for example, to the system to visualize the free blocks it has recognized: in this case, beside a verbal answer of the kind "There are N free blocks", into another window will appear the reconstructed scene with the vertices of the above N blocks differently colored.

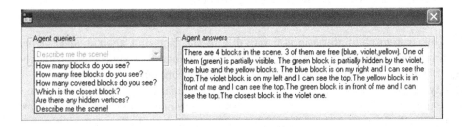

Fig. 4. Agent dialog interface

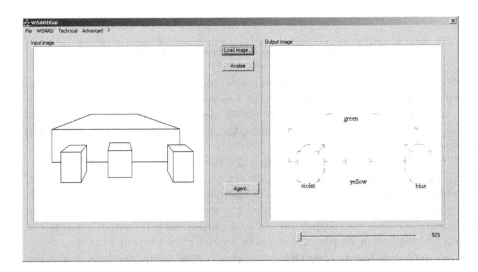

Fig. 5. Neural dialog interface

A query like: "Describe me the scene!" will produce the verbal answer reported in Fig. 4, while in another window will be shown the mental images of the elements of the blocks involved in the explanation (see Fig. 5).

5 Conclusions

It is a widely shared view that visual perception and imagery share various underlying mechanisms [19][20]. This basic view is reflected in the fact that visual object classification and generation (the latter one is a sort of imagery function performed by the system, since a picture of the object is generated from a verbal stimulus only) are performed by the same modules of the whole hybrid system, because these are capable of appropriate forms of bidirectional behaviour, as shown in one of the previous sections.

The proposed system has been tested in various situations giving encouraging results. In fact, the set of recognizable images belonging to a block world is quite wide and the system gives satisfactory explanations. This fact suggest a possible development in the direction of a "virtual scene navigation" system. Since Agent WiSARD is able to describe a scene, it means that it could virtually move in such an environment. Adding new plans for transforming co–ordinates and symbols, Agent WiSARD could be able to virtually navigate the scene. When the information is no longer sufficient to let the navigation continue, Agent WiSARD could ask the neural module to scan again the new environmental image (from the agent position point of view) in order to have an up to date "mental image" of the scene.

References

1. Burattini E., Coraggio P., De Gregorio M., Ripa B., (2003), "Agent Wisard go and catch that image", in *Proc. First IAPR–TC3 Workshop*, Florence, Italy, 89–95.
2. Burattini E., Coraggio P., De Gregorio M., (2003), "Agent Wisard: a hybrid system for reconstructing and understanding two–dimensional geometrical figures", *Design and Application of Hybrid Intelligent Systems*, Abraham *et al.* eds, 887–896.
3. Burattini E., De Gregorio M., Tamburrini G., (2001), "Pictorial and Verbal Components in AI Explaination", in *Vision: The Approach of Biophysics and Neuroscience*, C. Musio ed., World Scientific, 471–475.
4. Cawsey A., (1995), "Developing an explanation component for a knowledge-based system", *Expert Systems with Applications*, **8**, 527–531.
5. Sun R., Bookman L.A. (eds.), (1995), *Computational Architectures Integrating Neural and Symbolic Processes*, Kluwer, Dordrecht.
6. Sun R., Alexandre F., (1997), *Connectionist–Symbolic Integration*, Lawrence Erlbaum, Hillsdale, NJ.
7. De Gregorio M., (1996), "Integrating Inference and Neural Classification in a Hybrid System for Recognition Tasks", *Mathware & Soft Computing*, **3**, 271–279.
8. Burattini E., De Gregorio M., Tamburrini G., (1998), "Generating and classifying recall images by neurosymbolic computation", in *Proc. Second European Conference on Cognitive Modelling*, Nottingham, UK, Ritter and Young Eds., 127–134.

9. Aleksander I., Thomas W., Bowden P., (1984), "WiSARD, a radical new step forward in image recognition", *Sensor Rev*, 120–124.
10. De Gregorio M., (1997), "On the reversibility of multidiscriminator systems". *Technical report 125/97*, Istituto di Cibernetica CNR, Pozzuoli, Italy.
11. Burattini E., De Gregorio M., Tamburrini G., (2000), "Mental imagery in explaination of visual object classification", *VIth Brazilian Symposium on Neural Networks*, Rio de Janeiro, Brazil, 137–143.
12. Rao M. S., Georgeff M. P., (1992), "An abstract architecture for rational agents", *Proceedings of Knowledge Representation and Reasoning*, C. Rich, W. Swartout, and B. Nebel Eds, 439–449.
13. Staffa M., (2004), "A BDI Agent Acting in a Block World", *Laurea Thesis in Informatics*, Università degli Studi di Napoli "Federico II", Italy.
14. Fukushima K., (1980), "Neocognitron: A self-organizing Neural Network Model for a Mechanism of Pattern Recognition Uneffected by Shift in Position", *Biological Cybernetics*, **36**, Springer–Verlag, 193–202.
15. Fukushima K., Miyake S., (1982), "Neocognitron: a new algorithm for pattern recognition tolerant of deformations and shifts in position", *Pattern Recognition*, **36**, No. 6, Pergamon Press Ltd, 455–469.
16. Aleksander I., Morton H. (1990), *An introduction to Neural Computing*, Chapman & Hall Eds.
17. Coraggio P., (2003), "ProAgent: a language for designing and implementing BDI agents. A case of study: an agent for recognizing and classifying geometrical figures", *Laurea Thesis in Physics*, Univ. degli Studi di Napoli "Federico II", Italy.
18. Burattini E., De Gregorio M., (1994), "Qualitative Abduction and Prediction; Regularities over Various Expert Domains", *Inf. and Decision Technologies*, **8**, 471–481.
19. Kosslyn S. M., (1994), *Image and Brain*, MIT Press Cambridge MA, 54–60.
20. Ishai A., Sagi D., (1995), "Common mechanisms of visual imagery and perception.", in *Science*, **268**, 1772–1774.

One Generalization of the Naive Bayes to Fuzzy Sets and the Design of the Fuzzy Naive Bayes Classifier

Jiacheng Zheng[1] and Yongchuan Tang[2]

[1] College of Economics, Zhejiang University,
Hangzhou, Zhejiang Province, 310027, P. R. China
zjcheng@zj.com
[2] College of Computer Science, Zhejiang University,
Hangzhou, Zhejiang Province, 310027, P. R. China
yongchuan@263.net

Abstract. Despite its unrealistic independence assumption, the Naive Bayes classifier is remarkably successful in practice. In the Naive Bayes classifier, all variables are assumed to be nominal variables, it means that each variable has a finite number of values. But in large databases, the variables often take continuous values or have a large number of numerical values. So many researchers discussed the discretization (or partitioning) for domain of the continuous variables. In this paper we generalize the Naive Bayes classifier to the situation in which the fuzzy partitioning for the variable domains instead of discretization is taken. Therefore each variable in the Fuzzy Naive Bayes classifier can take a linguistic value represented by a fuzzy set. One method for estimating the conditional probabilities in the Fuzzy Naive Bayes classifier is proposed. This generalization can decrease the complexity for learning optimal discretization, and increase the power for dealing with imprecise data and the large databases. Some well-known classification problems in machine learning field have been tested, the results show that the Fuzzy Naive Bayes classifier is an effective tool to deal with classification problem which has continuous variables.

1   Introduction

During the past several years, many techniques dealing with uncertain information have attracted the attention of researchers and engineers. The uncertain information may include randomness, fuzziness. How to represent the uncertain information and how to reason using uncertain information are the key problems of the artificial intelligence research field. The uncertainty due to randomness can be dealt with by the theory of probability (subjective or objective), Bayesian belief networks (BN) are the normal knowledge representation for randomness. Because of its mathematics foundation, the Bayesian belief network has been an important tool to reason on the uncertain situation. For instance, it has been used in expert systems [6], classification systems [3] etc.. On the other hand, the uncertainty due to the fuzziness can be dealt with by the theory of fuzzy set, the theory of fuzzy set provides an appropriate framework to describe the fuzziness of human knowledge. For instance, Fuzzy logic control, fuzzy classification, fuzzy clustering, fuzzy intelligent decision have succeeded in many applications. These

J. Mira and J.R. Álvarez (Eds.): IWINAC 2005, LNCS 3562, pp. 281–290, 2005.

two theories can live side by side in providing tools for uncertainty analysis in complex, real-world problems [1, 7, 8, 11, 12, 13, 14]. Many researchers have discussed this cooperation, L.A. Zadeh firstly provided the concept of fuzzy event and its probability [14]. The term "fuzzy Bayesian Inference" was introduced by [13] meaning the generalization of Bayesian Statistics to fuzzy data, the more deep discussion about this was presented in [12].

Bayesian belief networks are powerful tools for knowledge representation and inference under uncertainty, they were not considered as classifiers until the discovery that Naive Bayes, a very simple kind of Bayesian belief network that assumes the variables are independent given the classification node, are surprisingly effective. In Naive Bayes, all variables are assumed to be nominal variables meaning each variables has a finite number of values. But in practice, in the large databases many attributes often have a large number of numerical values. In order to deal with the continuous variables, many approaches in machine learning were proposed by discretizing them [2, 4]. But this discretization or partitioning often causes the loss of the information. In this paper, we use fuzzy partitioning for the domains of the continuous variables instead of the discretization. So the Naive Bayes is generalized to the Fuzzy Naive Bayes in which each variable takes the linguistic values represented by the fuzzy sets on the domain of each variable. In the Fuzzy Naive Bayes, the computation and propagation of the probabilities have no difference with the classic Naive Bayes. One method for computing the conditional probabilities in the Fuzzy Naive Bayes from the observed data is proposed, some classification problems varying from several attributes and hundreds of data points to more than thirty attributes and thousands of data points are used to test the performance of the Fuzzy Naive Bayes classifier in this paper.

In what follows, we review the Bayesian belief network, and proceed to build our Fuzzy Naive Bayes classifier. We introduce one method to compute the conditional probabilities of each node in the Fuzzy Naive Bayes classifier and discuss its classification process. Finally, this Fuzzy Naive Bayes classifier is tested by some classification problems in machine learning field.

2   Bayesian Belief Network

A Bayesian belief network consists of a graphical structure that is augmented by a set of probabilities. The graphical structure is a directed, acyclic graph in which nodes represent domain variables which have a finite number of possible values. Prior probabilities are assigned to source nodes, and conditional probabilities are associated with arcs. In particular, for each source node x_i (i.e., a node without any incoming arcs), there is a prior probability function $P(x_i)$; for each node x_i with one or more direct predecessors π_i, there is a conditional probability function $P(x_i|\pi_i)$. That probability functions are represented in the form of explicit function tables called as the conditional probability tables (CPT). A general Bayesian belief network is represented as (V, A, P), where V is the set of variables (i.e., vertices or nodes). A the set of arcs between variables, and P the set of probabilities.

Bayesian belief network is capable of representing the probabilities over any discrete sample space, such that the probability of any sample point in that space can be computed from the probabilities in the Bayesian belief network. The key feature of Bayesian belief network is their explicit representation of the conditional independence among events. A Bayesian belief network represents a full join-probability space over the n event variables in the network, and it states that the join-probability $P(x_1, x_2, \ldots, x_n)$ can be factorized into the multiplication of the conditional probabilities of the variables given their respective parents, i.e.,

$$P(x_1, x_2, \ldots, x_n) = \prod_i P(x_i | \pi_i) \tag{1}$$

Let θ_{ijk} denote $P(x_i = j | \pi_i = k)$, where j is a value of variable x_i and k is a combination of the values of the parents of x_i. For convenience, we shall say that k is a value of π_i and call θ_{ijk} a *parameter pertaining to variable* x_i. And θ is the vector of all parameters θ_{ijk}. The parameter vector θ is to be estimated from a collection D of data cases $D_1, D_2, \ldots, D_l$ that are independent given θ. Each data case is a set of variable-value pairs. So the estimate θ^* of θ can be obtained by setting [5]

$$\theta_{ijk}^* = \frac{f(x_i = j, \pi_i = k)}{\sum_j f(x_i = j, \pi_i = k)} \qquad \text{for all } i, j, \text{ and } k. \tag{2}$$

where $f(x_i = j, \pi_i = k)$ stands for the number of data cases where $x_i = j$ and $\pi_i = k$ in D.

Although an arc from a node x to a node y frequently is used to express that x cause y, this interpretation of arcs in Bayesian belief networks is not the only one possible. For example, y may be only correlated with x, but not caused by x. Thus, although Bayesian belief networks are able to represent causal relationships, they are not restricted to such causal interpretations. In this regard, Bayesian belief networks can be viewed as a representation for probabilistic rule-based systems. Many researchers have studied the structure and the parameters learning problems of Bayesian belief network [9, 10], and its application [6].

When Bayesian belief network is applied to the classification problem, one of the most effective classifier is the so-called Naive Bayesian classifier. When represented as a Bayes network, it has the simple structure (V, A, P) proposed in Fig. 1. This classifier learns from observed data the conditional probability of each variable X_i given the class label C. Classification is then done by applying Bayes rule to compute the probability $P(C|X_1, \ldots, X_n)$ and then predicting the class with the highest posterior probability. This computation is feasible by making the strong assumption that the variables X_i are conditionally independent given the value of the class C.

In Naive Bayes classifier, the discriminant function is the class posterior probability function $P(C|X_1, \ldots, X_n)$, the classifier selects the class with the maximum discriminant function given the feature $X_1, \ldots, X_n$. Applying the Bayes rule the class posterior probability $P(C|X_1, \ldots, X_n)$ is given

$$
\begin{aligned}
P(C|X_1, \ldots, X_n) &= \frac{P(X_1, \ldots, X_n|C)P(C)}{P(X_1, \ldots, X_n)} \\
&= \frac{\prod_i P(X_i|C)P(C)}{P(X_1, \ldots, X_n)}.
\end{aligned} \tag{3}
$$

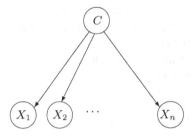

Fig. 1. The Naive Bayes classifier

Where $P(X_1, \ldots, X_n)$ is identical to all classes, so the discriminant function of the classifier is simplified as $P(X_1, \ldots, X_n|C)P(C)$ which also is called as class-conditional probability distribution.

In summary, Bayesian belief networks allow an explicit graphical representation of the probabilistic conditional dependencies and independencies among variables that may represent events, states, objects, propositions, or other entities. Generally, a Bayesian belief network greatly reduces the number of probabilities that must be assessed and stored (relative to the full join-probability space).

3   Fuzzy Naive Bayes Classifier

Handling the continuous variables is a key issue in machine leaning and pattern recognition, in real databases, the variables often are continuous. So in order to use the Naive Bayes classifier, one way is to discretize the domains of the variables. Here we partition the domain of each continuous variable into fuzzy regions. Therefore the Naive Bayes classifier is generalized to Fuzzy Naive Bayes classifier in which each variable is a linguistic variable taking the linguistic values. All linguistic values defined as fuzzy sets constructs a fuzzy partition of the domain of the continuous variable. It is clear that the Naive Bayes classifier is a specification of the Fuzzy Naive Bayes classifier.

Given the observed data set, eq. (2) provides a method to estimate the conditional probabilities of the Bayesian belief network. In the following we will consider the computation of the conditional probabilities of each variable in the Fuzzy Naive Bayes classifier.

In the Fuzzy Naive Bayes classifier, each variable takes the linguistic values which each linguistic value associates with a membership function. Let $\{A_j^i : j = 1, \ldots, n_i\}$ is the fuzzy partition of the domain of the variable X_i, and $\{C_i : i = 1, \ldots, m\}$ is the class label set of the class variable C. Therefore the observed data set D is partitioned into m subsets $D_i, i = 1, 2, \cdots, m$, in which each subset D_i includes the data elements having the form $\overline{X} = [X; C_i]$, and X is a n-dim vector $[x_1, x_2, \cdots, x_n]$. Here we provide a method to estimate the conditional probability as follows:

$$P(X_i = A_j^i | C = C_k) = \frac{\sum_{\overline{X} \in D_k} A_j^i(x_i)}{\sum_{j=1}^{n_i} \sum_{\overline{X} \in D_k} A_j^i(x_i)}, \tag{4}$$

where x_i is the i-th component of the vector $\overline{X} \in D_k$. It should be pointed out that eq. 4 is a direct generalization of eq. (2).

So we can use eq. (4) to compute the conditional probabilities in the Fuzzy Naive Bayes classifier. The following is the construction process of Fuzzy Naive Bayes,

1. Input the observed data $X = [x_1, \ldots, x_n]$, for each x_i, compute the membership degree $A_j^i(x_i), j = 1, \ldots, n_i$. Let A_i be the fuzzy set such that $A_i(x_i) = \bigvee_j A_j^i(x_i)$.
2. For each $A_i, i = 1, \ldots, n$, use eq. (3) to compute the conditional probability $P(C = C_k | X_1 = A_1, \ldots, X_n = A_n), k = 1, 2, \ldots, m$.
3. Output C_s as the class label of the data X in which $P(C = C_s | X_1 = A_1, \ldots, X_n = A_n) = \bigvee_k P(C = C_k | X_1 = A_1, \ldots, X_n = A_n)$.

The basic steps to construct a Fuzzy Naive Bayes classifier has been given. But how to determine the fuzzy partitioning for the domains of the variables is still an open problem. Because the Bayesian belief networks reflect the causal relationship of the current node and its parents. Like the construction of fuzzy if-then rules, the knowledge of the domain experts may be applied to partitioning the feature space. Of course the adapted tuning for the fuzzy partitioning needs more research.

4   Experiment Study

In this section we present some classification problems we have work with. All of database can be obtained in MLC++ datasets from UCI.

The first classification problem we have worked with is *Iris Plants Database* which was created by R.A. Fisher. This is perhaps the best known database to be found in the pattern recognition literature. Fisher's paper is a classic in the field and is quoted frequently nowadays. The data set contains 3 classes of 50 instances each, where each class refers to a type of iris plant. The number of instances in the Iris Plants Database is 150 (50 in each one of three classes), and the number of attributes is 4. Relevant information about it may be seen in Table 1.

Table 1. The relevant information of *Iris Plant Database*

C: C_1, Iris Setosa; C_2, Iris Versicolour; C_3, Iris Virginica						
Attribute	Type	Description	Min (cm)	Max (cm)	Mean (cm)	SD (cm)
X_1:	Continuous	Sepal Length	4.3	7.9	5.84	0.83
X_2:	Continuous	Sepal Width	2.0	4.4	3.05	0.43
X_3:	Continuous	petal Length	1.0	6.9	3.76	1.76
X_4:	Continuous	petal Width	0.1	2.5	1.20	0.76

The first attribute indicates sepal length, the second sepal width, the third petal length and the latter indicates petal width. All the attributes are measured in centimeters (cm). The class may be Iris Setosa, Iris Versicolor or Iris virginica. There are values for all the attributes in the Iris Plants Database. In the tests, we use 100 instance for computing the conditional probabilities of the Fuzzy Naive Bayes classifier, in which there are 35 instances labeled as Iris Setosa, 33 instances labled as Iris Versicolour and 32 instances labeled as Iris Virginica, and 50 instances (15 instances labeled as Iris Setosa, 17 instances labeled as Iris Versicolour and 18 instance labeled as Iris Virginica) for testing the Fuzzy Naive Bayes classifier.

We assume that the domain of the i-th feature is divided into n_i (here $n_i = 3$) fuzzy sets labeled as $A_1^i, A_2^i, \cdots, A_{n_i}^i$ for $i = 1, 2, 3, 4$. Though any type of membership functions (e.g., the triangle-shaped, trapezoid-shaped and bell-shaped) can be used for fuzzy sets, we employ the gauss-shaped fuzzy sets A_j^i, with the following membership function:

$$A_j^i(x_i) = \exp \frac{(x_i - a_j^i)^2}{-2(\sigma_j^i)^2}. \tag{5}$$

So the membership function is determined by two parameters (a, σ).

The classification results of the Fuzzy Naive Bayes classifier are compared with the given classes, and discrepancies arising from mismatch between the given classes and the achieved classes are reported in Table 2. For Iris Plant database, the discrepancies between the actual classes and the achieved classes is very few (a total of 7 in 100 data points for training and a total of 5 in 50 data points for testing).

Table 2. Classification of Iris Plant Database

	Train (100 points) Success rate 0.93		Test (50 points) Success rate 0.90	
	Achieved class label $C_1\ C_2\ C_3$		Achieved class label $C_1\ C_2\ C_3$	
Actual C_1	35		15	
Class C_2	33		17	
Label C_3	7	25	5	13

The second classification problem we have worked with is *Pima Indians Diabetes Database* which was created by Vincent Sigillito of Johns Hopking University. The data set contains 2 classes, Diabetes and No-Diabetes. The number of attributes is 8, all numeric-valued, the relevant information of *Pima Indians Diabetes Database* is in table 3. The total number of instances is 768. we use 512 instances (345 No-Diabetes; 167 Diabetes) to construct the Fuzzy Naive Bayes classifier, and 256 instances (155 No-Diabetes; 101 Diabetes) to test the performance of the final Fuzzy Naive Bayes classifier.

Many users have used different algorithms to test this classification problem, in the website of MLC++ their performances have been listed, see Table 4.

In our study we still employ gauss-shaped membership function for the fuzzy sets. And the fuzzy partitioning for the domian of each variable is an evenly fuzzy partitioning. In Table 5, we list the experimental results for different fuzzy partitions. Comparing the Table 4 and Table 5 shows that the performance of our Fuzzy Naive Bayes classifier

Table 3. Relevant information of *Pima Indians Diabetes Database*

Attribute	Type	Name	Min	Max	Mean	SD
		C: C_1, No-Diabetes; C_2, Diabetes				
X_1	Continuous	Number of times pregnant	0	17	3.8	3.4
X_2	Continuous	Plasma glucose concentration at 2 hours in an oral glucose tolerance test	0	199	120.9	32.0
X_3	Continuous	Diastolic blood pressure	0	122	69.1	19.4
X_4	Continuous	Triceps skin fold thickness	0	99	20.5	16.0
X_5	Continuous	2-Hour serum insulin	0	846	79.8	115.2
X_6	Continuous	Body mass index	0	67.1	32.0	7.9
X_7	Continuous	Diabetes pedigree function	0.084	2.42	0.5	0.3
X_8	Continuous	Age (years)	21	81	33.2	11.8

Table 4. List of the different algorithms applied to *Pima Indians Diabetes Database*

Algorithm	Success Rate Train	Test	Time Train	Test	Algorithm	Success Rate Train	Test	Time Train	Test
LogDisc	78.09	77.7	31	7	C4.5	86.92	73.0	12	1
Dipol92	?	77.6			IndCart	92.14	72.9	18	17
Discrim	78.01	77.5	27.3	6	BayTree	?	72.9		
Smart	82.27	76.8	314	?	LVQ	?	72.8		
Radial	?	75.7			Kohonen	?	72.7		
Itrule	?	75.5			Ac2	100	72.4	648	29
BackProp	?	75.2			NewId	100	71.1	10	10
Cal5	76.8	75.0	40	1	Cn2	98.98	71.1	38	3
Cart	77.31	74.5	61	2	Alloc80	71.24	69.9	115	?
Castle	73.97	74.2	29	4	KNN	100	67.6	1	2
QuaDisc	76.28	73.8	24	6	Default	?	65.0		
Bayes	76.07	73.8	2	1					

Table 5. Classification results of the Fuzzy Naive Bayes classifier applied to *Pima Indians Diabetes Database*

		The number of fuzzy sets of each variable					
		2	3	4	5	6	7
Success	Train	0.6816 (349/512)	0.7402 (379/512)	0.7715 (395/512)	0.7559 (387/512)	0.7617 (390/512)	0.7910 (405/512)
Rate	Test	0.6133 (157/256)	0.6758 (173/256)	0.7148 (183/256)	0.7266 (186/256)	0.7148 (183/256)	0.6953 (178/256)

is satisfactory and the improvement might be achieved by tuning the fuzzy partition of each variable.

The third classification problem we have worked with is a *Satellite Image Database*. This database consists of the multi-spectral values of pixels in 3×3 neighborhoods in a satellite image, and the classification associated with the central pixel in each neighborhood. The aim is to predict this classification, given the multi-spectral values. The original database was generated from Landsat Multi-Spectral Scanner image data, *Satellite Image Database* was generated taking a small section (82 rows and 100 columns) from the original data. One frame of Landsat imagery consists of four digital images of the same scene in different spectral bands. Two of these are in the visible region (corresponding approximately to green and red regions of the visible spectrum) and two are in the (near) infra-red. Each pixel is a 8-bit binary word, with 0 corresponding to black and 255 to white. The spatial resolution of a pixel is about $80m \times 80m$. Each image contains 2340×3380 such pixels. *Satellite Image Database* is a (tiny) sub-area of a scene, consisting of 82×100 pixels. Each line of data corresponds to a 3×3 square neighborhood of pixels completely contained within the 82×100 sub-area. Each line contains the pixel values in the four spectral bands of each of the 9 pixels in the 3×3 neighborhood and a number indicating the classification label of the central pixel. *Satellite Image Database* was divided into two sets, one is training set containing 4435 data points, the other is testing set containing 2000 data points. There are 36 attributes (= 4 spectral bands $\times 9$ pixels in neighborhood), each attribute can take a large number of possible values ranging from 0 to 255, and the decision class number is 6, Table 6 describes some relevant information of the classes.

Table 6. Class information of *Satellite Image Database*

Class	Description	Train	Test
C_1	red soil	1072 (24.17%)	461 (23.05%)
C_2	cotton crop	479 (10.80%)	224 (11.20%)
C_3	grey soil	961 (21.67%)	397 (19.85%)
C_4	damp grey soil	415 (9.36%)	211 (10.55%)
C_5	soil with vegetation stubble	470 (10.60%)	237 (11.85%)
C_6	7 very damp grey soil	1038(23.40%)	470 (23.50%)

In each line of data the four spectral values for the top-left pixel are given first followed by the four spectral values for the top-middle pixel and then those for the top-right pixel, and so on with the pixels read out in sequence left-to-right and top-to-bottom. Thus, the four spectral values for the central pixel are given by attributes 17,18,19 and 20.

In the website of MLC++ the test results of many other machine learning methods applied to *Satellite Image Database* have been given, see Table 7.

Like the classification problem *Pima Indians Diabetes Database* we have worked with, in the design of the Fuzzy Naive Bayes classifier, the fuzzy partition of each variable is still an evenly fuzzy partition and each fuzzy set is gauss-shaped. we have achieved different performances in different fuzzy partitions, in our study when the number of fuzzy sets of each variable is 5 the best performance is achieved, see Table 8. In this situation our test results are close to the algorithm Koholen, but the Fuzzy Naive Baye classifier needn't the burden of iterative learning process. The improvement might be achieved by studying the distribution of the fuzzy sets.

Table 7. Test results of other machine learning methods applied to *Satellite Image Database*

Algorithm	Success Rate Train	Test	Time Train	Test	Algorithm	Success Rate Train	Test	Time Train	Test
KNN	91.1	90.6	2105	944	Cal5	87.8	84.9	1345	13
LVQ	?	89.5			QuaDisc	89.4	84.5	276	93
Dipol92	?	88.9			Ac2	?	84.3	8244	17403
Radial	88.9	87.9	723	74	Smart	87.7	84.1	83068	20
Alloc80	96.4	86.8	63840	28757	LogDisc	88.1	83.7	4414	41
IndCart	98.9	86.2	2109	9	Cascade	?	83.7		
Cart	92.1	86.2	348	14	Discrim	85.1	82.9	68	12
BackProp	88.8	86.1	54371	39	Kohonen	89.9	82.1	12627	129
BayTree	?	85.3			Castle	?	80.6		
NewId	93.3	85.0	296	53	Bayes	71.3	71.3	56	12
Cn2	98.6	85.0	1718	16	Default	24.0	24.0		
C4.5	95.7	85.0	449	11					

Table 8. Test results of the Fuzzy Naive Bayes classifier applied to *Satellite Image Database*

	The number of fuzzy sets of each variable			
	3	4	5	6
Success Train	0.6207 (2753/4435)	0.7202 (3194/4435)	0.8059 (3574/4435)	0.7869 (3490/4435)
Rate Test	0.6085 (1217/2000)	0.7165 (1433/2000)	0.8090 (1618/2000)	0.7770 (1554/2000)

In our three experimental studies, the Fuzzy Naive Bayes classifier has been applied to the simple classification problem having a few attributes and hundreds of data points, the medium classification and the large classification having more attributes and thousands of data points , the test results manifest that the Fuzzy Naive Bayes classifier can adapt to the different kinds of classification problems. And the results show that the classifier can be improved. One improvement might consist of the studying the distribution of the fuzzy sets for the domain of each variable.

5   Conclusion

In order to deal with the large databases in which many attributes have a large number of numeric values or are continuous variables, we provide the Fuzzy Naive Bayes classifier which uses the fuzzy partition of the each continuous variable instead of the discretization. Based on this fuzzy partition, we propose a method to estimate the conditional probabilities in Fuzzy Naive Bayes classifier.

Several testing have been done and the results have been studied in-depth using three data bases, the *Iris Plants Database*, the *Pima Indians Diabetes Database* and the *Satellite Image Database*. The number of the attribute varies from 4 to more than 30, and number of data points varies from hundreds to thousands. The test results show that the Fuzzy Naive Bayes classifier is a very effective classifier. One possible way to improve the performance of Fuzzy Naive Bayes classifer is taking into account the distribution of the fuzzy sets of each variable.

Acknowledgements

This work has been supported by Zhejiang Province Natural Science Foundation (No. Y104225), Department of Education of Zhejiang Province (Grant No. 20040115).

References

1. Didier Dubois and Henri Prade, Random sets and fuzzy interval analysis, *Fuzzy Sets and Systems* **42** (1991) 87-101.
2. F. EL-MATOUAT, O. COLOT, P. VANNOORENBERGHE and J. LABICHE, From Continuous To Discrete Variables For Bayesian Network Classifiers, *Proceedings of 2000 IEEE international conference on systems, man and cybernetics*, Nashville, Tennessee, USA, October 2000, 2800-2805.
3. N. Friedman, D. Geiger, and M. Goldszmit, Bayesian network classifiers, *Machine Learning*, vol. 29, 1997, 131-163.
4. N. Friedman and M. Goldszmidt, Discretization of continuous attributes while learning Bayesian networks, in *Proceedings of the Thirteenth International Conference on Machine Learning*, (Morgan Kaufmann, San Francisco, CA, 1996), 157-165.
5. Gernot D. Kleiter, Propagating imprecise probabilities in Bayesian networks, *Artificial Intelligence* **88** (1996) 143-161.
6. Gernot D. Kleiter, Bayesian diagnosis in expert systems, *Artificial Intelligence* **54** (1992) 1-32.
7. Hung T. Nguyen, Fuzzy sets and probability, *Fuzzy Sets and Systems* **90** (1997) 129-132.
8. John A. Drakopolulos, Probabilities, possibilities, and fuzzy sets, *Fuzzy Sets and Systems* **75** (1995) 1-15.
9. Pedro Larranaga, Cindy M. H. Kuijpers, Roberto H. Murga, and Yosu Yurramendi, Learning Bayesian Network structures by searching for the best ording with genetic algorithms, *IEEE transactions on systems, man, and cybernetics-part A: systems and humans*, **vol. 26, No. 4**, July 1996, 487-493.
10. Radford M. Neal, Connectionist learning of belief networks, *Artificial Intelligence* **56** (1992) 71-113.
11. Ronald R. Yager,Dimitar P. Filev, Including Probabilistic Uncertainty in Fuzzy Logic Controller Modeling Using Dempster-Shafer Theory, *IEEE transactions on systems, man and cybernetics*, **vol. 25, No. 8**, Augest 1995, 1221-1230.
12. Sylvia Fruhwirth-Schnatter, On fuzzy bayesian inference, *Fuzzy Sets and Systems* **60** (1993) 41-58.
13. R. Viertl, Is it necessary to develop a fuzzy Bayesian inference? in: R. Viertl (Ed.), *Probability and Bayesian Statistics* (Plenum Publishing Company, New York, 1987) pp. 471-475.
14. L.A.Zadeh, Probability measures of fuzzy events, *J.Math. Analysis and Appl.*, **10** (1968),421-427.

Towards a Methodology to Search for Near-Optimal Representations in Classification Problems

Manuel del Valle[1], Beatriz Sánchez[1,2],
Luis F. Lago-Fernández[1,3], and Fernando J. Corbacho[1,3]

[1] Escuela Politécnica Superior,
Universidad Autónoma de Madrid, 28049 Madrid, Spain
[2] Telefónica Investigación y Desarrollo,
C/ Emilio Vargas 6, 28043 Madrid, Spain
[3] Cognodata Consulting, C/ Caracas 23, 28010 Madrid, Spain

Abstract. This paper provides a first step towards a methodology that allows the search for near-optimal representations in classification problems by combining feature transformations from an initial family of basis functions. The original representation for the problem data may not be the most appropriate, and therefore it might be necessary to search for a new representation space that is closer to the structure of the problem to be solved. The outcome of this search is critical for the successful solution of the problem. For instance, if the objective function has certain global statistical properties, such as periodicity, it will be hard for methods based on local pattern information to capture the underlying structure and, hence, afford generalization capabilities. Conversely, once this optimal representation is found, most of the problems may be solved by a linear method. Hence, the key is to find the proper representation. As a proof of concept we present a particular problem where the class distributions have a very intricate overlap on the space of original attributes. For this problem, the proposed algorithm finds a representation based on the trigonometric basis that provides a solution where some of the classical learning methods, e.g. multilayer perceptrons and decision trees, fail. The methodology is composed by a discrete search within the space of basis functions and a linear mapping performed by a Fisher discriminant. We play special emphasis on the first part. Finding the optimal combination of basis functions is a difficult problem because of its non-gradient nature and the large number of possible combinations. We rely on the global search capabilities of a genetic algorithm to scan the space of function compositions.

1   Introduction

Classical methods for pattern classification are based on the existence of statistical differences among the distributions of the different classes. The best possible situation is perfect knowledge of these distributions. In such a case, Bayes classification rule gives the recipe to obtain the best possible solution. In real problems,

J. Mira and J.R. Álvarez (Eds.): IWINAC 2005, LNCS 3562, pp. 291–299, 2005.
© Springer-Verlag Berlin Heidelberg 2005

however, class distributions are rarely available because the number of patterns is generally small compared with the dimensionality of the feature space. To tackle this problem many techniques of density estimation have been developed, both parametric and non-parametric [2]. When density estimation fails, we have a variety of supervised learning algorithms, such as neural networks [1] or support vector machines [9], that try to find a non-linear projection of the original attribute space on to a new space where a simple linear discriminant is able to find an acceptable solution.

Let us assume a particular classification problem in which, when looking at the original attribute space, we observe an almost complete overlap among the class distributions. Following the Bayes rule, we see that for any point in this attribute space, the probabilities of belonging to any of the classes are all equal. Must we conclude that there is no solution to the problem better than choosing the class randomly?

When dealing with a problem like this one, a supervised learning algorithm, like a backpropagation neural network, will try to exploit the local differences that exist in the examples of each class, assuming that these differences are generalizable and are not due to noise. This usually leads to very complex solutions, in general difficult to interpret, and that eventually may not be able to generalize. This kind of algorithms have the additional problem of local minima, derived from their intrinsic non-linearity. Functional Link Networks (FLNs) [6] avoid this problem by using a simple linear projection and relaying the non-linearity to the operation of input units. Using polynomials as input units, these networks can, at least theoretically, approximate almost any function. Sierra et al. [8] have shown that the use of FLNs, combined with a genetic algorithm that evolves the polynomial terms, can produce very compact, effective and interpretable architectures.

Following their work, we hypothesize that if there exists a function that discriminates among the different classes for a given classification problem, there must exist a suitable basis in which this function has a simple and compact expression. So solving such a problem can be reduced to finding the most appropriate basis or representation for the input data (with respect to the classification target). Once this representation is found, a linear discriminant will suffice to find a simple and compact solution. We propose an expansion of the work in [8] that incorporates other bases apart from polynomials. We use a genetic algorithm to perform both variable selection and search in the transformation space, and a Fisher discriminant that performs the final linear projection. We show that this approach is able to solve problems where other methods fail to find a solution, even when the overlap is so large that there are no apparent statistical differences among the classes. This overlap may be due simply to the fact that the original representation of data is not well suited to the problem. Actually, it is well known that many classification problems are solved only after the application of some "intelligent" transformations provided by a domain "expert". Here we want to go a step closer into the automatic selection of these intelligent transformations, by allowing the algorithm to search for the optimal basis.

The paper is organized as follows. In section 2 we present, as a proof of concept, a classification problem with a high degree of overlap between classes, whose solution appears to be quite difficult for any standard classification method. Section 3 summarizes the performance of different classical methods on this problem. Section 4 shows that the selection of an appropriate representation for the input data makes the solution much simpler. In section 5 we propose a general method that searches the space of transformations for near-optimal representations. Finally, section 6 contains the conclusions and discussion.

2   A Particular Problem

Let us consider the classification problem of figure 1. It consists of two classes (A and B) in a two-dimensional space represented by the variables x and y. There are 1000 patterns of class A (circles in the figure) and 1000 patterns of class B (crosses in the figure). At first sight one would say that this classification problem is impossible to solve, since the two classes apparently follow the same probability distribution, a bidimensional uniform distribution in the interval $[0 \leq x \leq 100, 0 \leq y \leq 100]$. In fact, statistical tests for mean and variance corroborate this hypothesis.

The next section shows the results obtained by different classification methods when trying to solve this problem. None of the tested strategies achieves a successful result. The difficulty they are confronting is due to the high overlap

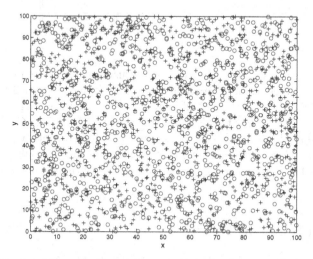

Fig. 1. Input patterns for a problem with two classes and two attributes x and y. The problem data consist of 1000 patterns of class A (circles) and 1000 patterns of class B (crosses). Apparently the two classes follow the same (uniform) distribution in the considered interval, and classical methods will have great difficulties to deal with this problem

between the two classes. Note that for an absolute class overlap, even the best (Bayes) class estimator fails. But it may happen that below this apparent class mixing there is a hidden structure that these methods are not able to discover when just using the original input space. It is clear that, if we trust standard methods, we must conclude there is no solution. But we could be quite far from the truth.

3   When Classical Methods Fail

When traditional classification methods were applied to the problem of section 2, no solutions with acceptable error rates were obtained. Three different methods have been tried: a multilayer perceptron trained with backpropagation, a decision tree trained with the C4.5 algorithm, and an evolutionary FLN with the polynomial basis. The results are summarized in table 1, which shows the best error rates obtained with each method. In no case we observed results better than approximately a 47% error on the test set. That is, none of the tested methods performs much better than simply selecting the classes randomly.

Table 1. Comparison of performances of various classification methods on the problem of section 2

Algorithm	Train Error %	Test Error %
Backpropagation	50.2	52.7
C4.5	48.7	49.8
EFLN	45.7	46.8

The backpropagation algorithm was tested using networks of one single hidden layer, with different number of hidden units (ranging from 3 to 10) with a sigmoidal activation function. Different values for the learning rate between 0.01 and 0.3 were tried. For the decision tree, we used Quinlan's C4.5 algorithm [7] with probabilistic thresholds for continuous attributes, windowing, a gain ratio criterion to select tests and an iterative mode with ten trials. Finally, the evolutionary FLN was trained as described in [8], with polynomials of up to degree 3.

4   Does a Solution Exist?

Let us now analyze the example problem in depth. In table 2 some examples of class A and class B patterns are shown. A closer study of the data in the table may reveal a relation between each pattern and its class. For all class A patterns the integer parts of x and y are either both odd or both even. However, for class B patterns there is always one odd and one even integer part. The class

Table 2. Some example of patterns from both classes. Can you find the relation between each pattern and its class?

Class A		Class B	
x	y	x	y
82.76	48.55	91.94	12.50
13.11	87.86	23.04	82.04
68.82	26.78	4.28	65.18
1.33	37.82	33.73	70.57
22.27	30.63	14.97	5.63
17.16	63.33	82.12	97.40
41.69	69.59	10.93	87.28
67.94	3.92	94.11	29.86
82.76	48.55	91.94	12.50

distribution is in fact, although we could not observe it from figure 1, like a chess board in which class A patterns occupy black squares and class B patterns occupy white ones. This is obvious in figure 2 (left), where we zoom in the area $[20 \leq x \leq 30, 20 \leq y \leq 30]$ and plot reference lines for ease of interpretation.

Hence, it is clear that there exists a solution to the problem. Nevertheless, standard methods fail to find it since they are not expressing the problem in an appropriate form (basis) to solve it. In figure 2 (right) we show one possible transformation that is well suited to the present problem. Instead of using the original attributes x and y, we used the transformations $sin(\pi x)$ and $sin(\pi y)$. We can see that in this new basis the problem is much easier to solve, and none of the methods considered in the previous section would have any difficulties to find the solution. The problem we face now is how to discover the appropriate basis. In the next section we explore a possible method.

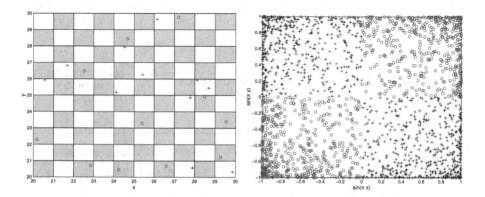

Fig. 2. Left. Zoom in of the region $[20 \leq x \leq 30, 20 \leq y \leq 30]$ of figure 1. Right. Representation of the problem data using the transformed attributes $sin(\pi x)$ and $sin(\pi y)$

5 Proposed Approach

Our approach follows on the work developed by [8] with the EFLN algorithm. In this regard our methodology involves transformations of the input space variables searching for a better approximation of the target function. These new transformed variables define the input to a linear classification method, which in turn will provide the quality measure over the selected transformation. These new sets of input space variables are evolved using a standard genetic algorithm (GA) [5], which will select the best subsets of transformed variables by using the linear classifier error rate as the fitness criterion. Consequently our algorithm can be viewed as a wrapper method [3].

There are two main differences between the EFLN method and the algorithm presented here. First, we do not limit the transformations to the polynomial basis, but we expand the representation capabilities by adding the trigonometric basis. Second, the linear projection is performed by a Fisher discriminant, instead of a linear neural network. The proposed algorithm includes feature construction as well as feature selection. For the first task, it combines different bases of transformation (e.g. polynomial and trigonometric). Feature selection is performed by the application of the GA. All the process is schematized in figure 3.

The algorithm is able to generate a rich set of new features through the generator function

$$F(x_1, x_2, ..., x_n) = \prod_{i=1}^{n} x_i^{a_i} T_i(b_i \pi x_i) \tag{1}$$

where the x_i represent the original input variables, a_i and b_i are integer coefficients, and T_i is a trigonometric function (a sine or a cosine). The coefficients a_i are the exponents applied to the original variables to get polynomial terms, whilst the coefficients b_i deal with trigonometric terms. This generator function can provide either pure polynomial or trigonometric terms, or combinations of both.

As previously mentioned, a genetic algorithm is used for feature selection since an exhaustive search over the new input spaces would be computationally

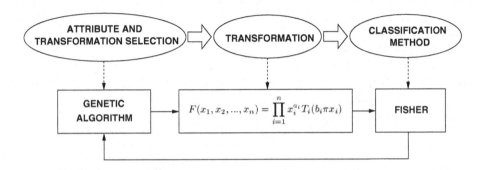

Fig. 3. Schematics of the overall methodology

too costly and would not scale properly on the number of input variables. The genetic operators for mutation and crossover are included from the standard library [5], the mutation probability is set to 15%, and elitism is selected to save the best individuals from each evolved population. The initial population is randomly generated. We have selected the most general GA options, so that the influence of the GA on the overall algorithm performance will not be a decisive factor. The coding scheme for every gene is based on the generator function described above. The aforementioned coefficients a_i and b_i are included in the gene, as well as n additional bit variables, t_i, which determine the kind (sine or cosine) of trigonometric function employed on the new feature.

Each chromosome has a variable number of genes, which generate the new input variables for the Fisher discriminant. Then the Fisher projection is performed, and the classification error on the training and validation sets are computed. These errors define the *fitness* of the chromosome.

An schematic overview of the algorithm follows:

1. *Initialize the first population of chromosomes randomly, and set other values for the GA such as the number of iterations or the mutation probability.*

2. *For each evolution iteration:*
   (a) For each chromosome:
       i. Generate new features based on the generator function and the original attributes.
       ii. Evaluate each chromosome to get the fitness value using the Fisher Linear Discriminant classification error on the training and validation data sets.
   (b) Select the lowest error chromosomes for the next iteration.
   (c) Generate a new population applying genetic operators and the chromosomes selected in (b).
3. *Evaluate the most accurate chromosome on the test data set.*

Finally, we applied this algorithm to the problem described in Section 2. Different trials converged fastly (in no more than 50 iterations) to classification errors close to 0% in test. This implies that the algorithm is discovering the intrinsic trigonometric nature of the problem. As an example, the following chromosome was obtained with a 0.9% error:

$$\begin{pmatrix} xy\sin(2\pi x)\sin(2\pi y) \\ xy\sin(\pi x)\sin(\pi y) \\ 0 \\ \sin(\pi x)\sin(\pi y) \\ xy\sin(\pi x)\sin(\pi y) \\ x\cos(2\pi x) \\ xy\sin(2\pi x)\sin(2\pi y) \end{pmatrix}$$

with the resulting Fisher projection given by the vector:

$$\begin{pmatrix} 0 & 0 & 0 & -9.5 & 0 & 0 & 0 \end{pmatrix}$$

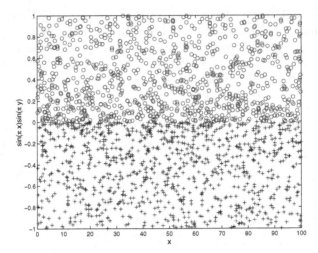

Fig. 4. Plot of $sin(\pi x)sin(\pi y)$ vs x for the patterns of the problem of section 2. The final transformation discovered by the proposed algorithm allows for a linear separation of the two classes

So the final transformation reached by the algorithm is $-9.5 sin(\pi x)sin(\pi y)$. Note that the Fisher projection is ignoring all the terms in the chromosome except the fourth. As shown in figure 4, this transformation allows for a linear separation of the two classes.

6   Conclusions

This paper presents a proof of concept for the construction of near-optimal problem representations in classification problems, based on the combination of functions selected from an initial family of transformations. The selection of an appropriate transformation allows the solution of complex nonlinear problems by a simple linear discriminant in the newly transformed space of attributes.

Work on progress includes the introduction of a more extensive family of basis functions that will allow for the construction of a wider repertoire of problem representations. Additionally, mechanisms to control the combinatorial explosion in the space of representations and the complexity of solutions will be analyzed.

Additional advantages of the proposed method are that a closer, more compact problem representation usually allows for easier model interpretation [8], and, hence, a deeper understanding of the structure and mechanisms underlaying the problem under study. Related work on the extraction of hidden causes [4], which provide the generative alphabet, will be farther explored.

Acknowledgements

We want to thank P. Hoelgaard and A. Sierra for very interesting comments and discussions, and Ministerio de Ciencia y Tecnología for financial support (BFI2003-07276).

References

1. Bishop, C.M.: Neural Networks for Pattern Recognition. Oxford Univ. Press (1995)
2. Duda, R.O., Hart, P.E., Stork, D.G.: Pattern Classification. John Wiley and Sons (2001) 84–214
3. Kohavi, R., John, G.H.: Wrappers for Feature Subset Selection. Artificial Intelligence **97** (1–2) (1997) 273–324
4. Lago-Fernández, L.F., Corbacho, F.J.: Optimal Extraction of Hidden Causes. LNCS **2415** (2002) 631–636
5. Levine, D.: Users Guide to the PGAPack Parallel Genetic Algorithm Library. T.R.ANL-95/18 (1996)
6. Pao, Y.H., Park, G.H., Sobajic, D.J.: Learning and Generalization Characteristics of the Random Vector Functional Link Net. Neurocomputing **6** (1994) 163–180
7. Quinlan, J.R.: C4.5: Programs for Machine Learning. Morgan Kaufmann Publishers Inc. (1992)
8. Sierra, A., Macías, J.A., Corbacho, F.: Evolution of Functional Link Networks. IEEE Trans. Evol. Comp. **5** (1) (2001) 54–65
9. Vapnik, V.N.: Statistical Learning Theory. John Wiley and Sons (1998)

Playing a Toy-Grammar with GCS

Olgierd Unold

Institute of Engineering Cybernetics,
Wroclaw University of Technology,
Wyb. Wyspianskiego 27, 50-370 Wroclaw, Poland
olgierd.unold@pwr.wroc.pl
http://www.ict.pwr.wroc.pl/~unold

Abstract. Grammar-based Classifier System (GCS) is a new version of Learning Classifier Systems in which classifiers are represented by context-free grammar in Chomsky Normal Form (CNF). Discovering component of the GCS and fitness function were modified and applied for inferring a toy-grammar, a tiny natural language grammar expressed in CNF. The results obtained proved that proposed rule's fertility improves performance of the GCS considerably.

1   Introduction

The problem of learning (inferring) context-free languages (CFLS's) is worth considering for both practical and theoretical reasons. Practical applications include pattern recognition and speech recognition. From a theoretical standpoint, the problem poses an interesting challenge in that, there are serious limitations to learning them. Finally, successful language learning algorithms model how humans acquire language.

There are very strong negative results for the learnability of context-free grammar (CFG). The main theorems are that it is impossible to evolve suitable grammar (each of the four classes of languages in the Chomsky hierarchy) only from positive examples [8], and that even the ability to ask equivalence queries does not guarantee exact identification of CFL in polynomial time [2]. Effective algorithms exist only for regular languages, thus construction of algorithms that learn context-free grammar is critical and still open problem of grammar induction [9].

The approaches taken have been to provide learning algorithms with more helpful information, such as negative examples or structural information; to formulate alternative representation of CFG's; to restrict attention to subclasses of CFL's that do not contain all finite languages; and to use Bayesian methods [19]. Grammar induction can be considered as a difficult optimization task. Evolutionary approaches are probabilistic search techniques especially [1] suited for search and optimization problems where the problem space is large, complex and contains possible difficulties like high dimensionality, discountinuity and noise.

Many researchers have attacked the problem of grammar induction by using evolutionary methods to evolve (stochastic) CFG or equivalent pushdown au-

J. Mira and J.R. Álvarez (Eds.): IWINAC 2005, LNCS 3562, pp. 300–309, 2005.

tomata [28], [27], [13], [7], [21], [17], [20], [14], [23], [15], [16], [24], but mostly for artificial languages like brackets, and palindromes. The survey of the non-evolutionary approaches for context-free grammatical inference one can find in [19].

In this paper we examine CFL inference using Grammar-based Classifier System (GCS) that take texts as input, where a text is a sequence of strings over the same alphabet as the CFL's being learned (a text contains both strings in and not in the target language). GCS [25] is a new model of Learning Classifier System (LCS) which represents the knowledge about solved problem in Chomsky Normal Form productions. As Wyard already pointed out [27], the approach in which the population consists of rules instead of complete grammars (employed successfully in LCS), might prove to be useful in grammatical inference. In such an algorithm, a population rule's fitness is determined by scoring its ability to correctly analyze the examples in conjunction with the other rules of population. A advantage of this approach is that in a population of rules the value of different grammar rules is evaluated only once, as opposed to a population of grammars, in which a single rule might appear in many different grammars.

In spite of intensive research into classifier systems in recent years [18] there is still slight number of attempts at evolving grammars using LCS. Bianchi in his work [4] revealed, on the basis of experiments with bracket grammars, palindromes and toy-grammar, higher efficiency of classifier system in comparison with evolutionary approach. Cyre [6] inducted a grammar for subset of natural language using LCS but comparison to his results is hard since usage of corpora protected by trademarks (DMA driver's American patents). GCS tries to fill the gap also bringing grammar induction issues up. As was shown in [25], GCS achieves better results than Bianchi's system with reference to artificial grammars. GCS has been tested on the natural language corpora as well, and it provided comparable results to the pure genetic induction approach proposed in [3], but in a significantly shorter time.

The paper improves the results presented in [25] by modification the discovery component of GCS and fitness function, and investigates some characteristics. All experiments were conducted to infer so called toy-grammar, a small natural language grammar expressed as CFG.

Architecture of learning classifier system is presented in second paragraph. Third section contains description of GCS preceded by short introduction to context-free grammars. Fourth paragraph shows new ideas incorporated in GCS's model, and fifth paragraph some selected experimental results, whereas sixth section is a short summary.

2   Learning Classifier System

Learning Classifier Systems proposed by J. Holland [10] exemplify the promising model from among many methods and algorithms of machine learning [5], both when we are talking about its simplicity, flexibility and efficiency.

LCS learns by interacting with an environment from which it receives feed-back in the form of numerical reward. Learning is achieved by trying to maximize the amount of reward received. There are many models of LCS and many ways of defining what a Learning Classifier System is. All LCS models, more or less, comprise four main components:

1. A finite population of condition-action rules (classifiers), that represent the current knowledge of system;
2. The performance component, which governs the interaction with the environment;
3. The reinforcement component (called credit assignment component), which distributes the reward received from the environment to the classifiers accountable for the rewards obtained;
4. The discovery component responsible for discovery better rules and improving existing ones through a genetic algorithm.

Classifiers have two associated measures: the prediction and the fitness. Prediction estimates the classifier utility in terms of the amount of reward that the system will receive if the classifier is used. Fitness estimates the quality of the information about the problem that the classifier conveys, and it is exploited by the discovery component to guided evolution. A high fitness means that the classifier conveys good information about the problem and therefore it should be reproduced more trough the genetic algorithm. A low fitness means that the classifier conveys little or no good information about the problem and therefore should reproduce less.

On each discrete time step the LCS receives as input the current state of the environment and builds a match set containing the classifiers in the population whose condition matches the current state. Then, the system evaluates the utility of the actions appearing in the match set; an action is selected from those in the match set according to certain criterion, and sent to the environment to be performed. Depending on the current state and on the consequences of action, the system eventually receives a reward. The reinforcement component distributes reward among the classifiers accountable of the incoming rewards. This can be either implemented with an algorithm specifically designed for the Learning Classifier Systems (e.g. bucket brigade algorithm [11] or with an algorithm inspired by traditional reinforcement learning methods (e.g. the modification of Q-learning [26]). On a regular basis the discovery component (genetic algorithm) randomly selects, with the probability proportional to their fitness, two classifiers from the population. It applies crossover and mutation generating two new classifiers.

3   Grammar-Based Classifier System

GCS operates similar to the classic LCS systems but differs from them in (i) representation of classifiers population, (ii) scheme of classifiers' matching to the environmental state, (iii) methods of exploring new classifiers.

3.1   Representation

Primary goal, the GCS was developed for, was natural language grammar inference, expressed in context-free form. As a result the population of classifiers consists of context-free grammar rules. A context-free grammar is usually summarized as:

$$G = (V_N, V_T, P, S)$$

where:
P productions are in the form of $A \rightarrow a$ and $A \in V_N$ and $\alpha \in (V_T \cup V_N)^*$,
S is a starting symbol,
V_N is a set of nonterminals,
V_T is a set of terminals.
System evolves only one grammar according to the so-called Michigan approach. In this approach each individual classifier - or grammar rule in GCS - is subject of the genetic algorithm's operations. All classifiers (rules) form a population of evolving individuals. In each cycle a fitness calculating algorithm evaluates a value (an adaptation) of each classifier and a discovery component operates only on a single classifier.

3.2   Matching

Automatic learning context-free grammar is realized with so-called grammatical inference from text [8]. According to this technique system learns using a training set that in this case consists of sentences both syntactically correct and incorrect. Grammar which accepts correct sentences and rejects incorrect and is able to classify unseen so far ones from a test set is an anticipated result. Cocke-Younger-Kasami (CYK) parser, which operates in $\Theta(n^3)$ time [29], is used to parse sentences from the sets. It uses context-free grammar in Chomsky Normal Form [12].

Chomsky Normal Form (CNF) allows only production rules in the form of $A \rightarrow a$ or $A \rightarrow BC$, where $A, B, C \in V_N, a \in V_T$. This is not a limitation actually because every context-free grammar can be transformed into equivalent CNF.

Environment of classifier system is substituted by an array of CYK parser. The array of size $m \times n$, where m represents number of rows and n represents length of considered sentence, stores complete history of sentence's parsing. If there is a nonterminal symbol in cell $[i, j]$ then this symbol derives a part of sequence beginning from position i and j-length. If the parsing ends with starting symbol S in cell $[m, 1]$ it means that considered grammar is able to derive whole sentence. In every single step parser a) generates every possible right side of the rule that matches current state of parsing, b) matches the right sides to the available production rules, c) inserts left side of every rule that matches into the appropriate cell in an array. If there is more then one rule in a cell it means that there are many possible derivation trees for this (part) of sentence.

A value of adaptation (fitness) is assigned for each rule as soon as parsing of every sentence from a set is finished. The fitness value is expressed as:

$$f_c = \begin{cases} \dfrac{w_p \cdot U_p}{w_n \cdot U_n + w_p \cdot U_p} & \text{for } U_N + U_P \neq 0 \\ f_0 & \text{for } U_N + U_P = 0 \end{cases} \tag{1}$$

where:

U_P – number of uses of rule while parsing correct sentence,

U_N – number of uses of rule while parsing incorrect sentence,

f_0 – fitness of classifier that wasn't used in parsing,

w_p, w_n – coefficients (commonly used settings are 1 and 2).

Fitness value is used by genetic algorithm while searching for new classifiers.

The following function f_G is applied to evaluate fitness of each grammar. In the equation, PS is the positive set of sentences, NS is the negative set of sentences, P is the number of positive sentences parsed by grammar and N is the number of negative strings parsed.

$$f_G = \frac{P + N}{|PS + PN|} \cdot 100\% \tag{2}$$

3.3   Discovery Component

GCS uses two techniques that explore space of all possible classifiers - just like many other classifiers systems. First of them is genetic algorithm and the second is covering.

Genetic algorithm in GCS works on a population of classifiers like in other CS but because of the different representation it operates only on production rules in form of $A \rightarrow BC$. System uses roulette-wheel or random selection (chosen in the options), classic crossover and mutation, and crowding technique in order to keep diversity in population [1]. Genetic operators are launched with given probability once analyzing of the train set ends.

Covering works regardless of genetic algorithm and during trains set analysis. It adds productions that allow continuing of parsing in the current state of the system. In GCS there are following sorts of covering:

1. terminal covering: a production rule in the form of $A \rightarrow a$ is created when system finds unknown (new) terminal symbol while parsing,
2. one-length covering: a production rule in the form of $S \rightarrow a$ is created for one-length, correct sentences,
3. two-length covering: a production rule in the form of $S \rightarrow AB$ is created if productions $A \rightarrow a$ and $B \rightarrow b$ exist in the population and there is two-length correct sentence,
4. full-covering: a production rule in the form of $S \rightarrow AB$ is created if symbols A and B can be derived and the last cell in the CYK array is considered and there is a correct sentence currently parsed,
5. aggressive-covering: a production rule in the form of $C \rightarrow AB$ is created if symbols A and B can be derived and there is a correct sentence currently parsed.

4   New Ideas

In [25] the set of experiments on bracket grammars, palindromes, toy-NL grammar, and natural language corpora was presented. It was observed that while learning natural language corpora fitness graph shows sudden changes of the fitness value. The most probable reason of this is strong cooperative nature of grammar production rules. Deletion or modification of a rule can deactivate a huge set of connected productions. This can decrease overall grammar's fitness. On the other hand creation or proper modification of existing rule can activate new set of rules, and dramatically increase overall fitness. Modifying discovery component could be one of the solutions to this problem. Discovery component could look at the rule's position at the derivation tree (rule's *fertility*) and more carefully remove rules that may be important to the parsing process.

According to the concept of the rule's fertility we introduce new formula for fitness value of rule:

$$ f = \frac{w_c \cdot f_c + w_f \cdot f_f}{w_c + w_f} \tag{3} $$

where:
f_c – "classic" fitness of classifier expressed by (1),
w_c, w_f – coefficients,
f_f – normalized fitness of classifier's fertility expressed as:

$$ f_f = \frac{p - d - f_{min}}{f_{fmax} - f_{fmin}} \tag{4} $$

where:
p – (*profit*) sum of credits of the classifier scored while parsing correct sentence,
d – (*debt*) sum of credits of the classifier scored while parsing incorrect sentence,
f_{fmin}, f_{fmax} – minimal / maximal credits in the set of classifiers.
The classifier receives the specific credit (equal *renounced amount factor * base amount*) from each rule in the derivation tree placed below. The terminal rule is rewarded by constant value (so-called *base amount*).

5   Playing a Toy-Grammar

Modified GCS was tested on the tiny natural language grammar called toy-grammar, that can be described by the following rules in CNF:
S → np vp
np → det n
np → np pp
pp → prep n
vp → v np
vp → vp pp
where S is a start symbol (sentence), *np* is a noun phrase, *vp* is a verb phrase, *det* is a determinant, *n* is a noun, *pp* is a preposition phrase, *prep* is a preposition and *v* is a verb.

GCS was trained using train set with 40 examples (20 correct and 20 incorrect ones). All results presented here are average of 50 independent runs.

In the first set of experiments the proper merit of the fertility coefficient w_f was searched. Fifth measures of performance were used. To calculate the average *fitness* of the grammar, first we have to find the average fitness for each generation (over all runs), and then average obtained results. The positive competence (*Competence*) of a grammar is the percentage of all positive sentences in the test corpus that it successfully parses, the negative competence (*Ncompetence*) is the percentage of all negative (incorrect) sentences being successfully parsed. Average competence and average negative are calculated similar to average fitness. *First 80%* is the number of generations needed to reach 80% average fitness. We measured maximal of the average fitness (*Max*) as well.

In order to put the curves of all measures in one figure, we multiply *Ncompetence* by 100, and divide *First 80%* by 5. Figure 1 shows the measures versus the fertility coefficient w_f. Experiment was performed with following settings: generations 500, crossover probability 0.1%, mutation probability 0.85%, population consisted of maximal 18 classifiers where 10 of them were created randomly in the first generation, base amount 1, renounced amount factor 0.5, $w_c = 1$. The plots of maximal fitness, average fitness, and average competence are similar in courses. One can observe the saddle for the fertility coefficient equals 3 or 4, for the coefficient greater than 9 the values of measures decrease considerably. GCS reaches 80% of average fitness in the smallest number of generations for fertility between 6 and 9. For the fertility equals 2 and 10, GCS accepts the largest amount of incorrect sentences. To summarize above results: it seems that the most promising merit for the fertility coefficient should be taken from the range 6 to 9.

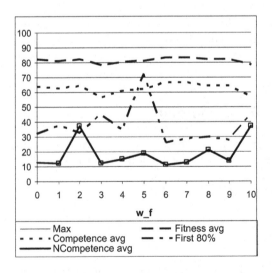

Fig. 1. The influence of the fertility coefficient (w_f)

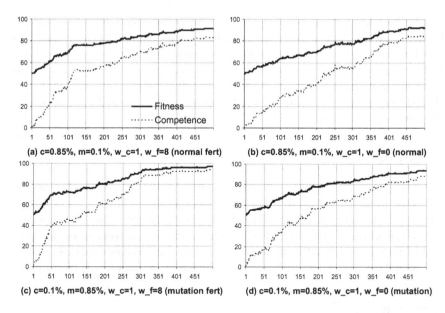

Fig. 2. The influence of mutation (m), crossover (c), and fertility (w_f)

Table 1. Statistics merits for plots from Fig. 2

Pairs of values	Fitness avg.	Compet. avg.	First 80%	Max
normal	74.7	49.6	327	92.3
normal fert	79	58.1	237	91.35
mutation	78.6	57.2	220	94
mutation fert	83.4	66.8	191	97.2

The aim of the second set of experiments was to compare the impact of mutation and crossover on GCS performance. Runs were conducted with two pairs of values: mutation $m = 0.1$, crossover $c = 0.85$ (Fig. 2a and 2b), and $m = 0.85$, $c = 0.1$ (Fig. 2c and 2d). Additionally, fertility was switched on (Fig. 2a and 2c) and off (Fig. 2b and 2d). Some statistics were drawn from algorithms curves (Tab. 1).

A large mutation rate has an interesting effect. The fitness and competence increases faster at the beginning, for example for 201 generation fitness reaches 70.25% and competence 40.5% without mutation (Fig. 2b), but with mutation 78.2% and 56.5% respectively (Fig. 2d)! Reinforcement with fertility gives much better results, for 201 generation fitness gains 77.9% and competence 56.1% without mutation (Fig. 2a), and 80% and 60.2% respectively with mutation (Fig. 2c). Experiments confirmed the conclusion known in the literature [22], that mutation works most efficiently in small populations, and proved important influence of fertility. Adding fertility to the experiment without large mutation rate (Fig. 2a) gives similar results as in experiment with large mutation rate but without fertility (Fig. 2d). Both curves of fitness and curves of competence look alike, the synthetic measures are almost the same (compare rows 2 and 3 from Tab. 1). The best results one can observe in experiment with large mutation and fertility (see row 4 from Tab. 1). It is noteworthy that in this case the curves of fitness and competence stabilize on the certain level when the population is approaching the optimum.

6 Summary

Some extensions to the grammar-based classifier system that evolves population of classifiers represented by context-free grammar productions rules in normal form were presented. Introduced coefficient of rule's fertility and new formula of rule's fitness seem to be promising modification of discovering module. Obtained results proved that fertility can improve performance of the GCS. Further experimentation to determine the optimal algorithm parameters is warranted.

References

1. Arabas, J.: Wyklady z algorytmow ewolucyjnych. WNT, Warszawa (2001) (in Polish)
2. Angluin, D.: Queries and concept learning. Machine Learning 2(4) (1988) 319-342
3. Aycinena, M., Kochenderfer, M.J., Mulford D.C.: An evolutionary approach to natural language grammar induction. Final project for CS224N: Natural Language Processing. Stanford University (2003)
4. Bianchi, D.: Learning Grammatical Rules from Examples Using a Credit Assignement Algorithm. In: Proc. of The First Online Workshop on Soft Computing (WSC1), Nagoya (1996) 113-118
5. Cichosz, P.: Systemy uczace sie. WNT, Warszawa (2000) (in Polish)
6. Cyre, W.R.: Learning Grammars with a Modified Classifier System. In: Proc. 2002 World Congress on Computational Intelligence, Honolulu Hawaii (2002) 1366-1371
7. Dupont, P.: Regular Grammatical Inference from Positive and Negative Samples by Genetic Search. In: Grammatical Inference and Application, Second International Colloquium ICG-94, Berlin, Springer (1994) 236-245
8. Gold, E.: Language identification in the limit. Information Control 10 (1967) 447-474
9. de la Higuera, C.: Current trends in grammatical inference. In: Ferri, F.J. at al (eds.): Advances in Pattern Recognition. Joint IAPR International Workshops SSPR+SPR'2000, LNCS 1876, Springer (2000) 28-31
10. Holland, J.: Adaptation. In: Rosen, R., Snell, F.M. (eds.): Progress in theoretical biology. New York, Plenum (1976)
11. Holland, J.: Escaping Brittleness: The possibilities of General-Purpose Learning Algorithms Applied to Parallel Rule-Based Systems. In: Michalski, R.S. at al. (eds.): Machine Learning, an artificial intelligence approach. vol. II. Morgan Kaufmann (1986) 593-623
12. Hopcroft, J.E., Ullman, J.D.: Introduction to Automata Theory, Languages, and Computation. Addison-Wesley Publishing Company. Reading Massachusetts (1979)

13. Huijsen, W.: Genetic Grammatical Inference: Induction of Pushdown Automata and Context-Free Grammars from Examples Using Genetic Algorithms. M.Sc.-thesis, Dept. of Computer Science, University of Twente, Enschede, The Netherlands (1993)
14. Kammeyer, T.E., Belew, R.K.: Stochastic Context-Free Grammar Induction with a Genetic Algorithm Using Local Search. Technical Report CS96-476, Cognitive Computer Science Research Group, Computer Science and Engineering Department, University of California at San Diego (1996)
15. Keller, B., Lutz, R.: Learning Stochastic Context-Free Grammars from Corpora Using a Genetic Algorithm. In: Proceedings International Conference on Artificial Neural Networks and Genetic Algorithms ICANNGA-97 (1997)
16. Korkmaz, E.E., Ucoluk, G.: Genetic Programming for Grammar Induction. In: Proc. of the Genetic and Evolutionary Conference GECCO-2001, San Francisco Ca, Morgan Kaufmann Publishers (2001)
17. Lankhorst, M.M.: A Genetic Algorithm for the Induction of Nondeterministic Pushdown Automata. Computing Science Reports CS-R 9502, Department of Computing Science, University of Groningen (1995)
18. Lanzi, P.L., Riolo, R.L.: A Roadmap to the Last Decade of Learning Classifier System Research. In: Lanzi, P.L., Stolzmann, W., Wilson, S.W. (eds.): Learning Classifier Systems. From Foundations to Applications, LNAI 1813, Springer Verlag, Berlin (2000) 33-62
19. Lee, L.: Learning of Context-Free Languages: A Survey of the Literature. Report TR-12-96. Harvard University, Cambridge, Massachusetts (1996)
20. Losee, R.M.: Learning Syntactic Rules and Tags with Genetic Algorithms for Information Retrieval and Filtering: An Empirical Basis for Grammatical Rules. In: Information Processing & Management (1995)
21. Lucas, S.: Context-Free Grammar Evolution. In: First International Conference on Evolutionary Computing (1994) 130-135
22. Mhlenbein, H., Schlierkamp-Voosen, D.: Analysis of Selection, Mutation and Recombination in Genetic Algorthms. In: Banzhaf, W., Eeckman, F.H. (eds.): Evolution as a Computational Process, LNCS 899, Springer, Berlin (1995) 188-214
23. Smith, T.C., Witten, I.H.: Learning Language Using Genetic Algorithms. In: Wermter, S., Rilo, E., Scheler, G. (eds.): Connectionist, Statistical, and Symbolic Approaches to Learning for Natural Language Processing, LNAI 1040 (1996)
24. Unold, O.: Context-free grammar induction using evolutionary methods, WSEAS Trans. on Circuits and Systems, 3 (2) (2003) 632-637
25. Unold, O., Cielecki, L.: Grammar-based Classifier System. In: Kacprzyk, J., Koronacki, J. (eds.) First Warsaw International Seminar on Intelligent Systems. EXIT Publishing House, Warsaw (2004) (in print).
26. Wilson, S.W.: Classifier Fitness Based on Accuracy. Evolutionary Computation 3 (2) (1995) 147-175
27. Wyard, P.: Context Free Grammar Induction Using Genetic Algorithms. In: Belew, R.K., Booker, L.B. (eds.): Proceedings of the Fourth International Conference on Genetic Algorithms, San Diego, CA. Morgan Kaufmann (1991) 514-518
28. Zhou, H., Grefenstette, J.J.: Induction of Finite Automata by Genetic Algorithms. In: Proceedings of the 1986 International Conference on Systems, Man and Cybernetics (1986) 170-174
29. Younger, D.: Recognition and parsing of context-free languages in time n3. University of Hawaii Technical Report, Department of Computer Science (1967)

A Genetic Approach to Data Dimensionality Reduction Using a Special Initial Population

M. Borahan Tümer and Mert C. Demir

Marmara University Faculty of Engineering, Göztepe Campus,
34722 Kadiköy, İstanbul, Turkey
{bora, mcdemir}@eng.marmara.edu.tr

Abstract. Accurate classification of data sets is an important phenomenon for many applications. While multi-dimensionality to a certain point contributes to the classification performance, after a point, incorporating more attributes degrades the quality of the classification. In a pattern classification problem, by determining and excluding the least effective attribute(s) the performance of the classification is likely to improve. The task of the elimination of the least effective attributes in pattern classification is called "data dimensionality reduction (DDR)". DDR using Genetic Algorithms (DDR-GA) aims at discarding the less useful dimensions and re-organizing the data set by means of genetic operators. We show that a wise selection of the initial population improves the performance of the DDR-GA considerably and introduce a method to implement this approach. Our approach focuses on using information obtained *a priori* for the selection of initial chromosomes. Our work then compares the performance of the GA initiated by a randomly selected initial population to the performance of the ones initiated by a wisely selected one. Furthermore, the results indicate that our approach provides more accurate results compared to the purely random one in a reasonable amount of time.

Keywords: data dimensionality reduction, feature extraction, genetic algorithms, attribute ranking, attribute quality.

1   Introduction

Pattern classification and recognition have been popular research subjects since longer than three decades. Availability of large data sets with a growing number of attributes has provided, on one hand, more efficient and higher performance pattern classifiers to a certain extent. On the other hand, increasing dimensionality has called for more complex classifier architectures that, beyond a point, suffer from the *curse of dimensionality*. Hence, one faces the problem of optimizing the number of attributes that offers a classifier performing satisfactorily well while it avoids the involvement of non-contributory attributes. The term data dimensionality reduction is the name of this optimization process that is widely applied to multi-dimensional data sets. Two basic approaches to data dimensionality reduction are feature (attribute) selection and feature extraction.

J. Mira and J.R. Álvarez (Eds.): IWINAC 2005, LNCS 3562, pp. 310–316, 2005.

Feature extraction is about obtaining a new (and smaller) set of features from the original features where the new features provide a better separability among the classes than do the original features [1]. Principal component analysis (PCA) [8, 10] and linear discriminant analysis (LDA) [3, 5, 9] are an unsupervised and a supervised technique in respective order that employ feature extraction. The transformation of original features to the generated set of features involves the computation of the eigenvectors of the covariance matrix of original features. Combined with concerns on how well PCA or LDA maintains the originality of data, the above transformation makes the feature selection a preferable alternative to feature extraction. In feature selection, a subset of original features is sought that contains the least number of features with the highest contribution to the classification performance. To find the optimal subset of features, an exhaustive search would be necessary. Exhaustive search requires exponential time and becomes infeasible for large number of features. Instead, genetic algorithms (GAs) are frequently employed [4, 7, 12] as an adaptive search mechanism for a suboptimal solution in data dimensionality reduction. GAs in data dimensionality reduction are popular with their polynomial time requirements. However, the random nature of the GA operators makes the algorithm sensitive to the starting point (i.e., initial population).

In this work, we present a GA for data dimensionality reduction with an intelligently selected initial population. Here, features that contribute more to the classification are determined by the attribute ranking procedure described in [5]. This ranking information is utilized to determine the probabilities of inclusion for the attributes in the initial population. We compare the performance of the GA initiated by a pure random initial population to the performance of the one initiated by a wisely selected initial population. Our experiments show that this approach outperforms the Genetic Algorithm initiated by a purely random population.

In determining the fitness of the attribute set selected, we perform a classification using this selected subset of features and calculate the accuracy and cost of the classification. We employ feed-forward neural networks [2] with supervised learning to classify the Letter Recognition data set [11].

This paper is organized as follows. The implemented attribute ranking method is discussed and our genetic model is introduced in Section 2. In Section 3, we give the results of the experiments and a brief discussion concludes the paper.

2   The Genetic Model

Genetic algorithms are search and optimization techniques based on natural selection and evolution mechanisms [6]. Imitating the learning mechanisms in living organisms, GA methods are applied to a given group of data called *population*.

A *chromosome* in the population is a binary string where a *gene* with value 1 stands for an attribute used in the current classification and a one with value 0 for an unused one. To reproduce the next generation we used *elitism* and scattered *recombination* as well as the mutation operator. To select the fit individuals into the next generation, we used the probabilities of inclusion originating from the attribute ranking.

2.1   Attribute Ranking

Although there are differences, techniques for the attribute ranking problem are commonly based on vector distances of patterns in the whole data set. The technique in [5] that uses the *separability correlation measure* (SCM) is one such technique, that is used in this work to generate initial populations.

Separability correlation measure combines two measures: *class separability measure* and *attribute-class correlation measure*. Class separability measure evaluates the ratio of intra-class distance to inter-class distance, while the latter estimates the correlation between the changes in attributes and the resulting changes in class labels.

2.2   Fitness Function

In order to define the fitness of individuals, the specifications of the desired solution must be determined. To evaluate the fitness of the chromosome, we use a function that incorporates two measures. First, we consider the classification accuracy. We employ feed-forward neural networks learning with supervision as the classifier; hence, the accuracy of the classification is defined by the average distance, e, between the observed and the desired output. Second, we include the complexity of the classifier, which is proportional to d^2 due to the nature of the neural network architecture. We represent the complexity, c, of the classifier as, $c = \left(\frac{d}{16}\right)^2$, in terms of the number of attributes, d, with normalization. Our goal is to *minimize* both of these measures, which we combine into a product, as in (1).

$$f(e,c) = e * (0.5 + c/2)^{0.04} \qquad (1)$$

Due to the nature of the error and the complexity terms, further transformation was necessary in order to level the impacts of these two factors.

The Letter Recognition data set [11] is considered as the case study. This data set consists of 20000 letter instances with 16 generated attributes. In order to find the most important attributes, the classical exhaustive search method requires experimenting $2^{16}-1$ different combinations (excluding the chromosome at which all genes are zero, i.e. no attributes are selected to perform the classification). Usually, an exhaustive search is infeasible due to the prohibitively large solution space. Therefore, a genetic algorithm is designed to find a "fit" solution to the problem.

3   Results and Conclusions

In this section the mechanism of the genetic model and the experiments are presented.

The main objective of the solution to the DDR problem is optimizing the initial multi-dimensional data set by representing it with a decreased number of dimensions. While employing the genetic algorithm approach to DDR we used the following two approaches to generate the initial solution of the genetic algorithm: The first one starts with a purely random (uniformly selected) initial space of the dimension subsets. The second set of experiments, on the other hand, starts with initial populations that are formed by the additional knowledge, namely the importance ranking of the dimensions found by SCM. Except this, all other steps of the two genetic algorithms are identical.

All experiments have been setup and run in JAVA. Feed-forward neural networks with four layers and an input layer have been used in the experiments. The input layer consists of 16 elements that hold the positive integer feature values for the presented pattern. Each of the three hidden layers consists of eight processing elements. Finally the output layer has a single node that generates an output signal ranging within $[0, 1]$. The values of the output node are normalized into this interval to represent one of the alphabet letters in respective order. The letter "A" is represented by $[0, \frac{1}{26})$, "B" by $[\frac{1}{26}, \frac{2}{26})$, "C" by $[\frac{2}{26}, \frac{3}{26})$, and so on until the last letter "Z" which is represented by $[\frac{25}{26}, 1)$. We have conducted experiments with one, two and three hidden layers with various combinations of the numbers of processing elements in each hidden layer. We have based our selection of the neural network architecture on the results of these experiments. Hyperbolic tangent sigmoid function given in (2) is employed as the activation function in all processing elements except the single node in the output layer.

$$o(net) = \frac{e^{net} - e^{-net}}{e^{net} + e^{-net}} \tag{2}$$

The output node simply transmits the value of its input with no further modifications, i.e., $o(net) = net$. To train the neural network, the Levenberg-Marquardt backpropagation scheme [2] has been used.

The genetic model was run 20 times and the best final chromosome case is presented. Out of these 20 runs, 10 of them were run with uniformly generated initial populations, and 10 with initial populations generated by the SCM method. The special selections favor the attributes that have higher importance found with SCM. Since SCM does not find probabilities but just weights, these weights are normalized to create a selection probability vector between 0.03 and 0.97. To select the special initial population, we generate a random vector of 16 dimensions and if a random element of this vector is less than the selection probability, then the corresponding attribute is selected in the initial population. The reason we are normalizing to $[0.03, 0.97]$ and not to the usual $[0,1]$ is to include the possibility of the worst attribute to be selected and of the best attribute not to be selected.

We chose the uniform mutation operator with the mutation probability set to 1%. In this model, each bit is selected (with 1% probability) for mutation, and the content of the selected bit is replaced by a 0 or 1 with equal probabilities. The elite count is set to 5 and the crossover ratio is 90%, meaning that each generation is crowded by 90% using crossover. Our experiments showed that among the possible crossover methods, scattered crossover worked best for this model. As for the other parameters of the genetic algorithm, we set the population size to 100 and run each GA model for 40 generations.

The only difference of these two approaches is in the way initial populations are generated. As mentioned above, there is an additional knowledge embedded in the second solution methodology. The importance ranking of the attributes, which are calculated by SCM formula, specifies the impact of discarding a single attribute to the classification performance. In the scope of this chapter these two solution methodologies will be experimented.

Although, both approaches apply the same standard steps (representation, recombination, mutation, selection) of the GA, they do not provide resembling results. As predicted, the knowledge-based solution ends up with better results compared to the purely random initial solution.

We have evaluated the success of the GA methods by the best chromosome generated by the final population. The best chromosome generated by the purely random initial population has a fitness value of 13.01 whereas the best chromosome obtained by the GA initiated by the SCM method has a fitness value of 12.80. It should also be noted that these are the best results obtained out of 10 different GA runs for each method. The progress of the runs are given in Figures 1 and 2. In both figures we observe that the average fitness values approach towards the best fitness value closely. This occurs slightly sooner in experiments starting with a purely random initial population than that with a wisely selected initial population. However, the worst (i.e., maximum) fitness values fluctuate less especially towards the final generations in the experiments with a wisely selected initial population compared with those with a purely random initial population.

In this work, we present an evolutionary solution for data dimensionality reduction problem applied to a letter recognition data set. Due to the nature of the problem, dimensionality reduction has a very big, discrete solution set. To find the optimal solution, one would need to exhaustively search this large solution space (in our case, this corresponds to $2^{16} - 1$ classification experiments

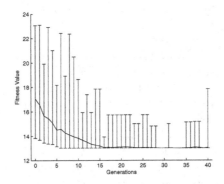

Fig. 1. Best, worst, and average scores of GA run with uniformly generated initial population

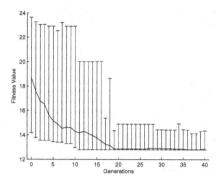

Fig. 2. Best, worst, and average scores of GA run with initial population generated via SCM

using a specific neural network architecture), which proves infeasible. Therefore, we have designed a genetic algorithm based model.

We believe that both the stability of the approach of the average fitness value towards the best one and the rather monotone reduction of the worst fitness values with increasing number of generations depend on the restricted and wise selection of the initial population. Further, not to turn off the possibility for finding a promising individual among the less fit ones while working with fitter individuals (i.e., to explore the less fit individuals while exploiting the fitter ones)[13], we used, in all experiments, a slightly higher mutation rate of 1% than that commonly used in experiments (i.e., 0.1%-0.5%).

References

1. E. Alpaydin. *Introduction to Machine Learning.* MIT Press, 2004.
2. C. M. Bishop. *Neural networks for pattern recognition.* Oxford University Press, 1995.
3. K.D. Bollacker and J. Ghosh. Mutual information feature extractors for neural classifiers,. *IEEE Int. Conf. Neural Networks*, 3:1528–1533, 1996.
4. Damian R. Eads, Steven J. Williams, James Theiler, Reid Porter, Neal R. Harvey, Simon J. Perkins, Steven P. Brumby, and Nancy A. David. A multimodal approach to feature extraction for image and signal learning problems. In *Proceedings of SPIE*, 2004.
5. X. Fu and L. Wang. Data Dimensionmality Reduction with Application to Simplify RBF Network Sutructure and Improving Classificaiton Performance. *IEEE Transactions on Systems, Man and Cybernetics: Part B*, 33(3):399–409, June 2003.
6. D.E. Goldberg. *Genetic algorithms in search, optimization and machine learning.* Addison-Wesley, 1989.
7. Jennifer Hallinan and Paul Jackway. Simultaneous evolution of feature subset and neural classifier on high-dimensional data. In *Conference on Digital Image Computing and Applications*, 1999.
8. N. Kambhatla and T. K. Leen. Dimension reduction by local principal component analysis. *Neural Computation*, 9:1493–1516, 1997.

9. I. Kononenko. Estimating Attributes: Analysis and Extensions of RELIEF. *Proc. Eur. Conf. Machine Learning*, pages 171–182, 1994.
10. E. Lopez-Rubio, J. M. Ortiz de Lazcano-Lobato, J. Munoz-Perez, and J. Antonio Gomez-Ruiz. Principal components analysis competitive learning. *Neural Computation*, 16(11):2459 – 2481, 2004.
11. P.M. Murphy and D. Aha. UCI repository of machine learning databases. [http://www.ics.uci.edu/ mlearn/MLRepository.html], 1994.
12. V. Ramos and F. Muge. Less is more: Genetic optimisation of nearest neighbour classifiers. In F. Muge, C. Pinto, and M. Piedade, editors, *10th Portuguese Conference on Pattern Recognition*, pages 293–301. Technical University of Lisbon, March 1988.
13. R. S. Sutton and Andrew G. Barto. *Reinforcement Learning: an Introduction*. MIT Press, 1999.

Engineering Optimizations via Nature-Inspired Virtual Bee Algorithms

Xin-She Yang

Department of Engineering, University of Cambridge,
Trumpington Street, Cambridge CB2 1PZ, UK
xy227@eng.cam.ac.uk

Abstract. Many engineering applications often involve the minimization of some objective functions. In the case of multilevel optimizations or functions with many local minimums, the optimization becomes very difficult. Biology-inspired algorithms such as genetic algorithms are more effective than conventional algorithms under appropriate conditions. In this paper, we intend to develop a new virtual bee algorithm (VBA) to solve the function optimizations with the application in engineering problems. For the functions with two-parameters, a swarm of virtual bees are generated and start to move randomly in the phase space. These bees interact when they find some target nectar corresponding to the encoded values of the function. The solution for the optimization problem can be obtained from the intensity of bee interactions. The simulations of the optimization of De Jong's test function and Keane's multi-peaked bumpy function show that the one agent VBA is usually as effective as genetic algorithms and multiagent implementation optimizes more efficiently than conventional algorithms due to the parallelism of the multiple agents. Comparison with the other algorithms such as genetic algorithms will also be discussed in detail.

1   Introduction

Nature inspired algorithms based on the swarm intelligence and the self-organized behaviour of social insects can now solve many complex problems such as the travelling salesman problem and the rerouting of traffic in a busy telecom network [1, 2]. This type of algorithms is only a fraction of biology-inspired algorithms. In fact, biology-inspired algorithms form an important part of computational sciences that are essential to many scientific disciplines and engineering applications. These biologically inspired algorithms include genetic algorithms, neural networks, cellular automata and other algorithms ([9, 10, 12]). However, substantial amount of computations today are still using the conventional methods such as finite difference, finite element, and finite volume methods. New algorithms are often developed in the form of the hybrid combination of biology-derived algorithms with the conventional methods, and this is especially true in the field of engineering optimizations. Engineering problems with optimization objectives are often difficult and time consuming, and the applications of nature

J. Mira and J.R. Álvarez (Eds.): IWINAC 2005, LNCS 3562, pp. 317–323, 2005.

or biology inspired algorithms in combination with the conventional optimization methods have been very successful in the last several decades.

Biology-derived algorithms can be applicable to a wide variety of optimization problems [12]. For example, the optimization functions can have discrete, continuous or even mixed parameters without any a priori assumptions about their continuity and differentiability. Thus, the evolutionary algorithms are particularly suitable for parameter search and optimization problems [4, 13]. In addition, they are suitable for parallel implementation. However, evolutionary algorithms are usually computationally intensive, and there is no absolute guarantee for the quality of the global optimizations. Besides, the tuning of the parameters can be very difficult for any given algorithms. Furthermore, there are many different evolutionary algorithms and the best choice of a particular algorithm depends on the type and characteristics of the problems concerned. In this paper, we will present a virtual bee algorithm and use this new algorithm to study the function optimization problems arising naturally from engineering applications.

2   Engineering Optimization

2.1   Engineering Optimization

Many problems in engineering and other disciplines involve optimizations that depend on a number of parameters, and the choice of these parameters affects the performance or objectives of the system concerned. The optimization target is often measured in terms of objective or fitness functions in qualitative models. Engineering design and testing often require some iteration process with parameter adjustment. Optimization functions are generally formulated as:

$$\text{Optimize: } f(\mathbf{x}), \qquad \text{Subject to: } g_i(\mathbf{x}) \geq 0, \qquad i = 1, 2, ., N.$$

where $\mathbf{x} = (x_1, x_2, , x_n) \in \Omega$ (parameter space). The optimization can be either expressed as maximization or more often as minimization [4]. As the space of parameter variations is usually very large, systematic adaptive searching or optimization procedures are required. In the past several decades, researchers have developed many optimization algorithms. Examples of conventional methods are hill-climbing, gradient methods, random walk, simulated annealing and heuristic method etc. The examples of evolutionary or biology inspired algorithms are genetic algorithms, photosynthetic method, neural network, and many others [9, 6]. In next section, we will briefly outline the genetic algorithm for engineering optimizations.

2.2   Genetic Algorithms

Genetic algorithm (GA) is a very effective and powerful method that has been widely used in engineering optimizations. GA is a model or abstraction of biological evolution, which includes the following operators: crossover, mutation, inversion and selection [9]. This is done by the representation within a computer of a population of individuals corresponding to chromosomes in terms of a set of character strings, then the evolution of individuals in the population through the crossover and mutation of the string from parents, and the selection or survival according to their fitness.

For the optimization of a function using genetic algorithms, one simple way is to use the simplest GA with a fitness function: $\phi = A - y$ with A being the large constant and $y = f(x)$, thus the objective is to maximize the fitness function and subsequently minimize the objective function $f(x)$. However, there are many different ways of defining a fitness function. For example, we can use the individual fitness assignment relative to the whole population $\phi(x_i) = f(x_i)/\sum_{i=1}^{N} f(x_i)$, where x_i is the phenotypic value of individual i, and N is the population size [13, 6].

3   Virtual Bee Algorithms

3.1   VBA Algorithms

Algorithms based on social insects and swarm intelligence begin to show their power and effectiveness in many applications. A swarm is a group of mobile agents such as bees that are liable to interact or communicate in a direct or indirect manner in their local environment. For example, when a bee find a food source and successfully bring some nectar back to the hive, it communicates by performing the so-called 'wangle dance' so as to recruit more other bees to go to the food source. The neighbouring bees seem to learn the distance and direction from the dance [7, 8]. As more and more bees forage the same source, it becomes the favoured path. Based on these major characteristics, scientists have developed several powerful algorithms [1, 11]. If we only use some of the nature or behaviour of bees and add some new characteristics, we can devise a class of new algorithms. In the rest of this paper, we will first describe the main procedure of our virtual bee algorithm (VBA) and then apply it to solve function optimizations in engineering.

The VBA scheme starts with a troop of virtual bees, each bee randomly wonders in the phase space and in most case, the phase space can be simply a 1-D or 2-D space. The main steps of the virtual bee algorithm for function optimizations are: 1) creating a population of multi-agents or virtual bees, each bee is associated with a memory bank with several strings; 2) encoding of the objectives or optimization functions and converting into the virtual food; 3) defining a criterion for communicating the direction and distance in the similar fashion of the fitness function or selection criterion in the genetic algorithms; 4) marching or updating a population of individuals to new positions for virtual food searching, marking food and the direction with virtual wangle dance; 5) after certain time of evolution, the highest mode in the number of virtual bees or intensity/frequency of visiting bees corresponds to the best estimates; 6) decoding the results to obtain the solution to the problem.

The procedure can be represented as the following pseudo code:

```
// Create a initial population of virtual bees A(t)
// Encode the function f(x,y,...) into virtual food/nectar
   Initial Population A(t);
   Encode f(x,y) |-> F(x,y);
// Define the criterion for communicating food location with others
   Food F(x,y) |-> P(x,y)
// Evolution of virtual bees with time
   t=0;
   while (criterion)
   // March all the virtual bees randomly to new positions
      t=t+1;
      Update A(t);
   // Find food and communicate with neighbouring bees
      Update F(x,y), P(x,y);
   // Evaluate the encoded intensity/locations of bees
      Evaluate A(t), F(x,y), P(x,y)
   end while
// Decode the results to obtain the solution
   Decode  S(x,y,t);
```

3.2   Function Optimization and Comparison with GA

The virtual bee algorithm has some similarity with genetic algorithms, but it has multiagents that work independently and thus it is much more efficient than the genetic algorithms due to the parallelism of the multiple independent bees. To test this, we first implement the VA algorithm to solve the optimization of the generalized De Jong's test function [3],

$$f(x) = \sum_{i=1}^{n} x^{2\alpha}, \quad |x| \le r, \quad \alpha = 1, 2, ..., m. \quad (1)$$

where α is a positive integer and r is the half length of the domain. This function has a minimum of $f(x) = 0$ at $x = 0$. For the values of $\alpha = 3$, $r = 256$, and $n = 50$, the results of optimization of this test function are shown in Figure 1. The left figure shows the optimization of the test function using the virtual bee algorithm (marked with dots) and its comparison with the results obtained by genetic algorithm (solid curve). The best estimate obtained is 0.1194. We can see that the new VA algorithm is much efficient than the GA method. The figure on the right shows the two different sets of results using 5 bees (dotted) and 20 bees (marked with diamonds). As the multi-bees work almost independently in a parallel manner, the set of 20 bees is much faster than the set of 5 bees in obtaining the best estimate of about 0.0016616.

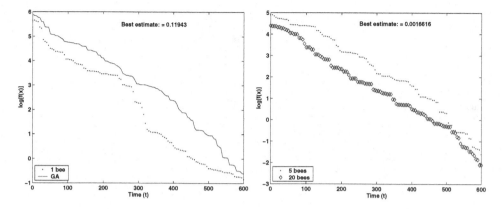

Fig. 1. Function optimization using virtual bee algorithms. Multi-bees are more effective than a single bee in approaching the best estimates $f(x) \to 0$

3.3   Multi-peaked Functions

The function we just discussed is relatively simple in the sense that it is single-peaked. In reality, many functions are multi-peaked and the optimization is thus multileveled. Keane studied the following bumpy function in the multi-peaked optimization problem [14],

$$f(x,y) = \frac{\sin^2(x-y)\sin^2(x+y)}{\sqrt{x^2+y^2}}, 0 < x, y < 10. \qquad (2)$$

The optimization problem is to find (x,y) starting $(5,5)$ to maximize the function $f(x,y)$ subject to: $x + y \leq 15$ and $xy \geq 3/4$. This problem makes the optimization difficult because it is nearly symmetrical about x=y, and thus the peaks occur in pairs but one is bigger than the other. In addition, the true maximum is $f(1.593, 0.471) = 0.365$ which is defined by a constraint boundary. Although the properties of this bumpy function make difficult for most optimizers and algorithms, the genetic algorithms and other evolutionary algorithms perform well for this function and it has been widely used as a test function in the genetic algorithms for comparative studies of various evolutionary algorithms or in the multilevel optimization environment [5]. Figure 2 shows the surface variation of the multi-peaked bumpy function (the left picture). We have used 40 bees in parallel to solve the optimization problem of this bumpy function. After t=500 runs, the concentration distribution of the virtual pheromone laid by the randomly moving virtual bees is shown on the right in Figure 2. We can see that the pheromone concentration is overlapping well with the contour of the function, thus the location of the pheromone focus is the location of the optimal solution.

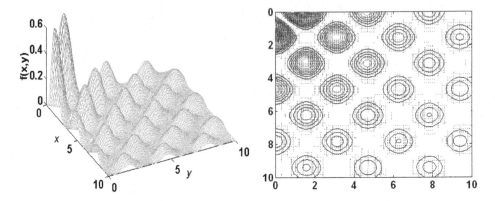

Fig. 2. Multi-peaked function optimization and the virtual pheromone concentration laid by 40 virtual bees

4   Conclusions

By simulating the swarm interactions of social honey bees, we have developed a new virtual bee algorithm (VAA) to solve the function optimizations. For the functions with two-parameters, we have used a swarm of virtual bees that wonder randomly in the space. These bees leave the virtual pheromone when they find some target food corresponding to the encoded value of the functions. The solution for the optimization problem can be obtained from the concentration of the virtual pheromone. Comparison with the other algorithms such as genetic algorithms suggests that virtual bee algorithms work more efficiently due to the parallelism of the multi-bees. As many engineering optimizations have multi-peaked functions, the new algorithms can be expected to have many applications in engineering. In addition, there are many challenges in the virtual bee algorithms. Further research can focus on the efficiency of different encoding/decoding of objective functions, updating rules, evaporation of the long term pheromone, and the parallel implementation technique of the virtual bee algorithms.

References

1. Bonabeau E., Dorigo M., Theraulaz G.: Swarm Intelligence: From Natural to Artificial Systems. Oxford University Press, (1999)
2. Bonabeau E. and Theraulaz G.: Swarm smarts, Scientific Americans. March, (2000)73-79.
3. De Jong K.: Analysis of the Behaviour of a Class of Genetic Adaptive Systems. PhD thesis, University of Michigan, Ann Arbor, (1975).
4. Deb K.: Optimization for Engineering Design: Algorithms and Examples, Prentice-Hall, New Delhi. (1995).

5. El-Beltagy M. A., Keane A. J.: A comparison of various optimization algorithms on a multilevel problem. Engineering Applications of Artificial Intelligence, **12** (1999) 639-654.
6. Flake G. W.: The Computational Beauty of Nature: Computer Explorations of Fractals, Chaos, Complex Systems, and Adaptation, Cambridge, Mass.: MIT Press. (1998).
7. von Frisch K.: The Dance Language and Orientation of HoneyBees, Harvard University Press, (1967).
8. Gordon D. M.: The organization of work in social insect colonies. Nature, **380** (1996) 121-124.
9. Holland J.: Adaptation in Natural and Artificial Systems, University of Michigan Press, Ann Anbor. (1975).
10. Mitchell, M.: An Introduction to Genetic Algorithms, Cambridge, Mass.: MIT Press. (1996).
11. Nakrani S. and Tovey C., On honey bees and dynamic server allocation in Internet hosting centers. Adaptive Behavior, **12** (2004) 223-240.
12. Goldberg D. E. Genetic Algorithms in Search, Optimization and Machine Learning, Reading, Mass.: Addison Wesley. (1989).
13. Jenkins W. M.: On the applications of natural algorithms to structural design optimization. Engineering Structures, **19** (1997) 302-308.
14. Keane A. J.: Genetic algorithm optimization of multi-peak problems: studies in convergence and robustness. Artificial Intelligence in Engineering, **9** (1995) 75-83.

Solving Partitioning Problem in Codesign with Ant Colonies

Mouloud Koudil, Karima Benatchba, Said Gharout,
and Nacer Hamani

Institut National de formation en Informatique,
BP 68M, 16270, Oued Smar, Algeria
{m_koudil, k_benatchba}@ini.dz
http://www.ini.dz

Abstract. Partitioning problem in codesign is of great importance since it can widely influence the characteristics of the system under design. The numerous constraints imposed by the environment and/or the underlying target architecture, in addition to its NP-Completeness makes the problem hard to solve. This paper introduces an automatic partitioning approach inspired by the collective behavior of social insects such as ants, which are able to find the shortest path from their nest to a food source.

1 Introduction

Real-time embedded systems are invading more and more our daily life. In fact, domains such as medicine (surgical robots, exploration tools, pacemakers, hearing prosthesis) take benefit from the advances in this domain. These systems are a mixture of hardware ad software: a software application is ran on a hardware architecture made of processors of various natures: hardware processors, software set-instruction processors, etc. The best approach for mixed system design is called codesign (concurrent-design). Codesign consists of a succession of steps. The major one is partitioning that is the process of determining the parts of the system that must be implemented in hardware and those parts that are to be in software [1]. This task is of critical importance since it has a big impact on final product cost/performance characteristics [2]. Any partitioning decision must, therefore, take into account system properties. It must also include several constraints related to the environment, implementation platform and/or system functionality requirements. Partitioning is known to be an NP-Complete problem. The only feasible way to solve it is, therefore, the use of heuristics since exhaustive methods take a prohibitive execution time to find the best solution.

This paper introduces an automatic partitioning approach that uses ant colony approach which is a meta-heuristic allowing to find the best solution to hard-to-solve optimization problems.

Section two reviews some of the reported partitioning approaches. The third one introduces the proposed partitioning approach. Section four presents ant colonies and the way they can solve partitioning problems, while the last one lists some experiments and results.

J. Mira and J.R. Álvarez (Eds.): IWINAC 2005, LNCS 3562, pp. 324–337, 2005.

2   Previous Work

Among the works that try to automatically solve partitioning problem, some of them use the exact approaches such as d'Ambrosio and al. [3], Pop and al. [4] works apply a branch and bound algorithm to map the tasks onto architectural components. Approached methods allow getting one (or many) solutions in an "acceptable" time. There are mainly two kinds of heuristics: the methods dedicated to the problem to solve, and the general heuristics, that are not specific to a particular problem. One of the dedicated methods is the "bottom-up" approach, also called "constructive approach" which is very popular [2], [5]. The partitioning strategy introduced in [6] combine a greedy algorithm, with an extern loop that takes into account global measures. In VULCAN, Gupta et De Micheli [7] use an approach where all the tasks are first affected to hardware. The tasks are progressively moved to software, using a greedy approach. The main advantage of applying specific approaches is that they are "tailored" for the given problem.

The general heuristics are not dedicated to a particular type of problems, and are widely used in other research fields, consisting of NP-Complete problems. This class also includes algorithms starting with an initial solution (often randomly chosen), that is iteratively improved. The different solutions are compared, using a cost function. The advantage of this type of heuristics is that it is possible to use cost functions that are arbitrary chosen and easy to modify. They also allow achieving a solution in a short time. Their drawback is that it is impossible to guarantee that then achieved solution is the optimum. Among the general heuristics, there are: variable-depth search methods such as variants of Kernighan-Lin (KL) migration heuristics (used in [8]), Knapsack Stuffing algorithm (used in [9]), hill-climbing method (used in [10]), simulated annealing (used in [11])

The works of Eles and Peng [12] compare two heuristics for automatic partitioning: the first one, based on simulated annealing, and the other one on taboo search. Other approaches exist, such as [13] that is based on a min-cut procedure. In addition, integer linear programming formulations are proposed for hardware/software partitioning, by [14].

Recent works have been published in partitioning area [15], [16], [17], [18], which tend to prove that the problem is still opened.

3   The Proposed Solution for Partitioning

When the size and complexity of the problem rise, it becomes difficult, for a human being, to apprehend all the details, and manual resolution of the problem becomes intractable. This is the reason why an environment called PARME [19] was designed and implemented, allowing the user to test partitioning heuristics. It offers the opportunity to study the parameter values, as well as the strategies to use, according to the type of problem.

In this paper, we introduce the tests performed with ant colonies, using PARME. Such tests allow tuning the parameters and the strategies of the algorithm. The results are compared with those achieved with Genetic Algorithms.

3.1   Representation of a Partitioning Solution

Partitioning can be seen as the mapping of all the application entities onto the different processors of the target architecture. Solving this affectation problem means trying to find a solution that minimizes a cost function defined by the user. The representation proposed for this mapping problem is the following:

Let *Nbe* be the number of entities of the application under design and *Nbp* the number of processors of the target architecture. The coding technique consists of creating a vector of Nbe cells, in which each entry corresponds to an entity number. The entities are sorted in the increasing order. Each vector cell contains the number of the processor to which the corresponding entity is affected during partitioning.

Example: The following representation corresponds to a solution of a partitioning problem with 4 entities (Nbe=4) mapped on 2 processors (Nbp=2).

Entity	$\lvert 0\lvert 1\lvert 2\lvert 3\rvert$
Processor	$\lvert 0\lvert 1\lvert 0\lvert 1\rvert$

This means that the entities 1 and 3 are mapped on processor Number 1; Entities 0 and 2 are mapped on processor Number 0.

3.2   The Cost Function

The partitioning algorithm is guided by a cost function that allows evaluating the quality of a given solution. It takes into account different cost constraints (hardware and software space), performance constraints (particular object execution times, global application time) and communication (because information exchange between different application entities is often a bottleneck). The characteristics taken into account in our approach are thus: space, execution time and communication.

A cost function is used to evaluate the quality of generated solutions. Unlike many partitioning approaches reported, the works presented in this paper take into account different cost constraints (hardware and software space), performance constraints (particular object execution times, global application time) and communication (because information exchange between different application entities is often a bottleneck). The characteristics taken into account in our approach are thus: space, execution time and communication. Weights are associated to each characteristic to allow according differentiating the relative importance of each parameter. The cost function is detailed in [17].

4   Solving Partitioning with Ant Colonies

The behavior of insects (in particular ants), as well as their organization, has always interested researchers. They wanted to understand how do ants to find the shortest path between their nest and a food source. The resolution of this problem, which is relatively complex, calls for a certain organization and a collective work. Ants can solve this problem in a collective manner, based on a particular communication means: "pheromone" which is a volatile chemical product they leave as they go by. This substance allows them to locate their path later, and also to attract other ants. The ants modify the pheromone concentration during their walk, and thus dynamically modify their environment.

4.1   Ant Colony Optimization (ACO) Meta-heuristic

Dorigo and al. [20] proposed an algorithm for solving the trading salesman problem that was inspired by ant cooperation to find the shortest path between their nest and a food source. Several algorithms, based on this principle have been developed to solve optimization problems [21], [22], [23], [24], etc.

```
PROCEDURE ACO_Meta_heuristic()
    While (not stopping criterion) do
        Program activities:
            Ant activity;
            Pheromone Evaporation;
            Demon actions optional;
        End Program activities
    End do
End.
```

Fig. 1. ACO Meta-heuristic

The ACO (Ant Colony Optimization) is a meta-heuristic that helps to design algorithms trying to solve optimization problems [20]. It uses three mechanisms that are: ant activity, pheromone evaporation and centralized actions of a demon that can be optional (fig. 1).

Ant Activity. Each ant builds a solution by scanning the research space, which is represented as a graph. There are as much solutions as ants that are launched to solve the problem. At the beginning, ants are randomly placed on the edges of the graph (initial states) that represent the allocation of values to variables. Then they move in their neighborhood to build their solution. The choice of the following edge to visit, among a certain number of neighbors (candidate list), is made according to "a stochastic local search policy". This policy is based on information that is local to the ant, values of pheromone trails that are local to the visited nodes, and constraints that are specific to the problem. Pheromone trails can be associated to components, or connections. When an ant moves towards a new node, it may add a quantity of pheromone on it, or on the vertex leading to that node. This quantity depends on the quality of the solution. When an ant finishes building its solution, it may update the pheromone rate on the path it has taken. It is then removed from the system.

Pheromone Effect: The more a path is used by ants, the more the pheromone rate on it tends to increase. The method for implementing pheromone trails is important ant has an effect on the behavior of the method and its performance. By modifying the pheromone rates on their path, ants modify dynamically the environment and influence the decision of the other ants. Pheromone evaporation allows avoiding a premature convergence of the algorithm towards a local optimum (this problem is often encountered with optimization methods).

Pheromone Update: The ant may put down pheromone on its way at two different moments:

- During the building of the solution (online step by step);
- At the end of the building operation. In this case, the ant retraces its steps ant puts down pheromone on the way it took (online delayed).

Pheromone Evaporation. During their walk, ants tend to choose paths that have the highest pheromone rate. However, this can lead to gathering the ants on the same regions of the search space and thus, and cause a convergence towards the same solution. This is why a mechanism called "evaporation" allows decreasing the pheromone rates, leading therefore to explore new regions, and slow down the convergence.

Demon Action. In order to improve ACO performance, a component called "demon" is integrated. It has a global view on the search state, acting on the environment. It may intensify the search by adding extrapheromone on the promising paths, or on the contrary diversify the search by steering it towards new regions. This is called "offline update".

Candidate List. When the problem to solve presents a big neighborhood, a candidate list can be used by ants to limit the number of neighbors to take into account. This latter can improve the performance of the algorithm. These lists contain subsets of the current state neighbors. They allow to greatly reducing the search space.

4.2   Ant System

The first algorithm implemented (called Ant System: AS) [25] allows to better understand the principle of the algorithms based on ants. This algorithm was designed to solve the traveler salesman problem that consist of finding the shortest path connecting two towns, each town being visited only once. This problem can easily be represented as a graph G (N, A) where N is the set of nodes representing the towns to visit, and A the set of vertices illustrating the distance between two towns. The principle of AS is the following: during several iterations, a set of ants is launched to solve this problem. Each ant (k) builds its solution in N steps. At the beginning of each iteration (t), the ants are randomly located on the edges of the graph (towns). An initial value TAU0 is placed on the vertices during the initialization phase.

Each ant covers the graph to build a solution, by making some choices and adding pheromone on its way (online step by step or online delayed). When the ant (k) is in one town (i), it must choose a town (j) among the list of towns it has not visited yet (J_i^k).

This choice is made on the basis of the heuristic information NUij = 1/dij (the reverse of the distance between the two towns i and j), and the pheromone information, according to the following rule [26]:

$$
P_{ij}^k(t) = \begin{cases} \dfrac{(TAU_{ij}(t))^A * NU_{ij}^B}{\sum_{l\in J}^k (TAU_{il}(t))^A * NU_{ij}^B} & If\, j \in J_i^k \\ 0 & Else \end{cases}
\tag{1}
$$

Where: TAUij(t) is the pheromone rate on node i;

J_i^k is the set of towns in the neighborhood of i, and that are not visited yet by ant k; A and B are the adjustment parameters of the heuristic specific to the problem versus pheromone. They are used to weight the magnitude of pheromone trail TAU and the attraction ; NUij is the value of the heuristic function that is the reverse of the distance between towns i and j in this case. A traveling towards a town j depends on A and B parameters. If A = 0, the traveling is made according to a value NUij that is to say that the nearest town is selected. The pheromone does not influence the decision of the ant. On the other hand, if B = 0, the values of pheromone trails decide the ant k to take a particular path. A and B allow to intensify or diversify the search according to their values. Once an ant k has built its solution, it adds on the path it followed a quantity of pheromone that depends on the quality of its solution. This quantity is defined as follows:

$$
\Delta TAUk_{ij}(t) = \begin{cases} \dfrac{Q}{L^k(t)} & If\, f(i,j) \in T^k(t) \\ 0 & Else \end{cases}
\tag{2}
$$

Where $T_{(t)}^k$ is the distance covered by the ant k at the instant t; $L_{(T)}^k$ is the length of the tour and Q a fixed parameter.

The following ants are influenced by these trails during the choice of the towns to visit through the different iterations. Before launching new ants searching for the optimal solution, pheromone evaporation is performed according to the following formula: $TAU_{ij}(t + 1) = (1 - RHO).TAU_{ij}(t) + \Delta TAU_{ij}(t)$ Where: $\Delta TAU_{ij}(t) = \sum_{k=1}^m \Delta TAU_{ij}^k(t)$;m is the number of ants; and (1- RHO) the pheromone evaporation coefficient. The algorithm of (fig. 2) describes the steps of an ant system.

Algorithm: Ant System
    For t = 1 to tmax
        For each ant k = 1 to m
            Randomly choose a town
            For each not visited town.
                Choose a town j, in the list J_i^k of the remaining towns
            End For
            Put a trail $\Delta TAU_{ij}^k(t)$ on the path $T^k(t)$
        End For
        Evaporate the trails
    End for
End

Fig. 2. ACS steps

4.3   Problem Representation

Let us consider a graph G=(C,L) associated to a discrete optimization problem.
The solution to this problem can be expressed as the feasible paths of G. The aim
of ACO algorithm is to find a path (a sequence) of minimal cost while respecting
the constraints O[27].

The ants of the colony collectively solve the problem by coding the collected
information (during their walk on the graph) as artificial pheromone. The vertices may have heuristic values NUij, giving information on the problem.

For example, in the TSP (traveling Salesman Problem), NUij is associated
to the inverse of the distance between the towns i and j [28]. Pheromone and
heuristic information is used to determine the probabilistic rules.

Graphical Modeling of Partitioning. In order to be able to solve partitioning
problem using ACO, it first must be modeled as CSP problem (Constraint Satisfaction Problem). X = E0, E1, , ENbe-1, D(Ei) = P0, P1, , ENbp-1, C represents the partitioning constraints. We proposed to consider the graph edges
as combinations of each entity with each processor of the system under design.
Each couple (Entity N, Processor N) is coded to obtain an integer that represents a given entity and the associated processor. The coding function used is:
Code:(0...Nbe)*(0...Nbp) → (0...Nbe*Nbp-1)

(ne, np))→ Code(ne, np) = ne*Nbp+n

Example: (fig. 3) If we have to map four entities E0, E1, E2 et E3 on two
processors P0 and P1, we get the following codes: Code(0,0) = 0*2+0 = 0;
Code(0,1) = 0*2+1 = 1; Code(1,0) = 1*2+0 = 2; Code(1,1) = 1*2+1 = 3;
Code(2,0) = 2*2+0 = 4; Code(2,1) = 2*2+1 = 5; Code(3,0) = 3*2+0 = 6;
Code(3,1) = 3*2+1 = 7; The path found 0, 3, 4, 7 represents the solution (0,0),
(1,1), (2,0), (3,1).

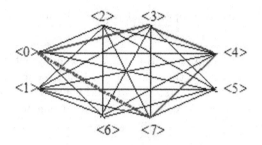

Fig. 3. Graph of the example

4.4   Ant Colony System for Partitioning

The main idea behind ACS [20] is to model the problem as a search for the best path in a particular graph. The ant behavior is determined by defining the starting state, the ending conditions, the building rules, the pheromone update rules and the demon actions [20]. Each solution S is a vector of integers. Each entry in the vector (corresponding to an entity number) contains the number of the processor the entity has been mapped onto.

```
Begin
  Initialize pheromone;
  For (i=1 to Max_Iter) do
    For (each ant) do
      Generate randomly an initial solution S0;
      Build a solution Sk;
      Apply the delayed online update of the pheromone;
    End;
    Determine the best solution found in this iteration;
    Apply the offline update of the pheromone;
  End;
End.
```

Fig. 4. ACS algorithm

The ACS algorithm for partitioning tries to improve the initial solutions that are generated. Each solution S is an array of integers. Each position in the array (corresponding to an entity number) contains the number of the processor to which it has been assigned.

Pheromone Utilization. The choice of the graph element to which pheromone is associated (components or connections) is an important point in the algorithm.

a. Pheromone on components: the first possibility is to put pheromone on the nodes (components). In this case, the pheromone quantity is proportional to the wish of getting a particular assignment in the solution (taking into account some characteristics of a given solution). The state transition rule used by an ant k, called pseudo-random proportional rule can be found in [29]. An ant k randomly generates q.

- If q is less or equal to q0, then the following component to choose (node to visit) will be the one, in the neighborhood, that contains the maximum quantity of product (between pheromone and heuristic) : intensification.
- Else (q greater than q0) it will be chosen according to a probability Pk (diversification).

Where q0 is a parameter of the algorithm, ranging between 0 and 1, and determining the relative importance between exploitation and exploration.

b. Pheromone on connections: another possibility to associate pheromone is to put it on connections. The quantity of pheromone on the connection between two components is thus proportional to the advantage of having the two corresponding assignments in the same solution. For an ant k that moves from a node r to another node u, the pheromone update rule remains the same.

c. Pheromone on connections with sum: the last possibility takes into account the dependence of an assignment with the ones already done. The ant k randomly generates q.

- If q is less or equal to q0, then the following component to take will be: the component in the neighborhood whose sum of products between pheromone and heuristic of the connections with all the visited components is maximal.
- Else (q greater than q0) it will be chosen with a probability Pk.

Other Parameters. Building a new solution: This strategy consists of making each ant start from a complete initial solution S0 which is randomly generated. At each step, a new solution Sk is generated by applying a flip on the current one. Flipping means changing the processor to which a selected entity was previously affected. At each iteration, the entity X_i to flip is chosen with a probability $P^k_{X_i}$ using the state transition rule:

If q $\leq$ q0 Then

$$P^k_{X_i}(t) = \begin{cases} 1 & if(i,j)=Argmax_{(i,j)\in Jk}\{TAU_{ij}(t)*[NU_{ij}(t)]^B\} \\ 0 & Else \end{cases} \tag{3}$$

Else (q greater than q0) Then

$$P^k_{X_i}(t) = \frac{TAU_{ij}(t) * [NU_{ij}(t)]^B}{\sum_{Xl\in Jk} TAU_{lm}(t) * [NU_{lm}(t)]^B} \tag{4}$$

Where $X_{ij} = X_{ik}$ at the instant t ; $k \in 0,1,,Np-1$ with Np the number of processors in the system.

The increase of q0 value leads to a concentration of the search on the components of the best solutions. The contrary promotes the exploration of other areas in the search space.

- Pheromone evaporation: evaporation is applied by the following rule:
  TAUij = (1-RHO)*TAUij.

- Online delayed update: adding pheromone to a solution S* follows the rule:
  TAUij = TAUij + RHO*(cost(S*)/total_number_of_components).

- Offline update is applied by the rule:
  TAUij = TAUij + RHO*(cost(Best_Solution)/cost(S*)).

- Properties of the heuristic used by the state transition rule: in order to solve partitioning with ACO, the following heuristic proceed as follows: if the objective function is F=(Space + Time + Communication), then for a couple entity, processor (e, p):
  Heuristic(e, p) = the cost obtained by the entities already assigned + The space that e occupies on p + The execution time of e on p + The cost of the communication between e and the entities that are already assigned.

5   Tests and Results

The following tests have been performed to illustrate the execution times obtained with partitioning alone. All the test have been performed on an Intel Pentium IV 2,8 GHz with a 256 Mb RAM. Notice that the execution times concern only the effective processing time of the algorithm, excluding display and Input/Output times.

5.1   Presentation of the Benchmark Used

The tests we performed using PARME concerned several benchmarks of various sizes, but for space reasons, we will focus on the most interesting one: it corresponds to a real program implementing a Genetic Algorithm (GA) used in PARME. The objective is to speed up this program by mapping it on a parallel architecture. This benchmark contains 10 entities that must be run on a target architecture made of two heterogeneous processors (a classical instruction-set processor, and a fast hardware one). It is very difficult to solve, since it has only two feasible solutions.

   The application of the exact method to the refined benchmark is possible, since the number of solutions is relatively low (1024). It is nevertheless clear that the application of a manual method becomes impossible. The results achieved by the exact method confirm the complexity of the problem: there exist only two feasible solutions to this problem. The first one, at the 382nd iteration, gives the following cost: 233343; the second one is the optimum, achieved at the 944th iteration. It gives the following cost: 214189.5. Only one optimal solution can be found over the 1024 possible combinations, that is to say, a probability of 0.00097 to get this optimum.

   In the following, some of the results obtained by the ant colony approach are presented. Numerous simulations have been carried out, but for space reasons, we will list only some of the most significant. The execution times are compared with those obtained with GA.

5.2   Execution of Ant Colonies

Table 1 illustrates some of the simulations that have been carried out, using ant colonies, with different sets of parameters. Pheromone was stored on nodes.

A cost of "X" means that no feasible solution was found. By "feasible solution", we mean a solution that does not exceed the maximum space available on each processor. Since the benchmark is hard, only an important size of the ant colony (not less than 150 ants) with a maximum candidate list (size of the neighborhood) and a number of generations at least equal to 50, could get the optimal solution. The algorithm stops after 50 generations.

Table 1. Parameters and results using Ant colonies

Col	Nit	Ng	Can	TAU0	RHO	A	B	q0	Cost	It	Time(s)
200	10	50	20	0.1	0.1	0.01	1	0.8	**214189.5**	27	**13.14**
100	10	50	10	0.1	0.1	0.01	1	0.8	233343.0	8	1.26
150	10	50	20	0.1	0.1	0.01	1	0.8	**214189.5**	13	**5.97**
90	10	50	20	0.1	0.1	0.5	1	0.8	233343.0	10	1.45
100	50	50	10	0.1	0.1	0.5	1	0.8	**214189.5**	13	**9.24**
200	8	50	20	0.1	0.1	0.01	1	0.8	X	X	X
200	10	30	20	0.1	0.1	0.01	1	0.8	X	X	X
150	50	50	10	0.1	0.1	0.5	1	0.8	**214189.5**	**2**	**2.92**

Col: colony size ; **Nit**: Number of iterations; **Ng**: Number of generations; **Can**: Size of the candidate list; It: iteration where the best solution was found; Time: Time to find the best solution. **q0**: intensification/diversification parameter; **TAU0**: initial value of pheromone; **RHO** : pheromone evaporation rate; **A**: influence of pheromone; **B**: influence of heuristics.

5.3   Comparing with the Results Obtained with Genetic Algorithms

It was also possible to get the optimum using another meta-heuristic: the Genetic Algorithms (GA) [26]. However, since the GAs are known to prematurely converge, the only way to get the best solution was to use a strategy called "sharing" [26]. This strategy allows diversifying the population, avoiding the GA to get trapped in a local optimum. But the cost is a much greater time consumption: 7,34 seconds of average time to get the best solution with ant colonies, versus 16.63 seconds with GAs. Notice that the benchmark used is rather small (only 1024 possible solutions). It is easy to imagine that the difference will be much more significant when benchmarks are of bigger size.

NG: Number of Generations; **S**: Sigma (Sharing); **A**: Alpha (Sharing); **C**: Crossover probability and replacement technique; **M**: Mutation probability and

Table 2. Simulations performed with Genetic Algorithms

NG	S	A	C	M	EC	It	ET(s)
60	1	0.01	0.8/CRP	0.1/CRP	214189.5	7	12.96
100	5	0.01	0.9/CRP	0.09/CRP	214189.5	8	15.97
100	1	0.01	0.7/CRP	0.9/CRP	214189.5	12	20.95

replacement technique; **NBS**: N Best Selection (The N best individuals are selected); **CRP**: Children Replace Parents; **EC**: Elite solution Cost; **It**: Iteration number; **ET**: Execution Time (sec.).

6   Conclusion

This paper presents a new algorithm that tries to solve partitioning in codesign by applying the properties of ant colonies in order to overcome the drawbacks of other meta-heuristics: early convergence of GA, cycles in taboo search, slowness of simulated annealing The ant colonies used in this paper are a perfect example of the mutual cooperation that can take place between technology and natural phenomena. Inspired from the social behavior of insects, ant colonies proved to be a good meta-heuristic that gave better results than other approaches we have tested such as taboo search or genetic algorithms (GA). The best solution was found in 100% of the experiments, in a shorter time than the one consumed by GA.

References

1. Kumar S. Aylor J.H. Johnson B. and Wulf W.A., "The Codesign of Embedded Systems", Kluwer Academic Publishers, 1996.
2. De Micheli G. and Gupta R., "Hardware/Software Co-design", Proceedings of the IEEE, Vol.85, N3, pp.349-365, 1997.
3. D'Ambrosio J.G. and Hu X., "Configuration-level hardware/software partitioning for real-time embedded systems", Proceedings of the Third International Workshop on Hardware/Software Codesign, pp.34-41, 1994.
4. Pop P., Eles P. and Peng Z., " Scheduling driven partitioning of heterogeneous embedded systems", Dept. of Computer and Information Science, Linkping University, Sweden, 1998.
5. Hou J. and Wolf W., "Process partitioning for distributed embedded systems", Proc CODES'96, pp.70-75, Pittsburgh, USA, 1996.
6. Kalavade A. and Lee E.A., "The extended partitioning problem: hardware/software mapping and implementation-bin selection", Proceedings of the Sixth International Workshop on Rapid Systems Prototyping, Chapel Hill, NC, June 1995.
7. Gupta R.K., "Co-Synthesis of Hardware and Software for digital embedded systems", Amsterdam: Kluwer, 1995.
8. Olukotun K.A., Helaihel R., Levitt J and Ramirez R., "A Software-Hardware cosynthesis approach to digital system simulation", IEEE Micro, pp.48-58, Aug. 1994.

9. Jantsch A., Ellervee P., Oberg J. and Hemani A., "A case study on hardware/software partitioning", Proceedings IEEE Workshop on FPGAs for Custom Computing Machines, pp.111-18, 1994.
10. Gajski D.D., Narayan S., Ramachandran L. and Vahid F., "System design methodologies: aiming at the 100h design cycle", IEEE Trans. on VLSI Systems, Vol.4, N1, pp.70-82, March 1996.
11. Hartenstein R., Becker J. and Kress R., "Two-level hardware/software partitioning using CoDe-X", IEEE Symposium and Workshop on Engineering of Computer-Based Systems, March 1996.
12. Eles P., Peng Z., Kuchinski K. and Doboli A., "System Level Hardware/Software Partitioning Based on Simulated Annealing and Tabu Search", Design Automation for Embedded Systems, Kluwer Academic Publisher, Vol.2, N1, pp.5-32, 1997.
13. Chou P., Walkup E.A. and Boriello G., "Scheduling strategies in the cosynthesis of reactive real-time systems", IEEE Micro, Vol.14, N4, pp.37-47, Aug. 1994.
14. Nieman R. and Marwedel P., "Hardware/Software partitioning using integer programming", in Proc. EDTC, pp.473-479, 1996.
15. Chatha K.S. and Vemuri R., "MAGELLAN: Multiway Hardware-Software Partitioning and Scheduling for Latency Minimization of Hierarchical Control-Dataflow Task Graphs", Proceedings of 9th International Symposium on Hardware-Software Codesign (CODES 2001), April 25-27, Copenhagen, Denmark, 2001.
16. Bolchini C., Pomante L., Salice F. and Sciuto D., "H/W embedded systems: online fault detection in a hardware/software codesign environment: system partitioning", Proceedings of the International Symposium on Systems Synthesis, Vol. 14 September 2001.
17. Koudil M., Benatchba K., Dours D., "Using genetic algorithms for solving partitioning problem in codesign", Lecture Notes in Computer Science, Springer-Verlag, Vol. 2687, pp. 393-400, June 2003.
18. Noguera J. and Badia R.M., "A hardware/software partitioning algorithm for dynamically reconfigurable architectures", Proceedings of the International Conference on Design Automation and Test in Europe (DATE'01), March 2001.
19. Benatchba K., Koudil M, Drias H., Oumsalem H. Et Chaouche K., "PARME un environnement pour la rsolution du problme Max-Sat", CARI'02, Colloque Africain sur la Recherche en Informatique, OCT. 2002.
20. Dorigo M., "Optimization, Learning and Natural Algorithms", PhD thesis, Politecnico di Milano, Italy, 1992.
21. Costa D. and Hertz V., "Ants can Colour Graphs", Journal of the Operational Research Society", 48:295-305, 1997.
22. Solnon C., "Solving Permutation Constraint Satisfaction Problems with Artificial Ants", Proceedings of the 14th European Conference on Artificial Intelligence, pp 118-122. IOS Press, Amsterdam, The Netherlands, 2000.
23. Benatchba K., Admane L., Koudil M. and Drias H. "Application of ant colonies to data-mining expressed as Max-Sat problems", International Conference on Mathematical Methods for Learning, MML'2004, Italy, June 2004.
24. Admane L., Benatchba K., Koudil M. and Drias H., "Evolutionary methods for solving data-mining problems", IEEE International Conference on Systems, Man & Cybernetics, Netherlands, October 2004.
25. Colorni A., Dorigo M., Maniezzo V., "Distributed Optimization by Ant Algorithm", Proceedings of the First European Conference on Artificial Life, pp. 134-142. MIT Press, Cambridge, MA, 1992
26. Bonabeau E., Dorigo M. and ThRaulaz G., "Swarm Intelligence: From Natural to Artificial Systems", Oxford University Press, New York, 1999.

27. Dorigo M., Dicaro G., "Ant Colony Optimisation : A new meta-heuristic", IEEE 1999.
28. Sttzle T. and Dorigo M., "ACO algorithms for the traveling salesman broblem", in K. Miettinen, M. M. Mkel, P. Neittaanmki, and J. Periaux, editors, John Wiley and Sons, 1999.
29. Dorigo M., and Gambardella L. M., "Ant Colony system : A cooperation learning approch to the traveling salesman problem", IEEE Trans. Evol. Comp., 1(1):53-66, 1997.

A Neuromimetic Integrated Circuit for Interactive Real-Time Simulation

Sylvain Saïghi, Jean Tomas, Yannick Bornat, and Sylvie Renaud

IXL Laboratory, UMR 5818 CNRS, ENSEIRB-University Bordeaux 1,
351 cours de la Libération, 33405 Talence, France
saighi@ixl.fr

Abstract. This paper presents a new analog neuromimetic Integrated Circuit. This IC includes tunable analog computation cores, which are based on Hodgkin-Huxley formalism, calcium channel and calcium-dependent potassium channel. The analog computation cores compute in real-time biologically realistic models of neurons and they are tuned by built-in digital functions. Several topologies are possible to reproduce different neural activity like fast spiking or regular spiking. Those activities are presented to illustrate the diversity of models simulated by this IC.

1 Introduction

Since the first silicon neuron from M. Mahowald and R. Douglas in 1991 [1], research groups have developed and used analog neuromimetic integrated circuits to address fundamental neurosciences questions. Such devices emulate and therefore allow a detailed analysis of activity patterns of single neurons or small networks. When based on biophysical models, the circuits provide a precise temporal replica of the neurons electrical activity. In this paper, we consider devices where the models are computed in analog mode and in real-time. The variations of the signal are then continuously computed, while their dynamics strictly fits the biological neurons ones. The applications of such circuits have been notably detailed and discussed in [2], [3] and [4].

Two approaches can be identified when designing those custom circuits: in the first one, an integrated circuit (IC) is fabricated to fit a specific model card (set of parameters), and will be used to study a single class of neurons. In that case, more silicon neurons can be integrated on a single chip, and applications generally address network and synaptic modulation questions [5]. For the second approach, the IC receives inputs to set the chosen model card. It is then used as a simulation tool where the user can access and tune the models parameters, building its proprietary neuron and network adapted to its application.

The IC presented here has been designed according to the second approach. It is specified to accept a wide range of model parameters, which correspond to realistic neurons diversity. We will show that it can precisely emulate different types of neurons, characterized by specific activity patterns. Then we will discuss

J. Mira and J.R. Álvarez (Eds.): IWINAC 2005, LNCS 3562, pp. 338–346, 2005.
© Springer-Verlag Berlin Heidelberg 2005

some applications using this IC. In particular, we will present a methodology to study the parameters variations influence on the membrane electrical activity.

2   Biological Model

We chose to exploit the Hodgkin-Huxley formalism as the design basis for the IC. The main advantage of this formalism is that it relies on parameters, which are biophysically realistic, by the way of a conductance-based expression of the neural activity. The electrical activity of a neuron is the consequence of the ionic species diffusion through its membrane. This activity is characterized by a membrane potential, which is the voltage difference between the outside and the inside of the cell. Ions flow through the cell membrane through ion-specific channels, generating specific ionic currents. A reverse potential is associated to each ionic species, according to the difference between the intracellular and extracellular concentrations. The fraction of opened ion-specific channels determines the global conductance of the membrane for that ion. This fraction results from the interaction between time and voltage dependent activation and inactivation processes.

The Hodgkin-Huxley formalism [6] provides a set of equations and an electrical equivalent circuit (Fig. 1) that describe these conductance phenomena. The current flowing across the membrane is integrated on the membrane capacitance, following the electrical equation (1),

$$C_{mem} \cdot \frac{\mathrm{d}V_{mem}}{\mathrm{d}t} = -\sum I_{ion} + I_s \qquad (1)$$

where V_{mem} is the membrane potential, C_{mem} the membrane capacitance and I_s an eventual stimulation or synaptic current.

I_{ion} is the current passing through one channel type, and is given by (2), in which g_{max} is the maximal conductance value, m and h respectively represent activation and inactivation term, which are the dynamic functions describing the permeability of membrane channels to this ion. V_{equi} is the ion-specific reverse potential and p, q are integers. Figure 1 displays an electrical circuit schematic that follows the voltage-current relationships given by (1) and (2).

$$I_{ion} = g_{max} \cdot m^p \cdot h^q \cdot (V_{mem} - V_{equi}) \qquad (2)$$

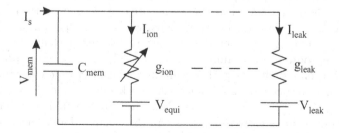

Fig. 1. Neuron electrical equivalent circuit

According to kinetic function (3), m converges to its associated steady-state value m_∞, which is a sigmoid function of V_{mem} (4). The time constant for the convergence is τ_m. In (4) V_{offset} is the activation sigmoid offset and V_{slope} the activation sigmoid slope.

$$\tau_m \cdot \frac{\mathrm{d}m}{\mathrm{d}t} = m_\infty - m \tag{3}$$

$$m_\infty = \frac{1}{1 + \exp(-\frac{V_{mem} - V_{offset}}{V_{slope}})} \tag{4}$$

The inactivation variable h follows same dynamics than the activation variable m, but the steady-state value calculation is done by modifying the minus sign between brackets in a plus sign.

The Hodgkin-Huxley primary equations describe sodium, potassium and leak channels with respectively in (2) $p = 3$ and $q = 1$; $p = 4$ and $q = 0$; $p = 0$ and $q = 0$. These channels are responsible of action potential generation. For more complex activity patterns, like bursting or action potentials discharge with adaptation phenomena, additional channels such as calcium and calcium-dependent potassium have to be taken into account. The calcium channel is described like the previous ones, but with several possible values for p and q according with (2) [7]. To keep our initial objective and model various neural activity, we chose p=[1;2] and q=[0;1].

The potassium channel dynamics also depends on internal variables, such as the calcium concentration. The calcium concentration can be computed following (5). The resulting value is introduced in (6) to evaluate the steady state activation value. For the calcium-dependent potassium channel, we define here $p = 1$ and $q = 0$.

$$\tau_{Ca} \frac{\mathrm{d}[Ca^{2+}]}{\mathrm{d}t} = I_{Ca^{2+}} - [Ca^{2+}] \tag{5}$$

$$m_\infty = \frac{[Ca^{2+}]}{[Ca^{2+}] + [Ca_0]} \cdot \frac{1}{1 + \exp(-\frac{V_{mem} - V_{offset}}{V_{slope}})} \tag{6}$$

3   Silicon Integration

Ionic channels calculation necessitates mathematical operations; a library of elementary analog circuits computing those operations has been developed and validated [8]. Those mathematical operations are defined in generic mode; to be exploited them for ionic channels operations, we tune the parameters V_{offset}, V_{slope}, τ_m, V_{equi} and g_{max}. We retained for our model the five channels described earlier: leakage, sodium, potassium, calcium and calcium-dependent potassium. Their computation circuitry is gathered in an analog electronic core. Ionic current generators can then be represented as block diagrams of function modules. For example, with potassium dependence towards calcium considered as a weighting of potassium steady-state value (6), the calcium-dependent potassium block diagram is as shown in Fig. 2.

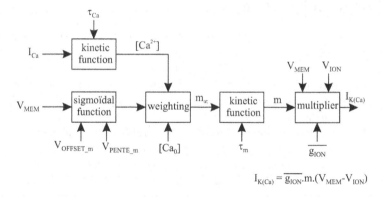

$$I_{K(Ca)} = \overline{g_{ION}}.m.(V_{MEM}-V_{ION})$$

Fig. 2. Block diagram of the $I_{K(Ca)}$ ionic current generator

We integrated a circuit including two analog computation cores, together with 8 synapses for each core (see Fig. 3). The different parameters are stored on-chip, using dynamic analog memories. These analog memories, based on integrated capacitors, were designed to store the 158 analog parameters that are necessary for the 2 analog cores. The memory cells array is driven by an external ADC, which sequentially refreshes the analog parameters values. This technique allows the dynamic modification of one or more parameters, even during the running of the simulation. One modification necessitates three refreshing cycles ($\leq 5\,ms$). To program the parameter memories we use another 3 bits bus (Clock, Reset and Data) and an analog bus (Parameters values).

To define which channels are used to compute the electrical activity, the experimenter chooses a topology before the simulation starts. This topology is stored on built-in dynamic digital memories and programmed by another 3 bits bus (Clock, Reset and Data).

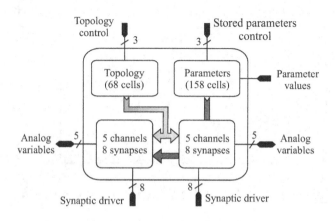

Fig. 3. Integrated Circuit block diagram

4   Circuit and System

The chip has been designed in full-custom mode with a BiCMOS SiGe 0.35m technology process from *austriamicrosystems* (AMS) under Cadence environment. We used bipolar transistors for the integration of the steady-state value and the computation of power functions. Figure 4 is a microphotograph of the integrated circuit named *Pamina*. Ionic channels, synapses, topology and analog memory cells can be identified on the figure. *Pamina* integrates around 19000 MOS transistors, 2000 bipolar transistors and 1200 passive elements; its area is $4170 \times 3480 \mu m^2$. Ionic channels, synapses and analog memory cells are designed in full-custom mode (71% of the components) whereas digital cells for topology are from the *austriamicrosystems*'s library. Optimized analog layout procedures, like common-centroïd, have been used to implement critical structures and harden it to technological process mismatch and variations [9].

A complete computer-based system was built to exploit the IC (see Fig. 5). The user defines the neuron characteristics he wants to model, using interface software running on the computer. Theses characteristics include the ionic channels choice between sodium, potassium, leakage, calcium and calcium-dependent potassium and the parameters values of each channel. These data are sent to the IC through the analog and digital buses described in paragraph 3. The analog computation core simulates in real-time the membrane potential, which is digitized through an analog digital converter and sent back to the computer for display and/or storage.

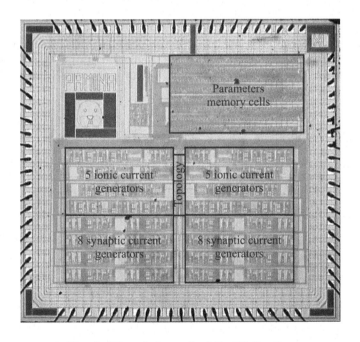

Fig. 4. Microphotograph of the IC *Pamina*

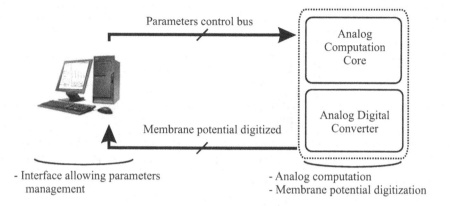

Parameters control bus

Analog
Computation
Core

Analog Digital
Converter

Membrane potential digitized

- Interface allowing parameters
  management

- Analog computation
- Membrane potential digitization

Fig. 5. Structure of the complete simulation system

5   Results and Application

The first result presented is a simulation of the activity of a 4-conductances neuron (sodium, potassium, leak and calcium). We can observe in Fig. 6, from bottom to top, plots of the stimulation current, the calcium current and the membrane voltage. Before the stimulation, the artificial neuron is silent. When the stimulation current is applied, the neuron starts oscillations and the calcium current increases, which produces the oscillations frequency raising. When the stimulation pulse stops, the oscillations are maintained during a period due to the calcium current presence. The decrease of the calcium current slows down the oscillations frequency, which enforces the calcium current decrease. Finally, The neuronactivity stops.

We then added to the precedent neuron model the calcium-dependent potassium channel. We can observe in Fig. 7 the stimulation current (bottom plot) and in the top the neuron electrical activity (top plot). When the stimulation current starts, the neuron begins to oscillate and activates the calcium channel. The calcium channel activates in turn the calcium-dependent potassium channel. The calcium conductance tends to increases the oscillation frequency whereas the calcium-dependent potassium conductance tends to decrease it. The calcium-dependent potassium effect finally predominant and the oscillation frequency decreases, whereas the stimulation current is still present. When the stimulation current stops, the calcium current is not strong enough to keep the oscillations; then we observe a hyperpolarization of the membrane while the calcium-dependent potassium channel is still activated. The calcium channel finally inactivates, inducing the inactivation of the calcium-dependent potassium channel, and the membrane potential returns to its resting state.

The same IC has been used with a different set of model parameters, programmed through the software/hardware interface, and stored on-chip in the analog memory cells. Whereas the initial values are fixed before the simulation

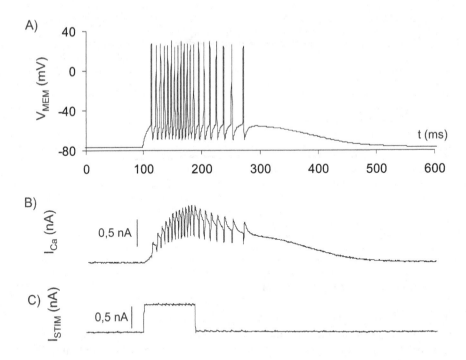

Fig. 6. A simulation of a 4 conductances neuron

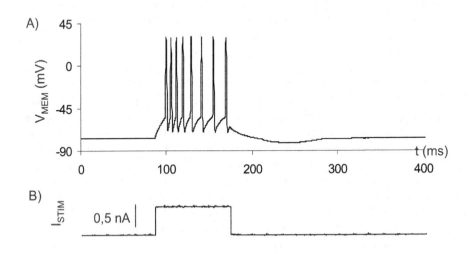

Fig. 7. A simulation of a 5 conductances neuron

starts, it is still possible to modify dynamically one or more values during the
simulation. This procedure is illustrated in figure 8: the oscilloscope screen cap-

Fig. 8. Effect of the dynamic modification of a parameter during a simulation

tures display the electrical activity a 4-conductances neuron, measured on the IC output. From A) to C), the maximal conductance value for the calcium channel ($g_{Ca\,max}$) is increased. In A) $g_{Ca\,max}$ is the smallest; the neuron electrical activity stops with the stimulation. In B), C) the oscillations are maintained during a period increasing with $g_{Ca\,max}$.

Changing the parameter value has been preformed without stopping or resetting the simulation; individual parameter influence is therefore easier to visualize. With the same principle, we can modify any other parameter. Combining both parameter dynamic tuning capability and membrane electrical potential digitization allows us to use such a real-time system to explore more precisely the different interplays of model parameters and neuron activity.

6   Conclusion

We presented in this paper a novel mixed neuromimetic integrated circuit. The chip simulates in real-time electrical activities of biologically realistic neuron models. The chip is organized around a tunable analog computation core, analog memory cells, and a communication digital circuitry. The neuron model parameters values are tunable via a software/hardware interface. The results presented illustrate the diversity of the activity pattern possibly emulated by a single IC. Using the system, it is then possible to study in real-time the influence of parameters in conductance-based neuron models. Integrated synapses, not described here, will allow the extension of the experimental principle to small networks.

References

1. M. Mahowald and R. Douglas, "A silicon neuron", Nature, vol. 345, pp 515-518, 1991.
2. G. Le Masson, S. Renaud-Le Masson, D. Debay and T. Bal, "Feedback inhibition controls spike transfer in hybrid thalamic circuits", Nature, vol. 417, pp. 854-858, 2002.
3. M. F Simoni, G. S. Cymbalyuk, M. E. Sorensen, R. L. Calabrese and S.P. DeWeerth, "A multiconductance Silicon Neuron with Biologically Matched Dynamics", IEEE Trans. Bio. Eng., vol. 51, NO 2, pp. 342-354, 2004.

4. G. Indiveri, "A Neuromorphic VLSI Device for Implementing 2-D Selective Attention Systems", IEEE Trans. Neural Network, vol. 12, NO 6, pp 1455-1463, 2001.
5. L. Alvado et al., "Hardware computation of conductance-based neuron models", Neurocomputing, vol. 58-60, pp. 109-115, 2004.
6. A.L. Hodgkin, .F. Huxley, "A quantitative description of membrane current and its application to conduction and excitation in nerve", Journal of physiology, vol. 117, pp. 500-544, 1952.
7. C. Koch, Biophysics of Computation, Oxford University Press, 1999, pp. 212-231.
8. S. Saïghi, J. Tomas, L. Alvado, Y. Bornat, S. Renaud - Le Masson, "Silicon Integration of Biological Neurons Models", DCIS 2003, Ciudad Real, Spain, November 19-21 2003, ISBN 84-87087-40-X, pp. 597-602.
9. A.Hastings, "The art of analog layout", Prentice Hall, 2001.

A FPGA Architecture of Blind Source Separation and Real Time Implementation

Yong Kim and Hong Jeong

Department of Electronic and Electrical Engineering,
POSTECH, Pohang, Kyungbuk, 790-784, South Korea
{ddda, hjeong}@postech.ac.kr
http://isp.postech.ac.kr

Abstract. Blind source separation(BSS) of independent sources from their convolutive mixtures is a problem in many real-world multi-sensor applications. However, the existing BSS architectures are more often than not based upon software and thus not suitable for direct implementation on hardware. In this paper, we present a new VLSI architecture for the blind source separation of a multiple input mutiple output(MIMO) measurement system. The algorithm is based on feedback network and is highly suited for parallel processing. The implementation is designed to operate in real time for speech signal sequences. It is systolic and easily scalable by simple adding and connecting chips or modules. In order to verify the proposed architecture, we have also designed and implemented it in a hardware prototyping with Xilinx FPGAs.

1 Introduction

Blind source separation is a basic and important problem in signal processing. BSS denotes observing mixtures of independent sources, and, by making use of these mixture signals only and nothing else, recovering the original signals. In the simplest form of BSS, mixtures are assumed to be linear instantaneous mixtures of sources. The problem was formalized by Jutten and Herault [1] in the 1980's and many models for this problem have recently been proposed.

In this paper, we present K. Torkkola's feedback network [2, 3] algorithm which is capable of coping with convolutive mixtures, and T. Nomura's extended Herault-Jutten method [4] algorithm for learning algorithms. Then we provide the linear systolic architecture design and implementation of an efficient BSS method using these algorithms. The architecture consists of forward and update processor.

We introduce in this an effficient linear systolic array architecture that is appropriate for VLSI implementation. The array is highly regular, consising of identical and simple processing elements(PEs). The design very scalable and, since these arrays can be concatenated, it is also easily extensible. We have designed the BSS chip using a very high speed integrated circuit hardware description language(VHDL) and fabricated Field programmable gate array(FPGA).

J. Mira and J.R. Álvarez (Eds.): IWINAC 2005, LNCS 3562, pp. 347–356, 2005.
© Springer-Verlag Berlin Heidelberg 2005

2   Background of the BSS Algorithm

In this section, we assume that observable signals are convolutively mixed, and present K. Torkkola's feedback network algorithm and T. Nomura's extended Herault-Jutten method. This method was used by software implementation by Choi and Chichocki [5].

2.1   Mixing Model

Real speech signals present one example where the instantaneous mixing assumption does not hold. The acoustic environment imposes a different impulse response between each source and microphone pair. This kind of situation can be modeled as convolved mixtures. Assume n statistically independent speech sources $s(t) = [s_1(t), s_2(t), \ldots, s_n(t)]^T$. There sources are convolved and mixed in a linear medium leading to m signals measured at an array of microphones $x(t) = [x_1(t), x_2(t), \ldots, x_m(t)]^T (m > n)$,

$$x_i(t) = \sum_p \sum_{j=0}^{n} h_{ij,p}(t) s_j(t-p), \quad for \ \ i = 1, 2, \ldots, m, \tag{1}$$

where $h_{ij,p}$ is the room impulse response between the jth source and the ith microphone and $x_i(t)$ is the signal present at the ith microphone at time instant t.

2.2   Algorithm of Feedback Network

The feedback network algorithm was already considered in [2, 6]. Here we describe this algorithm. The feedback network whose ith output $y_i(t)$ is described by

$$y_i(t) = x_i(t) + \sum_{p=0}^{L} \sum_{j \neq i}^{n} w_{ij,p}(t) y_j(t-p), \quad for \ \ i, j = 1, 2, \ldots, n, \tag{2}$$

where $w_{ij,p}$ is the weight between $y_i(t)$ and $y_j(t-p)$. In compact form, the output vector $y(t)$ is

$$\begin{aligned} y(t) &= x(t) + \sum_{p=0}^{L} W_p(t) y(t-p), \\ &= [I - W_0(t)]^{-1} \{ x(t) + \sum_{p=1}^{L} W_p(t) y(t-p) \}. \end{aligned} \tag{3}$$

The learning algorithm of weight W for instantaneous mixtures was formalized by Jutten-Herault algorithm [1]. In [4], the learning algorithm was the extended Jutten-Herault algorithm and proposed the model for blind separation where observable signals are convolutively mixed. Here we describe this algorithm. The learning algorithm of updating W has the form

$$W_p(t) = W_p(t-1) - \eta_t f(y(t)) g(y^T(t-p)), \tag{4}$$

where $\eta_t > 0$ is the learning rate. One can see that when the learning algorithm achieves convergence, the correlation between $f(y_i(t))$ and $g(y_j(t-p))$ vanishes. $f(.)$ and $g(.)$ are odd symmetric functions. In this learning algorithm, the weights are updated based on the gradient descent method. The function $f(.)$ is used as the signum function and the function $g(.)$ is used as the 1st order linear function because the implementation is easy with the hardware.

3   Systolic Architecture for a Feedback Network

We present a parallel algorithm and architecture for the blind source separation of a multiple input mutiple output(MIMO) measurement system. The systolic algorithm can be easily transformed into hardware. The overall architecture of the forward process and update is shown first and then follows the each detailed internal structure of the processing element(PE).

3.1   Systolic Architecture for Forward Process

In this section, we introduce the architecture for forward processing of the feed-back network. The advantage of this architecture is spatial efficiency, which accommodates more time delays for a given limited space. The output vector of the feedback network, $\boldsymbol{y}(t)$ is

$$\boldsymbol{y}(t) = [\boldsymbol{I} - \boldsymbol{W}_0(t)]^{-1}\{\boldsymbol{x}(t) + \sum_{p=1}^{L}\boldsymbol{W}_p(t)\boldsymbol{y}(t-p)\}. \tag{5}$$

Let us define $\boldsymbol{C}(t) = [\boldsymbol{I} - \boldsymbol{W}_0(t)]^{-1}$ where the element of $\boldsymbol{C}(t) \in \boldsymbol{R}^{n\times n}$ is $c_{ij}(t)$.

$$\boldsymbol{y}(t) = \boldsymbol{C}(t)\{\boldsymbol{x}(t) + \sum_{p=1}^{L}\boldsymbol{W}_p(t)\boldsymbol{y}(t-p)\},$$

$$= \boldsymbol{C}(t)\boldsymbol{x}(t) + \sum_{p=1}^{L}\hat{\boldsymbol{W}}_p(t)\boldsymbol{y}(t-p). \tag{6}$$

Applying (6) and the above expressions to (2), we have

$$y_i(t) = \sum_{j=1}^{n}c_{ij}(t)x_j(t) + \sum_{p=1}^{L}\sum_{j=1}^{n}\hat{w}_{ij,p}(t)y_j(t-p), \quad for \ \ i=1,2,\ldots,n. \tag{7}$$

Let us define the cost of the pth processing element $f_{i,p}(t)$ as

$$f_{i,0}(t) = 0,$$

$$f_{i,p}(t) \equiv f_{i,p-1}(t) + \sum_{j=1}^{n}\hat{w}_{ij,p}(t)y_j(t-p), \quad for \ \ p=1,2,\ldots,L. \tag{8}$$

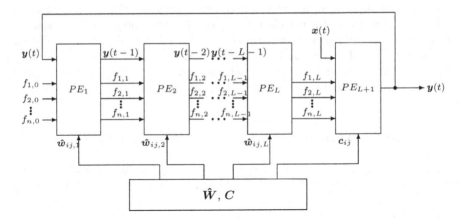

Fig. 1. Linear systolic array for the feedback network

Combining (7) and (8), we can rewrite the output $\boldsymbol{y}(t)$ as

$$y_i(t) = \sum_{j=1}^{n} c_{ij}(t)x_j(t) + f_{i,L}(t), \quad for \ \ i = 1, 2, \ldots, n. \qquad (9)$$

We have constructed a linear systolic array as shown in Fig. 1. This architecture consists of $L+1$ PEs. The PEs have the same structure, and the architecture has the form of a linear systolic array using simple PEs that are only conneted with neighboring PEs and thus can be easily scalable with more identical chips.

During $p = 1, 2, \ldots, L$, the pth PE receives three inputs $y_j(t - p)$, $\hat{w}_{ij,p}(t)$ and $f_{i,p-1}(t)$. Also, it updates PE cost $f_{i,p}(t)$ by (8). The $L + 1$th PE calculates

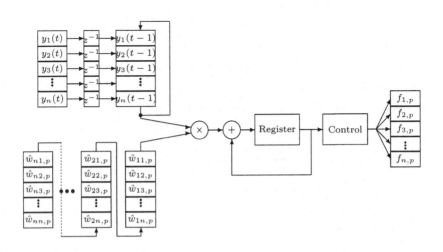

Fig. 2. The internal structure of the processing element

$y(t)$ according to (9) using inputs $x(t)$ and $c_{ij}(t)$. In other words, we obtain the cost of PE using recursive computation. With this method, computational complexity decreases plentifully. The computational complexity for this architecture is introduced in the last subsection.

The remaining task is to describe the internal structure of the processing element. Fig. 2 shows the processing element. The internal structure of the processing element consists of signal input part, weight input part, calculation part, and cost updating part. The signal input part and weight input part consist of the FIFO queue in Fig. 2, and take signal input $y(t)$ and weight $w(t)$ respectively. The two data move just one step in each clock. As soon as all inputs return to the queue, the next weight is loaded. When all the weights are received, the next input begins. The calculation part receives inputs $y(t)$ and w at two input parts, then updates PE cost $f_{i,p}(t)$ according to (8). This part consists of adder, multiplier, and register.

3.2   Systolic Architecture for the Update

This section presents the architecture that updates the weights. This architecture also consists of the processing elements that operate in the same method. We define the processing element which the used in the update as the Update Element(UE).

Efficient implementation in systolic architecture requires a simple form of the update rule. The learning algorithm of updating has the form

$$w_{ij,p}(t) = w_{ij,p}(t-1) - \eta_t f(y_i(t))g(y_j(t-p)),$$
$$i, j = 1, 2, \cdots, n, (i \neq j) \quad p = 1, 2, \cdots, L. \tag{10}$$

In this architecture, the function $f(.)$ is used as the signum function $f(y_i(t)) = sign(y_i(t))$ and the function $g(.)$ is used as the 1st order linear function $g(y_j(t)) = y_j(t)$ because the implementation is easy with the hardware.

Fig. 3 shows the systolic array architecture of the update process. If the number of signals is N, then the number of rows D is $(N^2+N)/2$. All arrays have the same structure and all weignts can be realized by using $y(t)$ simultaneously.

In a row in Fig. 3, if the number of PE is L, the number of columns is $2L+1$. In other words, the architecture of the update consists of $D \times (2L+1)$ UEs.

The cost of (d,p)th UE has the form

$$u_{d,p}(t) = u_{d,p}(t-1) - \eta_t f(y_i(1/2(t-p-L)))y_j(1/2(t+p-L)),$$
$$d = 1, 2, \cdots, D. \tag{11}$$

Fig. 4 shows the internal structure of the UE of the feedback network. The processing element performs simple fixed computations resulting in a simple design and low area requirements. The pth UE receives two inputs y_i and y_j, then one input becomes $f(y_i)$. The cost of UE is added to the accumulated cost of the same processor in the previous step.

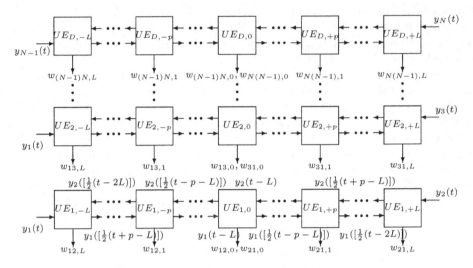

Fig. 3. Linear systolic array for update

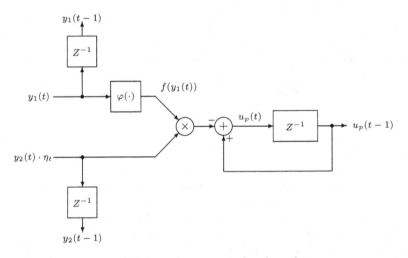

Fig. 4. The internal structure of update element

3.3 Overall Architecture

The system configuration is shown in Fig. 5. We observe a set of signals from an MIMO nonlinear dynamic system, where its input signals are generated from independent sources. The minimization of spatial dependence among the input signals results in the elimination of cross-talking in the presence of convolutively mixing signals.

We assumed that auditory signals from n sources were mixed and reached n microphones far from the sources. The forward process of the feedback network algorithm uses two linear arrays of $(L + 1)$ PEs. (nL) buffers are required for

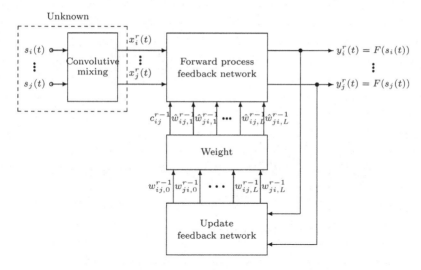

Fig. 5. Overall block diagram of feedback network architecture

output $\boldsymbol{y}(t)$,nL buffers for cost of PE, and (DnL) buffers for weight $\boldsymbol{W}(t)$. Since each output $\boldsymbol{y}(t)$ and weight $\boldsymbol{W}(t)$ are B_y bits and B_w bits long, a memory with $O(nLB_y + DnLB_w)$ bits is needed. Each PE must store the partial cost of B_p bit and thus additional $O(nLB_p)$ bits are needed. As a result, total $O(nL(B_y + B_p + DB_y))$ bits are sufficient.

The update of the feedback network algorithm uses a linear array of $D(2L+1)$ UEs. $2D(2L+1)$ buffers are required for output $\boldsymbol{y}(t)$ and $D(2L+1)$ buffers for cost of UE. If UE stores the partial cost of B_u bits, total $O(DL(4B_y + 2B_w))$ bits are sufficient.

4   System Implementation and Experimental Results

The system is desiged for an FPGA(Xilinx Virtex-II XC2V8000). The entire chip is designed with VHDL code and fully tested and error free. The following experimental results are all based upon the VHDL simulation. The chip has been simulated extensively using ModelSim simulation tools. It is designed to interface with the PLX9656 PCI chip. The FPGA design implements the following architecture:

- Length of delay: L=50
- The number of input source: n=4
- The buffer size for learning: 200 samples
- The learning rate: $\eta_t = 10^{-6}$

As a performance measure, Choi and Chichocki [5] have used a signal to noise ratio improvement,

$$SNRI_i = 10 \log_{10} \frac{E\{(x_i(k) - s_i(k))^2\}}{E\{(y_i(k) - s_i(k))^2\}}. \tag{12}$$

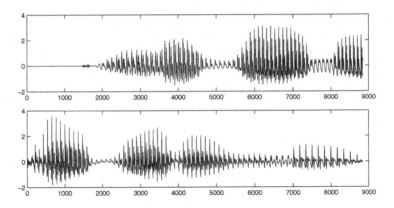

Fig. 6. Two original speech signals

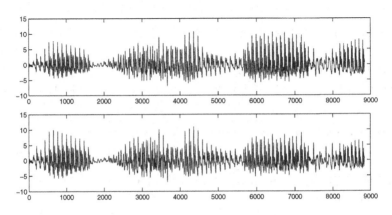

Fig. 7. Two mixtures of speech signals

Table 1. The experimental results of SRNI with noisy mixtures

	SNR(signal to noise ratio)$= 10\log_{10}(s/n)$					
	clean	10dB	5dB	0dB	-5dB	-10dB
SNRI1(dB)	2.7061	2.6225	2.4396	2.3694	1.9594	0.1486
SNRI2(dB)	4.9678	4.9605	4.7541	4.6506	3.9757	3.2470

Two different digitized speech signals $s(t)$, as shown in Fig. 6, were used in this simulation. The received signals $x(t)$ collected from different microphones and recovered signals $y(t)$ using a feedback network are shown in Fig. 7, 8. In this case, we obtained $SNRI_1 = 2.7061$, and $SNRI_2 = 4.9678$.

We have evaluted the performace of the feedback network, given noisy mixtures. Table. 1 shows the experimental results of SRNI with noisy mixtures. The performance of this system measures were scanned for the SNR from -10dB to 10dB with an increment of 5dB. The system has shown good performance in high SNR (above 0dB only).

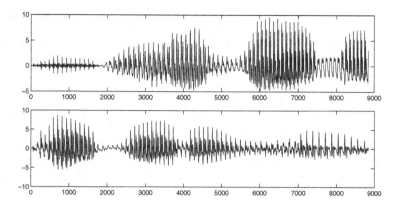

Fig. 8. Two recovered signals using the feedback network

5   Conclusion

In this paper, the systolic algorithm and architecture of a feedback network have been derived and tested with VHDL code simulation. This scheme is fast and reliable since the architectures are highly regular. In addition, the processing can be done in real time. The full scale system can be easily obtained by the number of PEs, and UEs. Our system has two inputs but we will extend it for N inputs.

Because the algorithms used for hardware and software impelmentation differ significantly it will be difficult, if not impossible, to migrate software implementations directly to hardware implementations. The hardware needs different algorithms for the same application in terms of performance and quality. We have presented a fast and efficient VLSI architecture and implementation of BSS.

References

1. C. Jutten and J. Herault, Blind separation of source, part I: An adaptive algorithm based on neuromimetic architecture. *Signal Processing* 1991; vol.24, pp.1-10.
2. K. Torkkola, Blind separation of convolved sources based on information maximization. *Proc. IEEE Workshop Neural Networks for Signal Processing* 1996; pp.423-432.
3. Hoang-Lan Nguyen Thi, Christian Jutten, Blind Source Separation for Convolutive Mixtures. *Signal Processing* 1995; vol.45, pp.209-229.
4. T. Nomura, M. Eguchi, H. Niwamoto, H. Kokubo and M. Miyamoto, An Extension of The Herault-Jutten Network to Signals Including Delays for Blind Separation. *IEEE Neurals Networks for Signal Processing* 1996; VI, pp.443-452.
5. S. Choi and A. Cichocki, Adaptive blind separation of speech signals: Cocktail party problem. *in Proc. Int. Conf. Speech Processing (ICSP'97)* August, 1997; pp. 617-622.
6. N. Charkani and Y. Deville, Self-adaptive separation of convolutively mixed signals with a recursive structure - part I: Stability analysis and optimization of asymptotic behaviour. *Signal Processing* 1999; 73(3), pp.255-266.

7. K. Torkkola, Blind separation of delayed source based on information maximaization. *Proc. ICASSP*, Atlanta, GA, May, 1996; pp.7-10.

8. K. Torkkola, Blind Source Separation for Audio Signal-Are we there yet? *IEEE Workshop on Independent Component Analysis and Blind Signal Separation* Aussois, Jan, 1999; France .

9. Aapo Hyvarinen, Juha Karhunen and Erkki Oja, *Independent Component Analysis,* New York, John Wiley & Sons, Inc., 2001.

10. K.C. Yen and Y. Zhao, Adaptive co-channel speech separation and recognition. *IEEE Tran. Speech and Audio Processing* 1999; 7(2),pp. 138-151.

11. A. Cichocki, S. Amari, and J. Cao, Blind separation of delayed and convolved signal with self-adaptive learning rate. *in NOLTA,,* tochi, Japan, 1996; pp. 229-232.

12. T. Lee, A. Bell, and R. Orglmeister, Blind source separation of real world signals. *in ICNN,*

Description and Simulation of Bio-inspired Systems Using VHDL–AMS

Ginés Doménech-Asensi, José A. López-Alcantud,
and Ramón Ruiz-Merino

Universidad Politécnica de Cartagena, Dpto. de Electrónica,
Tecnología de Computadoras y Proyectos, c/ Dr. Fleming,
sn Cartagena 30201, Spain

Abstract. In this paper we explain the usefulness of VHDL–AMS, a
hardware description language oriented for description and simulation of
mixed signal circuits and processes. Its range of application includes not
only electrical circuits, but also combined physical domains like thermal,
mechanical or fluid dynamics. In this paper we preset two examples of
neuronal applications described and simulated using VHDL-AMS.

1 Introduction

The use of specific simulator tools for research in neurosciences and in the interplay between biological and artificial systems requires a dedicated effort to
develop and debug such tools. However, for certain applications, the use of a
generic simulator can be useful because of the time expended with the simulation
tool is negligible in comparison with the time dedicated to the research itself.

In this paper we explain the usefulness of VHDL–AMS, a hardware description language oriented for description and simulation of mixed signal circuits
and processes. VHDL–AMS [1], specified in the standard 1076.1 is defined as
a superset of original VHDL [2]. This hardware description language is able to
model mixed signal systems and its range of application includes not only electrical circuits, but also combined physical domains like thermal, mechanical or
fluid dynamics.

Modern hardware description languages support the description of both behavior and structure of a given process. This feature is inherited by VHDL–
AMS, which together with the capabilities above explained, makes it a general
description language rather than a hardware description one. However it scope
is limited to processes which can be modeled using lumped parameters than
can be described by ordinary differential and algebraic equations, which may
contain discontinuities. Modeling of neural mechanism frequently use this kind
of equations to describe individual cell or cell populations dynamics, as well as
to model interfaces between biological and artificial systems. This is the main
reason, VHDL–AMS can be an useful tool to describe and model such systems.

This paper is organized as follows: an introduction to the problem has been
presented in this section. Section 2 describes the fundamentals of VHDL–AMS

J. Mira and J.R. Álvarez (Eds.): IWINAC 2005, LNCS 3562, pp. 357–365, 2005.

and its usefulness for neural modeling. In section 3 we present two examples of neural applications. The first one is the FitzHugh–Nagumo neuron model and the second one is the interface between a motor control system and a limb in plain movement. Finally, section 4 summarizes the conclusions obtained from the use of this language.

2   VHDL–AMS

This description language is limited to processes which can be modeled using lumped parameters and than can be described by ordinary differential and algebraic equations, which may contain discontinuities. Basically, each model is described using a pair *entity–architecture*. The *entity* defines the interfaces between this model and the environment and the *architecture* describes its functionality, either in a structural way or in a behavioral fashion.

Interfaces can be *generic* or *ports*. Generic interfaces are used to define certain parameters of an architecture which do not change during simulation. For instance gain in an amplifier or capacity in a membrane cell. Port interfaces are the interfaces themselves. They are used to define the links between the model or architecture and the environment. In an amplifier, the input and the output are port interfaces, and in a biological cell, potassium and sodium channels could be defined as such interfaces.

There are several kind of port interfaces, but those called *terminals* are specially recommended when describing systems with conservative semantics, like electrical systems following Kirchoff's laws. This feature is interesting when describing electrical interaction between a cell membrane and the environment. Associated to a terminal we can find *branch quantities*, which represent the unknowns in the equations describing the system. These quantities can be *across quantities* or *through quantities*. The first ones are used to express effort like effects, like voltage, pressure or temperature while the second ones represent a flow like effect like current, fluid flow or heat flow.

Once an *entity* has been defined, next step is the architecture description. Here, there are two options: structural descriptions or behavioral descriptions. First ones simply describe links between models in a lower hierarchy level. Second ones do not imply any knowledge about a model structure, but allow powerful resources to describe complex processes behavior. In this case, differential equations and algebraic equations are expressed using *simple simultaneous statements*, which are all computed and simulated in a concurrent way with independence of the position in which they appear in the code.

VHDL–AMS includes additional types of *simultaneous statement*, used to define conditional clauses. For instance, a *simultaneous if statement* is used to evaluate sets of other *simultaneous statements* depending on its conditional clause. An important statement is the *break statement* used to define discontinuities to the analog solver when performing simulations with such limitations. These sentences are used also to set initial conditions for ordinary differential equations.

3   Examples

We are going to show the capability of VHDL–AMS to model neural applications using two examples. The first one describes the neuron model proposed by FitzHugh–Nagumo and the second one a simplified motor control system connected to a limb in horizontal movement.

3.1   FitzHugh–Nagumo Model

FitzHugh–Nagumo model [3] describes dynamic behavior of a cell membrane excited by an input current I. This model, which is shown in figure 1 is in fact a simplification of Hodgkin–Huxley [4] model and comprises two ionic channels, for sodium and potassium, a membrane capacity, and a leakage current.

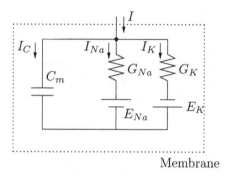

Fig. 1. Fitz–Hugh Nagumo model

In this figure, C_m represents the cell membrane, g_K and g_{Na} identify the ionic channel conductances and I_L defines the leakage current. Equations which model this circuit are shown below:

$$C\frac{du}{dt} = I(t) - I_K - I_{Na} \tag{1}$$

$$\frac{dI_K}{dt} = V - RI_K + V_O \tag{2}$$

The neuron is modeled using a pair *entity–architecture*, as shown below. The *entity* describes the interface between the neuron and the environment, which in this case is the cell membrane. This membrane has a certain potential and through it flows a given current. The membrane is thus modeled using a *terminal* of nature electrical:

```
LIBRARY IEEE;
USE ieee.math_real.all;
USE ieee.electrical_systems.all;
ENTITY nagumo IS
```

```
  PORT (terminal membrana: electrical);
END ENTITY nagumo;

ARCHITECTURE behavior OF nagumo IS
  QUANTITY v ACROSS i TROUGH membrane TO electrical_ref;
  QUANTITY INa,IK:real;
  CONSTANT R: real := 0.8;
  CONSTANT L: real := 1.0;
  CONSTANT C: real := 0.1;
  CONSTANT VO: real := 0.7;
BEGIN
  C*v'dot==(-INa-IK+i);
  L*IK'dot==V+VO-R*IK;
  INa==v**3.0-v;
END ARCHITECTURE behavior;
```

In the code above, the *architecture* describes the behavior of the cell itself. First we declare two *quantities*, v and i which represent the membrane voltage and membrane current. Two additional quantities of type real are declared to model current in the potassium channel and in the sodium channel. Finally, four constants are defined, to implement the resistance, inductance, capacity and threshold voltage of the cell. Figure 2 shows this VHDL–AMS model.

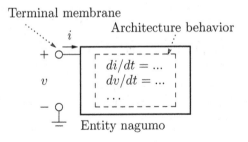

Fig. 2. FitzHugh–Nagumo VHDL–AMS model

For the modeling itself, we have used three *simple simultaneous statements*, one for each equation. The first one describes evolution of membrane voltage in time and the second one describes variation of current in the potassium channel. The last statement calculates the value of current in the sodium channel. Figure 3 shows a simulation of the Fitz–Hugh Nagumo model.

In this simulation we have excited the neuron with a current pulse train. The voltage in the membrane obtained corresponds to equations (1) and (2).

3.2   Motor Control Model

In this example we explain the modeling of a motor control model [5]. This model comprises a limb in planar horizontal movement, two agonist muscles

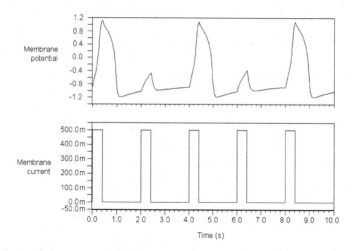

Fig. 3. Simulation of FitzHugh–Nagumo model

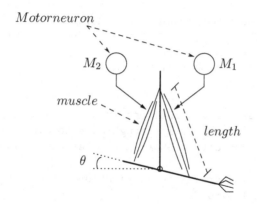

Fig. 4. Motor–control model

and two agonist motorneurons. Figure 4 shows the general structure of a limb controlled by a pair of agonist muscles. Equations which drive limb behavior are summarized below. Firstly, the dynamics of the limb is defined by:

$$\frac{d^2}{dt^2}\theta = \frac{1}{I_m}(F_1 - F_2 - n\frac{d}{dt}\theta) \tag{3}$$

where θ is the rotation angle, I_m is the inertia of the limb, n represents dynamic friction and F_i are forces coming from agonist muscles.

For the calculation of muscles length we use a simple trigonometric relation, given by:

$$L_1 = \sqrt{cos^2\theta + (20 - sin\theta)^2} \tag{4}$$

$$L_2 = \sqrt{cos^2\theta + (20 + sin\theta)^2} \tag{5}$$

Muscle forces are calculated as:

$$F_i = k * (L_i + C_i - L_0) \tag{6}$$

where k is a constant and L_0 is the resting length of the muscle. C_i represents the contraction state of the muscle, and is defined by:

$$\frac{d}{dt}C_i = (1 - C_i)M_i - C_i - [F_i - 1]^+ \tag{7}$$

In this expression, $[\ ]^+$ means the maximum between $[\ ]$ and cero. M_i is the activation value of the motorneuron, given by:

$$\frac{d}{dt}M_i = (1 - M_i) * A_i - (M_i + 1, 6) * 0, 2 \tag{8}$$

In this model we can make a difference between the limb itself, including the pair of agonist muscles, and the neuronal circuitry. So, we can assign a VHDL–AMS entity for each one of these submodels.

VHDL–AMS description of the motorneuron is shown below:

```
ENTITY motorneuron IS
    PORT (terminal axon,input: electrical);
END ENTITY motorneuron;

architecture complete of motorneurona is
    quantity activation across I through axon to electrical_ref;
    quantity excitation across entrada to electrical_ref;
BEGIN
    activation'dot==(B-activation)*activation
        -(excitation+1.6)*0.2;
END architecture complete;
```

The *entity* includes a *port* whose interfaces are two *terminals* of electrical nature: the input to the neuron and the axon (output). The *architecture* models cell behavior by means of a *simple simultaneous statement* which describes a ordinary differential equation. This equation drives motorneuron dynamic behavior. Description of the limb contains again an *entity* and an *architecture*. The *entity* is described below:

```
ENTITY limb IS
    PORT (terminal insertion1,insertion2: electrical;
          terminal elbow: rotational);
END ENTITY limb;
```

This *entity* contains a *port* interface to which there are associated three *terminals*. The first ones are referred to muscle to bone insertion and the third one models the elbow. The *architecture* is:

```
ARCHITECTURE complete OF limb IS
  QUANTITY omega ACROSS torque THROUGH elbow TO rotational_ref;
  QUANTITY M1 ACROSS incision1 TO electrical_ref;
  QUANTITY M2 ACROSS incision2 TO electrical_ref;
  QUANTITY theta: real:=0.0;
  QUANTITY omega_ini: real:=0.0;
  QUANTITY force1, force2: real:=0.0;
  QUANTITY length1, length2: real:=0.0;
  QUANTITY contraction1, contraction2: real:=0.0;
  CONSTANT IM: real := 1.00;
  CONSTANT n: real := 1.00;
  CONSTANT K: real := 1.00;
BEGIN
  omega'dot*IM==force1-force2-n*omega;
  theta'dot==omega;
  torque==0.0;
  length1==sqrt( (cos(theta))**2.0 + (20.0-sin(theta))**2.0  );
  length2==sqrt( (cos(theta))**2.0+(20.0+sin(theta))**2.0);
  force1==K*(length1-20.025+contraction1);
  force2==K*(length2-20.025+contraction2);
  IF force1>1.0 USE
     contraction1'dot==((1.0-contraction1)*M1-contraction1)
        -(force1-1.0);
  ELSE
     contraction1'dot==((1.0-contraction1)*M1-contraction1);
  END USE;
  IF force2>1.0 USE
     contraction2'dot==((1.0-contraction2)*M2-contraction2)
        -(force2-1.0);
  ELSE
     contraction2'dot==((1.0-contraction2)*M2-contraction2);
  END USE;
END ARCHITECTURE COMPLETE;
```

The architecture contains the *quantities* and *constant* declarations and the list of statements which model the system. We have used seven *simple simultaneous statements* to describe limb behavior. Two first statements describe limb dynamics, defined in (3) . Then, sentences four and five calculate muscle length, while sentences six and seven calculate muscle force, described in equations (5) to (6).

The *if* clauses contain sentences which model the muscle contractile state (7). These clauses are called *simultaneous if statements* and are the type of conditional statements explained in section 2. When simulating this system, if the condition in the *if statement* is true, the *simultaneous statement* specified

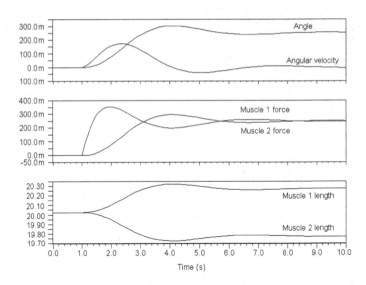

Fig. 5. Simulation of motor control model

after reserved word *use* is evaluated. If condition is false, then the statement after *else* is computed.

Figure 5 shows the simulation of the motor control model. For this simulation we have used two voltage sources as excitation for both motorneurons.

Simulation shows the evolution of the motor–control system for an asymmetric excitation of 1.0 and 0.0 units at 1 s simulation time. The excitation has been applied to both motorneurons through their interface port "input". The motorneuron activation states, are driven through interface port "axon", which are linked to limb port interface "insertion". This produces an initial force in muscle 1 which starts rotation in the elbow. Then, muscles length is modified, and a reaction force appears in muscle 2. System stabilizes approximately after nine seconds.

4   Conclussions

We have presented a tool for description and modeling of neural systems able to be described in terms of ordinary differential and algebraic equations. We have proved the utility of VHDL–AMS describing two types of biological systems: a membrane cell and a motor–control system. In both cases, description has been done in a clear and fast way, focusing our efforts in the systems rather than in programming specific code for such simulations. Although this is a general purpose tool, oriented to hardware description modeling, VHDL–AMS language includes resources to describe systems with enough flexibility and accuracy to be used for biologically inspired systems and processes.

References

1. Design Automation Standards Committee of the IEEE Computer Society.: IEEE Standard VHDL Language Reference Manual (integrated with VHDL–AMS changes). IEEE Std 1076.1 1999
2. Design Automation Standards Committee of the IEEE Computer Society.: IEEE Standard VHDL Language Reference Manual. IEEE Std 1076 1993 IEEE Standar 1076–1993
3. Fitz–Hugh, R.: Impulses and physiological states in theoretical models of nerve membrane. Byophis. J. **177** (1961)
4. Hodgkin, A. L. and Huxley, A. F.: A Quantitative Description of Membrane Current and its Application to Conduction and Excitation in Nerve" Journal of Physiology **117** 500–544 (1952)
   Fitz–Hugh, R.: Impulses and physiological states in theoretical models of nerve membrane. Byophis. J. **177** (1961)
5. Bullock, D., Contreras–Vidal, J. L., Grossberg, S. Inertial load compensation by a model spinal circuit during single joint movement Tech. Report CAS/CNS–95–007

Transistor-Level Circuit Experiments Using Evolvable Hardware

Adrian Stoica, Ricardo Zebulum, Didier Keymeulen, and Taher Daud

Jet Propulsion Laboratory,
California Institute of Technology
taher.daud@jpl.nasa.gov

Abstract. The Jet Propulsion Laboratory (JPL) performs research in fault tolerant, long life, and space survivable electronics for the National Aeronautics and Space Administration (NASA). With that focus, JPL has been involved in Evolvable Hardware (EHW) technology research for the past several years. We have advanced the technology not only by simulation and evolution experiments, but also by designing, fabricating, and evolving a variety of transistor-based analog and digital circuits at the chip level. EHW refers to self-configuration of electronic hardware by evolutionary/genetic search mechanisms, thereby maintaining existing functionality in the presence of degradations due to aging, temperature, and radiation. In addition, EHW has the capability to reconfigure itself for new functionality when required for mission changes or encountered opportunities. Evolution experiments are performed using a genetic algorithm running on a DSP as the reconfiguration mechanism and controlling the evolvable hardware mounted on a self-contained circuit board. Rapid reconfiguration allows convergence to circuit solutions in the order of seconds. The paper illustrates hardware evolution results of electronic circuits and their ability to perform under 280C temperature as well as radiations of up to 175kRad.

1 Introduction

The Jet Propulsion Laboratory (JPL) performs research in fault tolerant, long life, and space survivable electronics for the National Aeronautics and Space Administration (NASA). JPL has been involved in Evolvable Hardware (EHW) technology research for the past several years. EHW can bring some key benefits to spacecraft survivability, adaptation to new mission requirements and mission reliability. The idea behind evolutionary circuit synthesis/design and EHW is to employ a genetic search/optimization algorithm that operates in the space of all possible circuits and determines solution circuits that satisfy imposed specifications, including target functional response, size, speed, power, etc. In a broader sense, EHW refers to various forms of hardware from sensors and antennas to

J. Mira and J.R. Álvarez (Eds.): IWINAC 2005, LNCS 3562, pp. 366–375, 2005.

complete evolvable space systems that could adapt to changing environments and, moreover, increase their performance during their operational lifetime. In a narrower sense here, EHW refers to self-reconfiguration of transistor-level electronic hardware by evolutionary/genetic reconfiguration mechanisms. EHW can help preserve existing circuit functionalities, in conditions where hardware is subject to faults, aging, temperature drifts and high-energy radiation damage. The environmental conditions, in particular the extreme temperatures and radiation effects, can have catastrophic impacts on the spacecraft. Interstellar missions or extended missions to other planets in our solar system with lifetimes in excess of 50 years are great challenges for on-board electronics considering the fact that presently, all commercial devices are designed for at most a 10-year lifespan. Further, new functions can be generated when needed (more precisely, new hardware configurations can be synthesized to provide required functionality). Finally, EHW and reconfigurable electronics provide additional protection against design mistakes that may be found after launch. Design errors can be circumvented during the mission either by human or evolutionary driven circuit reconfiguration. One clear example of a space exploration area where EHW can directly provide benefits is the extreme-environment operation of electronics for in-situ planetary exploration. It may require electronics capable of operating at low temperatures of -220C for Neptune or moon, (-235C for Titan and Pluto) and also high temperatures, such as above 470C needed for operation on the surface of Venus. Terrestrial applications may include combustion systems, well logging, nuclear reactors, and automotive industry requiring high temperature operation and dense electronic packages. In this paper we propose the use of reconfigurable chips, which allow for a large number of topologies to be programmed in-situ, allowing adaptation to extreme temperatures and ionizing radiation. The experiments presented here illustrate hardware recovery from degradation due to extreme temperatures and radiation environments. A 2nd generation reconfigurable chip, the Field Programmable Transistor Array (FPTA-2) integrated circuit, developed at JPL, is used in these experiments. We separately subjected the chips to high temperatures and ionizing radiations using JPL facilities. Measurement results show that the original functionality of some of the evolved circuits, such as half-wave rectifiers and low-pass filters, could be recovered by again using the Evolutionary Algorithm that altered the circuit topologies to lead to the desired solutions. The Evolutionary Algorithms thus control the state of about 1500 switches, which alter the circuit topology. Using a population of about 500 candidate circuits and after running the Evolutionary process for about 200 generations, the desired functionality was recovered.

The paper is organized as follows. Section 2 reviews main aspects of EHW including reconfigurable devices and reconfiguration mechanisms for hardware evolution on a EHW testbed. Section 3 describes experiments on the evolutionary design of analog circuits. Section 4 describes experiments on electronic survivability through evolution, including evolutionary recovery at extreme temperatures and ionizing radiation. Concluding remarks are given in Section 5.

2   Evolvable Hardware

Presently, the evolutionary search for a circuit solution is performed either by software simulations [1] or directly in hardware on reconfigurable chips [2]. However, software simulations take too long for practical purposes, since the simulation time for one circuit is multiplied by the large number of evaluations required by evolutionary algorithms. In addition, the resulting circuit may not be easily implemented in hardware, unless implementation constraints are imposed during evolution. Hardware evaluations can reduce by orders of magnitude the time to get the response of a candidate circuit, potentially reducing the evolution time from days to seconds [2]. Hardware evaluations commonly use commercial re-configurable devices, such as Field Programmable Gate Arrays (FPGA) or Field Programmable Analog Arrays (FPAA)[3]. These devices, designed for several applications other than EHW, lack evolution-oriented features, and in particular, the analog ones are sub-optimal for EHW applications.

2.1   Evolution-Oriented Reconfigurable Architectures

Many important aspects of evolution-oriented reconfigurable architectures (EORA) must be considered to best support the EHW. The granularity of the programmable chip is an important feature. A first limitation of commercial FPGAs and FPAAs is their coarse granularity. From the EHW perspective, it is interesting to have programmable granularity, allowing the sampling of novel architectures together with the possibility of implementing conventional architectures. The optimal choice of elementary block type and granularity is task dependent. From the point of view of experimental work in EHW, it appears that the reconfigurable hardware based on elements of the lowest level of granularity is a good choice to build. Virtual higher-level building blocks can be considered by imposing programming constraints. EORA should also be transparent, thereby allowing analysis and simulation of the evolved circuits. They should also be robust enough not to be damaged by any bit-string configuration existent in the search space, potentially sampled by evolution. Finally EORA should allow evolution of both analog and digital functions.

With the granularity in mind, Field Programmable Transistor Array (FPTAs) chips were designed at JPL and particularly targeted for EHW experiments. The first two versions of the FPTA (FPTA-0 and FPTA-1) relied on a cell with 8 transistors interconnected by 24 switches [3]. They were used to demonstrate intrinsic evolution of a variety of analog and digital circuits, including logical gates, trans-conductance amplifiers, computational circuits, etc. The newer version, FPTA2, is a second-generation reconfigurable mixed-signal chip consisting of an array of cells, each with 14 transistors connected through 44 switches. The chip is able to map different building blocks for analog processing, such as two- and three-stage OpAmps, logarithmic photo-detectors, or Gaussian computational circuits. Figure 1 shows the details of the FPTA-2 cell. As shown, these cells can be programmed at the transistor level. The chip architecture consists of an 8x8 matrix of re-configurable cells. The chip can receive 96 analog/digital

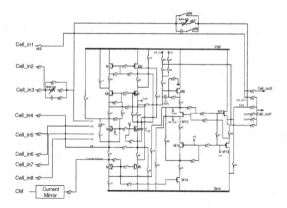

Fig. 1. FPTA-2 cell topology with transistors M1 thru M14, connected through 44 switches. The chip has an 8x8 array of cells

inputs and provide 64 analog/digital outputs. Each cell is programmed through a 16-bit data bus/9-bit address bus control logic that provides an addressing mechanism to download the bit-string of each cell. A total of 5000 bits is used to program the whole chip.

2.2   Evolutionary Reconfiguration Mechanisms

The main steps of evolutionary synthesis are illustrated in Figure 2. The genetic search in EHW is tightly coupled with a coded representation that associates each circuit to a "genetic code" or chromosome. The chromosomes are converted into circuit models for evaluation in SW (extrinsic evolution) or into control bitstrings downloaded to programmable hardware (intrinsic evolution). The simplest representation of a chromosome is a binary string, a succession

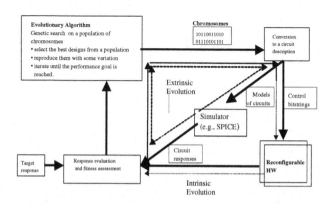

Fig. 2. Main steps for the evolutionary synthesis of electronic circuits showing extrinsic and intrinsic evolution paths

of 0s and 1s that encode a circuit. Each bit of the binary sequence refers to a particular switch location. The status of the switches (ON or OFF) determines a circuit topology and consequently a specific response. Thus, the topology can be considered as a function of switch states, and can be represented by a binary sequence, such as "1011", where a '1' is associated to a switch turned ON and a '0' to a switch turned OFF. More details can be found in [3].

2.3 Evolvable Hardware Testbed

A Stand-Alone Board Level Evolvable System (SABLES), developed for autonomous portable experiments, is a stand-alone platform integrating the FPTA-2 and a digital signal processor (DSP) chip that implements the Evolutionary Programming (EP) code. The system is stand-alone and is only connected to the PC for the purpose of receiving specifications and communicating back the results of evolution for analysis.

The evolutionary algorithm, implemented in a DSP that directly controlled the FPTA-2 provided with fast internal communication ensured by a 32-bit bus operating at 7.5MHz. Details of the EP were presented in Ref [2]. Over four orders of magnitude speed-up of evolution was obtained on the FPTA chip compared to SPICE simulations on a Pentium processor (this performance figure was obtained for a circuit with approximately 100 transistors; the speed-up advantage increases with the size of the circuit).

3 Evolution Experiments of Analog Circuits

The first demonstration of SABLES was reported in Ref. [2]. A half-wave rectifier circuit was evolved in about 20 seconds after processing a population of 100 individuals running for 200 generations. The testing of candidate circuits was

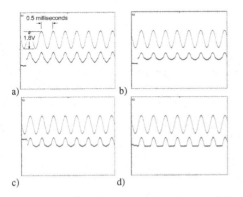 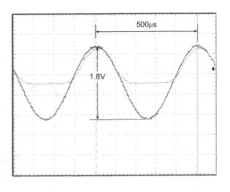

Fig. 3. Evolution of a halfwave rectifier showing the response of the best individual of generation a) 1, b) 5, c) 50 and finally the solution at generation d) 82. The final solution, which had a fitness value less than 4500, is illustrated on the right

performed for an excitation input of 2kHz sine wave of amplitude 2V. A computed rectified waveform of this signal was considered as the target. The fitness function rewarded those individuals exhibiting behavior closer to target (using a simple sum of differences between the response of a circuit and target) and penalized those farther from it. In this experiment only two cells of the FPTA were allocated. Figure 3 displays snapshots of evolution in progress, illustrating the response of the best individual in the population over a set of generations. Fig. 3 a) shows the response of the best individual of the initial population, while the subsequent ones (Fig. b, c, and d) show the best after 5, 50 and 82 generations respectively. The final solution response is shown on the right. The evolution of other analog circuits is shown in [4].

4 Experiments Under Extreme Environment

4.1 Experimental Testbeds

Testbed for evolution at high temperatures: A high temperature testbed was built to achieve temperatures exceeding 350C on the die of the FPTA-2 while staying below 280C on the package. It was necessary to keep the package temperature below 280C so as not to destroy the interconnects and preserve package integrity. Die temperatures were kept below 400C to make sure die attach epoxy does not soften and that the crystal structure of the aluminum core does not degrade. To achieve these high temperatures the testbed includes an Air Torch system. The Air Torch forces out hot compressed air through a small hole within a temperature-resistant ceramic, protecting the chip. The temperatures were measured by attaching thermocouples to the die and the package.

High-energy electron radiation chamber: In the case of the radiation experiments, the radiation source used was a high-energy (1MeV) electron beam obtained using a Dynamitron accelerator. The electrons are accelerated in a small vacuum chamber with a beam diameter of 8". The flux in the chamber was $4x10^9$ [electrons/(sec-cm^2)], which is around 300 rad/sec. Below we describe experiments for evolutionary recovery of the functionality of the following circuits: (1) Half-wave rectifier at 280C temperature; (2) Low-pass filter at 230C temperature; and (3) Half-wave rectifier at 175kRads.

4.2 Half-Wave Rectifier on FPTA-2 at 280C

The objective of this experiment was to recover functionality of a half wave rectifier for a 2kHz sine wave of amplitude 2V using only two cells of the FPTA-2 at 280C. The fitness function does a simple sum of error between the target function and the output from the FPTA. The input was a 2kHz excitation sine wave of 2V amplitude, while the target waveform was the rectified sine wave. The fitness function rewarded those individuals exhibiting behavior closer to target (by using a sum of differences between the response of a circuit and the target) and penalized those farther from it. The output must follow the input during the positive half-cycle but stay constant at a level half-way between the rails (1V)

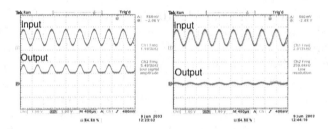

Fig. 4. Input and output waves of the half-wave rectifier. On the left we show the response of the circuit evolved at 27C. On the right we show the degraded response of the same circuit when the temperature was increased to 280C

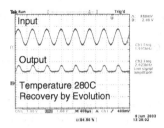

Fig. 5. The response shows the recovery for the half-wave rectifier circuit at 280C following successful evolution

during the negative half-cycle. After evaluation of 100 individuals, they were sorted according to fitness and a 9% (elite percentage) portion was set aside, while the remaining individuals underwent crossover (70% rate), (either among themselves or with an individual from the elite), followed by mutation (4% rate). The entire population was then reevaluated. In Figure 4 the left graph depicts response of the evolved circuit at room temperature whereas the right graph shows degraded response at high temperature. Figure 5 shows the response of circuit obtained by running evolution at 280C, whereby we can see that the functionality has been recovered.

4.3   Low-Pass Filter on FPTA-2 at 230C

The objective of this experiment was to recover the functionality of a low-pass filter using ten cells of the FPTA-2 chip. The fitness function given below performs a sum of errors between the target function and the output from the FPTA in the frequency domain.

Given two tones at 1kHz and 10kHz, the circuit after evolution is to have at the output only the lowest frequency tone (1kHz). This evolved circuit demonstrated that the FPTA-2 is able to recover the functionality of the active filter circuit with some gain at 230C. Figure 6 shows the response of the evolved filter at room temperature and degradation at 230C. Figure 7 shows the time response of the recovered circuit evolved at 230C.

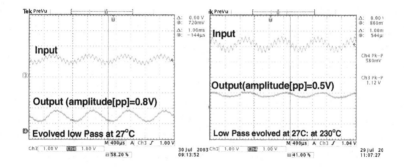

Fig. 6. Low-pass filter response. The graph displays the input and output signals in the time domain when the FPTA-2 was used for evolution at room temperature (left) and at 230C temperature (right)

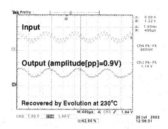

Fig. 7. Recovered Low Pass Filter at 230C

At room temperature, the originally evolved circuit provided a gain of 3dB at 1kHz and a roll-off of -14dB/dec. When the temperature was increased to 230C, the roll-off went to -4dB/dec and the gain at 1kHz fell to -12dB. In the recovered circuit at high temperature the gain at 1kHz increased back to 1dB and the roll-off went to -7dB/dec. Therefore the evolved solution at high temperature was able to restore the gain and to partially restore the roll-off.

4.4   Half Wave Rectifier at 175krads

This experiment was to evaluate the recovery of a half-wave rectifier after the FPTA-2 was subjected to radiation. Figure 8(a) illustrates the response of a previously evolved rectifier after the chip was exposed to a radiation dose of 50 krads. It can be observed that the circuit response was not affected by radiation. After exposure to radiation of up to 175Krad the rectifier malfunctions as the output response is identical to that of the input as shown in Figure 8(b). When the evolutionary mechanism was activated, the correct output response was recovered and retained as shown in Figure 8(c).

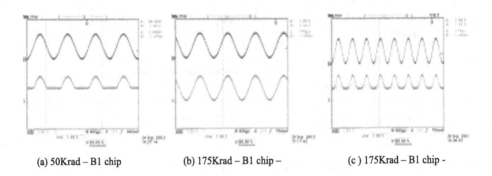

<table>
<tr><td>(a) 50Krad – B1 chip</td><td>(b) 175Krad – B1 chip –</td><td>(c) 175Krad – B1 chip -</td></tr>
</table>

Fig. 8. Response of the Rectifier circuit at (a) 50kRads, (b) after being radiated to 175kRads resulting in deterioration through loss of rectification, followed by (c) recovery through Evolution

5   Conclusions

The above experiments illustrate the power of EHW to synthesize new functions and to recover degraded functionality due to faults or extreme temperatures. A mechanism for adapting a mixed analog reconfigurable platform for high temperature and radiation induced faults was presented. Different experiments were carried out which exercised the reconfigurable device up to 280C and 175Krad radiation dosages demonstrating that the technique is able to recover circuit functionality such as those of rectifiers and filters.

Acknowledgements

The research described in this paper was performed at the Jet Propulsion Laboratory, California Institute of Technology and was sponsored by the Air Force Research Laboratory (AFRL) and the National Aeronautics and Space Administration (NASA).

References

1. A. Stoica, D. Keymeulen, R. Zebulum: Evolvable Hardware Solutions for Extreme Temperature Electronics, Third NASA/DoD Workshop on Evolvable Hardware, Long Beach, CA, July, 12-14, 2001, pp.93-97, IEEE Press.
2. A. Stoica, R. S. Zebulum, M.I. Ferguson, D. Keymeulen, V. Duong. Evolving Circuits in Seconds: Experiments with a Stand-Alone Board Level Evolvable System. 2002 NASA/DoD Conference on Evolvable Hardware, Alexandria, VA, July 15-18, 2002, pp. 67-74, IEEE Press.

3. A. Stoica, R. Zebulum, D. Keymeulen, R. Tawel, T. Daud, A. Thakoor. Reconfigurable VLSI Architectures for Evolvable Hardware: from Experimental Field Programmable Transistor Arrays to Evolution-Oriented Chips. IEEE Transactions on VLSI, IEEE Press, Vol 9, 227-232, 2001.
4. R.S. Zebulum, A. Stoica, D. Keymeulen, M.I Ferguson, V. Duong, V. Vorperian. Automatic Evolution of Tunable Filters Using SABLES. Proceedings of the International Conference on Evolvable Systems (ICES) Norway, pp. 286-296, March, 2003.

An Electronic Reconfigurable Neural Architecture for Intrusion Detection

F. Ibarra Picó, A. Grediaga Olivo, F. García Crespi, and A. Camara

Department of computer science - University of Alicante,
Apartado 99 Alicante 03080. Spain
ibarra@dtic.ua.es
http://www.ua.es/tia

Abstract. The explosive growth of the traffic in computer systems has made it clear that traditional control techniques are not adequate to provide the system users fast access to network resources and prevent unfair uses. In this paper, we present a reconfigurable digital hardware implementation of a specific neural model for intrusion detection. It uses a specific vector of characterization of the network packages (intrusion vector) which is starting from information obtained during the access intent. This vector will be treated by the system. Our approach is adaptative and to detecting these intrusions by using a complex artificial intelligence method known as multilayer perceptron. The implementation have been developed and tested into a reconfigurable hardware (FPGA) for embedded systems. Finally, the Intrusion detection system was tested in a real-world simulation to gauge its effectiveness and real-time response.

1 Introduction

Intrusion Dection System (IDS) is an important component of defensive measures protecting computer systems. It has been an active field of research for about two decades [12]. When an IDS is properly deployed, it can provide warnings indicating that a system is under attack. It can help users alter their installation's defensive posture to increase resistance to attack. In addition, and IDS can serve to confirm secure configuration and operation of other security mechanism such us firewalls. So, IDS's objective is to characterize attack manifestation to positively identify all true attacks without falsely identifying any attacks [13].

Accompanying our growing dependency on network-based computer systems is an increased importance of protecting our information systems. Intrusion detection, the process of identifying and responding to malicious activity targeted at computing and networking resources [11], is a critical component of infrastructure protection mechanisms. A natural tendency in developing an IDS is trying to maximize its technical effectiveness. This often translates into IDS vendors attempting to use brute force to correctly detect a larger spectrum of intrusions than their competitors. However, the goal of catching all attacks has proved to be a major technical challenge. After more than two decades of research and development efforts, the leading IDS's still have marginal detection

J. Mira and J.R. Álvarez (Eds.): IWINAC 2005, LNCS 3562, pp. 376–384, 2005.

rates and high false alarm rates, especially in the face of stealthy or novel intrusions. This goal is also impractical for IDS deployment, as the constraints on time (i.e., processing speed) and resources (both human and computer) may become overwhelmingly restrictive. An IDS usually performs passive monitoring of network or system activities rather than active filtering (as is the case with Firewalls). It is essential for an IDS to keep up with the throughput of the data stream that it monitors so that intrusions can be detected in a timely manner. A real-time IDS can thus become vulnerable to overload attacks [9]. In such an attack, the attacker first directs a huge amount of malicious traffic at the IDS (or some machine it is monitoring) to the point that it can no longer track all data necessary to detect every intrusion. The attacker can then successfully execute the intended intrusion, which the IDS will fail to detect. Similarly, an incident response team can be overloaded by intrusion reports and may be forced to raise detection and response thresholds [10], resulting in real attacks being ignored. In such a situation, focusing limited resources on the most damaging intrusions is a more beneficial and effective approach. A very important but often neglected facet of intrusion detection is its cost-effectiveness, or cost-benefit trade-off.

Artificial neural networks (ANNs) are a form of artificial intelligence, which have proven useful in different areas of application, such as pattern recognition [1] and function approximation/prediction. Multilayer Perceptron is an artificial neural model for training networks of simple neuron like units connected by adjustable weights to perform arbitrary input/output mappings. Patterns are presented by setting the activation set of "input layer" units. Activation flows forward to units in hidden layers via adjustable weights, eventually reaching the output layer. A supervised training is necessary before the use of the neural network. A highly popular learning algorithm called back-propagation is used to train this neural network model (Hayking, 1999). One trained, the perceptron can be used to solve classification of the traffic packets and detect intruder access.

The proposal to palliate the main limitation of the traditional IDS and their inability to recognize attacks lightly modified regarding the patterns with those that carries out the comparisons, it is the one of using neural nets . By means of this proposal it is sought to prove if by means of using a neural net, it is possible to carry out the distinction among packages that represent normal flow in a computers net, and packages that represent attacks. Concretely we will carry out the experimentation with a multilayer perceptron.

Finally, we use reconfigurable architecture (FPGA) to implementation of the final system so we can reach fast response in high traffic conditions and real-time response. An FPGA is a programmable device in which the final logic structure can be directly configured by the end user for a variety of applications. In its simplest form an FPGA consist of an array of uncommitted elements that can be programmed or interconnected according to a user's specification. In our system we use a Xilinx V1000 FPGA to implement the final neural IDS.

The research is supported by the project "Design of a reconfigurable architectures and Mapping Algorithms for vision" (CICYT: 2001-2004).

2   Intrusion Detection Systems

As malicious intrusions into computer systems have become a growing problem, the need for accurately detecting these intrusions has risen. Many methods for detecting malicious intruders (firewalls, password protected systems) currently exist. However, these traditional methods are becoming increasingly vulnerable and inefficient due to their inherent problems. As a result, new methods for intrusion detection that are not hampered by vulnerability and inefficiency must be developed (Eskin et al, 2001). This research sought to design such a detection method through the use of a multilayer perceptron algorithm.

Traditional systems in place for intrusion detection primarily use a method known as "fingerprinting" to identify malicious users. Fingerprinting requires the compilation of the unique traits of every type of attack on a computer system. Each generated fingerprint is first added to the attack database of a detection system and then compared to all subsequent user connections for classification as either a malicious or normal connection (Lane, 1998). This trait compilation is typically accomplished through human analysis by the creators of the system. The resulting fingerprint updates must then be manually installed on each individual system in use (Eskin,2001; Stolfo, 2000). There are several inherent problems with this method: a system must first be compromised by an attack for a fingerprint to be generated; a separate fingerprint is required for each different type of attack; and as the number of fingerprints grows, more computer resources must be allocated to detection, degrading overall system performance. In addition, to gain protection from new attacks, there is a significant waiting period from the time a new attack is first reported to the time that a fingerprint is generated. During this waiting period, a system is left vulnerable to the new attack and may be compromised. Moreover, in extreme scenarios, a fingerprint-based system may be unable to allocate all required resources to detect attacks because of the number of fingerprints, resulting in undetected attacks (Lee et al, 2001).

As an alternate solution for protecting computers from malicious users, a model-based Intrusion Detection System (IDS) may be used. Instead of using a fingerprinting method of user classification, an IDS compares learned user characteristics from an empirical behavioural model to all users of a system. User behavior is generally defined as the set of objective characteristics of a connection between a client (e.g., a user's computer) and a server. Using a generalized behavioural model is theoretically more accurate, efficient, and easier to maintain than a fingerprinting system. This method of detection eliminates the need for an attack to be previously known to be detected because malicious behavior is different from normal behavior by nature (Sinclair et al, 1999). Also, a model based system uses a constant amount of computer resources per user, drastically reducing the possibility of depleting available resources. Furthermore, while actual attack types by malicious users may vary widely, a model-based IDS does not require the constant updates typical of fingerprint-based systems because the characteristics of any attack against a system will not significantly change throughout the lifetime of the system because attacks are inherently different from normal behavior (Eskin et al, 2001; Lee et al, 2001; Sinclair et al, 1999). In previous research, the options for model generation have been to base it on normal users or to base it on malicious users (Eskin et al, 2001). Models based on normal users, known as Anomaly Detection models, use an empirical behavioural model of a normal user and classify any computer activity that does not fit this model as malicious. Models based on malicious users are known as Misuse Detection models. These models look for a pattern of malicious behavior, and behavior that fits this model is classified as malicious (Eskin et al, 2001). In this research, neither model was explicitly specified, allowing the genetic algorithm to generate the best model. An Intrusion Detection System must first be able to detect malicious user connections. Our proposal combines both methods.

2.1   Intrusion Vector

The first thing that it is necessary to think about when training a net neuronal, is what data they will contain the samples that it uses as entrances. This belongs since the most important decision that can take, to her it will depend in great measure the distinction capacity that it can acquire the net neuronal once concluded the training. For the election of these data, they have taken as reference the data that a traditional IDS uses, concretely 'Snort', the one which, apart from using numerous data characteristic of the packages TCP, UDP, or ICMP, it also considers the content of the own package. It is since something that outlines a serious problem, in a principle that this was the most important fact to distinguish among a dangerous inoffensive and other package, but it is very complicated to introduce the content of the data of the package in a net neuronal. Finally it has been opted to take out four characteristics of the same one in form statistic of probability of the four more frequent characters which appear in it. The IVD components are showed in Table 1.

Table 1. Intrusion Vector Data (IVD)

Head IP	TCP	UDP
Port origin	Reserved Flag 1	Size data + head
Port destination	Reserved Flag 2	**ICMP**
Protocol (TCP/UDP/ICMP)	Urgent Pointer Flag OR	type
Type of Service (COUGH)	Acknowledge Flag TO	Code
Size of the Head (Iplen)	Push Function Flag P	O. Protocol
Total Size (Dmglen)	Reset Connection Flag R	**DATA CODE**
Reserved Bit (RB)	Syncronize Flag S	Most frecuent
Don't Fragment Bit (DF)	END Flag F	Second more frecuent
Fragment Bit Lives (MF)	Size of Window (Win)	Third more frecuent
Number of options IP	Size Head (Tcplen)	Fourth more frecuent
	Number of options	

3   The Multilayer Perceptron Model

The most popular Neural Network is the Multi-Layer Perceptron (MLP) trained using the error back propagation algorithm [3]. An input vector (IVD) is presented to the neural network which determines an output. The comparison between the computed and desired output (class attribute) provides an output error. This signal is used by the learning algorithm to adapt the network parameters. A Hardware implementation of MLP has been built for solving the classification problem for the IVD.

The perceptron network can easily cope with complex classification problems. The method of ensuring it captures the right IVD is not controlled by a set of rules but by a learning process. In many respects this learning process is rather similar to the way the brain learns to distinguish certain patterns from others.

Multilayer perceptrons (MLPs) are feedforward neural networks trained with the standard backpropagation algorithm. They are supervised networks so they require a desired response to be trained. They learn how to transform input data into a desired response, so they are widely used for pattern classification. With one or two hidden layers, they can approximate virtually any input-output map. They have been shown to approximate the performance of optimal statistical classifiers in difficult problems. Most neural network applications involve this model.

MLP learning process is done by way of presenting the network with a training set composed of input patterns (intrusion vector) together with the required response pattern (classification). By present we mean that a certain pattern is fed into the input layer of the network. The net will then produce some firing activity on its output layer which can be compared with a 'target' output. By comparing the output of the network with the target output for that pattern we can measure the error the network is making. This error can then be used to alter the connection strengths between layers in order that the network's response to the same input pattern will be better the next time around. The main purpose of the first layer is just to deliver the input signals for all the neurons of the hidden layer. As the signals are not modified by the first layer neurons (the neurons do not have arithmetical operations), the first layer can be represented by a single

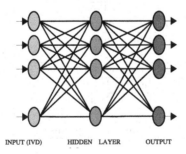

Fig. 1. MLP topology (input/hidden/output layers)

set of busses in the hardware model. The IVD has 29 elements (Figure 1), so the first layer has 29 neurons.

For our case the MLP is trained by backpropagation algorithm to distinguish among inoffensive packages and packages that represent attacks to a host in a computer network. In his work we use a workstation to train the net and later we put the weights into the FPGA architecture for recognition purpose.

4   FPGA Architecture of the IDS

The digital architecture proposed here is an example of a reconfigurable computing application, where all stages of the MLP algorithm in recognition mode reside together on the FPGA at once. Neural networks in general, work with floating-point numbers. Working with floating-point numbers in hardware is a difficult problem because the arithmetic operations are more complex that with integer numbers. Further more, the dedicated circuits for floating point operation are more complex, slower, and occupy a larger chip area that integer number circuits. However, to achieve best results we use floating point IEEE 754 Single Precision (Exponent is 8 bits and has a bias of 127. Mantissa is 23 bits not including the hidden 1) to code the intrusion vector and the parameters of the neural model.

The Architecture of the neural network was designed with parallel Processor Element (EP). The EP is the elemental element in neurocomputing and the whole system is built with multiple interconnected EP. Each EP performs weighted accumulation of inputs (intrusion vector) and evaluation of output levels with a certain activation function. Figure 2 shows the internal structure of an EP which has a 32 bit multiplier (MULT), and adder (ADD), an accumulator register (AC), look-up table for non-linear filter implementation and shift-register for the internal weights and the network pattern (intrusion vector). The final IDS has seven EP which can operate in parallel (six in the hidden layer and one in the output level).

The high level design of the IDS has been defined using Handel-C [110] using the development environments FPGA advantage and DK1 to extract EDIF

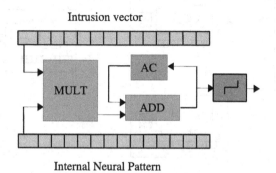

Fig. 2. Block Diagram of an PE

files. All designs have been finally placed and routed in a XilinxVirtex V1000 FPGA, using the development environment Xilinx ISE (Integrated Software Enviroment).

5   Comparative Results

The obtaining the data, it is necessary to gather so much data of inoffensive packages, as of dangerous packages, to separate them, and to make sure that was significant. To obtain the dangerous packages two machines they are used, one from which the attacks rushed using the scanner ' Nessus' and another from which those packages were captured using the traditional IDS ' Snort', in total 433 dangerous packages were obtained. To obtain the inoffensive packages, a real net of computers is used, that is to say, where the habitual packages flow in any company or corporation where it seeks to settle an IDS. Finally the study has been carried out using a small departmental net. The fact that the address IP origin and the destination are the same one in all the packages it doesn't rebound in the obtained results, since that information one doesn't keep in mind. Between Windows and Linux 5731 inoffensive packages were obtained.

In our experiments we proceed separating a group of samples of those obtained to train, and using them for test. Concretely 80% of packages has been chosen for training, and 20% for test, as much for the inoffensive packages as for the obtained dangerous packages. The results will measure them in terms of percentage of successes obtained in the test packages that will be: dangerous or normal traffic packets. Figure 4 shows the quick learn of the intrusion vector patterns in the MPL and table 2 shows the final test results.

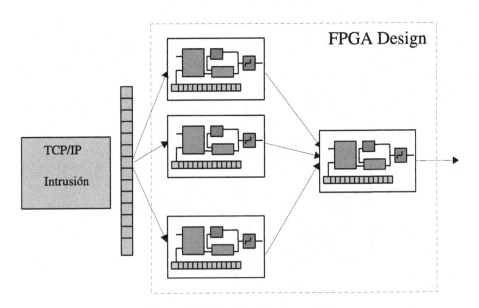

Fig. 3. IDS Classifier Implementation

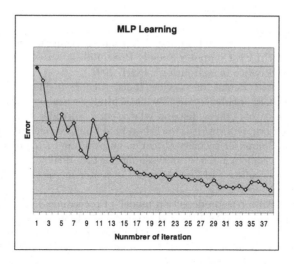

Fig. 4. Learning of the intrusion patterns error evolution

Table 2. Final packet IDS results

Packet type	Training	TCP/IP network Test	Correct Classification Probability
Normal	4585	1146	0.985395
Intrusión	346	87	0.977273

6   Conclusions

One limitation of current modeling techniques is that when cost metrics change, it is necessary to reconstruct new cost-sensitive models. In our approach, it is built dynamic models which can learn new intrusion patterns. These techniques will help reduce the cost of static models due to changes in intra-site or metrics. Our system can operate in real-time response and has good response in detection of intrusion. In future work, we will also study how to incorporate uncertainty of cost analysis due to incomplete or imprecise estimation, especially in the case of anomaly detection systems, in the process of cost-sensitive modeling. We will also perform rigorous studies and experiments in a real-word environment to further refine our system parameters.

Acknowledgement

The work has been supported by CICYT TIC2001-2004 Diseño de una arquitectura reconfigurable para algoritmos.

References

1. M. D. Abrams, S. Jajodia, H.J. Podell. "Information Security: An Integrated Collection of Essays" IEEE Computer Society Press 1997
2. C. P. Pfleeger. "Security in computing". PHH, 1997.
3. CERT. Distributed denial of service tools. CERT Incident Note IN-99-07, http://www.cert.org/incident_notes/IN-99-07.html, December 1999
4. CERT. Results of the distributed-systems intruder tools workshop, http://www.cert.org/reports/dsit_workshop-final.html, November 1999
5. J.P. Anderson. Computer security threat monitoring and surveillance. Technical report, James P. Anderson Co. Fort Washington, PA, 1980
6. S. Kumar. "Classification and Detection of Computer Intrusion" PhD thesis, Purdue University, August 1995
7. D. E. Denning "An intrusion-detection model". Proceedings IEEE Symposium on Security and Privacy, pp 118-131 Oakland, CA May 1986
8. K.L. Fox, R. Henning, J. Reed, "A Neural Network Approach Towards Intrusion Detection". Proceedings of the 13th NationalComputer Security Conference. pp 125-134 Washington, DC, October 1990.
9. V. Paxson. Bro: A system for detecting network intruders in real-time. In Proceedings of the 7th USENIX Security Symposium, San Antonio, TX, 1998.
10. F. Cohen. Protection and Security on the Information Superhighway. John Wiley & Sons, 1995.
11. E. Amoroso. Intrusion Detection: An Introduction to Internet Surveillance, Correlation, Traps, Trace
12. Back, and Response. Intrusion.Net Books, 1999.
13. D.E. Denning "An intrusion detection model". IEEE Trans. Software Eng, vol, se-13, no 2, feb 1987, pp 222-23.
14. R. Durst et al., "Testing an evaluation computer intrusion detection systems" Comm ACM, vol 42, no 7, 1999, pp 53-61.

Construction and VHDL Implementation of a Fully Local Network with Good Reconstruction Properties of the Inputs

Joël Chavas[1], Demian Battaglia[2], Andres Cicuttin[3], and Riccardo Zecchina[4]

[1] ISI Foundation, Viale San Severo 65, I-10133 Torino, Italy
`chavas@isiosf.isi.it`
[2] SISSA, via Beirut 9, I-34100 Trieste, Italy
`battagli@sissa.it`
[3] ICTP-INFN MLab Laboratory, via Beirut 31, I-34100 Trieste, Italy
`cicuttin@ictp.trieste.it`
[4] ICTP, strada della Costiera 11, I-34100 Trieste, Italy
`zecchina@ictp.trieste.it`

Abstract. This report shows how one can find a solution to the K-SAT equations with the use of purely local computations. Such a local network, inspired by the Survey Propagation equations driven by an external input vector, potentially has an exponential number of attractors. This gives the network powerful classification properties, and permits to reconstruct either noisy or incomplete inputs. It finds applications from bayesian inference to error-correcting codes and gene-regulatory networks, and its local structure is ideal for an implementaion on FPGA. Here we write its algorithm, characterize its main properties and simulate the corresponding VHDL code. One shows that the time of convergence towards a solution optimally scales with the size of the network.

1 Introduction

The fast development of components efficiently performing parallely simultaneous computations (FPGAs, CPLDs) increases the interest of computations making massively use of local parallelism, such as computations on graphs [8].

Belief propagation algorithms are well known algorithms which permit, by the use of purely local messages, to converge towards the exact distribution of *a posteriori* probabilities in tree-like graphicals models, or towards a good approximation even in loopy networks [11].

But when the graph is too constrained, such classical local algorithms fail to converge towards the desired distribution of probabilities.

The graphical model K-SAT (described below), recently studied and solved using techniques of statistical mechanics, can be viewed as a toy-model of such a constrained graph. The Survey Propagation equations are a generalization of the Belief Propagation equations which permit to solve the K-SAT formulas in a range of parameters in which the graph is highly constrained, and in which the Belief Propagation equations fail [9, 10, 4].

J. Mira and J.R. Álvarez (Eds.): IWINAC 2005, LNCS 3562, pp. 385–394, 2005.

But while the Survey Propagation equations use fully local computations, the decimation based on these and which permits to converge to one particular solution of the K-SAT problem uses a global criterion [10, 4], and is as such not suitable for low-level parallelization.

In the present article, one builds a fully local network based on the Survey Propagation equations, the dynamic of which lets it converge towards a desired instance of the K-SAT equations. We then propose an algorithm and its VHDL code, which will permit the implementation of this network on a FPGA. To conclude, we will then shortly review the fields in which such an implementation can be applied.

2   Solving and Properties of the K-SAT Equations

2.1   K-SAT Formulas

The K-SAT formula consists of N Boolean variables $x_i \in \{0, 1\}$, $i \in \{1, ..., N\}\}$, with M constraints, in which each constraint is a clause, which is the logical OR ($\vee$) of the variables it contains or of their negations. A clause is written as

$$(z_{i_1} \vee ... \vee z_{i_r} \vee ... \vee z_{i_K}) \tag{1}$$

where $z_{i_r} = x_{i_r}$ (resp. $\bar{x}_{i_r}$) if the variable is directed (resp. negated) in the clause. The problem is to find an assignment (if it exists) of the $x_i \in \{0, 1\} = \{$directed,negated$\}$ which is such that all the M clauses are true. We define the energy E of a configuration $\mathbf{x} = (x_1, ..., x_N)$ as the number of violated clauses.

2.2   The Solution Space Becomes Divided into an Exponential Number of Clusters

When the number of constraints $M = \alpha N$ is small, the solutions of the K-SAT formulas are distributed close one to each other over the whole N−dimensional space, and the problem can be solved by the use of classical local search algorithms. When α is included in a narrow region $\alpha_d < \alpha < \alpha_c$, the problem is still satisfiable but the now limited solution phase breaks down in an exponential number of clustered components. Solutions become grouped together into clusters which are fart apart one from the other.

2.3   Survey Propagation Equations with External Inputs

Recently, statistical physics methods derived from spin glass physics (making use of the cavity method) have permitted to derive closed set of equations, which allow, after decimation, the retrieval of a large number of different solutions belonging to different clusters [9, 10, 4, 2, 3].

The principles of the method are schemed on Fig. 1. The message $\eta_{a \to i}$ represents the probability that the clause a sends a warning onto the variable i, $i.e.$ the probability that i is forced to be fixed at a value which satisfies a. As the

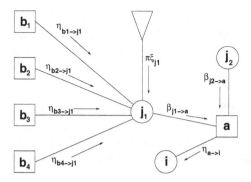

Fig. 1. Factor graph representation of the K-SAT formula. A constraint (called a clause) is represented by a square ; the variables are represented by a circle

clause a is an OR function, it corresponds to the probability that both j_1 and j_2 do not satisfy a (resp. $\beta_{j_1 \to a}$ and $\beta_{j_2 \to a}$). If one assumes independence between j_1 and j_2 when the link (the "cavity") $a - i$ is absent, this writes

$$\eta_{a \to i} = \prod_{j \in V(a) \backslash i} \beta_{j \to a} , \tag{2}$$

in which $V(a)$ means "all the variables belonging to the clause a".

One then writes without further derivation the dependency of $\beta_{j \to a}$ (the probability that j does not satisfy a) on the $\eta_{b \to j}$ (see Ref.[4, 10] for further details) :

$$\beta_{j \to a}^{\pm} = \frac{\Pi_{j \to a}^{\pm}}{\Pi_{j \to a}^{\pm} + \Pi_{j \to a}^{\mp} + \Pi_{j \to a}^{0}} , \tag{3}$$

$\beta_{j \to a}^{+}$ (resp. $(-)$) meaning that j is negated (directed) in the clause a. And :

$$\Pi_{j \to a}^{\pm} = \left[1 - \lambda \left(1 - \pi \delta_{\xi_j, \pm} \right) \prod_{b \in V_{\pm}(j) \backslash a} \left(1 - \eta_{b \to j} \right) \right] \cdot$$

$$\cdot \left(1 - \pi \delta_{\xi_j, \mp} \right) \prod_{b \in V_{\mp}(j)} \left(1 - \eta_{b \to j} \right) \tag{4}$$

$$\Pi_{j \to a}^{0} = \lambda \left(1 - \pi \delta_{\xi_j, +} \right) \left(1 - \pi \delta_{\xi_j, -} \right) \prod_{b \in V(j) \backslash a} \left(1 - \eta_{b \to j} \right)$$

in which $V^{+}(j) \backslash a$ means "all the clauses in which j is directed, except the clause a". By putting the real parameter λ to 0, one retrieves the well known Belief Propagation equations. Last, these equations include the external input field imposed onto the variable i, which has an intensity π and a direction $\xi_i \in \{-1, 0, 1\}$. The fact that $\pi \xi_i = \pm \pi$ means that an *a priori* probability π of assuming the value ± 1 is being assigned to the variable i. If, on the other hand, ξ_i equals 0, the variable i is *a priori* left unconditioned.

The Equations (2, 3) form a closed set of equations between the $\eta_{a \to i}$, called "Survey Propagation (SP) equations with external inputs". Iterating these equations permits to find the fixed point $\eta_{a \to i}^*$, probability that, among all the clusters of solutions, a sends a warning to i.

2.4   Temporary Local Field

At each sweep, to perform the local updates of the intensity of the external inputs, as will be explained below, one needs to compute the local temporary field for each variable i, *i.e.* the temporary tendency of the variable i to be equal to $1^{(+)}$, $0^{(-)}$ or left unconstrained.

The tendency of the variable i to be equal to $1^{(+)}$ (resp. $0^{(-)}$) then reads:

$$W_i^{(\pm)} = \frac{\hat{\Pi}_i^{\pm}}{\hat{\Pi}_i^{\pm} + \hat{\Pi}_i^{\mp} + \hat{\Pi}_i^0} \, . \tag{5}$$

in which $\hat{\Pi}_i^{\pm,0}$ is equivalent to $\Pi_i^{\pm,0}$ of Equ. 4, but by now considering all the inputs including the clause a.

3   Presentation of the Network

3.1   A Fully Parallel Decimation of the K-SAT Equations

We shall now introduce the important corollary of the SP algorithm with external forcing field, which allows the retrieval of solutions close to any N-dimensional point $\boldsymbol{\xi}$ in $ln(N)$ time when implemented on a fully distributed device (SP-parallel). On each variable i, the direction of the forcing field is fixed at the external input value $\xi_i \in \{-1, 1\}$, while the intensity π_i of its external forcing field is now regularly updated : it equals π with probability 1 (resp. with probability p) when the temporary local field is aligned (resp. unaligned) with the external forcing ; its value is 0 otherwise. In words, if, locally, the network wants to align with the external forcing, one further stabilizes it in this direction ; if the network really does not want, one still tries to drive it in that direction (so that one obtains a solution as close as possible from the external forcing), but not too much (with a probability $p < 1$) in order not to generate contradictions.

Good properties of convergence are obtained by updating the direction of the forcing field once every two steps, *i.e.* once every two parallel updates of all the etas. At the end of a unique convergence, using a right and formula-dependent choice of π and p, most variables are completely polarized, and a solution of the K-SAT formula is finally found by fixing each variable in the direction of its local field. As the update of the forcing is reminiscent of the Perceptron learning, one calls such an algorithm the "Perceptron-like algorithm" : it is an efficient distributed solver of the K-SAT formulas in the hard-SAT phase.

3.2   "Perceptron-Like Decimation" Algorithm

INPUT: the K-SAT formula, an external input vector; a maximal number of iterations t_{max}, a requested precision ϵ, and the parameters π and p for the update of the external forcing

OUTPUT: if it has converged before t_{max} sweeps: one assigment, close from the input vector, which satisfies all clauses.

0. At time $t = 0$: for every edge $a \rightarrow i$ of the factor graph, randomly initialize the cavity bias $\eta_{a \rightarrow i}(t = 0) \in [0, 1]$. The intensities of the input forcing field are initially set to $\pi_i = 0$.
1. For $t = 1$ to $t = t_{max}$:
   1.1 update parallelely the $\eta_{a \rightarrow i}(t)$ on all the edges of the graph, using subroutine CBS-UPDATE.
   1.2 half of the times, update parallelely the intensity $\pi \in [0, 1]$ for all variables x_i, using subroutine FORC-UPDATE.
   1.3 If $|\eta_{a \rightarrow i}(t) - \eta_{a \rightarrow i}(t-1)| < \epsilon$ on all the edges, the iteration has converged and generated $\eta^*_{a \rightarrow i} = \eta_{a \rightarrow i}(t)$: GOTO label 2.
2. If $t = t_{max}$ return UN-CONVERGED. If $t < t_{max}$ return the satisfying assignment which is obtained by fixing the boolean variable x_i parallel to its local field : $x_i = sign\{W_i^{(+)} - W_i^{(-)}\}$.

Subroutine CBS-UPDATE($\eta_{a \rightarrow i}$).
INPUT: Set of all $\eta_{b \rightarrow j}$ arriving onto each variable node $j \in V(a) \setminus i$
OUTPUT: new value for the $\eta_{a \rightarrow i}$.

1. For every $j \in V(a) \setminus i$, compute the values of $\Pi^{\pm}_{j \rightarrow a}, \Pi^{\mp}_{j \rightarrow a}, \Pi^0_{j \rightarrow a}$ using Eq. (4).
2. Compute $\eta_{a \rightarrow i}$ using Eq. (2, 3).

Subroutine FORC-UPDATE(π_i).
INPUT: Set of all cavity bias surveys arriving onto the variable node i, including the forcing field (π_i, ξ_i).
OUTPUT: new value for the intensity π_i of the additional survey.

1. Compute the local fields $W_i^{(+)}, W_i^{(-)}, W_i^{(0)}$ using Eq.5.
2. Compute $\pi_i : \pi_i = \pi$ with probability $P(\pi_i = \pi) = p + (1 - p) \times \theta(W_i^{(\xi_i)} - W_i^{(-\xi_i)})$, $\pi_i = 0$ otherwise.

3.3   Choice of the Network

The results of the experiments presented below are performed with $K5$ "regular" formulas, in which : first, the number of clauses where any given variable appears negated or directed is kept strictly equal ; second, the number of clauses to which belongs a given variable (the connectivity of the variable) is kept strictly constant for all variables (here one chooses $c = 84$, $i.e.$ $\alpha = 16.8$).

The choice has been motivated by the good reconstruction properties of such graphs when used to implement "lossy data compression" [2, 3]. Except if stated otherwise, ones takes networks of $N = 10000$ variables, above which size convergence properties don't dramatically change (Fig. 7).

4   Characterization of the Network

4.1   Distance Between the Input Forcing and the Output Solution

One presents to the network an input vector i. The network, driven by i, progressively converges towards a stable solution σ_o, which is also the output of the network.

The optimal intensity π_{opt} and the optimal probability of flip p_{opt}, for which one obtains in average the closest solution σ_o from the input vector i, are respectively equal to 0.52 and 0.72. For such choices of the parameters, the average Hamming distance $d(i, \sigma_o) = \frac{1}{2N}(N - \sum_{i=1}^{N} i_i \sigma_{o,i})$ between the input vector and the corresponding solution equals 0.314, which is slightly better (but in agreement with) the value 0.329 found by performing the classical serial decimation [2, 3].

4.2   Time of Convergence

At each sweep (*i.e.* at each parallel update of all etas), one constructs a temporary assignment σ_t in which each coordinate $\sigma_{t,i} = sign(W_i^+ - W_i^-)$ is fixed parallel to the temporary local field. Fig. 2 represents the evolution respectively of the energy E of the temporary assignment σ_t and of the Hamming distance $d(i, \sigma_t)$ between the input vector and σ_t. The algorithm converges towards a solution after an average of 350 sweeps.

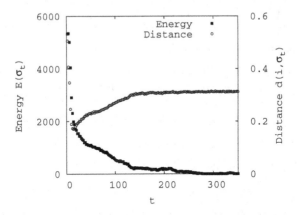

Fig. 2. Time of convergence towards a solution, by imposing a random forcing input

4.3   Stability of the Network - Resistance to Noise

The stability of the network is examined by taking a solution σ_o of the K-SAT, and forcing the network with a noisy input vector i increasingly distant from the solution σ_o. One recovers a solution σ which tends to belong to the same cluster as σ_o (Fig. 3). In the language of the attractor network, the critical distance $d_c = 0.265$, for which the cluster converges in half of the cases towards the initial cluster $Cl(\sigma_o)$ is a measure of the average radius of the basin of

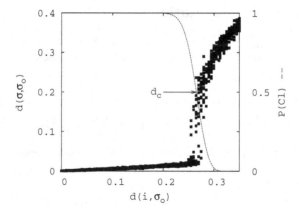

Fig. 3. Stability of the cluster of solutions $Cl(\sigma_o)$ to noise. Forcing the system with the input i increasingly distant from a solution σ_o permits to recover a solution σ. $P(Cl)$ is the probability that σ belongs to the same cluster as σ_o

attraction of the cluster $Cl(\sigma_o)$. Conversely, the curve P(Cl) of the probability of convergence towards the same cluster $Cl(\sigma_o)$ as a function of the distance $d(\sigma, \sigma_o)$ between the initial solution and the output solution is a measure of the sharpness of the boundary of the basis of attraction of $Cl(\sigma_o)$.

4.4   Associative Properties of the Network

The associative properties of the network have been used (in the context of serialized computation) to build "lossy data compressors" [2, 3]. By presenting a truncated part of the solution, one recovers after convergence a whole solution from the same cluster. Without further optimization, and by choosing an intensity $\pi = 0.99$ and a probability of flip $p = 0.99$, Fig. 4 shows that 18% of the solutions is enough to recover a whole solution from the same cluster.

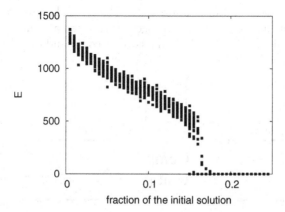

Fig. 4. Stability to missing inputs. One forces the system with a fraction of the initial solution σ_o. E is the energy of the assignment found at the end of the convergence

5   Electronic Implementation

The previous algorithm is translated into a VHDL code, in view of a FPGA implementation. All calculations are performed synchronously.

5.1   Structure of the VHDL Code

At the descriptive level : a component called *graph* is linked to the external word (which can be a computer), receiving the vector i as input, and providing the corresponding solution σ_o as output. *Graph* links together components called *computational-unit*, each of them performing the calculations involving directly a given variable. If two variables i and j belong to the same clause in the K-SAT formula, then the two *computational-unit* representing respectively each of the variables are linked by a signal (Fig. 5).

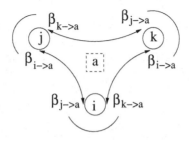

Fig. 5. Scheme of the transmission of the $\beta_{j\to a}$ when COMMAND reads "01"

Main input/output lines of "computational-unit" representing the variable j are drawn in Fig. 6. The INPUT-BETA-IA line transmits the value $\beta_{i\to a}$ from the variable i belonging to the same clause a as j. The OUTPUT-BETA-JA line transmits to all variables belonging to clause a the value $\beta_{j\to a}$. The EPS line is not compulsory : it transmits the information about local convergence.

Moreover, all components are wired together through a common clock signal (CLK), an enable signal (ENABLE) and a two-bit command line (COMMAND).

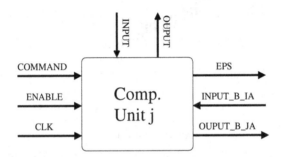

Fig. 6. Main input/output of the "computational-unit" component

At the behavioral level, each unit constantly repeats four synchronized operations, determined by the value of the COMMAND line. When COMMAND equals :

- "00" : calculate $\beta_{j\to a}$ using Eq. (3), for all clauses a to which belongs j
- "01" : transmit $\beta_{j\to a}$
- "10" : read $\beta_{i\to a}$, calculate $\eta_{a\to j}$ using Eq. 2
- "11" : update intensity of the forcing π_i

5.2   Bit Representation of the Reals

For space optimization, the reals $(\beta_{j\to a}, \eta_{a\to j} \in [0,1]^2$ have to be represented by the possibly smallest number of bits. Convergence can be consistently achieved if the reals are encoded in 12 bits, and if the update of the forcing is performed once every two updates of the etas.

5.3   Logarithmic Dependence of the Time of Convergence on the Size of the Graph

Fig. 7 shows that the time to converge to a solution depends logarithmically on the size of the graph : $t \simeq 37 \times ln(N)$ for $N \geq 2000$.

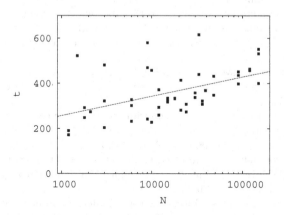

Fig. 7. Logarithmic dependence of the time of convergence on the size of the graph. The fit is performed for $N \geq 2000$. Graphs until a size of 3000 inc. have been done using ModelSim and the VHDL code. Beyond, one used a C code performing the same task

For such constrained graphs, information from local messages has to propagate through the whole graph before convergence. This can't be done faster than in a logarithmic time. Thus, this time of convergence to find a K-SAT solution on a distributed device is, in terms of scaling, the best possible.

6 Conclusion

One has built a network and written its VHDL code which permit to find solutions of the K-SAT formulas in $ln(N)$ time, the best achievable scaling on a distributed device. A physical interpretation of alternative methods of parallelization will be described in detail in another manuscript [5]. This algorithm or its derivatives, which define a local network with an exponential number of attractors, has applications in error-correcting codes [2, 3] and gene-regulatory networks [6]. Its powerful reconstruction properties are also likely to be used in probabilistic inference for constrained graphs [8], artificial intelligence and theoretical neuroscience [1, 7].

Acknowledgments

We warmly thank Alexandre Chapiro and Maria Liz Crespo for useful discussion and technical advises. We thank Alfredo Braunstein for discussions, Michele Leone and Andrea Pagnani for comments on the manuscript.

References

1. Amit,D.,J.: Modeling brain function: the world of attractor neural network. Cambridge University Press (1992)
2. Battaglia, D., Braunstein,A., Chavas,J., Zecchina,R.: Source coding by efficient selection of ground states. Submitted to Phys.Letters (2004) cond-mat/0412652
3. Battaglia, D., Braunstein, A., Chavas,J., Zecchina,R.: Exact probing of glassy states by Survey Propagation. Proceedings of the conference SPDSA-2004,Hayama, Japan (2004)
4. Braunstein,A., Mézard,M., Zecchina,R.: Survey propagation: an algorithm for satisfiability. Preprint, to appear in Random Structures and Algorithms (2002) cs.CC/0212002
5. Chavas,J., Furtlehner,C.,Mézard,M.,Zecchina,R.: Survey-propagation based decimation through purely local computations. J. Stat.Phys. (2005) in preparation
6. Correale, L., Leone, M., Pagnani, A., Weigt, M., Zecchina, R.: Complex regulatory control in boolean networks. Sumitted to Phys.Letters (2004) cond-mat/0412443
7. Latham,P.E., Deneve, S., Pouget,A.: Optimal computation with attractor networks. J. Physiol. Paris **97** (2003) 683–694
8. Madsen, M., L., Jensen,F.,V.: Parallelization of inference in bayesian networks.
9. Mézard,M., Parisi,G., Zecchina, R.: Analytic and algorithmic solution of random satisfiability problems. Science **297** (2002) 812
10. Mézard,M., Zecchina,R.: The random K-satisfiability problem : from an analytic solution to an efficient algorithm. Phys. Rev. Letters E **66** (2002) cs.CC/056126
11. Yedidia,J.S., Freeman,W.T., Weiss,Y.: Generalized belief propagation. Advances in Neural Information Processing Systems **13** (2001) 689–695

Reconfigurable Hardware Implementation of Neural Networks for Humanoid Locomotion

Beatriz Prieto, Javier de Lope, and Darío Maravall

Department of Artificial Intelligence, Faculty of Computer Science,
Universidad Politécnica de Madrid, Campus de Montegancedo,
28660 Madrid
{bprietop, jdlope, dmaravall}@dia.fi.upm.es

Abstract. An artificial neural network (ANN) is a parallel distribution of linear processing units arranged as layers. Parallelism, modularity and dynamic adaptation are computational characteristics associated with networks. These characteristics support FPGA implementation of networks, because parallelism takes advantage of FPGA concurrency, and modularity and dynamic adaptation benefit from network reconfiguration. The most important aspects of FPGA implementation of neural networks are: the benefits of reconfiguration, the representation of internal data and implementation issues like weight precision and transfer functions. This paper proposes a number of internal data formats that optimize the network precision and a way of implementing sigmoid transfer functions to make the most of FPGA implementation.

1 Introduction

It was an old dream to have a personal robot looking like a human [1]. One of the main features of humanoid robots are biped walking in which the research has focused on getting the robot to maintain stability as it walks in a straight line. Usually complex analytical and hard computing methods are used for this purpose, but a different, bio-inspired approaches can also be materialized [2].

The method is based on the definition of several sets of joint and Cartesian coordinates pairs, which are generated using the humanoid robot direct kinematics [3]. These data sets are named *postural schemes*, and, for example, one of them includes the positions involved in the execution of a single robot step. For each postural scheme, a different ANN is used, and the respective network will be selected depending on the posture to be achieved. A two-layer backpropagation network is trained separately with the respective generated data and is used to learn a representative set of input-target pairs, which has been introduced in elsewhere [4]. The topology of the ANN is as follows: the input layer has 3 neurons, one for each element of the Cartesian coordinates of the free foot, the output layer has 12 neurons, associated to the joint coordinates of the humanoid robot, and, after trying out several configurations, the hidden layer contains no more than 6 neurons.

J. Mira and J.R. Álvarez (Eds.): IWINAC 2005, LNCS 3562, pp. 395–404, 2005.
© Springer-Verlag Berlin Heidelberg 2005

In this paper, we introduce the hardware implementation of the artificial neural networks used by the postural scheme method. The use of this technology provides an improved autonomy to the humanoid robot, making be able the use of specific postural schemes when needed thanks to the FPGA reconfigurability.

The rest of this paper is organized as follows. We first present the current problems surrounding the FPGA implementation of neural networks and then go on to explain a range of design solutions. This is followed by a description of how the network has been implemented and the evaluation of the results achieved by the implemented neural network.

2   Hardware Implementation of ANN

The implementation of networks with a large number of neurons is still a challenging task, because the most of the neural network algorithms are based on arithmetic operations such as multiplications. Multipliers are very expensive operators to implement on FPGAs, as they take up quite a lot of silicon area. Using FPGA reconfigurability, there are strategies that efficiently and inexpensively improve the implementation of networks.

ANN were first implemented on FPGAs over ten years ago. Since then, a lot of research and applications related to neural networks have been developed using FPGA-based approaches. The implementations can be classed according to the following criteria: purpose of reconfiguration and data representation.

Many of the implementations that have been realized take advantage of FPGA reconfigurability. Depending on the use to which reconfiguration is put, there are several implementation approaches. For instance, sometimes FPGAs are used for prototyping due to they can be rapidly reconfigured an unlimited number of times. The first network implementation was an implementation of a connectionist classifier and is, also, a clear example of prototyping using device reconfiguration to generate tailored hardware for each classifier application [5].

A second implementation approach is the design density enhancement, which exploits run-time and parallel reconfiguration features. On the one hand, the sequential steps taken by the network algorithm can be time-multiplexed to divide the algorithm into executable steps [6, 7]. Another option is to divide the neural network into specialized modules with a set of operands. This technique is known as *dynamic constant holding* and it can be used for implementing multilayer perceptrons [8].

Another important topic is the topology adaptation, which refers to the fact that FPGAs can be reconfigured dynamically, enabling the implementation of neural networks with modifiable topologies. The algorithm called FAST (*Flexible Adaptable-Size Topology*) [9] that is used for implementing two-layered neural networks, is an example of this kind of application.

As we commented previously, the second criterion to class the FPGA implementation of neural networks is the data representation. A lot of the research into the FPGA implementation of networks is based on network training with integer weights, skirting the problems of working with real numbers in FPGAs.

There are special training algorithms that constrain the weights to powers of two or sums of powers [10]. The main advantage with this is that multipliers are replaced by shift and add registers. Some floating-point designs have been tested, but some authors [11] conclude that, despite FPGA and design tool sophistication, neural networks cannot be feasibly implemented on a FPGA with an acceptable level of precision. Some designers have opted for the bit-stream arithmetic solution, based on random bit streams that represent real numbers. This method reduces the use of multipliers, but severely limits the network learning and problem-solving ability.

3   Comments on Weight Precision and Transfer Functions

A lot of research into FPGA implementations of neural networks comes to the same conclusion, namely, that the most important implementation-related decisions to be taken are weight precision selection and transfer function implementation.

Neural networks usually use floating-point weights. Obviously, this involves using floating-point numbers within the network. From the hardware viewpoint, however, this is much more complicated than working with integers. Additionally, floating-point circuits are slower, more complex and have larger silicon area requirements than integer circuits. A very commonly used solution to simplify the design is to convert real numbers into integers, which leads to a significant loss of precision. High precision leads to fewer errors in the final implementation, whereas low precision leads to simple designs, high speed and reductions in area and power consumption.

The implementation of transfer functions raises problems as to how to represent the arithmetical operations. The most commonly used solutions are: piecewise linear approximation and look-up tables. The *piecewise linear approximation* describes a combination of lines, and authors use it as an approximation to implement a multilayer perceptron [12]. Using *look-up tables*, examples of values from the center of the function are entered in a table. The regions outside this area are represented by the area boundaries, because the values-based approximation is considered a straight line. This is how the function is approximated. The more examples there are, the bigger the area will be. Consequently, a decision will have to be made as to what the minimum accepted precision is depending on the occupied silicon area.

4   Data Representation

As we explained previously, we are using the artificial neural networks as a method for calculating the final foot positions of an humanoid robot. For this reason, the inputs for the implemented neural networks are Cartesian coordinates— the desired final position of a foot— and the outputs it returns are angles as radians —the values for the joints—. Therefore, floating-point numbers need to be used, which can be considered as the common case in the most of applications.

This is where the first important constraint materializes: real numbers cannot be synthesized. Consequently, a numerical format needs to be designed that can represent real numbers and can, at the same time, be synthesized.

We use a 16-bit sign-magnitude binary format, allocating the 6 most significant bits (including the sign bit) to represent the integer part and the following 10 to represent the decimal part. This gives a representation range of $(-32, 32)$ with a precision of $1/2^{10}$.

Previously we used this format in another design, but the integer part was represented in 2's Complement rather than sign-magnitude format. This format is more efficient, because the FPGA is capable of working in 2's Complement. The main problem with this format is that 2's Complement cannot represent -0. Therefore, none of the values between $(-1, 0)$ are represented.

However, sometimes is recommended to raise the precision, usually when values between $(0, 1)$ must be represented. Thanks to the possibility to define the precision in a variable way, a special format is also used for this sort of situations, which raises precision significantly. This is an 18-bit format, all of which are allocated to the decimal part, getting a precision of $1/2^{18}$.

5   Transfer Functions

The implemented networks are multilayer. The most commonly used transfer functions in such networks are logarithmic and tangential sigmoid:

$$f_l(n) = \frac{1}{1 + e^{-n}} \tag{1}$$

$$f_t(n) = \frac{e^n - e^{-n}}{1 + e^{-n}} \tag{2}$$

As the tool used does not synthesize floating-point numbers, implementation of the sigmoid transfer functions comes up against several problems: the number e cannot be raised to a real number, nor can divisions be performed, since the result of dividing one number by a higher one is always zero, because real numbers are not permitted.

Our first idea for solving this problem was to search a function that approximates e^n and e^{-n}. To this end, we examined the division-based Taylor approximation function, shown in (3).

$$e^x = 1 + x + \frac{x^2}{2!} + \frac{x^3}{3!} + \ldots + \frac{x^n}{n!} \tag{3}$$

This solution was rejected, because floating-point divisions cannot be performed, as explained above. Therefore, we decided to use splines to approximate the function by means of polynomials.

5.1   Logarithmic Sigmoid Function

Splines are used to approximate the function, using a polynomial in each curve piece. This polynomial should be of a small degree, because the greater the degree is, the larger the number of multiplications is and the higher the FPGA area requirements are. A polynomial of degree 3 is calculated that approximates the function quite precisely, the interpolation spline being as follows:

$$
s_l(n) = \begin{cases}
0 & \text{if } n < -5 \\
0.0466\,n^3 + 0.0643\,n^2 + 0.3064\,n + 0.5131 & \text{if } -5 \le n < -1.5 \\
-0.0166\,n^3 + 0.249\,n + 0.5 & \text{if } -1.5 \le n < 1.5 \\
0.0466\,n^3 - 0.0643\,n^2 + 0.3064\,n + 0.5131 & \text{if } 1.5 \le n < 5 \\
1 & \text{if } n \ge 5
\end{cases} \tag{4}
$$

The case of a very simple two-layered network, with two neurons in the first layer and two in the output layer, was analyzed. In each network, we implemented the XOR logical function, getting a mean error of 0.0044. In this manner, we managed to interpolate the transfer function very accurately, there is a small loss of precision at the inflection points, exactly in −5 and 5.

5.2   Tangential Sigmoid Function

As in the above case, the aim is to find a polynomial of the least possible degree that interpolates the function curve. As in the logarithmic case, we get an approximation interpolation spline of degree 3.

$$
s_t(n) = \begin{cases}
-1 & \text{if } n < -2.5 \\
0.0691\,n^3 + 0.4849\,n^2 + 1.1748\,n & \text{if } -2.5 \le n < 0 \\
0.0691\,n^3 - 0.4849\,n^2 + 1.1748\,n & \text{if } 0 \le n < 2.5 \\
1 & \text{if } n \ge 2.5
\end{cases} \tag{5}
$$

To evaluate the errors for all approximations, the same test case was run for all the error studies. The mean error (0.0185) is very small and, as in the above case, we find that there is a small loss of precision at the inflection points, −2.5 and 2.5 in this case.

6   FPGA Network Design

We are considering neural networks composed of three layers: an input layer, a hidden layer and an output layer. The transfer functions applied in the intermediate layer are sigmoid (logarithmic or tangential) functions, and the transfer function applied in the output layer is linear.

For design purposes, the networks were divided into three layers, which do not exactly match the layers enumerated above. The Input Layer is responsible for operating on the inputs using hidden layer weights and thresholds calculated by network training. The Intermediate Layer is in charge of applying the transfer function to the outputs of the input layer to get the normalized inputs for the output layer. As we discuss previously only the logarithmic and tangential sigmoid functions have been implemented. The Output Layer operates on the module inputs using the output layer weights and thresholds calculated by training the network.

As we can see, the input and output layers perform the same operation. The main difference lies in the weights and thresholds each layer uses to perform this operation.

6.1   Input and Output Layers

The input and output layers are modules responsible for using weights and thresholds to operate on the input signals and get results that can be used as input for the intermediate layer (in the case of the input layer) and network output (in the case of the output layer).

As both layers are alike, they are composed of the same blocks, as illustrated in Fig. 1. The *Operator "n"* block (bottom left) is responsible for reading a row of weights and thresholds from memory and then use these values to operate on the layer inputs. The *Layer Control* block (top center) is in charge of activating the above block, indicating the starting position of the weights and thresholds row to be read. Finally, the *Register Set* (right) guarantee layer output signals stabilization.

6.2   Intermediate Layer

In-between the input and output layers is a module in charge of applying the transfer functions to the outputs of the input layer so that they are normalized before they enter the output layer, that is, their values are bound within a small range. This range varies depending on the transfer function used. As we explained above the transfer functions to be implemented are logarithmic and tangential sigmoids.

A module is realized that implements the transfer function depending on the operator type, that is, is in charge of realizing the sigmoid approximation. This module is designed by means of a state machine and is called as many times

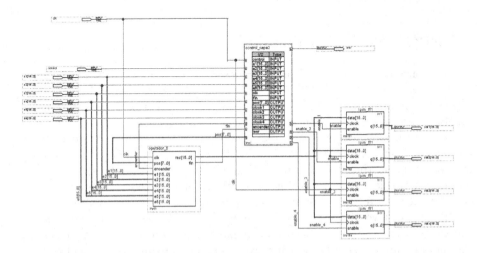

Fig. 1. Overall structure of the 6x4 output layer

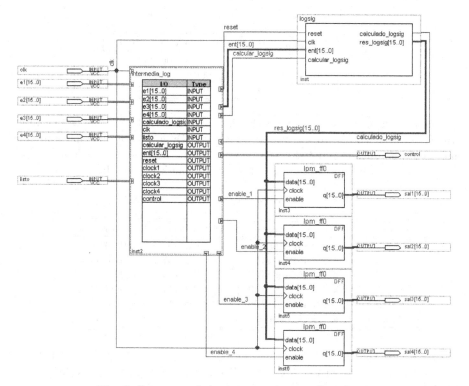

Fig. 2. Structure of the 4-neuron intermediate layer

as there are input layer outputs. Therefore, another state machine is realized to control these calls. The Fig. 2 shows the structure of a 4-neuron intermediate layer. As we can see, this intermediate layer is divided into several blocks. The *Operator* block (top right) is in charge of implementing the interpolation spline, appearing in designs as logarithmic or tangential sigmoids, depending on the transfer function to be applied, both having the same structure. The *Intermedia* block: controls the calls to the respective operator, and is called *intermediate_log* or *intermediate_tan* depending on the transfer function. Similarly to the previous cases, a *Register Set* (right bottom) is used to stabilize the output signals.

7   Experiments and Results

To verify the correct implementation of neural networks on FPGAs, we have selected one test network is selected and one data set. Then, the results are evaluated in each network layer this way to be able to study the error in each one and later get the mean network error.

In the following and as example, we will show how the network works with one data item, a tuple of 3D coordinates, which decimal and binary representation formats are shown in Table 1. The test network is composed of 3 inputs, 12 outputs and 6 neurons in the hidden layer.

Table 1. Data item for experimentation

Coordinate	Decimal Data Item	Binary Data Item
x	1.12	000001.0001111010
y	0.04	000000.0000101000
z	−0.99	100000.1111110101

The input layer is executed with the data shown in Table 1. The results and error for each of the layer outputs are shown in Table 2. The decimal number that should be output is listed first, followed by the binary result calculated by the network, the decimal equivalent of this result and, finally, the error.

Table 2. Results and error for the input layer

Ref Output	Binary Network	Decimal Network	Error
−2.7883	100010.1100100000	−2.78125	0.00705
−1.6107	100001.1001111100	−1.6210937	0.01039375
0.5447	000000.1001001100	0.57421875	0.02951875
0.3264	000000.0101000100	0.31640625	0.00999375
1.5848	000001.1001111100	1.62109375	0.03629375
−1.6096	100001.1010000000	−1.625	0.0154

To examine the cumulative error, the outputs calculated by the input layer are entered into the intermediate layer. The results and errors are evaluated as before and they are shown in Table 3.

Table 3. Results and error for the intermediate layer

Ref Output	Binary Network	Decimal Network	Error
0.058	000000.0000111110	0.060546875	0.002546875
0.1665	000000.0010101010	0.166015625	0.000484375
0.6329	000000.1010011000	0.6484375	0.0155375
0.5809	000000.1001011001	0.586914063	0.006014063
0.8299	000000.1101010110	0.833984375	0.004084375
0.166	000000.0010101001	0.165039063	0.000960938

Finally, the outputs of the intermediate layer are entered as inputs for the last layer. This way the total network error can be evaluated. The results and errors are shown in the Table 4.

Following the same procedure, the error analysis was run on a large data set, and the results were as follows. The mean error for the input layer is 0.013. This error is due, mainly, to format-induced error, because precision is $1/2^{10}$. The cumulative error of the intermediate and input layers is considerably lower (falling from 0.013 to 0.0008). Probably, we guess that sigmoid functions absorbs some of the error of the preceding layer. Finally, the mean network error, i.e. the cumulative error for the three network layers, is 0.03, that is, the network is precise to one hundredth of a unit.

Table 4. Results and error for the output layer

Ref Output	Binary Network	Decimal Network	Error
−0.2277	100000.0011110111	−0.241210998	0.013510938
−0.3504	100000.0100001110	−0.263671875	0.086728125
0.2102	000000.0011011000	0.2109375	0.0007375
0.1423	000000.0000111001	0.055664063	0.086635938
0.2274	000000.0011110111	0.241210938	0.013810938
−0.0012	100000.0000000010	−0.001953125	0.000753125
0.0016	000000.0000000100	0.00390625	0.00230625
−0.2296	100000.0011111001	−0.243164063	0.013564063
−0.4843	100000.0110001110	−0.388671875	0.095628125
0.4763	000000.0111010100	0.45703125	0.01926875
0.0088	000000.0001000101	0.067382813	0.058582813
0.2275	000000.0011110111	0.241210938	0.013710938

8   Conclusions and Future Work

Compared with the other neural network implementations mentioned earlier, the result of this work is quite satisfactory. A 16-bit binary format has been used, capable of representing floating-point numbers. This has solved one of the controversial aspects that other authors point out concerning FPGA implementations of neural networks. Thanks to this format, we have managed to implement quite a precise neural network design that does not have excessive FPGA area requirements.

Innovations have been introduced as regards transfer functions too. These functions were approximated in other implementations by means of look-up tables or piecewise linear approximations. In the implemented design, the functions were approximated non-linearly, resulting in a notably higher precision, again without FPGA area requirements being excessive.

Now, we are focusing our efforts for dividing the design into separate FPGAs to optimize operation, for example, by separating it by layers. Therefore, each layer can be loaded in a different FPGA making possible the use of smaller devices or, as counterpart, bigger neural networks. On the other hand, this network modularization can also be used over the same FPGA by means of reconfiguration. So, examining the advantages of FPGA reconfiguration and its impact on the implementation of the design presented here is another of our current works.

References

1. Kopacek, P. (2005) Advances in Robotics. Proc. of 10th Int. Workshop on Computer Aided Systems Theory, 272–274
2. De Lope, J., Maravall, D. (2004) A biomimetic approach for the stability of biped robots. Proc. of 5th Int. Conf. on Climbing and Walking Robots

3. De Lope, J., González-Careaga, R., Zarraonandia, T., Maravall, D. (2003) Inverse kinematics for humanoid robots using artificial neural networks. In R. Moreno-Díaz, F.R. Pichler (eds.) Computer Aided Systems Theory – EUROCAST-2003, LNCS 2809. Springer-Verlag, Berlin, 448–459

4. De Lope, J., Zarraonandia, T., González-Careaga, R., Maravall, D. (2003) Solving the inverse kinematics in humanoid robots: A neural approach. In J. Mira, J.R. Álvarez (eds.) Artificial Neural Nets Problem Solving Methods, LNCS 2687. Springer-Verlag, Berlin, 177–184

5. Cox, C.E., Blanz, E. (1992) Ganglion — A fast field-programmable gate array implementation of a connectionist classifier. IEEE J. of Solid-State Circuits, 28, 288–299

6. Eldredge, J.G., Hutchings, B.L. (1994) Density of enhancement of a neural network using FPGAs and run-time reconfiguration. Proc. of IEEE Workshop on Field-Programmable Custom Computing Machines, 180–188

7. Eldredge, J.G., Hutchings, B.L. (1994) RRANN: A hardware implementation of the back-propagation algorithm using reconfigurable FPGAs. Proc. of IEEE Int. Conf. on Neural Networks, 2097–2102

8. James-Roxby, P. (2000) Adapting constant multipliers in a neural network implementation. Proc. of IEEE Symposium on Field-Programable Custom Computing Machines, 335–336

9. Pérez-Uribe, A., Sánchez, E. (1996) FPGA implementation of an adaptable-size neural network. Proc. of 6th. Int. Conf. on Artificial Neural Network, 382–388

10. Marchesi, M., Orlandini, G., Piezza, F., Uncini, A. (1993) Fast neural networks without multipliers. IEEE Trans. on Neural Networks, 4, 53–62

11. Nichols, K.R., Moussa, M.A., Areibi, S.M. (2002) Feasibility of floating-point arithmetic in FPGA based artificial neural networks. Proc. of 15th Int. Conf. on Computer Applications in Industry and Engineering, 8–13

12. Wolf, D.F., Romero, R.A.F., Marques, E. (2001) Using embedded processors in hardware models of artificial neural networks. Proc. do Simposio Brasileiro de Automatao Inteligente, 78–83

An Associative Cortical Model of Language Understanding and Action Planning

Andreas Knoblauch, Heiner Markert, and Günther Palm

Department of Neural Information Processing, University of Ulm,
Oberer Eselsberg, D-89069 Ulm, Germany
{knoblauch, markert, palm}@neuro.informatik.uni-ulm.de

Abstract. The brain representations of words and their referent actions and objects appear to be strongly coupled neuronal assemblies distributed over several cortical areas. In this work we describe the implementation of a cell assembly-based model of several visual, language, planning, and motor areas to enable a robot to understand and react to simple spoken commands. The essential idea is that different cortical areas represent different aspects of the same entity, and that the long-range cortico-cortical projections represent hetero-associative memories that translate between these aspects or representations.

1 Introduction

When words referring to actions or visual scenes are presented to humans, distributed cortical networks including areas of the motor and visual systems of the cortex become active [1, 2]. The brain correlates of words and their referent actions and objects appear to be strongly coupled neuron ensembles in specific cortical areas. Being one of the most promising theoretical frameworks for modeling and understanding the brain, the theory of cell assemblies [3, 4] suggests that entities of the outside world (and also internal states) are coded in overlapping neuronal assemblies rather than in single ("grandmother") cells, and that such cell assemblies are generated by Hebbian coincidence or correlation learning. One of our long-term goals is to build a multimodal internal representation using several cortical areas or neuronal maps, which will serve as a basis for the emergence of action semantics, and to compare simulations of these areas to physiological activation of real cortical areas.

In this work we describe a cell-assembly-based model of several visual, language, planning, and motor areas to enable a robot to understand and react to simple spoken commands [5]. The essential idea is that different cortical areas represent different aspects (and correspondingly different notions of similarity) of the same entity (e.g., visual, auditory language, semantical, syntactical, grasping related aspects of an apple) and that the (mostly bidirectional) long range cortico-cortical projections represent heteroassociative memories that translate between these aspects or representations. Fig. 1 illustrates roughly the assumed locations and connections of the cortical areas actually implemented in our model.

J. Mira and J.R. Álvarez (Eds.): IWINAC 2005, LNCS 3562, pp. 405–414, 2005.

 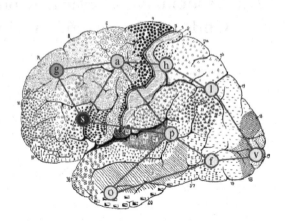

Fig. 1. Left: Robot on which the cortex model has been implemented to demonstrate a scenario involving understanding simple sentences as well as seeking and pointing to objects lying on a table. **Right:** Interaction of the different areas of the cortical model (v: visual, l: location, f: contour features, o:visual objects, h:haptic/proprioceptive, p:phonetics, s:syntactic, a:action/premotoric, g:goals/planning) and their rough localization in the human brain

This system is used in a robotics context to enable a robot to respond to spoken commands like "bot show plum" or "bot put apple to yellow cup". The scenario for this is a robot close to one or two tables carrying certain kinds of fruit and/or other simple objects (Fig. 1). We can demonstrate part of this scenario where the task is to find certain fruits in a complex visual scene according to spoken or typed commands. This involves parsing and understanding of simple sentences, relating the nouns to concrete objects sensed by the camera, and coordinating motor output with planning and sensory processing. The cortical model system can be used to control a robot in real time because of the computational efficiency of sparse associative memories [6, 7, 8].

2   Language, Finite Automata, Neural Networks and Cell Assemblies

In this section we briefly review the relation between regular grammars, finite automata and neural networks [9, 10, 11]. Regular grammars can be expressed by generative rules $A \rightarrow a$ or $B \rightarrow bC$ where upper case letters are variables and lower case letters are terminal symbols from an alphabet Σ. Regular grammars are equivalent to deterministic finite automata (DFA). A DFA can be specified by a set $M = (Z, \Sigma, \delta, z_0, E)$ where Z is the set of states, Σ is the alphabet, $z_0 \in Z$ is the starting state, $E \subseteq Z$ contains the terminal states, and the function $\delta : (Z, E) \rightarrow Z$ defines the state transitions. A sentence $s = a_1 a_2 \ldots a_n \in \Sigma^*$

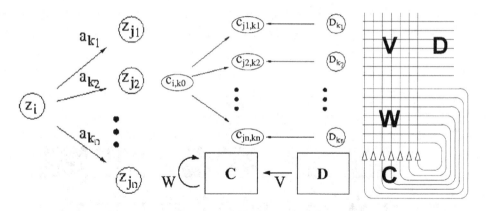

Fig. 2. Left: DFA; Middle: Neural Network; Right: Cell assemblies

(where Σ^* is the set of all words over the alphabet Σ) is said to be well formed with respect to the grammar if $\delta(...\delta(\delta(z_0, a_1), a_2), ..., a_n) \in E$.

DFAs can be simulated by neural networks [12,13]: E.g., it is sufficient to specify a simple model of recurrent binary neurons by $N = (C, D, W, V, c_{00})$, where C contains the local cells of the network, D is the set of external input cells, W and V are binary matrices specifying the local recurrent and the input connections (Fig. 2). The network evolves in discrete steps, where a unit is activated, $c_i(t) = 1$, if its potential $x_i(t) = (Wc(t-1) + Vd(t-1))_i$ exceeds threshold Θ_i, and deactivated, $c_i(t) = 0$, otherwise. A simple emulation of the DFA requires one neuron c_{jk} for each *combination* of state z_j and input symbol a_k, plus one neuron d_k for each input symbol a_k. Further we require synaptic connections $w_{il,jk} = d_{k,jk} = 1$ $(0 < l < |\Sigma|)$ for each state transition $(z_i, a_k) \mapsto z_j$. As threshold we use $\Theta_i = 1.5$ for all neurons. If at the beginning only a single neuron c_{0l} (e.g., $l = 0$) is active the network obviously simulates the DFA. However, such a network is biologically not very realistic since, for example, such an architecture is not robust against partial destruction and it is not clear how such a delicate architecture could be learned.

A more realistic model would interpret the nodes in Fig. 2 not as single neurons but as groups of nearby neurons which are strongly interconnected, i.e. local cell assemblies [3,6,14,15]. This architecture has two additional advantages: First, it enables fault tolerance since incomplete input can be completed to the whole assembly. Second, overlaps between different assemblies can be used to express hierarchical relations between represented entities. In the following subsection we describe briefly a cortical language model based on cell assemblies.

3   Cortical Language Model

Our language system consists of a standard Hidden-Markov-based speech recognition system for isolated words and a cortical language processing system which

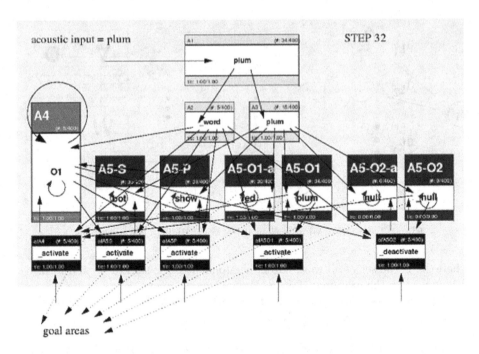

Fig. 3. Architecture of cortical language areas. The language system consists of 10 cortical areas (large boxes) and 5 thalamic activation fields (small black boxes). Black arrows correspond to interareal connections, gray arrows to short-term memory

can analyse streams of words detected with respect to simple regular grammars [16].

Fig. 3 shows the 15 areas of the language system. Each area is modeled as a spiking associative memory of 400 neurons [17, 8]. Binary patterns constituting the neural assemblies are stored auto-associatively in the local synaptic connections by Hebbian learning.

The model can roughly be divided into three parts. (1) Primary cortical auditory areas A1, A2 and A3. (2) Grammatical areas A4, A5-S, A5-O1a, A5-O1, A5-O2a, and A5-O2. (3) Relatively primitive "activation fields" af-A4, af-A5-S, af-A5-O1, and af-A5-O2 that subserve to coordinate the activation or deactivation of the grammar areas. When processing language, first auditory input is represented in area A1 by primary linguistic features (such as phonemes), and subsequently classified with respect to function in A2 and content in A3. The main purpose of area A4 is to emulate a DFA in a similar way as the neural network in Fig. 2. Fig. 4 shows the state graph of A4. Each node corresponds to an assembly representing a grammatical state, and each edge corresponds to a state transition stored in delayed recurrent hetero-associative connections of area A4. For example, processing of a sentence "Bot show red plum" would activate the state sequence S→Pp→OA1→O1→ok_SPO corresponding to expectation of processing of a subject, a predicate, an object or attribute, and finally an object.

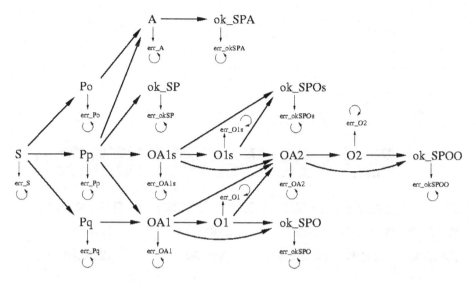

Fig. 4. Sequence assemblies stored in area A4 representing grammatical states. Each node corresponds to an assembly, each arrow to a hetero-associative link, each path to a sentence type. E.g., a sentence "Bot show red plum" would be represented by the sequence (S,Pp,OA1,O1,ok_SPO)

If the sentence was well formed with respect to the grammar, then the sequence terminates in an "ok_X" state, otherwise in one of the "err_X" states.

In our robot scenario it is not sufficient to decide if language input is grammatically well formed or not, but is also necessary to "understand" the sentence by transforming the word stream into an action representation. This is the purpose of areas A5-X which correspond to different grammatical roles or categories. In our example, area A5-S represents the subject "bot", A5-P the predicate "show" and A5-O1a and A5-O1 the object "red plum". Although establishing a precise relation to real cortical language areas of the brain is beyond the scope of this work [18, 19], we suggest that areas A1, A2, A3 can roughly be interpreted as parts of Wernicke's area, and area A4 as a part of Broca's area. The complex of the grammatical role areas A5 might be interpreted as parts of Broca's or Wernicke's area, and the activation fields as thalamic nuclei.

3.1   Example: Disambiguation Using Contextual Information

As discussed in section 2 our associative modeling framework is closely connected to finite state machines and regular languages in that we embed the automaton states in the attractor landscape of the associative neural networks. However, in contrast to the (purely symbolic) automata our state representations can express a similarity metric (e.g., by overlaps of different cell assemblies coding different states) that can be exploited in order to implement fault tolerance against noise

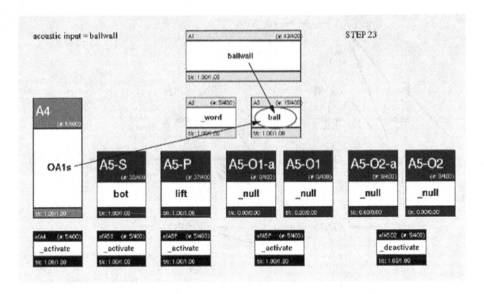

Fig. 5. Disambiguation using context: The example illustrates the states of the language areas after processing "bot lift" and then receiving noisy ambiguous acoustic input ("ballwall") which can be interpreted either as "ball" or as "wall". The conflict is solved by contextual information in areas A4 and A5-P representing the verb "lift" which expects a "small" object (such as a ball)

and to use contextual information to resolve ambiguities, e.g. by selecting the most probable interpretation.

The following example illustrates the ability of our model to resolve conflicts caused by noisy ambiguous input (see Fig. 5). After processing "bot lift" the primary auditory area A1 obtains noisy ambiguous input "ballwall" which can be interpreted either as "ball" or as "wall". The conflict is solved by contextual information in areas A4 and A5-P representing the previously encountered verb "lift" which expects a "small" object such as "ball", but not a large (non-liftable) object such as "wall". Thus contextual input from area A4 (where "OAs" represents a "small" object) biases the neural activity in area A3 such that the unambiguous representation "ball" is activated.

4   Action Processing

Our system for cortical planning, action, and motor processing can be divided into three parts (see Fig. 6). (1) The action/planning/goal areas represent the robot's goal after processing a spoken command. Linked by hetero-associative connections to area A5-P, area G1 contains sequence assemblies (similar to area A4) that represent a list of actions that are necessary to complete a task. For example, responding to a spoken command "bot show plum" is represented by

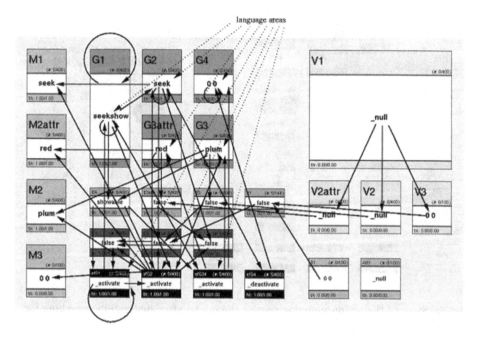

Fig. 6. Architecture of the cortical action model. Conventions as in Fig. 4

a sequence (seek, show), since first the robot has to seek the plum, and then the robot has to point to the plum.

Area G2 represents the current subgoal, and araes G3, G3attr, G4 represent the object involved in the action, its attributes (e.g., color), and its location, respectively. (2) The "motor" areas MX represent the motor command necessary to perform the current goal, and also control the low level attentional system. Area M1 represents the current motor action, and areas M2, M2attr, and M3 represent again the object involved in that action, its attributes, and its location. (3) Similar to the activation fields of the language areas, there are also activation fields for the goal and motor areas, and there are additional "evaluation fields" that can compare the representations of two different areas.

To illustrate how the different subsystems of our architecture work together, we describe a scenario where an instructor gives the command "Bot show red plum!", and the robot ("Bot") has to respond by pointing to a red plum located in the vicinity. To complete this task, the robot first has to understand the command as described in section 3, which activates the corresponding A5-representations. Activation in area A4 has followed the corresponding sequence path (see Fig. 4). Immediately after activation of the A5-representations the corresponding information is routed further to the goal areas where the first part of the sequence assembly (seekshow, pointshow) gets activated in area G1. Similarly, the information about the object is routed to areas G2, G3 and G3attr. Since the location of the plum is unknown, there is no activation in area G3. After checking semantics, the "seek" assembly in area G2 and the corresponding

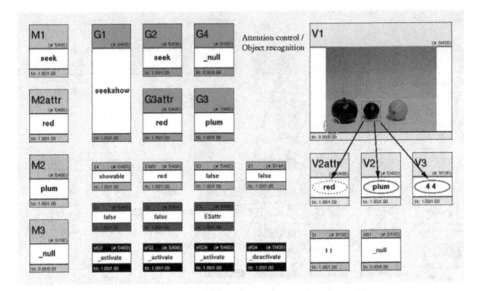

Fig. 7. State of the action planning part of the model after successfully searching for a red plum

representations in the motor areas MX are activated. This also triggers the attentional system which initiates the robot to seek for the plum. Fig. 7 shows the network state when the visual object recognition system has detected the red plum and the corresponding representations have been activated in areas V2, V2attr and V3. The control fields detect a match between the representations in areas V2 and G3, which initiates area G1 to switch to the next part "point" of the action sequence. The robot will then adjust its "finger position" represented in area S1 in order to point to the plum. The matching of the positions will be detected by the evaluation fields and this eventually activates the final state in G1.

5   Conclusion

We have described the implementation of a cell assembly-based model of cortical language and action processing on a robot [16, 5]. The model consists of about 40 neuron populations each modelled as a spiking associative memory containing many "local" cell assemblies stored in local auto-associative connections [17]. The neuron populations can be interpreted as different cortical and subcortical areas, where it is a long term goal of this project to establish a mapping of our "areas" into real cortex [1].

Although we have currently stored only a limited number of objects and sentence types, it is well known for our model of associative memory that the number of storable items scales with $(n/\log n)^2$ for n neurons [6, 7]. However, this is true only if the representations are sparse and distributed which is a design principle of our model. As any finite system, our language model can implement only regular languages, whereas human languages seem to involve context-sensitive grammars. On the other hand, also humans cannot "recognize" formally correct sentences beyond a certain level of complexity suggesting that in practical speech we use language rather "regularly".

Acknowledgments

We are grateful to Friedemann Pulvermüller for valuable discussions. This work has been supported by the MirrorBot project of the European Union.

References

1. Pulvermüller, F.: Words in the brain's language. Behavioral and Brain Sciences **22** (1999) 253–336
2. Pulvermüller, F.: The neuroscience of language: on brain circuits of words and serial order. Cambridge University Press, Cambridge, UK (2003)
3. Hebb, D.: The organization of behavior. A neuropsychological theory. Wiley, New York (1949)
4. Palm, G.: Cell assemblies as a guideline for brain research. Concepts in Neuroscience **1** (1990) 133–148
5. Fay, R., Kaufmann, U., Knoblauch, A., Markert, H., Palm, G.: Integrating object recognition, visual attention, language and action processing on a robot in a neurobiologically plausible associative architecture. accepted at Ulm NeuroRobotics workshop (2004)
6. Willshaw, D., Buneman, O., Longuet-Higgins, H.: Non-holographic associative memory. Nature **222** (1969) 960–962
7. Palm, G.: On associative memories. Biological Cybernetics **36** (1980) 19–31
8. Knoblauch, A.: Synchronization and pattern separation in spiking associative memory and visual cortical areas. PhD thesis, Department of Neural Information Processing, University of Ulm, Germany (2003)
9. Hopcroft, J., Ullman, J.: Formal languages and their relation to automata. Addison-Wesley (1969)
10. Chomsky, N.: Syntactic structures. Mouton, The Hague (1957)
11. Hertz, J., Krogh, A., Palmer, R.: Introduction to the theory of neural computation. Addison-Wesley, Redwood City (1991)
12. Minsky, M.: Computation: finite and infinite machines. Prentice-Hall, Englewood Cliffs, NJ (1967)
13. Horne, B., Hush, D.: Bounds on the complexity of recurrent neural networks implementations of finite state machines. Neural Networks **9(2)** (1996) 243–252
14. Palm, G.: Neural Assemblies. An Alternative Approach to Artificial Intelligence. Springer, Berlin (1982)
15. Hopfield, J.: Neural networks and physical systems with emergent collective computational abilities. Proceedings of the National Academy of Science, USA **79** (1982) 2554–2558

16. Knoblauch, A., Fay, R., Kaufmann, U., Markert, H., Palm, G.: Associating words to visually recognized objects. In Coradeschi, S., Saffiotti, A., eds.: Anchoring symbols to sensor data. Papers from the AAAI Workshop. Technical Report WS-04-03. AAAI Press, Menlo Park, California (2004) 10–16
17. Knoblauch, A., Palm, G.: Pattern separation and synchronization in spiking associative memories and visual areas. Neural Networks **14** (2001) 763–780
18. Knoblauch, A., Palm, G.: Cortical assemblies of language areas: Development of cell assembly model for Broca/Wernicke areas. Technical Report 5 (WP 5.1), MirrorBot project of the European Union IST-2001-35282, Department of Neural Information Processing, University of Ulm (2003)
19. Pulvermüller, F.: Sequence detectors as a basis of grammar in the brain. Theory in Bioscience **122** (2003) 87–103

Neural Clustering Analysis of Macroevolutionary and Genetic Algorithms in the Evolution of Robot Controllers

J.A. Becerra and J. Santos

Grupo de Sistemas Autónomos, Departamento de Computación,
Facultade de Informática, Universidade da Coruña
{ronin, santos}@udc.es

Abstract. In this work, we will use self-organizing feature maps as a method of visualization the sampling of the fitness space considered by the populations of two evolutionary methods, genetic and macroevolutionary algorithms, in a case with a mostly flat fitness landscape and low populations. Macroevolutionary algorithms will allow obtaining better results due to the way in which they handle the exploration-exploitation equilibrium. We test it with different alternatives using the self-organizing maps.

1 Introduction

When evolutionary methods are applied to automatically obtain robot controller behaviours, their interconnection or aspects of the robot morphology [8], the computer time requirements are high as each individual of the population must be checked in a real or simulated environment during an adequate number of steps or lifetime to assign it a reliable fitness. This implies the use of low populations in the evolutionary methods to obtain reasonable solutions in bearable amounts of time. In addition, in cases with mostly flat fitness landscapes, except for a very sparse distribution of peaks where fit individuals are located [1], such as the neural behaviour controllers we consider in that work, it is very important to concentrate adequately the computer effort in the search of good solutions, that is, to choice a good equilibrium balance between the exploration and exploitation phases.

In methods of simulated evolution such as genetic algorithms (GAs), selective pressure determines how fast the population converges to a solution. The more pressure, the faster the convergence, at the cost of increasing the probability of the solution found being suboptimal. The designer can choose, through the selection criterion parameters, to consider the evaluation of a large number of candidates throughout the search space or concentrate the search in the direction of a good, possibly suboptimal, solution. There are different avenues in the exploration-exploitation dilemma to concentrate the search in an adequate number of candidate solutions, not only one. A first possible approximation to

J. Mira and J.R. Álvarez (Eds.): IWINAC 2005, LNCS 3562, pp. 415–424, 2005.

this objective is the use of parallel evolutionary algorithms, with subpopulations that can cover different areas of the search space, at least at the beginning of the evolution process, as studied in [10].

Another possibility to avoid the problem of ill convergence is to obtain selection procedures that produce the desired clustering or concentration of the search efforts on the different candidate solutions of the fitness landscape. This "niching" process means, at the computational level, that these groups can be formed around each local fitness peak of the solution space. The most classical solution in this line is to consider the use of the so-called "crowding" operator [3]: when a new individual is generated, it replaces the most similar individual of the population, which prevents the possibility of having many similar individuals ("crowds") at the same time in the population. Thus, the key point in this approach seems to take into account some measurement of similarity among the individuals, as done, for example, by Menczer et al. [6], who have also used a local selection scheme for evolving neural networks in problems that require multi-criteria fitness functions.

Another more formal solution in the same line, followed in this work, is the one proposed by Marín and Solé [5]. The authors consider a new temporal scale, the "macroevolutionary" scale, in which the extinctions and diversification of species are modelled. The population is interpreted as a set of species that model an ecological system with connections between them, instead a number of independent entities as in classical GAs. The species can become extinct if their survival ratio with respect to the others is not higher than a "survival coefficient". This ratio measures the fitness of a species respect to the fitness of the other species. When species become extinct, a diversification operator colonizes the holes with species derived from those that survived or with completely new ones. The use of this method was studied in previous works [1][2], with average better results in terms of fitness with low populations respect to GAs.

The usual measurements of fitness progress across generations in evolutionary methods do not provide us with sufficient information to know how they are working in the exploration and exploitation of the fitness landscape. Because of that, in this work we will use self-organizing feature maps (SOMs) as a method of visually represent the sampling of the fitness space that is considered by the populations of the evolutionary methods. The SOMs create a translation between a high dimensional space and a lower one preserving the spatial topology, that is, close points in the original space are projected to nearby points in the map, with the same consideration with distant points, and without forgetting the important feature of generalization in neural network models. We will use 2-D maps with the original space being the gene space. In that manner, the SOMs will provide us with a useful tool to inspect that sampling when GAs and Macroevolutionary Algorithms (MAs) are used in the evolution of neural controllers that imply the commented mostly flat fitness landscapes.

2   Brief Description of Macroevolutionary Algorithms

Here, we summarize the model proposed by Marín and Solé [5], which explains the dynamics of an ecosystem based only on the relation between species. The individuals in the population are referred as species. They can survive or become extinct in each generation of the evolutionary process. The number of species is a constant. The relation between them is established by a matrix in which the term $W_{i,j}(t)$ represents the influence of species j on species i at time t, which is a continuous value in a given range. This influence is a measurement of the difference of the relative fitness of the two species, considering the distance between both in genotypic space:

$$W_{ij} = \frac{f(p_i) - f(p_j)}{|(p_i) - (p_j)|}$$

where $p_i = (p_{i1}, \ldots, p_{in})$ is the genotype of species i, with its parameters in a n-dimensional space. f represents the fitness of each species. Thus, the influence is the difference in fitness with a normalization factor that weighs the distance between the two.

Two operators are applied in each generation:

1. Selection operator: defines what species survive and what species become extinct. To determine this, the "state" of each individual is calculated as:

$$S_i(t+1) = \begin{cases} 1 \, if \, \sum_{j=1}^{P} W_{ij}(t) \geq 0 \\ 0 \, otherwise \end{cases}$$

that is, if the sum of the influences of a species relative to all the other species in the population is positive, the species survives, otherwise, it becomes extinct.

2. Colonization operator: it defines how the extinct species are replaced. The authors define a probability Γ to determine if a new solution p_n is generated. Otherwise, exploitation of surviving solutions takes place through "colonization". One of the surviving solutions, p_b, is chosen as a base to replace the extinct solution p_i, and the new species that replaces the extinct one is attracted toward p_b, in the following manner:

$$p_i(t+1) = \begin{cases} p_b(t) + \rho\lambda\,(p_b(t) - p_i(t)) \, if \, \xi > \Gamma \\ p_n \, if \, \xi \leq \Gamma \end{cases}$$

where ξ is a random number in [0,1], λ a random number in [-1,1], both with uniform distribution, ρ describes a maximum radius around the surviving solution and Γ controls the percentage of random individuals. This parameter may act as a temperature, because it can decrease in evolutionary time to perform a type of simulated annealing process. That is, when the temperature is low, the randomness is low, and, consequently, there is a tendency towards increased exploitation around the surviving individuals, and reduced exploration of new species. Thus, when using a macroevolutionary algorithm one can tweak with two parameters. On one hand, Γ determines what proportion of the species is randomly generated, that is, how much exploitation or exploration we perform in a given generation. On the other, one can modify ρ and thus juggle with the size of the attractor basin around p_b, that is, it permits deciding how the exploitation is carried out.

3   Test Case: Evolution of a Robot Wall Following Controller

As we commented before, the comparison between GAs and MAs will be carried out in a problem where robot controllers, made up of artificial neural networks, are the end product. This common problem implies a sparse fitness function and a large amount of processing per individual. In particular, all the examples presented here correspond to the evolution of a wall following controller for a Pioneer II robot.

The fitness evaluation is performed in a simulated environment where each individual (a candidate controller) runs for a given lifetime of 1000 steps. As a controller for the wall-following behaviour, we use a multilayer perceptron with eight inputs that correspond to eight sonar sensors, 6 hidden nodes, and two output nodes that specify the linear and angular velocity of the robot. The genetic representation is direct, that is, each genotype codifies the 76 real values of the neural weights, sigmoid slopes and neuron bias.

Evolution was carried out using two types of algorithms, with a population of 800 individuals distributed in 8 races, with 100 individuals each, and during 2000 generations. One of the algorithms was a pure genetic algorithm, with a 0.8 crossover probability and a 0.01 mutation probability. Crossover was a standard one-point crossover and mutation was random. The selection strategy employed was tournament with a window size of 4 individuals. The GA used a diagonal distribution for the initial population, as it has been shown to provide the best usage of resources [10]. This algorithm was taken as a standard in order to compare different aspects. The other one was a macroevolutionary algorithm in which ρ is 0.5 and Γ decreases linearly from 1 (in generation number 1) to 0 (in generation 1000) and then it keeps a constant value of 0 to end of the evolutionary process. In the two methods are used migrations between the 8 races. There are two types of migrations: local and global. The local migration has place each 40 generations and the best individual of each race is copied to neighbour races (neighbourhood is defined in one dimension, so every i race has two neighbours: $i-1$, $i+1$). The global migration happens each 80 generations and the best individual of each race is copied to every other race.

Figure 1 shows the evolution of average fitness and the fitness of the best individual across generations. The GA displays the classical evolution curve for the best individual, that is, very fast fitness increase at the beginning, even with the low pressure applied. There is a very fast exploitation at the beginning of the evolutionary process where the GA make use of the genetic material present in the initial population and recombines it to obtain the best possible individuals

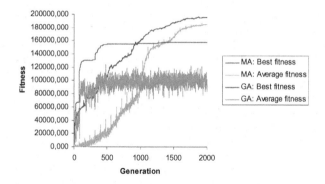

Fig. 1. Evolution of the fittest and average fitness for a GA and a MA

from this material, but then it gets stuck in a local maximum and it is barely able to improve the solution using mutation.

In the case of MAs, exploration takes place first, and exploitation comes into play slowly throughout the process as Γ decreases. In fact, due to the way in which exploitation is carried out in this type of algorithms, new genetic material arises also in the exploitation phase. This leads to a slower evolution in the beginning, but it is much more constant as shown in the figure. In fact, it usually leads to better results than GAs, especially in the low population cases, such as those with 800 individuals.

In the case of the average fitness, the GA average fitness is far below maximum fitness and it never converges to it. The search space in this problem could be described as a flat surface with sporadic high peaks [1]. This particular search space leads to most of the individuals in the population, resulting from crossover or random mutation, being quite poor when performing the desired task. In the MA case, this is different. Because of the way exploitation is carried out in this algorithm, individuals tend to concentrate in the high peaks as time progresses. In the next section, we will visually inspect these assertions with the neural clustering of the population in both types of algorithms.

4   Clustering Analysis with Macroevolutionary and Genetic Algorithms

To test the different behaviours used by the GA and the MA, in addition to the usual measurements of the best individual and average fitness evolution across generations, it would be interesting to have a method to determine the behaviour of the evolutionary methods in relation to the sampling of the search space. Obviously, we need a method to translate the n-dimensional space the evolutionary method must consider to a lower dimensional one where that information can be checked. Among the different alternatives, we have used the self-organizing feature maps, which perform a mapping from a continuous input space to a

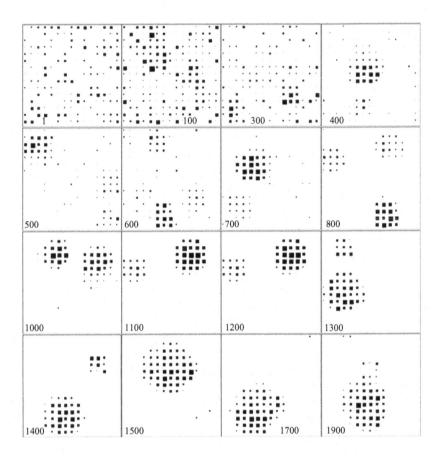

Fig. 2. Clustering of the population of race number 0 through generations with the MA. In the initial generations predominates the exploration phase that progressively decreases to play basically with exploitation

discrete output space preserving the topological properties of the input. The maps can provide us, in a visual manner, with the clustering effect of the input space, which could be interpreted in terms of exploration and exploitation of the evolutionary method.

The different trainings with the self-organizing maps have been done using the NeuroSolutions 4.2 environment [7]. The inputs to the feature map are the gene values, that is, the 76 genes that codify the neural controller weights. The maps have 15x15 nodes, with Gaussian bell as function of activation in its nodes. In all the cases it was used a final neighbourhood of two, as learning rate a value 0.001, 100 training epochs and a dot product based metric to determine the output winner in the map.

4.1   MA Population Clustering

Figure 2 shows the clustering results in the self-organizing map with Hinton diagrams, where the size of the squares represents the frequency of activation of each node in the 2-D self-organizing map. That is, it is a histogram of win frequencies. Each of the Hinton diagrams corresponds to a particular generation, from generation number 1 to generation 1900, using the populations of race number 0. In the first 300 generations, the individuals of the population are distributed in such a manner that practically covers the entire search space. In this stage predominates the exploration phase, with the introduction of new random individuals. Toward the end of evolution is clear a tendency to group the population in progressively less number of clusters, for example, from the three clusters in generation number 800, to the final and unique cluster in generations 1500-1900. We must take into account that the clusters can be formed in any position of the map, as it is has been trained in each generation with their particular population. To understand this tendency, we must remember that the introduction of new individuals is decreased with the generations, that is, there is a progressively exploitation of the good found solutions, in accordance with the comments previously done with the evolution of the fitness.

We can also use the maps to inspect if the different areas covered by the different races are or not the same. We can train a map with the individuals of a particular race, and, once trained, use it to check if the individuals of other races, in the same generation, correspond to the same clusters of the first one. One example of that is in figure 3. In generation number 800, the left figure represents the clustering once the map has been trained with 100 epochs with the 100 individuals of race number 0 in that generation. That is, as the previous cases, the Hinton diagram represents the frequency of the winner's outputs with the training set. The other three maps represent the frequency of the winner nodes, using the weights of the map that corresponds to race number 0, but tested with the individuals of other races (1, 6 and 7) in the same generation. The first map of race number 0, trained with other random initial weights, has the same three clusters shown in figure 2. The tests with the population of the other races show that their clusters "fall" in the same cluster areas of race number 0. For example, race number 1 is concentrating its population in two

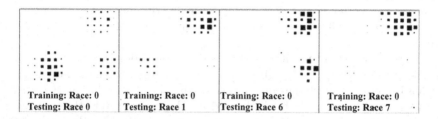

Fig. 3. Testing of the sampling in different races in a given generation (number 800). Populations of races number 1, 6 and 7 were tested with the map of race number 0 (upper left figure)

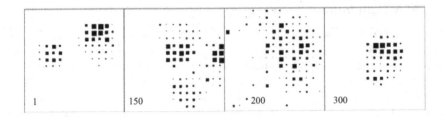

Fig. 4. Clustering of the population of race number 4 through generations with the GA. The exploration phase only takes place in the 200 first generations. From generation 300 to the end the GA is working with only one cluster

of the same areas of the solution space considered in race number 0, but with a greater concentration in the cluster that appears in the right top part. Race number 6 is concentrating its population in other two clusters also considered in race number 0, while race number 7 practically has all of its population in the right top cluster. That means that the different races, at generation 800, are exploring in the same areas, but with a different concentration of individuals, with one race that can be exploring an area that is not extensively explored by others. This fact is probably due to migrations.

4.2   GA Population Clustering

Figure 4 shows a sketch of the evolution of the clusters, trained, as in figure 2, only with the population in each generation. It is appreciated only in the first 200 generations the exploration phase, even with the low selective pressure applied: 4 individuals (4% of the total individuals in each race) in the window tournament, to select the one to procreate. From generation number 300 to the end, the GA is only exploiting around the individuals situated in the same cluster, and the fitness never reaches the MA values shown in figure 1.

4.3   Clustering Evolution Through Generations

Another possibility is to use SOMs, trained with all the individuals in all the generations, for a posterior visualization of the movement of the population across the fitness landscape. For example, in [9], the authors have used this alternative with a GA, for the evolution of neural networks for a classification problem.

We also have tested this alternative. Figure 5 now shows the evolution of the clusters of the population of a given race (number 0) when the self-organizing map has been trained with the population considered by the evolutionary method, the MA in this case. The individuals used for the training were those of the 8 races, taken with intervals of 50 generations, from generation 0 to the final generation number 2000, that proportionate an ample vision of the different genotypes the MA considers through the complete evolution. That is, figure 5 displays in particular generations the clusters that appear when the individuals of that race

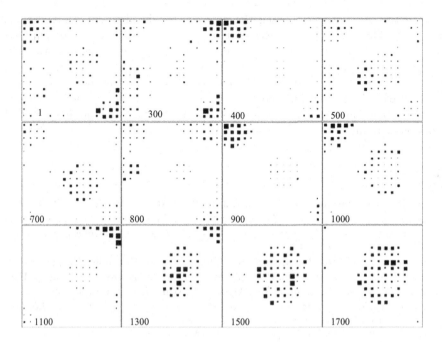

Fig. 5. Evolution of clusters of race number 0 through generations. The population is now tested with the map trained with all the individuals considered across generations

in a given generation are tested with the trained map. In that manner, the evolution of the clusters gives an idea of the race population progress across the fitness landscape.

There is a continuous movement of the clusters while there is an exploration component in the MA, which progressively decreases with generations. In the initial generations, new areas of the landscape are explored, but no one of the promising areas disappears thanks to the way exploitation is carried out selecting randomly one of the survival species. The introduction of new individuals in the migrations between races, individuals that can be in different zones of the search space and that consequently form clusters in other areas of the search space, as we previously shown, helps to explore new areas. Finally, from generation number 1300 to the end, the main cluster has not considerable movement, when predominates the exploitation phase.

5   Conclusions

We have checked the use of self-organizing feature maps to inspect how different evolutionary methods search across the fitness landscape. The analysis is useful when the work with low populations and mostly flat fitness landscapes is necessary, such as the one considered of a neural controller evolution.

With the original space being the gene space, the different visualizations with the SOMS allowed us to check different assumptions that we only suppose with the usual measurements of fitness evolution, from the clustering formed in each population, the test of the sampling of the landscape by different races, to the evolution of the populations across the landscape through generations. Finally, the used macroevolutionary algorithm has clearly shown a better balance between exploration and exploitation than a classical GA, with an easy control through few parameters.

References

1. Becerra, J.A., Santos, J. and Duro, R.J. (2002), "MA vs. GA in Low Population Evolutionary Processes with Mostly Flat Fitness Lanscapes", *Proceedings of 6th Joint Conference on Information Sciences - Frontiers in Evolutionary Algorithms (FEA 2002)*, pp. 626-630.
2. Becerra, J.A., Santos, J. and Duro, R.J. (2004), "Robot Controller Evolution with Macroevolutionary Algorithms", *Information Processing with Evolutionary Algorithms*. M. Graña, R.J. Duro, A. d'Anjou & P. Wang (Eds.), pp. 117-127.
3. De Jong, K.A. (1975), *An Analysis of the Behavior of a Class of a Genetic Adaptive Systems*, Ph. Thesis, University of Michigan, Ann Arbor.
4. Kohonen, T. (1984), "Self-Organization and Associative Memory", *Series in Information Sciences*, Vol. 8, Springer-Verlag, Berlín-Heildeberg-New York-Tokio; 2nd ed., 1988.
5. Marín, J., Solé, R.V. (1999), "Macroevolutionary Algorithms: A New Optimization Method on Fitness Landscapes", *IEEE Transactions on Evolutionary Computation* 3(4):272-286.
6. Menczer, F., Street, W.N., Degeratu, M. (2001), "Evolving Heterogeneous Neural Agents by Local Selection", *Advances in the Evolutionary Synthesis of Intelligent Agents*, M.J. Patel, V. Honavar and K. Balakrishnan (Eds.), MIT Press, 337-365.
7. *NeuroSolutions version 4.2*, NeuroDimension, Inc.
8. Nolfi, S., Floreano, D. (2000), *Evolutionary Robotics*, MIT Press.
9. Romero, G. Arenas, M.G., Castillo, P.A. and Merelo, J.J. (2003), "Visualization of Neural Net Evolution", *Computational Methods in Neural Modeling, Lecture Notes in Computer Science*, Vol. 2686, pp. 534-541.
10. Santos, J., Duro, R.J., Becerra, J.A., Crespo, J.L., and Bellas F. (2001), "Considerations in the Application of Evolution to the Generation of Robot Controllers", *Information Sciences* 133, pp. 127-148.

Induced Behavior in a Real Agent Using the Multilevel Darwinist Brain

F. Bellas, J.A. Becerra, and R.J. Duro

Grupo de Sistemas Autónomos,
Universidade da Coruña, Spain
{fran, ronin, richard}@udc.es

Abstract. In this paper we present a strategy for inducing a behavior in a real agent through a learning process with a human teacher. The agent creates internal models extracting information from the consequences of the actions it must carry out, and not just learning the task itself. The mechanism that permits this background learning process is the Multilevel Darwinist Brain, a cognitive mechanism that allows an autonomous agent to decide the actions it must apply in its environment in order to fulfill its motivations. It is a reinforcement based mechanism that uses evolutionary techniques to perform the on line learning of the models.

1 Introduction

When trying to deal with the problem of teaching a task to a robot through real time interaction, there are lots of variables that must be taken into account. In fact, the problem is normally driven by the designer and it is difficult to determine what has been autonomously learned and what has been induced by the limitations of the teaching process. As mentioned in [1] the representation reflects to a large extent the human designer's understanding of the robot task. In order to generalize the solutions obtained, the automatic design of robot controllers has become a very important field in autonomous robotics.

A natural approach to this problem is learning by demonstration (LBD) [2], but other methods such as gestures [3] or natural language [4] are also techniques that have been successfully applied. In LBD authors have tried to separate "how" to carry out a task from "what" to do in a task, that is, what the task is, focusing their research on systems that could be applied to different tasks. Within this field of learning by demonstration, relevant results have been achieved based on training and demonstration stages as in [5]. A technique that has been demonstrated as very appropriate in the automatic development of controllers is reinforcement learning [6], [7] where a robot (any agent in general) learns from rewards and punishments (which are basically some type of error signal) provided by a teacher in real time. These systems can be applied to different tasks (different "what") but they don't work if, for example, something fails in the robot (different "how").

A new approach to the automatic design process is called cognitive developmental robotics (CDR). As explained in [8], the key aspect of CDR is that

J. Mira and J.R. Álvarez (Eds.): IWINAC 2005, LNCS 3562, pp. 425–434, 2005.

the control structure should reflect the robot's own process of understanding through interactions with the environment.

In this work we present a solution in which the main objective is to complete a task driven by a motivation and not by the way the goal is achieved. Thus, if the environment or the agent itself changes, the way the task is carried out may also be modified but the agent readapts autonomously. The cognitive mechanism we have developed is called the Multilevel Darwinist Brain and in the next section we are going to provide an introduction of its concepts and operation.

2   The Multilevel Darwinist Brain

The Multilevel Darwinist Brain (MDB) is a Cognitive Mechanism that allows a general autonomous agent to decide the actions it must apply in its environment in order to fulfill its motivations. The main feature of the MDB is that the acquisition of knowledge is automatic, this is, the designers do not impose their knowledge on the system. In its development, we have resorted to bio-psychological theories [9], [10], [11] within the field of cognitive science that relate the brain and its operation through a Darwinist process. Each theory has its own features, as shown in the references, but they all lead to the same concept of cognitive structure based on the brain adapting its neural connections in real time through evolutionary or selectionist processes. This idea of Darwinism in the acquisition of knowledge is the basis for the development of the practical Cognitive Mechanism proposed.

2.1   Cognitive Model

To implement the MDB, we have used an utilitarian cognitive model [12] which starts from the premise that to carry out any task, a *motivation* (defined as the need or desire that makes an agent act) must exist that guides the behavior as a function of its degree of *satisfaction*. From this basic idea, several concepts arise:

- *Action*: set of commands applied to the actuators of the agent.
- *World model (W)*: function that relates sensory inputs in time t with sensory inputs in time t+1 after applying an action.
- *Satisfaction model (S)*: function that relates sensory inputs with the satisfaction the agent should perceive in the same instant of time according to the motivation for the agent.
- *Action-perception pair*: set of values made up by the sensorial inputs and the satisfaction obtained after the execution of an action in the real world. It is used when perfecting world and satisfaction models.

Two processes must take place in a real non preconditioned cognitive mechanism: models W and S must be obtained as the agent interacts with the world, and the best possible actions must be selected through some sort of internal optimization process using the models available at that time.

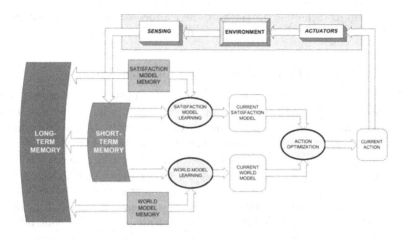

Fig. 1. Block diagram of the MDB

2.2   Constructing the MDB

The main difference of the MDB with respect to other model based cognitive mechanisms is the way the models are obtained and the actions planned from them. Its functional structure is shown in Fig. 1. The main operation can be summarized by considering that the selected action (represented by the *current action* block) is applied to the *environment* through the *actuators* obtaining new *sensing* values. These acting and sensing values provide a new action-perception pair that is stored in the action-perception memory (*Short-Term Memory* from this point forward). Then, the *model learning* processes start (for world and satisfaction models) trying to find functions that generalize the real samples (action-perception pairs) stored in the *Short-Term Memory*. The best models in a given instant of time are taken as *current world model* and *current satisfaction model* and are used in the process of *optimizing the action* with regards to the predicted satisfaction of the motivation. After this process finishes, the best action obtained is applied again to the *environment* through the *actuators* obtaining new *sensing* values.

These steps constitute the basic operation cycle of the MDB, and we will call it an iteration of the mechanism. As more iterations take place, the MDB acquires more information from the real environment (new action-perception pairs) so the models obtained become more accurate and, consequently, the action chosen using these models is more appropriate.

The block labeled *Long-Term Memory* stores those models that have provided succesful and stable results on their application to a given task in order to be reused directly in other problems or as seeds for new learning processes. Details about implementing the *Long-Term Memory* can be found in [13].

Summarizing, there are two main processes that must be solved in MDB: the learning of the best models predicting the contents of the short-term memory and the optimization of the action trying to maximize the satisfaction using the previously obtained models. In the way these processes are carried out lies the main difference of the MDB with respect to other mechanisms.

2.3   Learning of Models and Action Optimization

The model search process in the MDB is not an optimization process but a learning process (we seek the best generalization, which is different from minimizing an error function in a given instant t). Consequently, the search techniques must allow for gradual application as the information is known progressively and in real time. In addition, they must support a learning process through input/output pairs using an error function. To satisfy these requirements we have selected *Artificial Neural Networks* as the mathematical representation for the models and *Evolutionary Algorithms* as the most appropriate search technique.

Evolutionary techniques permit a gradual learning process by controlling the number of generations of evolution for a given content of the short-term memory. Thus, if evolutions last just a few generations per iteration, gradual learning by all the individuals is achieved. To obtain a general model, the populations of the evolutionary algorithms are maintained between iterations. Furthermore, the learning process takes place through input/output pairs using as fitness function the error between the predicted values provided by the models and the real values in each action-perception pair.

Strongly related to this process is the management of the Short Term Memory. The quality of the learning process depends on the data stored in this memory and the way it changes. The data stored in this memory (samples of the real world) are acquired in real time as the system interacts with the environment and, obviously, it is not practical or even useful, to store all the samples acquired in agent's lifetime. We have developed a replacement strategy for this memory that permits storing the most relevant samples for the best possible modelling. Details can be found in [12].

The search for the best action in the MDB is not a learning process because we are looking for the best possible action for a given set of conditions. That is, we must obtain the action whose predicted consequences given by the world model and satisfaction model result in the best predicted satisfaction. Consequently, for the actions, a simple optimization problem must be solved for which any optimization technique is valid.

The operation of the basic mechanism we have presented has been tested in real agents [12], so the objective of this paper is not to show the operation of the mechanism in terms of evolution of the models, action optimization or STM managment, but to study how it operates in a conceptually higher level problem.

3   Induced Behavior in a Real Robot

As commented before, the operation of the MDB is based on an external process of interaction with the world where the action-perception pairs are obtained, and an internal process where models are updated and actions selected from the current models. Both of these processes occur concurrently on the different world and satisfaction models present in the model bases. Consequently, even though, the motivation structure we have established leads the robot to maximize its "real" satisfaction in terms of rewards obtained from the teacher, internally, several satisfaction models may arise through interaction with the world that relate different sensory patterns to expected satisfactions. An example of this is the real satisfaction (teacher rewards) that can be directly modeled in terms of the relationship between actions and sounds for a given communication language through a world model that provides error values and a satisfaction model that provides the expected satisfaction. But the real satisfaction can also be modeled (if the teacher is consistent in its commands and rewards) in terms of the perceived distance decrease to the objective when performing an action (world model) and the related predicted satisfaction (satisfaction model). Obviously, if the teacher was consistent, any of these satisfaction models lead the robot to performing the same behavior, although they are not really modeling the same. One of them models the satisfaction obtained from following the teacher´s commands and has nothing to do with the task the robot is performing. The other does not take into account the teacher´s commands, but generates high satisfaction values when the behavior corresponds to the one the teacher wanted (reaching the object). This is what we call cross-induced behavior. Thus, if we take into account that any satisfaction model is only applicable when its inputs are present, and that the real satisfaction is obtained through rewards of the teacher, the MDB will use as satisfaction model the first one of these two when there is input from the teacher and when there is no input it will resort to the second one, that is, the induced satisfaction model.

Typical induced behaviors are based on the generation of a behavior pattern that arises from the task that is directly taught. In this example, the task is very simple: we want that a real agent (a robot) to reach an object. In the learning stage it must follow the commands that a teacher provides in order to reach the object. This teacher rewards or punishes the robot depending on if the action applied agrees with the command or not. The induced behavior we try to obtain appears from the fact that each time the robot applies the correct action, the distance to the object decreases. This way, once the teacher disappears, the robot can continue with the task because it just has to perform actions that reduce the distance to the object.

To carry out this example, we have designed the following experiment:

– First of all, the real agent we are going to use is a wheeled robot called Pioneer 2. It has a sonar sensor array as its main sensorial system and all the computation is carried out in a laptop that is placed on top of the robot.
– We place the robot in the central point of a simple real environment and the object it must detect in an area around the robot. In Fig. 2 (left) we display the experimental setup including the real robot and object.
– A teacher, external to the MDB, must make the robot reach the object frontally through a set of commands that guide its movements. Possible commands have been reduced to a group of seven which represent the se-

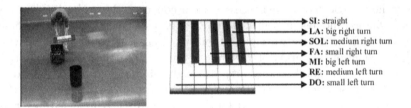

Fig. 2. Experimental setup (left) and translation of commands into musical notes (right)

OPERATION WITH TEACHER

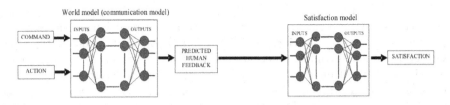

OPERATION WITHOUT TEACHER

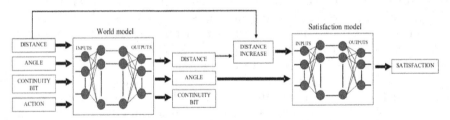

Fig. 3. World and satisfaction models for operation with and without teacher

mantics of the language used by the teacher to communicate with the robot. The sensor we use for communication purposes in the robot is a microphone, and the commands have been translated into musical notes. A possible translation is found in Fig. 2 (right) but it is not pre-established and we want the teacher to make use of any correspondence it wants.

- After perceiving a command, the robot can apply one of the following seven actions: 1-No turn, 2-Small right turn, 3-Medium right turn, 4-Large right turn, 5-Small left turn, 6-Medium left turn, 7-Large left turn. As we can see, the actions are in accordance with the possible commands.
- Depending on the degree of fulfilment of a command, the teacher must reward or punish the robot. To do this, we use a numerical value as a pain or pleasure signal that is introduced through a keyboard.

In figure 3 we display a schematic view of the two world model satisfaction model combinations that arise in this experiment. The first world model has two inputs (Fig. 3 top): command provided by the teacher and action applied; and one output: the predicted human feedback. As we can see, this model is related with robot-teacher communication, so we will call it *communication model* from this point forward. In this case, the satisfaction model is trivial because the satisfaction coincides with the output of the communication model, this is, the reward or punishment.

In the stage where the teacher is present, the communication model is the one used to select the action the robot must apply, while other world and satisfaction models are learnt in the background using the information provided by the action-perception pairs obtained by the robot through the interaction with the teacher and the environment. The second world model used has 4 inputs (Fig. 3 bottom): distance and angle to the object, continuity bit (necessary because of the symmetry of the sensed angles in this kind of circular robots) and the applied action; and 3 outputs: distance, angle to the object and continuity bit predicted after applying an action. The satisfaction model has two inputs: relative distance covered and angle after applying an action; and one output: the satisfaction value. The two operation modes are directly derived from the teaching process, because when the teacher is present the MDB can use direct sensorial information from the teacher (commands and rewards) while in the operation without teacher the MDB can use just the sensorial information obtained by the robot (distance and angle to the object). We can see this operation scheme as divided into two hierachical levels: when the teacher is present, the robot follows its commands and when he has gone it must use its own perceptions.

To evolve the models we have used a promoter based genetic algorithm [14] obtaining succesful results in the modelling of the action-perception pairs with multilayer perceptron artificial neural networks as individuals. The optimization of the action is very simple in this case as a reduced set of possible actions is considered, so we test all in the communication model (or in the world model depending on the operation mode) and select the one that provides the best satisfaction value.

To show the flexibility of this operation scheme, in Fig. 4 we have represented the evolution of the mean squared error provided by the best communication model during 2000 iterations. In the first 500 iterations the teacher provides commands using the encoding (language) shown in Fig. 2 (right). From iteration 500 to iteration 1000 the teacher stops providing commands and the robot uses the world and internal models. From iteration 1000 to iteration 1500 another teacher appears using a different language (different relation between musical notes and commands) and, finally, from iteration 1500 this second teacher dissapears too and the robot must use the world and satisfaction models again. There are no data in the operation without teacher because there are neither commands nor rewards and, consequently, there is no evolution of the communication model. As we can see in the figure, in the first 500 iterations the error decreases fast to below 10% which results in a very accurate prediction of the rewards. Consequently, the robot succesfully follows the commands of the teacher. What is really interesting in this graph, is that the behavior is the same from iteration

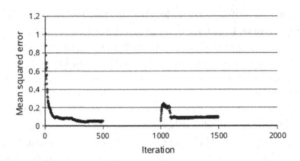

Fig. 4. Evolution of the mean squared error provided by the best communication model during 2000 iterations

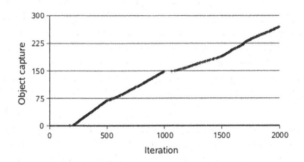

Fig. 5. Number of object captures through iterations

1000 to 1500, this is, when the second teacher appears. The iterations needed to follow his commands are more or less the same as in the previous case (related to the iterations needed to empty the action-perception pairs stored in the STM due to to the first teacher and fill it with the pairs of the second teacher) so the robot learns the new language and follows the commands quickly adapting itself to teacher characteristics. Finally, we want to remark that the learning of the world and internal models occurs in the same way with both teachers because the sensorial data used in these models (Fig. 3 bottom) are independent of the language selected by the teacher.

The main result from this experiment is shown on Fig. 5 that representes the number of object captures as the robot interacts with the world, taking into account that the teacher provides commands from until iteration 500 and from iteration 1000 to 1500. As shown in the figure, from this iteration on, the robot continues capturing the object in the same way the teacher had taught it, so we can say that the induced learning of the models has been successful. The decrease of the slope in the figure implies that the robot takes a large number of iterations (movements) to reach the object using the induced models. This is because it has learnt to decrease its distance to the object and not the fastest way to do it (these models aren't based on obedience but on distance increments). According

Fig. 6. Actions applied by the Pioneer 2 robot in the operation with (left) and without teacher (right)

to Fig. 4, when the second teacher appears there is a brief gap with no captures. In the first 200 iterations there are no captures because we have forced a random exploration stage to store relevant information in the STM. This was necessary due to the complexity of the world model motivated by the way the sonar sensor ring of the Pioneer 2 operates.

Fig. 6 displays a real execution of actions. In the left part, the robot is following commands by a teacher; in the right it is performing the behavior without any commands, just using its induced models. It can be clearly seen that the behaviour is basically the same although a little more inefficient.

Summarizing, this mechanism provides a way to teach a robot to do something where the world models and satisfaction models are separate. Thus, the satisfaction models determine the final behavior ("what") in terms of motivation, the world models represent the circumstances of the environment and the action optimization strategy will obtain the "how" from this combination. Thus, any teacher can teach any machine any behavior it can carry out through a consistent presentation of sensorial inputs and rewards/punishments and the system will obtain a model of the world and satisfaction with regards to the teacher commands and expected actions as well as any other induced world and satisfaction models that relate other inputs with the expected satisfaction.

4   Conclusions

In this paper we have presented an approach, that has been implemented through the Multilevel Darwinist Brain cognitive mechanism, for performing teaching-learning processes between humans and robots in a natural manner. The MDB allows for the induction of satisfaction models from reinforcement or carrot and stick teaching processes which can be used by the system to operate when no teacher is present. In addition, the separation of the different types of models in the architecture provides for a straightforward implementation of the trainable communication models required for the system to be able to learn to understand what the teacher means. An example with a real agent has been presented.

Acknowledgements

This work was funded by the MCYT of Spain through project VEM2003-20088-C04-01 and Xunta de Galicia through project PGIDIT03TIC16601PR.

References

1. Weng J., Zhang, Y.: Developmental Robots - A New Paradigm, Proceedings Second International Workshop on Epigenetic Robotics: Modeling Cognitive Development in Robotic Systems 94, pp. 163-174 (2003).
2. Bakker P., Kuniyoshi, Y.: Robot see, robot do: An overview of robot imitation, Autonomous Systems Section, Electrotechnical Laboratory, Tsukuba Science City, Japan (1996).
3. Voylesl, R. Khosla, P.: A multi-agent system for programming robotic agents by human demonstration. In Proc., AI and Manufacturing Research Planning Workshop (1998).
4. Lauria, S., Bugmann, G., Kyriacou, T., Klein, E.:Mobile robot programming using natural language, Robotics and Autonomous Systems, 38:171–181, (2002).
5. Nicolescu, M., Mataric, M.J.: Natural Methods for Robot Task Learning: Instructive Demonstration, Generalization and Practice, Proceedings, Second International Joint Conference on Autonomous Agents and Multi-Agent Systems, pp 241-248 (2003).
6. Schaal, S.: Learning from demonstration, Advances in Neural Information Processing Systems 9, pp.1040-1046, (1997).
7. Ullerstam, M.: Teaching Robots Behavior Patterns by Using Reinforcement Learning– How to Raise Pet Robots with a Remote Control, Master's Thesis in Computer Science at the School of Computer Science and Engineering, Royal Institute of Technology (2004).
8. Asada, M., MacDorman, K.F., Ishiguro H., Kuniyoshi, Y.: Cognitive Developmental Robotics As a New Paradigm for the Design of Humanoid Robots. Robotics and Autonomous System, Vol.37, pp.185-193 (2001).
9. Changeux, J., Courrege, P., Danchin, A.: A Theory of the Epigenesis of Neural Networks by Selective Stabilization of Synapses, Proc.Nat. Acad. Sci. USA 70, pp 2974-2978 (1973).
10. Conrad, M.:Evolutionary Learning Circuits. Theor. Biol. 46, pp 167-188 (1974).
11. Edelman, G.: Neural Darwinism. The Theory of Neuronal Group Selection. Basic Books (1987).
12. Bellas, F., Duro, R. J.: Multilevel Darwinist Brain in Robots: Initial Implementation, ICINCO2004 Proceedings Book (vol. 2), pp 25-32 (2004).
13. Bellas, F., Duro, R. J.: Introducing long term memory in an ann based multilevel darwinist brain, Computational methods in neural modeling, Springer-Verlag, pp 590-598 (2003).
14. Bellas, F., Duro, R.J.: Statistically neutral promoter based GA for evolution with dynamic fitness functions, Proceedings of the 2nd iasted international conference, pp 335-340 (2002).

Landscaping Model for Virtual Environment

Madjid Fathi and Ursula Wellen

University of Siegen, 57076 Siegen, Germany,
Faculty of Electrical Engineering and Computer Science,
Section of Knowledge Based Systems & Knowledge Management,
fathi@informatik.uni-siegen.de

1   Introduction

In this paper we discuss the development and management of Virtual Environment [1] – a virtual Environment for autonomous robotic agents – with Process Landscaping [9], a method to co-ordinate complex and (geographically) distributed processes like software development processes or business processes. It enables us to model and analyze distributed processes in a well-structured way at different levels of abstraction and with special focus on interfaces.

In [1] the authors present Virtual Environment to simulate autonomous robotic agents. It can incorporate various models of an environment for simulation as well as the agents under consideration, which can be trained by simulation. In many circumstances, several physical models and simulations need to be interconnected and have to take place simultaneously for a larger, more complex simulation to yield results. This requirement leads to the notion of distributed modeling and simulation. The architecture of Virtual Environment therefore serves as a framework of interaction and modeling, including discrete-event-system specifications (DEVS).

There are some desirable features of the proposed Virtual Environment simulation environment described in [1]. We are convinced, that Process Landscaping can support us to get these features and to improve therefore the existing release of Virtual Environment.

At a waste of more than $80 Billion per year for the development of software systems in the United States, it is more then necessary to create a high level model structured such as software process landscapes designed with Process Landscaping. This multi level model will be able to cover multiple agents' tasks and behavior in order to derive a suitable architecture with increased mission robustness and learning ability. Learning is one of the important aspects of multiagent robotics where robustness and performance is demanded in the face of environmental uncertainty.

This paper is structured as follows: We first introduce the Virtual Environment framework, its purpose and some of the already mentioned desirable features and improvement ideas for its next release (section 2). Afterwards we give an overview of Process Landscaping (section 3) and discuss how this method is able to support Virtual Environment's purposes (section 4). The paper finishes by summarizing the main results and by discussing some conclusions.

J. Mira and J.R. Álvarez (Eds.): IWINAC 2005, LNCS 3562, pp. 435–447, 2005.

2   Virtual Environment

The Virtual Environment architecture relies heavily on the structure defined by the DEVS (Discrete Event System) environment. It defines six different categories of DEVS models the simulation will be composed from:

- SimEnv (Simulation Environment) and SimMan (Simulation Management)
- control models
- agent models
- physics models
- terrain models and
- dynamic models.

Each of these models is referred to as a high-level model and is constructed from atomic and coupled models that are not high-level models.

At the heart of Virtual Environment, and the first category of the high-level models, is the SimEnv coupled model and the SimMan atomic model. SimEnv is the highest level coupled model in the simulation and is responsible for creating the instances of all of other models. It acts as the housing for all high-level models in the simulation.

SimMan is an atomic model where all other models in SimEnv are connected. It is responsible for coordinating messages between other high-level models, controlling the flow of time in the simulation and tracking information about the state of the agents in the simulation. All of the high-level models have at least one input and output port tied to SimMan. The only interconnection of other high-level models with each other occurs between an agent model and its control model. SimEnv and SimMan are objects that will be common to every simulation using the Virtual Environment architecture.

The second type of high-level models is the control model. Those models store the behavior algorithms that will be used to control the agent, physics, and terrain models. Agent control models do not directly instruct the agent's behaviors. Instead, they indirectly determine what an agent will do by controlling the agent's actuators. This will later cause a dynamic model to actually change the state of an agent.

The third type of high-level models is the agent model. Each of these models contains sensor models and actuator models. The sensor models contain the information about the environment that the agent is aware of. The actuators contain the information that dynamic models use to modify the agent and the environment.

The fourth type of high-level models is the physics model. These models are used to model the real world physical phenomena represented in the simulation. How the agent interacts with the environment and how agent sensors set their state is encapsulated within the physics models. Frequently in simulations, differential equations are used to explain physical phenomena. It is not the goal of Virtual Environment to act as a differential equation solver, however, a differential equation solver could be encapsulated within a DEVS model and be included in Virtual Environment as a physics model.

The fifth type of high-level models is the terrain model. The terrain models contain information about the layout of the environment in the simulation. They are the maps used by SimMan to determine what type of terrain the agents occupy.

The sixth and last type of high-level models is the dynamic model. These models are responsible for making changes to the agent models and the information about the agents that SimMan tracks. They make these changes based on the current state of the environment and each agent's actuators.

For all the high-level models, SimMan will act as a message liaison between the models. Each of the models is completely unaware of the fact that there are other high-level models aside from SimMan. For instance, the sensors on an agent need to be updated with information external to the agent. SimMan directs a request from the sensors to an appropriate physics model asking for the correct state for the sensors. This information is then sent from the physics model back to SimMan that will redirect the information back to the sensors. Neither the physics model nor the agent model knows of the existence of the other.

SimMan also controls the timing and sequence of events that occur in the simulation. This sequence is broken down into five phases.

In phase 1 SimMan checks to see if the termination conditions for the simulation are satisfied. If they are, the simulation halts. If not, SimMan proceeds to phase 2. In this phase, SimMan sends messages to each of the agent's sensors instructing them to update their state. The sensors will query SimMan for information from physics models and update their state. After all the sensors and physics models are done, SimMan will proceed to phase 3. In this phase a similar parlay of messages will occur between the control objects and their agent's activators. This phase will end after the control objects have updated the state of the actuators in the agents they control. During phases 4 and 5, the state of the agent's actuators will be analyzed by the dynamic models, which might in turn modify the state of the environment or the agents themselves. SimMan will repeat the Virtual Environment cycle until the termination conditions for the simulation have occurred.

3   Process Landscaping

Process Landscaping is a structured method to model and analyze complex and possibly distributed processes with special focus on interfaces between and within these processes. The latter may be geographically distributed business processes of any kind, but may also describe the activities of autonomous robotic agents interconnected with each other via geographically distributed different machines.

The method of Process Landscaping can be characterized as an integrated approach where both activities and their interfaces are treated as first class entities. The modeling result is a set of process models, called *process landscape*, depicting all core activities under consideration on different levels of abstraction. To attain this process landscape, the following steps have to be undertaken:

1. Identification of core activities and their interfaces
2. Refinement of activities to process models
3. Refinement of interfaces
4. Refinement of process models
5. Validation of the resulting process landscape

The differentiation between the terms *activity* and *process model* indicates an important feature of Process Landscaping: At the coarse-grained levels of a process landscape temporal and logical dependencies are not yet modeled. They only show *activities* and their input and output data but not the order of sequences in which the activities are carried out. If we specify the process landscape in more detail by adding this procedural information, we talk about *process models.* This modeling approach allowing incomplete and partial specification of process models conforms to Fugetta's requirement for process modeling languages (PML) [2].

The identification of core activities and their interfaces (step 1) results in a top-level process landscape where the existence of an interface between activities is indicated by bi-directional arrows. Such an interface has to be modeled every time when there is at least one information exchange between the (interconnected) activities. Both core activities and interfaces can be identified by requesting the corresponding process owners to all tasks and types of information to be exchanged.

Refinement of activities to process models (step 2) means to identify and model all (sub)activities belonging to an activity together with their temporal and logical dependencies. These (sub)activities have to be arranged within the process landscape in the order of sequences they have to be carried out. To keep a comprehensible overview of the process landscape under consideration, the hierarchical relations between all activities can be arranged additionally as a tree where the leaves represent refinements of superordinated activities.

Refinement of interfaces (step 3) means to identify all types of data to be exchanged via this interface together with the corresponding direction of data flow.

Finally, refinement of process models (step 4) means to describe activities already arranged in a process landscape on a more detailed level. For this refinement step it is important to mention that the resulting more detailed level always has to consider logical dependencies between the subactivities. Additionally, these subactivities have to be added to the tree mentioned above, representing now the leaves of the refined process landscape.

Steps 2 to 4 of the modeling steps can be executed in a flexible manner. The degree of refinement is also flexible and can vary individually for each activity within a process landscape. The resulting landscape has then to be validated (step 5) by walkthroughs with process owners where he/she decides whether both activities and interfaces are properly modeled.

Fig. 1 represents an abstract process model as a result of the Process Landscaping modeling approach. Rectangles filled with letters A to F represent activities at different levels of abstraction, where the different levels are indicated

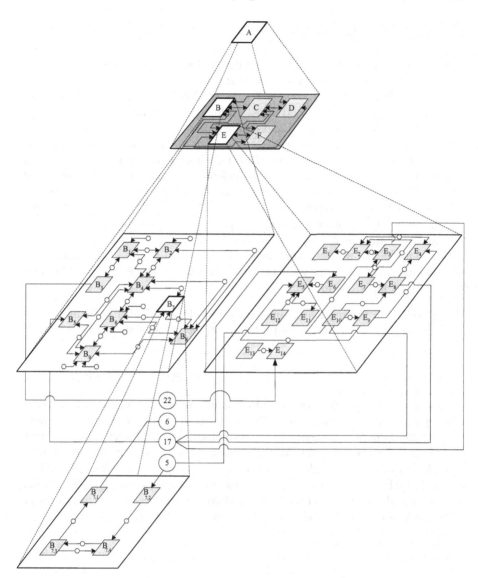

Fig. 1. Overall view of an abstract process landscape

by further rectangles surrounding sets of activities (see e.g. B_1 to B_9 and E_1 to E_{14} in Fig. 1) and connected to a superordinated activities by dotted lines. Levels, where activities are connected by bi-directional arrows do not yet consider procedural information. Every activity at those coarse-grained levels can be refined to process models (see e.g. activities $B_{7,2}$ to $B_{7,4}$ in Fig. 1), where refined interfaces are modeled as small circles, connecting activities by directed arrows. Circles 22, 6, 17 and the related directed arrows in Fig. 1 represent e.g. the refined interface between activities B and E.

The process landscape shown in Fig. 1 has been developed with focus on logical dependencies between activities. To focus on the distribution aspects of the modeled activities, Process Landscaping offers a restructuring algorithm to rearrange the *logical view* into a *locational view* of the process landscape by ordering all activities according to their location. This means that activities taking place at multiple locations are represented as multiples in the locational view. Communication features between different locations and within specific locations can then be better analyzed. To make clear whether we talk about communication between or within different locations, we talk about external and internal communication, correspondingly.

In [8] and [9] the analysis approach of Process Landscaping concerning specific communication features within a distributed landscape is discussed in more detail. In this paper, we focus on the

- application of the modeling approach to develop the logical view and
- restructuring algorithm to obtain the locational view of autonomous robotic agents.

These (main) features of Process Landscaping enable us to support Virtual Environment's purposes.

The following example shows how to restructure the logical view of a process landscape into its locational view. Fig. 2 illustrates part of component-based software development processes. It shows activities of superordinated activity *component engineering* posting each component to the quality management after its implementation (rectangle A) and release (rectangle B) in order to validate (rectangle QM) the software. Activity *error correction* receives an error report afterwards and decides whether the tested component has to be improved or whether it can be sent to activity *adding component to system*.

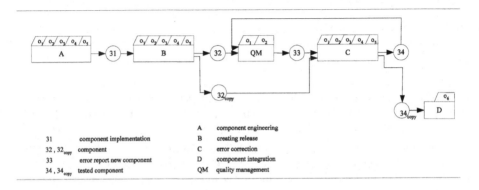

Fig. 2. Example process landscape in its logical view

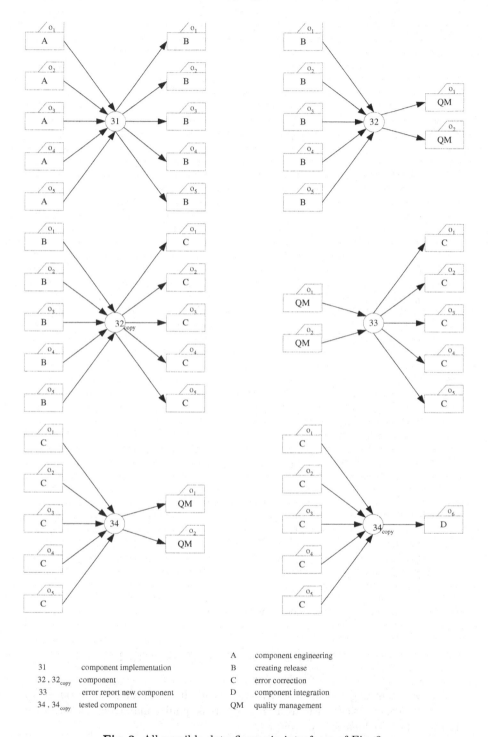

Fig. 3. All possible data flows via interfaces of Fig. 2

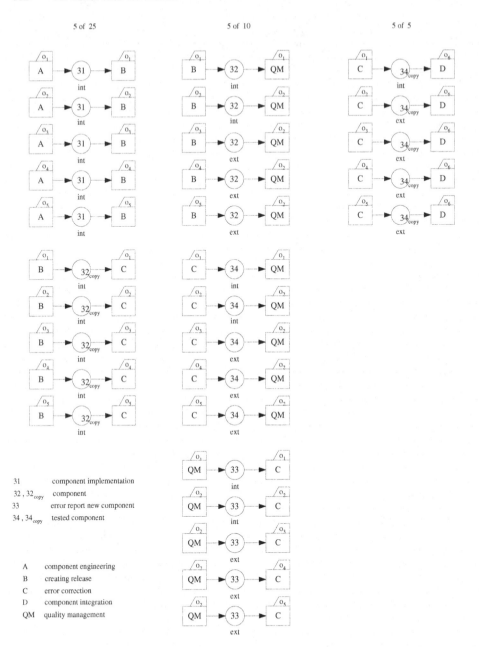

Fig. 4. Existing data flows via interfaces of Fig. 2

All component-engineering activities take place at five different locations o_1 to o_5 (see tabs at activities in Figure Fig. 2, whereas quality management activities only take place at locations o_1 and o_2.

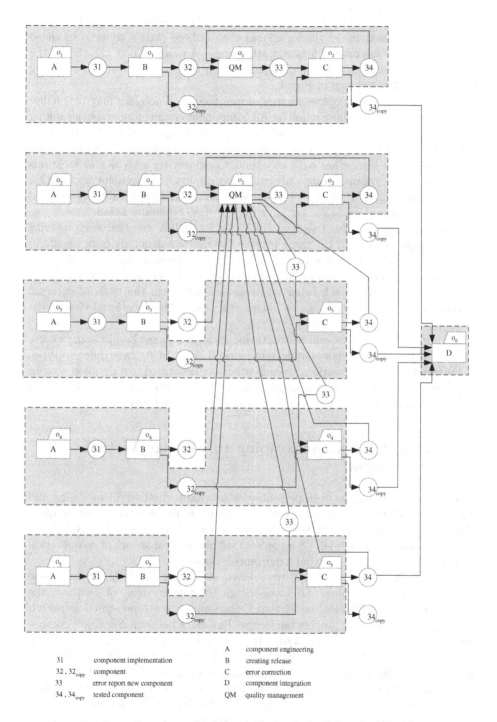

Fig. 5. Locational view of example process landscape of Fig. 2

Fig. 3 shows all possible data flows by relating each interface of the logical view with each activity of the locational view (where each activity is modeled as often as there are related location attributes). A modeler now can choose the subset of truly existing data flows depending on the modeled context, resulting in those data flows shown in Fig. 4.

Fig. 4 makes clear why the number of interfaces and accesses may be smaller than the number of all those possible: All component engineering activities taking place at location o_1 communicate only with quality management activities at the same location. This is the same for component engineering activities at location o_2, because they – together with component engineering activities at locations o_3, o_4 and o_5 – communicate only with the quality management at location o_2. Abbreviation *int* related to interfaces 31 to 34 in Fig. 4 indicates internal communication, abbreviation *ext* indicates external communication.

Connecting all process fragments of Fig. 4 in such a way that each activity with the same name and the same location attribute exists only once results in the process landscape shown in Fig. 5. External interfaces connecting different locations are arranged outside the gray-shaded locations o_1 to o_6. They connect the six locations with each other and consequently form the basis of external communication, which now can be better analyzed in this locational view of the process landscape.

To apply the modeling and restructuring steps of Process Landscaping we use different types of Petri nets as underlying formal basis. In [9] these different types of Petri nets and their relations to each other are described in a detailed way. In this paper we do not discuss the formal basis in detail but focus on the methodical concept of using Process Landscaping to manage Virtual Environment.

4   Using Process Landscaping to Manage Virtual Environment

Process Landscaping has been presented as a suitable method for modeling and analyzing distributed processes with special focus on interfaces and communication aspects on different levels of abstraction. Petri Nets as the underlying notation enable us to use DEVS as general simulation framework of co-operation and communication like Virtual Environmen does.

Fig. 6 shows how to design Virtual Environment as a distributed landscape with Process Landscaping. All components (atomic models) of *SimEnv*, the highest-level coupled model of Virtual Environment, are represented as activities at the top level of a process landscape. Relations between *SimEnv*'s components are indicated initially as bi-directional arrows. *SimMan* is the only activity connected via interfaces to all other activities. This fact already indicates the importance of SimMan: It is responsible for coordination of messages between other high-level models, controlling the flow of time, and tracking information about the state of the agents under consideration during simulation.

The advantage of modeling Virtual Environment as a process landscape is the use of only one model instead of three as presented in [1], figures 10, 11, 12.

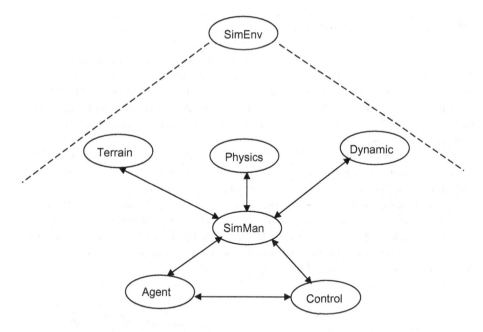

Fig. 6. Components of coupled model "SimEnv" - logical view

The relations between *SimEnv* and its atomic models are indicated at a course-grained level of a process landscape; more detailed information is shown at a more detailed level as shown in Fig. 7. There we consider the interfaces between *SimMan*, *Control* and the components of an agent – *Sensor* and *Actuator*. Petri nets ensure the consistency between the different levels of process landscape *SimEnv* as underlying formal basis, where refinements of sets and places are common and well defined [7], [9].

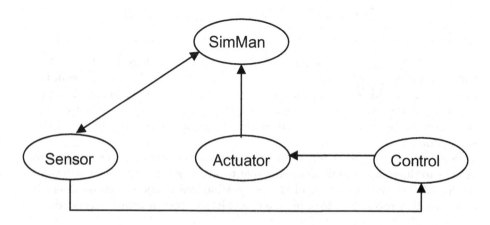

Fig. 7. Refinement of agent and its interfaces to SimMan and Control

Because Petri nets are used as discrete-event system specifications, process landscape *SimEnv* can be simulated to analyze control flow and data flow. In other words, we are able to analyze the initialization of *SimEnv*'s atomic models together with all coupling information and all simulation cycles including different states of *SimEnv*'s components until the termination conditions are made and the simulation ends.

One important feature of Virtual Environment we did not consider yet is the distribution of different robotic agents to a set of machines geographically distributed and connected in a hardware network. Whereas co-ordination tasks within a SimEnv-model can be analyzed in the logical view, amount of messages between SimEnv-models can be analyzed in the locational view of the process landscape under consideration (see Fig. 6). With the latter view we can identify models with excessive communication and arrange them on the same machine to optimize communication efficiency.

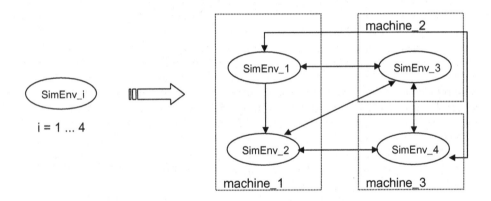

Fig. 8. From logical to locational view of SimEnv

Fig. 8 shows one distribution of different *SimEnv_i*, where i = 1, ..., 4, to three machines related to each other via a hardware network. Simulation of this distribution can identify optimization possibilities concerning interaction time and cost of the distributed SimEnv-models. The next step is then the rearrangement of the models among the different hardware locations and a further simulation run.

Summarizing the procedure to find the optimal distribution of several SimEnv-models to different machines, we are able to identify the best solution by simulation of the locational view of the process landscape under consideration. In other words, Process Landscaping offers a suitable way for a task, declared as a very difficult [1].

5   Conclusions

In this paper we have discussed the application of Process Landscaping to improve the existing release of Virtual Environment by considering some desirable features not yet implemented with the available release. We have shown e.g. how Process Landscaping can support us to find the optimal distribution of coupled models *SimEnv* among different computers. More details of the Process Landscaping Modeling approach can be found in [3] and [4]. The restructuring algorithm is further described along a software development example in [5].

Our future work will focus on the further development of the process landscape discussed above and on the implementation of an improved Virtual Environment.

References

1. Aly I. El-Osery, John Burge, Antony Saba, Mo Jamshidi, Madjid Fathi, Mohammed-R. Akbarzadeh-T., *Virtual Environment – A Virtual Laboratory for Autonomous Agents – SLA-Based Learning Controllers*, IEEE Transactions on Systems, Man and Cybernetics-Part B: Cybernetics, Vol. 32, No. 6, december 2002
2. A. Fugetta, Software Process: A Roadmap, in A. Finkelstein (ed.) *The future of Software Engineering, 22nd International Conference on Software Engineering*, pp 25-33, ICSE, Toronto, Kanada, 2000, ACM Press
3. V. Gruhn and U. Wellen, *Process Landscaping – eine Methode zur Geschäftsprozessmodellierung*, Wirtschaftsinformatik, 4:297-309, august 2000
4. V. Gruhn and U. Wellen, Structuring Complex Software Processes by "Process Landscaping", in R. Conradi (ed.) *Software Process Technology – Proceedings of the 7th European Workshop on Software Process Technology EWSPT*, pp. 138-149, Kaprun, Austria, february 2000, Springer Verlag, published as Lecture Notes in Computer Science LNCS 1780
5. U. Wellen, *Process Landscaping – Modeling and Analyzing Communication of Distributed Software Processes*, to be published in 2004
6. M. Fathi, *Distributed Robotic Agents*, NASA Autonomous Control Engineering (ACE) Center, Albuquerque, NM 87111, 2003
7. E. Smith. Principles of High-Level Net Theory, in W. Reisig and G. Rozenberg (eds.) Lectures on Petri Nets I: Basic Models – Advances in Petri Nets, pp. 174-210, Springer Verlag, 1998, published as Lecture Notes in Computer Science LNCS 1491
8. M. Stöerzel and U. Wellen, Modelling and Simulation of Communication between Distributed Processes, in M. Al-Akaidi *Proceedings of the Fourth Middle East Symposium on Simulation and Modelling MESM'2002*, pp. 156-169, Sharjah, U.A.E., september 2002, SCS European Publishing House, ISBN: 90-77039-09-0
9. Ursula Wellen, *Process Landscaping – Eine Methode zur Modellierung und Analyse verteilter Softwareprozesse* (PhD thesis), Shaker Verlag, 2003, ISBN 3-8322-1777-2

Sensitivity from Short-Term Memory vs. Stability from Long-Term Memory in Visual Attention Method

María T. López[1], Antonio Fernández-Caballero[1],
Miguel A. Fernández[1], and Ana E. Delgado[2]

[1] Universidad de Castilla-La Mancha,
Escuela Politécnica Superior, 02071 - Albacete, Spain
{mlopez, caballer, miki}@info-ab.uclm.es
[2] Universidad Nacional de Educación a Distancia,
E.T.S.I. Informática, 28040 - Madrid, Spain
adelgado@dia.uned.es

Abstract. In this paper a special focus on the relationship between sensitivity and stability in a dynamic selective visual attention method is described. In this proposal sensitivity is associated to short-term memory and stability to long-term memory, respectively. In first place, all necessary mechanisms to provide sensitivity to the system are included in order to succeed in keeping the attention in our short-term memory. Frame to frame attention is captured on elements constructed from image pixels that fulfill the requirements established by the user and gotten after feature integration. Then, stability is provided by including mechanisms to reinforce attention, in such a way that elements that accept the user's predefined requirements are strengthened up to be configured as the system attention centre stored in our long-term memory.

1 Introduction

The name dynamic selective visual attention (DSVA) embraces a set of image processing mechanisms for focusing vision on those regions of the image where there are relevant local space-time events. These DSVA mechanisms help find, using an active search process, the relevant information at each moment to perform the interaction task with the system [1], [2]. In this paper a special focus on the behavior of sensitivity and stability in our visual attention method is pursued. Sensitivity and stability are terms widely expressed in dynamic systems [3]. In systems associated to image sequences sensitivity and stability have also been explored due to their importance [4], [5]. Our intention is to introduce these concepts in dynamic visual attention, associating sensitivity to short-term memory and stability to long-term memory, respectively. Fig. 1 shows the block diagram that illustrates the two components of sensitivity and stability of the DSVA task as studied in this paper.

As also shown in Fig. 1, our solution to DSVA defines a model with two types of processes: bottom-up processes (based on the scene), which enable to

J. Mira and J.R. Álvarez (Eds.): IWINAC 2005, LNCS 3562, pp. 448–458, 2005.

extract features from the input image and allow to create the elements of interest; and top-down processes (based on the object) by means of which the features are integrated. The selection of the interest elements of the scene starts with setting some criteria based on the features extracted from the elements (Feature Extraction). This way, in first place, all necessary mechanisms to provide sensitivity to the system are included in order to succeed in capturing the attention. Frame to frame attention is derived (Attention Building) to elements constructed from image pixels that fulfill the requirements established by the user and gotten after a Feature Integration. On the other hand, stability has to be provided to the system. This has been achieved by including mechanisms to reinforce attention (Attention Reinforcement), in such a way that elements that accept the user's predefined requirements are strengthened up to be configured as the system attention centre. Thus, the relationship between sensitivity gotten in the Short-Term Memory (Attention Building) and stability obtained in the Long-Term Memory (Attention Reinforcement) is developed in our proposal.

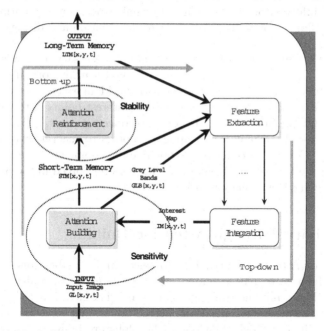

Fig. 1. DSVA block diagram with special emphasis on Attention Building and Attention Reinforcement

In previous works of our research team some methods based on image segmentation from motion have already been used. These methods are the permanency effect and the lateral interaction [6]. Based on the satisfactory results of these algorithms [7], [8], in this paper we propose to use mechanisms of charge and discharge together with mechanisms of lateral interaction to solve the fundamental aspects of sensitivity and stability in the DSVA task.

2   Short-Term and Long-Term Memory in DSVA Method

Short-term memory and long-term memory are expressions taken from cognitive psychology. Short-term memory (STM) -also called working memory or functional memory- is the cognitive system that allows keeping active a limited amount of information for a brief period of time [9], [10], [11], [12]. It was thought to have two functions: storing material that we have to recall in a few seconds and providing a gateway to long-term memory (LTM) [13]. LTM contrasts with STM in that information can be stored for extended periods of time. In standard theories of memory [13], [14], information can be stored in LTM only after it has been stored in STM, and even then, storage in LTM is a probabilistic event. Originally, it was proposed that the probability of storage in LTM is a function of the time an item was maintained in STM. More recently, Anderson [15] suggested that the probability of storage is a function of the number of times an item enters STM. LTM has a strong influence on perception through top-down processing. This is the process by which our prior knowledge affects how we perceive sensory information. LTM influences what aspects of a situation we pay attention to -allowing us to focus on relevant information and disregard what is not important [16].

In the DSVA method proposed in this paper all necessary mechanisms necessary to obtain a Short-Term Memory and a Long-Term Memory are explained. The mechanisms used to generate the Short-Term Memory endow the system of sensitivity, as it includes elements associated to interest points in the memory at each frame. But, the Short-Term Memory introduced is noisy, as blobs that are not of a real interest to the user may appear. In order to generate the Long-Term Memory, that is to say, in order to provide stability, some cues are included for inserting into the Long-Term Memory all elements reinforced in the Short-Term Memory through a persistency measure.

3   Sensitivity Through Attention Building

The purpose of Attention Building is to select and to label zones (blobs) of the objects (figures) to pay attention on. See, therefore, that after processing Attention Building, not the complete figures are classified, but each one of the blobs, understood as homogeneous connected zones that form the figures, are marked with different labels. Obviously, the blobs are built from image points that fulfill the requisites established by the guidelines of the observer (points of interest). Fig. 2 shows a process scheme for Attention Building. The output of Attention Building is precisely called Short-Term Memory. In our case, only blobs constructed in the Short-Term Memory will potentially form the figures of the system's Long-Term Memory.

In our proposal the blobs of the Short-Term Memory are built from the information provided through the so called Interest Map and from the input image divided into Grey Level Bands. The Interest Map is obtained by performing

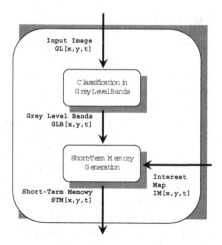

Fig. 2. Attention Building process scheme

feature integration, both of motion and shape features. For each image pixel, in the Interest Map the result of a comparison among three classes - "activator", "inhibitor" and "neutral"- is stored. The interest points are those points of the Interest Map labeled as "activator" points.

3.1 Classification in Grey Level Bands

Classification in Grey Level Bands transforms the 256 grey level input images into images with a minor number of levels. These new images are called images segmented into Grey Level Bands (GLB). The reason to working with grey level bands is twofold. (1) Some traditional methods of motion detection are based on image differencing. The noise level diminishes for little changes in grey level (or luminosity) of a same object between two consecutive images, when joining a range of grey levels into a single band. (2) On the other hand, a decrease of the computational complexity is achieved, bearing in mind the great parallelism used in the algorithms of the proposed model. We now calculate in parallel in the order of magnitude of grey level bands n, and not of grey levels N, where $N > n$.

The calculus of the grey level band of pixel $[x, y]$ at t, $GLB[x, , y, t]$, is expressed in Equation 1. As you may notice, this is just an easy scale transformation.

$$GLB[x, y, t] = \frac{GL[x, y, t] \cdot n}{GL_{max} - GL_{min} + 1} + 1 \tag{1}$$

where n is the number of grey level bands in which the image is split, GL_{max} is the maximum and GL_{min} are the minimum grey levels, respectively, of the input image.

3.2   Short-Term Memory Generation

The objective of Short-Term Memory Generation is firstly to select and to label (to classify numerically) image blobs associated to pixels of interest -pixels that possess dynamic features in predefined numerical intervals. Secondly, it eliminates the blobs whose shape features do not correspond with the pre-established ones. In order to achieve these aims, the images in Grey Level Bands are segmented into regions composed of connected points whose luminosity level belongs to a same interval (or grey level band) and to select only connected regions that include some "activator" point (or, point of interest) in the Interest Map. Each region or zone of uniform grey level is a blob of potential interest in the scene.

The idea consists in overlapping, as with two superposed transparencies, the Grey Level Bands image of the current frame (t) with the Interest Map image built at the previous frame ($t-1$). At t, only blobs of the Grey Level Bands image are selected where at least one point of interest fell at $t-1$ in the Interest Map. Nevertheless, not the total blob is taken; pixels that coincide with points of the Interest Map classified as "inhibitors" are eliminated. The computational model used to perform the preceding steps incorporates the notion of lateral interaction, which enables that the points of interest flood their zones of uniform grey levels whilst eliminating all points classified as "inhibitors". In order to achieve the aims of Short-Term Memory Generation, the processes shown in Fig. 3 are performed.

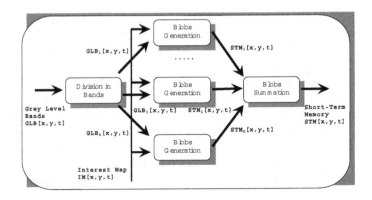

Fig. 3. Short-Term Memory Generation process scheme

Division in Bands. Division in Bands obtains from an image in grey level bands, $GLB[x,y,t]$, n binary images $GLB_i[x,y,t]$ (one image for each band). Each one of these images, $GLB_i[x,y,t]$, stores a value of 1 for a pixel whose grey level band is i and a 0 in the opposite case. That is to say (Equation 2):

$$GLB_i[x,y,t] = \begin{cases} 1, & \text{if } GLB[x,y,t] = i \\ 0, & \text{otherwise} \end{cases} \tag{2}$$

Blobs Generation. For each $GLB_i[x, y, t]$, the different connected regions that include an "activator" point in the Interest Map and that do not correspond to "inhibitor" points in the Interest Map are labeled. Thus, Blobs Generation gets and labels for each grey level band, pixels belonging to connected regions that include any "activator" point in the Interest Map but do not correspond with "inhibitor" points of the Interest Map. Its output, Short-Term Memory for Grey Level Band i, $STM_i[x, y, t]$, stores for each pixel the label corresponding to the generated blob if it belongs to the blob, or the value 0. Let us define $v_{activator}$ as the value given to the points of interest ("activators") of the Interest Map, $v_{neutral}$ as the value for the "neutral" points of the Interest Map, and $v_{inhibitor}$ as the value for the "inhibitor" points of the Interest Map. Let also NR be the number of rows of the image, and NC the number of columns of the image. Firstly, all points where $GLB_i[x, y, t] = 1$ are assigned an initial and provisional label value as shown in Equation 3:

$$STM_i[x, y, t] = \begin{cases} x * NC + y + 1, & \text{if } GLB_i[x, y, t] = i \wedge IM[x, y, t] = v_{activator} \\ NR * NC + 1, & \text{if } GLB_i[x, y, t] = i \wedge IM[x, y, t] = v_{neutral} \\ 0, & \text{otherwise} \end{cases}$$

(3)

where $IM[x, y, t]$ is the value of the Interest Map at pixel $[x, y]$. This value corresponds to $(x * NC + y + 1)$ when $IM[x, y, t] = v_{activator}$, to a greater value than $NR * NC$ when $IM[x, y, t] = v_{neutral}$ and to value 0 in the rest of the cases. In other words, if pixel $[x, y]$ belongs to the grey level band and corresponds to a point of interest of the Interest Map, the tentative value for it is a function of its proper coordinate. Now, if the pixel belongs to the grey level band but corresponds to a "neutral" point of the Interest Map, the provisional value given to it is a value greater than any possible value of the coordinate function. In any other case, the value is 0. This initial value assignment to all pixels serves to get an agreement in the labels of the blobs after a negotiation (consensus) period. The label value for each pixel $[x, y]$ is iteratively calculated as the minimum value of the proper value of the pixel and the value of its 8 surrounding pixels. Of course, there will only be collaboration among neighboring pixels that possess an initial value greater than 0. The iterative calculus up to obtaining a common value for all pixels of a same blob is shown in Equation 4.

$$STM_i[x, y, t] = min(STM_i[\alpha, \beta, t]), \forall [\alpha, \beta] \in [x \pm 1, y \pm 1]$$

(4)

whenever $0 \leq STM_i[\alpha, \beta, t] \leq NR * NC + 1$.

Thus, blobs are labelled with the ordinal corresponding to the point with the lowest coordinate (if taking as origin the superior left image pixel).

Blobs Summation. Lastly, Blobs Summation gets the Short-Term Memory, $STM[x, y, t]$, as the result of summing up all blobs computed at each of the n Short-Term Memories for Grey Level Band i, $STM_i[x, y, t]$, where $i = 1, 2, ..., n$. The final value for each pixel $[x, y]$ is the maximum value of the n $STM_i[x, y, t]$.

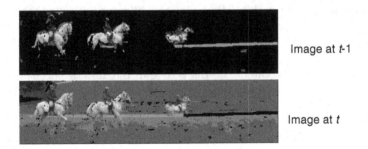

Image at *t*-1

Image at *t*

Fig. 4. Short-Term Memory for a couple of images

Only values that possess a label value less than $NC \cdot NR + 1$ are considered, as shown in Equation 5.

$$STM[x,y,t] = max_i STM_i[x,y,t]), \forall i \in [1..n] | STM_i[x,y,t] < NR*NC+1 \quad (5)$$

Notice that this maximum selection operation has to be performed for all elements of matrixes $STM_i[x,y,t]$ to obtain the corresponding element in a single matrix of blobs, $STM[x,y,t]$. This way all blobs of all grey level bands have been united and labeled with a common value. Fig. 4 shows the contents of the Short-Term Memory after processing all steps of Attention Building.

See also the sensitivity of the task through the contents of the Short-Term Memory in two consecutive frames as shown Fig. 4 taken from the picture "The Living Daylights". The input sequence has been captured by a camera in constant translational movement following the motion of the horse riders. Elements of the Short-Term Memory are composed of connected pixels that are not drawn in black color. The attention focus pursued in this case is the set of horses and horsemen. As you may notice, all interest elements are really detected. Nevertheless, other elements appear that neither are of the user's interest. The example shows the necessity for stability.

4   Stability Through Attention Reinforcement

The mechanisms used to generate the Short-Term Memory endow the system of sensitivity, as it includes elements associated to interest points ("activators") in the memory at each frame. Unfortunately, in the Short-Term Memory scene blobs whose shape features do not correspond to those defined by the observer may appear at a time instant t. This is precisely because their shape features have not yet been studied. But, if these blobs shape features really do not seem to be interesting for the observer, they will appear as "inhibitors" in $t+1$ in the Interest Map (now, in $t+1$, their shape features will have been obtained). And, this means that in $t+1$ they will disappear from the Short-Term Memory.

In order to obtain at each frame only blobs with the desired features, Attention Reinforcement performs an accumulative mechanism followed by a thresh-

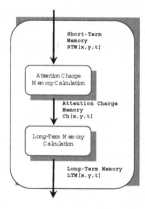

Fig. 5. Scheme for "Attention Reinforcement"

old. Accumulation is realized on pixels that have a value different from 0 (pixels that do not belong to labeled blobs) in the Short-Term Memory. The result of this accumulative process followed by a threshold offers as output the Long-Term Memory, $LTM[x, y, t]$. More concretely, pixels that appear with a value different from 0 in the Short-Term Memory reinforce attention, whilst those that appear with a value 0 diminish the attention value.

The process manages to keep activated in a stable way a set of pixels that belong to a group of objects (figures) of the scene that are interesting for the observer. Fig. 5 shows the decomposition of Attention Reinforcement into Attention Charge Memory Calculation and Long-Term Memory Calculation.

4.1   Attention Charge Memory Calculation

Attention Charge Memory Calculation performs an accumulative computation on the Short-Term Memory to get the Attention Charge Memory $Ch[x, y, t]$. The idea underlying Attention Charge Memory Calculation is that pixels that belong to a blob of the Short-Term Memory through time reinforce attention whilst all other ones decrease attention. The accumulative computation [1], [6], [7] takes the form of Equation 6, based on the more general charge/discharge accumulative computation mode [17].

$$Ch[x, y, t] = \begin{cases} \max(Ch[x, y, t-1] - D, Ch_{min}), \\ \quad \text{if } STM[x, y, t] = 0 \\ \min(Ch[x, y, t-1] + C, Ch_{max}), \\ \quad \text{if } NC * NR + 1 > STM[x, y, t] > 0 \end{cases} \quad (6)$$

where Ch_{min} is the minimum and Ch_{max} is the maximum value, respectively, that the values stored in the Attention Charge Memory can reach, and C and D are the charge increment and decrement, respectively, in the memory computation. The charge value $Ch[x, y, t]$ goes incrementing up to Ch_{max}, if pixel $[x, y]$ belongs to a blob of the Short-Term Memory, and goes decrementing down to Ch_{min} if the pixel does not. Charge value $Ch[x, y, t]$ represents a measure of the persistency of a blob in the Short-Term Memory on each image pixel $[x, y]$.

4.2   Long-Term Memory Calculation

Long-Term Memory Calculation produces, starting from the Attention Charge Memory, the points that configure the Long-Term Memory, labeling the figures obtained. The focus is in form of figures, obtained by the union of the connected blobs that have appeared successively in the Short-Term Memory and whose value in the Attention Charge Memory is greater or equal to a given threshold, θ. In the output, the label corresponding to the figure is stored; value 0 is assigned to all pixels that do not belong to any figure. Firstly, the Long-Term Memory at pixel $[x, y]$ is assigned an initial and provisional value (yet not agreed with the neighbours) corresponding to a function of the coordinate of the pixel, if the charge value overcomes threshold θ (see Equation 7):

$$LTM[x, y, t] = \begin{cases} x * NC + y + 1, & \text{if } Ch[x, y, t] > \theta \\ 0, & \text{otherwise} \end{cases} \quad (7)$$

Next, in an iterative way up to reaching a common value for all pixels of a same figure (by calculating the minimum value of each pixel and its 8 surrounding neighbours), a calculation is performed according to Equation 8:

$$LTM[x, y, t] = min(LTM[\alpha, \beta, t]), \forall [\alpha, \beta] \in [x \pm 1, y \pm 1] | 0 < LTM[\alpha, \beta, t] \quad (8)$$

Fig. 6 now shows the contents of the Long-Term Memory after processing all steps of Attention Reinforcement. Notice that we got the desired stability.

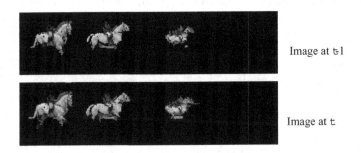

Image at t-1

Image at t

Fig. 6. Long-Term Memory for a couple of images

5   Conclusions

In this paper the relationship between sensitivity and stability in our particular DSVA method has been described. In the proposal sensitivity has been associated to Short-Term Memory and stability to Long-Term Memory, respectively. The generation of the Short-Term Memory, and hence the coming out of sensitivity, is related to task Attention Building, whereas Long-Term Memory is obtained after Attention Reinforcement, getting the desired stability to visual attention.

As described, Attention Building is achieved by means of two main steps, namely Classification in Grey Level Bands and Short-Term Memory Generation, getting as output figure blobs in the Short-Term Memory in a noisy way. On the other hand, Attention Reinforcement is divided into Attention Charge Memory Calculation and Long-Term Memory Calculation, and obtains persistent figures through time in a stable Long-Term Memory.

Acknowledgements

This work is supported in part by the Spanish CICYT TIN2004-07661-C02-01 and TIN2004-07661-C02-02 grants.

References

1. Fernández-Caballero, A., López, M.T., Fernández, M.A., Mira, J., Delgado, A.E., López-Valles, J.M.: Accumulative computation method for motion features extraction in active selective visual attention. LNCS **3368** (2004) 206–215
2. López, M.T., Fernández, M.A., Fernández-Caballero, A., Delgado, A.E.: Neurally inspired mechanisms for the active visual attention map generation task. LNCS **2686** (2003) 694–701
3. Oppenheim, A.V., Willsky, A.S., Nawab, S.H.: Signals and Systems, 2nd edition. Prentice-Hall Inc (1997)
4. Daniilidis, K., Spetsakis, M.: Understanding noise sensitivity in structure from motion. Aloimonos, Y. (ed.), Visual Navigation (1996) 61–88
5. Fermüller, C., Aloimonos, Y.: Algorithm-independent stability analysis of structure from motion. University of Maryland TR 3691 (1996)
6. Fernández-Caballero, A., Mira, J., Delgado, A.E., Fernández, M.A.: Lateral interaction in accumulative computation: A model for motion detection. Neurocomputing **50** (2003) 341–364
7. Fernández-Caballero, A., Fernández, M.A., Mira, J., Delgado, A.E.: Spatio-temporal shape building from image sequences using lateral interaction in accumulative computation. Pattern Recognition **36**:5 (2003) 1131–1142
8. Fernández-Caballero, A., Mira, J., Fernández, M.A., Delgado, A.E.: On motion detection through a multi-layer neural network architecture. Neural Networks **16**:2 (2003) 205–222
9. Baddeley, A.D., Hitch, G.J.: Short-Term Memory. Bower, G., (ed.), Recent Advances in Learning and Motivation **8** (1974)
10. O'Reilly, R.C., Braver, T.S., Cohen, J.D.: A biologically-based computational model of working memory. Miyake, A., Shah P. (eds.), Models of Working Memory: Mechanisms of Active Maintenance and Executive Control (1999) 375–411
11. Awh, E., Anllo-Vento, L., Hillyard, S.A.: The role of spatial selective attention in working memory for locations: evidence from event-related potentials. Journal of Cognitive Neuroscience **12** (2000) 840–847

12. Awh, E., Jonides J.: Overlapping mechanisms of attention and spatial working memory. Trends in Cognitive Sciences **5**(3) (2001) 119–126
13. Atkinson, R.C., R. M. Shiffrin, R.M.: Human memory: A proposed system and its control processes. Spence, K.W., Spence, J.T. (eds.), The Psychology of Learning and Motivation: Advances in Research and Theory **2** (1968)
14. Waugh, N., Norman, D.A.: Primary memory. Psychological Review **72** (1965) 89–104
15. Anderson, J.R.: The Architecture of Cognition. Harvard University Press (1983)
16. Winn, W., Snyder, D.: Cognitive perspectives in psychology. Jonassen, D.H. (ed.), Handbook of Research for Educational Communications and Technology (1996) 115–122
17. Mira, J., Fernández, M.A., López, M.T., Delgado, A.E., Fernández-Caballero, A.: A model of neural inspiration for local accumulative computation. LNCS **2809** (2003) 427–435

Visual Attention, Visual Saliency, and Eye Movements During the Inspection of Natural Scenes

Geoffrey Underwood[1], Tom Foulsham[1],
Editha van Loon[1], and Jean Underwood[2]

[1] School of Psychology, University of Nottingham,
Nottingham NG7 2RD, UK
geoff.underwood@nottingham.ac.uk
lpywtf@psychology.nottingham.ac.uk
editha.vanloon@nottingham.ac.uk
[2] Division of Psychology, Nottingham Trent University,
Nottingham NG1 4BU, UK
jean.underwood@ntu.ac.uk

Abstract. How does visual saliency determine the attention given to objects in a scene? Viewers' eye movements were recorded during the inspection of pictures of natural office scenes containing two objects of interest. According to the Itti and Koch algorithm one object had lower visual saliency relative to the other that was visually complex. We varied the purpose of picture inspection to determine whether visual saliency is invariably dominant in determining the pattern of fixations, or whether task demands can provide a cognitive override that renders saliency as of secondary importance. When viewers inspected the scene in preparation for a memory task, the more complex objects were potent in attracting early fixations, in support of a saliency map model of scene inspec-tion. When the viewers were set the task of search for the lower-saliency target the effect of the distractor was negligible, requiring the saliency map to be built with cognitive influences.

1 Introduction

How do we decide where to look first when shown an image of a natural scene? Itti and Koch [1] have developed a computational procedure for the determination of visual saliency of images such as photographs that also serves as a model of where attention should be directed when we look at those images. The model relies upon the low-level visual characteristics of the image to determine saliency and hence where attention should be directed, and in what order attention should be moved around the image. In the case of two-dimensional static image these low-level characteristics are colour, intensity and orientation, and with dynamic displays the relative motion of objects would also contribute to their saliency values. Separate saliency maps are first computed for each of these

J. Mira and J.R. Álvarez (Eds.): IWINAC 2005, LNCS 3562, pp. 459–468, 2005.
© Springer-Verlag Berlin Heidelberg 2005

characteristics, and the maps are then combined linearly to find the saliency peak using a winner-take-all network. A change in any of the three characteristics results in an increase in the saliency value assigned to that region of the picture. So, a picture of a blue ocean and blue sky with a white cloud in the sky will deliver high saliency values for the horizon (orientation change), and for the cloud (colour, intensity and orientation change). The model predicts that areas of high saliency should attract the viewer's attention, and that eye fixations should be directed first to the region of highest saliency, then to the area of next highest saliency, and so on. When attention has been directed towards an object, a process of 'inhibition of return' lowers the saliency value of that location, enabling attention to be directed to the next area of interest.

Parkhurst, Law and Niebur [2] have evaluated the Itti and Koch model against the performance of human viewers in a study in which a variety of images were shown. The images included colour photographs of natural scenes as well as computer-generated fractals, and each was displayed for 5 sec. Four participants were instructed to "look around at the images" while their eye movements were recorded. A modified model was evaluated in their study, such that central regions were given higher saliency values than peripheral locations. This modification takes account of the decline in visual sensitivity from the central visual field to the periphery of vision, and so gives priority to a central object over other objects with equal saliency but placed in the periphery. The model performed well, with the saliency value of the regions within an image predicting the order of fixations, especially for the first few fixations. The modified model, with greater weighting given to central regions over the periphery, was more successful than a model using a uniform visual field.

When viewers look around a picture with no specific goal, their attention and their eye movements are attracted to areas of high visual saliency. Nothdurft [3] has also suggested that saliency has effects in visual search. The task used in his experiments was to detect a single item that was unlike the other items in the display. The displays were formed of a large number of short lines of varying orientation, and a single item was made more salient than the others in a number of ways, including increasing the intensity and by introducing a short movement. Nothdurft concluded that high saliency in an item attracts focal attention and is responsible for the 'pop-out' effect whereby an item of different colour or different orientation will stand out from other items and will be found without detailed item-by-item scrutiny. Focal attention is then used to identify the detailed properties of the selected item. The question then arises as to whether searching a natural scene will show the same effects of saliency as Nothdurft's texture-like line displays, with a highly salient object inevitably attracting focal attention, or whether the nature of the task can override visual saliency and result in a highly salient object being ignored.

Visual search is clearly open to top-down cognitive influences. When we look around a kitchen searching for a bread knife, or a chopping board, we will direct our attention to the most likely locations rather than to the most visually salient objects in view, such as gleaming metal pans or rightly coloured bowls. Underwood, Jebbett and Roberts [4] have demonstrated the selective nature of searching for information in digitized photographs with an eye-tracking study. Viewers performed a sentence verification task in which they had to judge whether or not a sentence correctly described some aspect of a photograph of a natural scene. Sentences could appear before or after the picture. When the picture appeared first the task was essentially one of encoding as much detail as possible, not knowing the subject of the sentence. Participants characteristically made a large number of fixations over the whole picture, before indicating that they were ready to read the sentence. When participants were able to read the sentence first they were able to focus their attention on the subject of the sentence once the picture was displayed. They made fewer fixation overall, and their fixations were guided to the objects described in the sentence. For example, in a picture of road scene, which showed a car is parked near to a phone box, the sentence declared correctly that the car was parked next to a phone box, or, incorrectly that it was next to a mailbox. When the sentence was read before seeing the picture, viewers characteristically looked at the car, and then at the phone box. After very few fixations they would then press the response key to indicate that the sentence was true or false. Their search was determined by their knowledge rather than by the saliency map of the picture. Other modifications of the search by top-down factors has been reported in a range of tasks [5],[6],[7] with the suggestion that searches are influenced by relevance of objects to the specific task.

The two experiments here compare the attention given to objects in the same pictures when the task is changed from general encoding for a later memory test, to a search for an item defined by its natural category membership. This target item had lower visual saliency than a distractor, and so the experiments were used to answer the question of whether task demands can override the attentional pull of a high saliency object.

2   Experiment 1: Inspecting for Remembering

In this experiment viewers looked at a set of pictures in preparation for a memory test. They were given as long as they wanted to look at each picture, and their eye movements were recorded as they did so. Two objects of interest were present, although this was not indicated to the participants. A high-saliency object was designated as a "distractor' and a lower-saliency object as a "target". (These labels were used for consistency with Experiment 2, in which the participants would search for the target.) This task was close to the 'free-viewing" task used by Parkhurst et al. [2], and was expected to demonstrate a dominance for the high-saliency distractor object. To avoid the problem addressed by the Parkhurst et al. modified saliency model, namely, the decline in visual sensitivity in peripheral vision, all objects of interest were displayed at either 3 deg or 6 deg from the initial fixation point. These two eccentricities were used to determine the interference between competing objects that are nearer or further from each other.

2.1   Method

The images presented for inspection were 48 digital photographs of desktop office scenes. A further 8 photographs were used to give participants familiarity with the images and practice with task. Pictures subtended 11.6 deg by 15.4 deg when displayed on a colour monitor and viewed at a distance of 60 cm. Within each scene there was a piece of fruit that is designated here as a target although in no sense was it identified to the participants as having special significance. Different pieces of fruit were used in different pictures, and when two similar items were used, their orientation was changed between pictures. There was also a distractor object. Other objects such as papers, books, general desktop equipment and computers were visible. The target and distractor were of importance only to the experimental design, and the viewer's attention was not directed towards them in the task instructions. These objects were placed 3 deg or 6 deg from centre, with target and distractor on opposite sides of the picture. In addition, one third of the pictures had no distractor. There were 8 pictures in each of the 6 experimental conditions: target at 3 deg or at 6 deg, and distractor at 3 deg, at 6 deg, or absent.

Each picture was analysed using the Itti and Koch [1] saliency map program, to determine the saliency values of each object. The pictures used in the experiment all had the characteristic of the distractor object being the most salient object shown, and the target being the next most salient object.

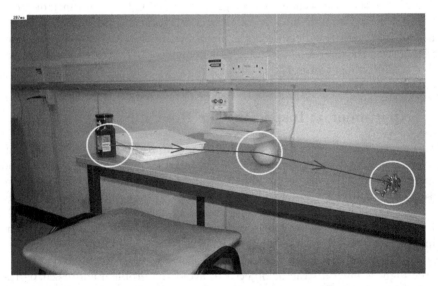

Fig. 1. One of the pictures used in the experiments. All of the pictures showed desktop office scenes, with a range of objects visible. Also shown here is the output of the Itti and Koch program that determines the visual saliency of the regions within the picture. In this case the most salient object is the food jar on the left, with the lemon being the next most salient object, and then the bunch of keys on the right

An example of one of the pictures used in the experiment is shown in Fig. 1. While participants looked at the pictures, their eye movements were recorded using an SMI EyeLink system recording eye positions at 250 Hz and with a spatial accuracy of 0.5 deg. A chin rest was used to ensure a constant viewing distance.

The participants were 20 undergraduates who were paid for their involvement. All volunteers had normal or corrected-to-normal vision, and all were paid for their participation. After calibrating the eye tracking equipment they were instructed that they would see a set of pictures, and that they would be tested for their memory of them at the end of the experiment. The 8 practice pictures were then shown, followed by a two-alternative forced-choice task in which two pictures were shown on the screen, with the task being to say which one of them had been shown previously. The 48 test pictures were then shown in a different random order for each participant, with each picture remaining on the screen until the participant pressed the space bar to indicate that they were ready for the next one.

2.2   Results

The number of eye fixations made prior to fixation of the target and the distractor were compared at the two eccentricities (3 deg and 6 deg) by separate analysis of variance for the two objects. The numbers of fixations made prior to inspection of the two objects are shown in Fig. 2, where it can be seen that there was earlier fixation of near distractors ($F = 20.66, df = 1, 19, p < .001$). The position of the target also influenced fixation of the distractor ($F = 23.08, df = 1, 19, p < .001$), with fewer fixations required before inspection of the distractor when the target was shown at 6 deg rather than 3 deg. The distractor was fixated earlier when it was positioned nearer to the initial point of fixation, and when the target was

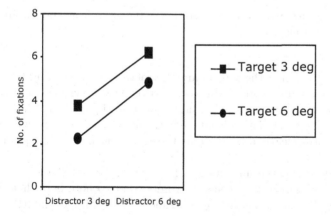

Fig. 2. Number of fixations made prior to inspection of the high-saliency distractor object (a manufactured object such as a food package) in Experiment 1

positioned furthest away. This analysis confirmed the effect of the target upon fixation of the higher-saliency distractor.

A second analysis was conducted on the number of fixations prior to fixation of the target object. These fixations are shown in Fig. 3. Target position was not a reliable factor ($F = 1.29, df = 1, 19, n.s.$), but distractor position did influence target inspection ($F = 8.52, df = 2, 38, p < .001$). Distractors were placed at 3 deg from the centre of the picture, at 6 deg, or absent altogether. Paired comparisons indicated that compared to the baseline condition of no distractor being present, there was an increase in the number of fixations when distractors were placed at 3 deg and at 6 deg. This analysis confirmed the effect of a high-saliency distractor object upon the time taken to attend to a lower-saliency object.

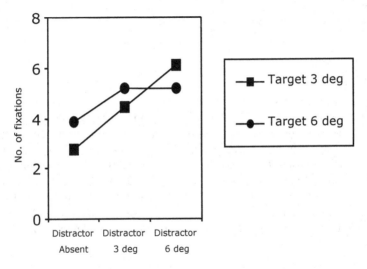

Fig. 3. Number of fixations made prior to inspection of the low-saliency target object (a piece of fruit) in Experiment 1

The course of inspection of the targets and distractors can be seen in Fig. 4, which compares the cumulative probabilities of fixation over the viewing interval. All targets and distractors are considered here, and the Fig. confirms that the distractors are fixated earliest. This advantage in the fixation of the high-saliency objects is apparent most clearly during the first few seconds of the display being available.

Experiment 1 confirmed the importance of visual saliency in the allocation of attention to objects shown in a scene. It is important to note here that the object designated as distractor has the highest saliency according to the Itti and Koch [1] model, and that the target had lower saliency. There was no special significance of these objects as targets and distractors for the viewers, who had the task of looking at the pictures in order to remember them for a memory test.

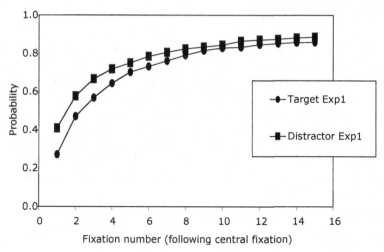

Fig. 4. Probability of target and distractor fixation as a function of ordinal fixation number in Experiment 1

The high-saliency distractor was fixated earlier than the second most salient object (see Fig. 4) in this experiment. The two objects also influenced each other's fixation. The inspection of the distractors was influenced by the proximity of the targets, and the inspection of the targets was influenced by the presence or absence of distractors.

3   Experiment 2: Inspecting for Search

Whereas in Experiment 1 the participants looked at the pictures in order to remember the scene, in this study the same pictures were used for a search task. Each time a picture was shown, the participants were required to judge whether or not a piece of fruit was present. The task now requires the viewers to look for a particular object, in the presence of a distractor that had higher visual saliency.

3.1   Method

The same pictures as were used in Experiment 1 were again used here, with 21 additional pictures that contained a high-saliency distractor, but no target object. The same equipment was used with 20 undergraduates. No participant had taken part in Experiment 1, all were paid volunteers, and all had normal or corrected-to-normal vision. The procedure was otherwise similar to Experiment 1, except that the instructions were to search for a piece of fruit in the picture, and to indicate its presence or absence by pressing one of two keyboard keys.

3.2   Results

Trials were analysed only where a target had been present in the picture. Distractors were fixated on only 20.4% of trials in Experiment 2 (in contrast with 84.8% in Experiment 1), and so no analysis was possible. The number of fixations prior to inspection of the target was again analysed as a function of the positions of the target and distractor, and the means are shown in Fig. 5. An analysis of variance indicated a reliable effect of target position ($F = 22.08, df = 1, 19, p < .001$), with fewer fixation being required for inspection of targets appearing 3 deg from the initial fixation. There was also an effect of distractors on the number of fixations made prior to target inspection ($F = 15.07, df = 2, 38, p < .001$), paired comparisons indicated that when the distractor was located at 3 deg from centre then it was more disruptive than when it was placed at 6 deg or when it was absent altogether.

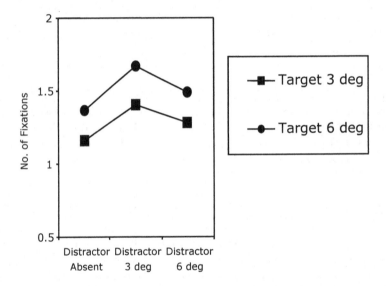

Fig. 5. Number of fixations made prior to inspection of the low-saliency target object (a piece of fruit) in Experiment 2

The cumulative probability of fixation of objects is shown in Fig. 6, for all targets and all distractors. The earlier fixation of the targets is in marked contrast with the target/distractor comparison from Experiment 1 (see Fig. 4). Targets are now fixated with the first fixation on 60.2% of trials and by the second fixation on 81.8% of trials. In Experiment 1 the targets had been fixated on only 47.1% of trials by the second fixation. In contrast the distractor was fixated on 18.2% of trials, whereas in Experiment 1 it was inspected 41% of the time with the first fixation.

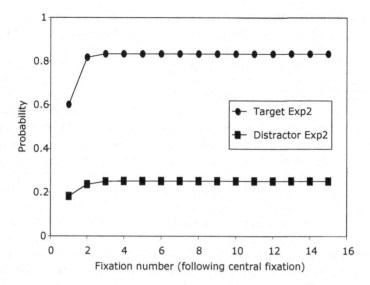

Fig. 6. Probability of target and distractor fixation as a function of ordinal fixation number in Experiment 2

4   Discussion

There are a number of striking contrasts in the results from the two experiments. When set the task of encoding a picture in preparation for a recognition memory test, viewers were strongly influenced by the visual saliency of the objects shown. Their eyes were attracted to objects that had colour, intensity and orientation features that distinguished them from their backgrounds. Viewers looked at the most salient region of a picture sooner than the second most salient region, and this relationship between the order of fixation and the saliency value confirms a result reported by Parkhurst et al. [2] in a task in which viewers were instructed simply to look around the picture. When the viewers in Experiment 2 were instructed to search for an example of a piece of fruit, and to indicate whether such a target item was present or not, then visual saliency was much less potent. The highly salient distractors that had attracted fixations in Experiment 1 were now only occasionally inspected at all. Viewers tended to look at the less-salient fruit without looking at the distractors, suggesting that the saliency values of the objects were being ignored during the search for the target. Top-down cognitive demands can override bottom-up visual saliency, and when the task is well-specified attention can be guided around the scene independently of the saliency map.

The dominance of visual saliency in the memory task was seen in mutual influences between the two objects of interest. The distractor was fixated sooner when the target was further away, and the target was inspected sooner when there was no distractor present. A distant distractor had as much interference as a near distractor, relative to there being no other salient object in the picture. In the search experiment this pattern was moderated. Targets were still inspected sooner when they were nearer to the initial point of fixation when the display started, and near distractors caused more interference than absent distractors, but distant distractors caused no more interference than absent distractors. Distant distractors could be ignored in this search task, whereas distant distractors in the memory task caused similar amounts of interference as near distractors.

The results are consistent with the predictions of the saliency map hypothesis for free viewing conditions [1],[2]. When searching for an object defined in terms of its category membership, however, visual saliency did not predict the locations of the first few fixations. Indeed, the most salient object, a non-target, was seldom fixated. The saliency map was not influential, if indeed it was constructed at all during the search task. If its construction is obligatory, then cognitive demands can exclude or minimise its input to the saccadic programming mechanism. Viewers guided their searches to locations where targets were likely to be placed, and to objects possessing features shared with members of the search category. Henderson, Weeks and Hollingworth [8] suggest that the saliency map is built with low-level visual features that guide the first few fixations. As the scene is inspected and the content understood, cognitive guidance can take over and the semantic content of the picture determines our inspection pattern. The present data suggest instead that the content of a picture can be understood without these preliminary fixations, and that cognitive guidance can be used to program the first saccadic movement.

References

1. Itti, L., Koch, C.: A saliency-based search mechanism for overt and covert shifts of visual attention. Vis. Res. **40** (2000) 1489–1506
2. Parkhurst, D., Law, K., Niebur, E.: Modeling the role of salience in the allocation of overt visual attention. Vis. Res. **42** (2002) 107–123
3. Nothdurft, H.-C.: Attention shifts to salient targets. Vis. Res. **42** (2002) 1287–1306
4. Underwood, G., Jebbett, L., Roberts, K.: Inspecting pictures for information to verify a sentence: Eye movements in general encoding and in focused search. Quart. J. Exp. Psychol. **57**A (2004) 165–182
5. Ludwig, C.J.H., Gilchrist, I.D.: Stimulus-driven and goal-driven control over visual selection. J. Exp. Psychol.: Hum. Perc. Perf. **28** (2002) 902–912
6. Lamy, D., Leber, A., Egeth, H.E.: Effects of task relevance and stimulus-driven salience in feature-search mode. J. Exp. Psychol.: Hum. Perc. Perf. **30** (2004) 1019–1031
7. Wolfe, J.M., Horowitz, T.S., Kenner, N., Hyle, M., Vasan, N.: How fast can you change your mind? The speed of top-down guidance in visual search. Vis. Res. **44** (2004) 1411–1426
8. Henderson, J.M., Weeks, P.A., Hollingworth, A.: The effects of semantic consistency on eye movements during complex scene viewing. J. Exp. Psychol.: Hum. Perc. Perf. **25** (1999) 210–228

Model Performance for Visual Attention in Real 3D Color Scenes

Heinz Hügli, Timothée Jost, and Nabil Ouerhani

Institute of Microtechnology,
University of Neuchâtel,
Rue A.-L. Breguet 2,
CH-2000 Neuchâtel, Switzerland
{Heinz.Hugli, Nabil.Ouerhani}@unine.ch

Abstract. Visual attention is the ability of a vision system, be it biological or artificial, to rapidly detect potentially relevant parts of a visual scene. The saliency-based model of visual attention is widely used to simulate this visual mechanism on computers. Though biologically inspired, this model has been only partially assessed in comparison with human behavior. The research described in this paper aims at assessing its performance in the case of natural scenes, i.e. real 3D color scenes. The evaluation is based on the comparison of computer saliency maps with human visual attention derived from fixation patterns while subjects are looking at the scenes. The paper presents a number of experiments involving natural scenes and computer models differing by their capacity to deal with color and depth. The results point on the large range of scene specific performance variations and provide typical quantitative performance values for models of different complexity.

1 Introduction

Visual attention is the ability of a system, be it biological or artificial, to analyze a visual scene and rapidly detect potentially relevant parts on which higher level vision tasks, such as object recognition, can focus. On one hand, artificial visual attention exists as the implementation of a model on the computer. On the other hand, biological visual attention can be read from human eye movements. Therefore, the research presented in this paper aims at assessing the performance of various models of visual attention by comparing the human and computer behaviors.

It is generally agreed nowadays that under normal circumstances human eye movements are tightly coupled to visual attention. This can be partially explained by the anatomical structure of the human retina. Thanks to the availability of sophisticated eye tracking technologies, several recent works have confirmed this link between visual attention and eye movements [1, 2, 3]. Thus, eye movement recording is a suitable means for studying the temporal and spatial deployment of visual attention in most situations.

J. Mira and J.R. Álvarez (Eds.): IWINAC 2005, LNCS 3562, pp. 469–478, 2005.

In artificial vision, the paradigm of visual attention has been widely investigated during the last two decades, and numerous computational models of visual attention have been suggested. A review on existing computational models of visual attention is available in [4]. The saliency-based model proposed in [5] is now widely used in numerous software and hardware implementations [6, 7] and applied in various fields.

However, and despite the fact that it is inspired by psychophysical studies, only few works have addressed the biological plausibility of the saliency-based model [8]. Parkhurst et al [9] presented for the first time a quantitative comparison between the computational model and human visual attention. Using eye movement recording techniques to measure human visual attention, the authors report a relatively high correlation between human attention and the saliency map, especially when the images are presented for a relatively short time of few seconds. Jost et al [10] run similar experiments on a much larger number of test persons and could measure the quantitative improvement of the model when chromaticity channels are added to the conventional monochrome video channels. Visual attention in 3D scenes was first considered in [11] and recently, a visual attention model for 3D was quantitatively analyzed in presence of various synthetic and natural scenes [12].

This paper presents a more global analysis, where the performance of a family of visual attention models in presence of 3D color scenes is evaluated. The basic motivation is to get insight into the contribution of the various channels like color and depth. Another motivation is to assess possible improvements when artificial visual attention is made more complex.

The remainder of this paper is organized as follows. Chapter 2 recalls basics of the saliency models. Chapter 3 presents the methods for acquiring the human fixation patterns and comparing them to the saliency map. Chapter 4 details the experiments and obtained results. A general conclusion follows in Chapter 5.

2   Saliency Models

The saliency-based visual attention [5] operates on the input image and starts with extracting a number of features from the scene, such as intensity, orientation chromaticity, and range. Each of the extracted features gives rise to a conspicuity map which highlights conspicuous parts of the image according to this specific feature. The conspicuity maps are then combined into a final map of attention named saliency map, which topographically encodes stimulus saliency at every location of the scene. Note that the model is purely data-driven and does not require any a priori knowledge of the scene.

2.1   Feature and Conspicuity Maps

From a scene defined by a color image (R, G, B) and a range image Z, a number of features F_j are extracted as follows:

Intensity feature $F_1 = I = 0.3 \cdot R + 0.59 \cdot G + 0.11 \cdot B$.

Four features F_2, F_3, F_4, F_5 for the local orientation according to the angles $\theta \in \{0^o, 45^o, 90^o, 135^o\}$.

Two chromaticity features F_6, F_7 based on the two color opponency components R^+G^- and B^+Y^- defined with the help of the yellow component Y as follows:

$$Y = \frac{R+G}{2} \qquad F_6 = \frac{R-G}{I} \qquad F_7 = \frac{B-Y}{I} \qquad (1)$$

Depth feature represented by a depth map $F_8 = Z$.

Each feature map is then transformed into its conspicuity map C_j which highlights the parts of the scene that strongly differ, according to the feature specificity, from their surroundings. The computation of the conspicuity maps noted $C_j = T(F_j)$ relies on the center-surround mechanism, a multiscale approach and a normalization and summation step during which, the maps from each scale are combined, in a competitive way, into the feature-related conspicuity map C_j.

2.2   Cue Maps

Given the nature of the different features, the model groups together conspicuities belonging to the same category and we define cue conspicuity maps for intensity (int), orientation (orient), chromaticity (chrom.) and range as follows:

$$\hat{C}_{int} = C_1; \quad \hat{C}_{orient} = \sum_{j \in \{2,3,4,5\}} N(C_j); \quad \hat{C}_{chrom} = \sum_{j \in \{6,7\}} N(C_j); \quad \hat{C}_{range} = C_8$$

$$(2)$$

where $N(.)$ is a normalization operator which simulates the competition between the different channels. A detailed description of the normalization strategy is given in [6].

2.3   Saliency Map

Finally, the cue maps are integrated, in a competitive manner, into a universal saliency map S as follows:

$$S = \sum_{cue} N(\hat{C}_{cue}) \qquad (3)$$

More specifically, in this study we work with three alternative saliency maps of in-creasing complexity, namely:

- A greyscale saliency map S_{grey} that includes intensity and orientation: $S_{grey} = N(\hat{C}_{int}) + N(\hat{C}_{orient})$.
- A color saliency map S_{color} that includes intensity, orientation and chromaticity: $S_{color} = N(\hat{C}_{int}) + N(\hat{C}_{orient}) + N(\hat{C}_{chrom})$.
- A depth saliency map S_{depth} that includes intensity, orientation, chromaticity and range: $S_{depth} = N(\hat{C}_{int}) + N(\hat{C}_{orient}) + N(\hat{C}_{chrom}) + N(\hat{C}_{range})$.

3   Comparing Computer and Human Visual Attention

The evaluation principle illustrated in figure 1 is based on the comparison of the computed saliency map with human visual attention. Under the assumption that under most circumstances, human visual attention and eye movements are tightly coupled, the deployment of visual attention is experimentally derived from the spatial pattern of fixations.

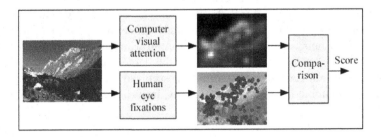

Fig. 1. Comparison of computer and human visual attention

3.1   Eye Movement and Fixation Pattern Recording

Eye movements were recorded with an infrared video-based tracking system (Eye-Link^{TM}). It has a temporal resolution of 250 Hz, a spatial resolution of 0.01^o, and a gaze-position accuracy relative to the stimulus position of $0.5^o -$ 1.0^o, largely dependent on subjects' fixation accuracy during calibration. As the system incorporates a head movement compensation, a chin rest was sufficient to reduce head movements and ensure constant viewing distance.

A considerable challenge of this research has been to record eye movements while a subject is watching a stereo image. It was made possible with the use of an autostereoscopic display. It avoids using glasses on the subject, which would prevent eye movement tracking. The images were presented in blocks of 10. Each image block was preceded by a 3×3 point grid calibration scheme. The images were presented in a dimly lit room on the autostereoscopic 18.1" CRT display (DTI 2018XLQ) with a resolution (in stereo mode) of 640×1024, 24 bit color depth, and a refresh rate of 85 Hz. Active screen size was 36×28.5 cm and viewing distance 75 cm, resulting in a viewing angle of 29×22°. Every image was shown for 5 seconds, preceded by a center fixation display of 1.5 seconds. Image viewing was embedded in a recognition task.

Eye monitoring was conducted on-line throughout the blocks. The eye tracking data was parsed for fixations and saccades in real time, using parsing parameters proven to be useful for cognitive research thanks to the reduction of detected microsaccades and short fixations (< 100 ms). Remaining saccades with amplitudes less than 20 pixels (0.75^o visual angle) as well as fixations shorter than 120 ms were discarded after-wards [10].

For every image and each subject i, the measurements yielded an eye trajectory T^i composed of the coordinates of the successive fixations f_k, expressed as image coordinates (x_k, y_k):

$$T^i = (f_1^i, f_2^i, f_3^i, ...)$$ (4)

3.2   Score s

The score s is used as a metric to compare human fixations and computer saliency maps. Also called chance-adjusted saliency by Parkhurst et al. [9], the score s corresponds to the difference of average values of two sets of samples from the computer saliency map $S(x)$. Formally:

$$s = \frac{1}{N} \sum_{f_k \in T} S(f_k) - \mu_S$$ (5)

The first term corresponds to the average value of N fixations f_k from an eye trajectory T^i . The second term μ_S is the saliency map average value. Thus the score measures the excess of salience found at the fixation points with respect to arbitrary points. If the human fixations are focused on the more salient points in the saliency map, which we expect, the score should be positive. Furthermore, the better the model, the higher the probability to reach the points with highest saliency and the higher this score should be.

4   Experiments and Results

The experimental process was divided into two parts. A first part is devoted to the measurement of visual attention induced by 2D images. A second part compares human visual attention in presence of 3D color scenes.

4.1   Dataset 2D

This dataset consists of 41 color images containing a mix of natural scenes, fractals, and abstract art images (see figure 2). Most of the images (36) were shown to 20 subjects. As stated above, these images were presented to the subjects for 5 seconds apiece, resulting in an average of 290 fixations per image.

4.2   Dataset 3D

This dataset consists of 12 3D scenes representing quite general natural scenes. Each scene is represented by a stereo image pair. Figure 3 presents sample images from this dataset. These image pairs were presented to 20 different subjects for 5 seconds apiece, resulting in an average of 290 fixations per image.

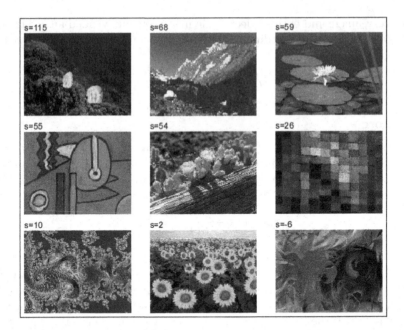

Fig. 2. Images from the dataset 2D, ranked by score for the color model

Fig. 3. Sample scenes from the dataset 3D

4.3   Performance in Presence of 2D Images

For all images of dataset 2D, we created a greyscale saliency map S_{grey} and a color saliency map S_{color}, both normalized to the same dynamic range. Then, a comparison of these two models with the whole set of human fixation patterns was performed in order to obtain the respective scores. Note that the score s was computed taking the first 5 fixations of each subject into account, since it has been suggested that, with regard to human observers, initial fixations are controlled mainly in a bottom-up manner [10].

Figure 4 shows the scores for the different individual images. The main observation here is that the resulting scores are widely spread in their value, covering the range [-7 .. 115]. The values show the model performance depends in a strong way on the kind of image. To illustrate these results and explain somehow these strong variations, we refer to figure 2 showing sample images from the dataset

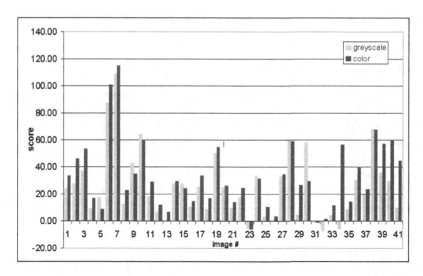

Fig. 4. Individual scores for images of dataset 2D, both for the greyscale and color models

2D. There, the images are ordered according to the score S_{color} obtained by each image. The image yielding the best results is top left. The score decreases from left to right and top to bottom. It is apparent that the images found on the top row generally contain few and strong salient features, such as the fish, the small house or the water lily. They yield the best results. On the other hand, images that lack highly salient features, such as the abstract art or the fractal images on the bottom row, result in much lower scores. Here, the model loses its effectiveness in the single image (out of 41) yielding a negative score.

Referring to performance of the models, it is expected that the color model performs better because it includes the additional chromaticity cue. We therefore expect the score for the color model to be at least as good as the score of the greyscale model. Although this is not true for all images it is the case for a majority of about 85% of the cases.

A general comparison is given in table 1 showing the estimated average model scores. The standard error was computed using the variance from both random picks and human fixations means. The main observation is that the color model fares better than the greyscale one. More specifically, the color model yields an average score 25.8% higher than the greyscale model. This underlines the

Table 1. Scores of the greyscale and color models

	score s
greyscale model S_{grey}	24.8 ± 1.2
color model S_{color}	31.2 ± 1.1

usefulness of the chromaticity cue in the model and goes toward assessing that this cue has a considerable influence on visual attention.

4.4   Performance in Presence of 3D Scenes

For all scenes of dataset 3D, we created a color saliency map S_{color} and a depth saliency map S_{depth} , both normalized to the same dynamic range. Then, a comparison of these two models with the whole set of human fixation patterns was performed in order to obtain the respective scores. The score s was computed as in previous experiments.

Figure 5 shows the scores for the 12 individual images. The main observation here is that the resulting scores are widely spread in their value [5 .. 76]. The effect is the same as in previous experiments and above comments keep their full validity here. It shows again that the model performance depends in a strong way on the kind of scene.

Referring to the model performance, table 2 presents the average scores s over the whole dataset, for both the color and the depth models. The standard error was computed as above. The main observation is that the depth model fares better than the color one. More specifically, the depth model yields an average score s that is 11.8% better than the color model. This general result underlines the usefulness of the depth channel in the model and goes toward assessing that depth contributes to the visual attention process.

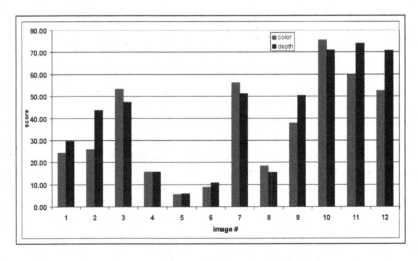

Fig. 5. Individual scores for images of dataset 3D, both for the color and depth models

Table 2. Scores of the color and depth models

	score s
color model S_{color}	36.2 ± 2.1
depth model S_{depth}	40.5 ± 2.1

5   Conclusion

The research described in this paper aims at assessing the performance of the saliency model of visual attention in the case of natural scenes. The evaluation is based on the comparison of computer saliency maps with human visual attention derived from fixation patterns while subjects are looking at the scenes of interest.

A new aspect of this research is the application to 3D color scenes. In this respect, this study provides for the first time quantitative performance comparisons of different models of visual attention, giving new insights to the contribution of some of its components, namely color and depth.

The experiments involved test persons watching at 3D scenes generated by stereo-vision. An autostereoscopic display was used so that stereo image pairs could be shown to the subjects while recording their eye movements. A first series of experiments refers to the performance in presence of color images. It involves 40 images of different kinds and nature. A second series refers to the performance in presence of color 3D scenes. It involves 12 scenes of different kinds and nature. The number of test persons is 20 in each case.

The eye saccade patterns were then compared to the saliency map generated by the computer. The comparison provides a score (s), i.e. a scalar that measures the similarity of the responses. The higher the score, the better the similarity and the better the performance of the attention model for predicting human attention.

The experiments provide scores covering a wide range of values, i.e. the range is [-5 .. 120] for the images and [5...75] for the 3D scenes. These large score variations illustrate the strong dependence on the kind of scenes: Visual attention of some scenes is very well predicted by the model, while the prediction is quite poor in some cases. These results confirm previous understanding of the model capability and earlier measurements on smaller datasets.

Beyond these large variations, the study shows significant performance differences between the three investigated models. The model performance increases with the model complexity. The performance is first increased when passing from the basic greyscale model to the color model. This is quantitatively assessed by a score increase of 25%. A further performance increase, assessed by a score increase of 11%, characterizes the model extension to depth.

The study therefore confirms the feasibility of a quantitative approach to performance evaluation and provides a first quantitative evaluation of specific models differing by their capacity to deal with color and depth.

Acknowledgements

This work was partially supported by the Swiss National Science Foundation under grant FN 64894. We acknowledge the valuable contribution of René Müri and Roman von Wartburg from University of Berne, Switzerland, who provided the eye saccade data.

References

1. A. A. Kustov and D.L. Robinson. Shared neural control of attentional shifts and eye movements. *Nature, Vol. 384, pp. 74-77*, 1997.
2. D.D. Salvucci. A model of eye movements and visual attention. *Third International Conference on Cognitive Modeling, pp. 252-259*, 2000.
3. C. Privitera and L. Stark. Algorithms for defining visual regions-of-interest: Comparison with eye fixations. *Pattern Analysis and Machine Intelligence (PAMI), Vol. 22, No. 9, pp. 970-981*, 2000.
4. D. Heinke and G.W. Humphreys. Computational models of visual selective attention: A review. *In Houghton, G., editor, Connectionist Models in Psychology*, in press.
5. Ch. Koch and S. Ullman. Shifts in selective visual attention: Towards the underlying neural circuitry. *Human Neurobiology, Vol. 4, pp. 219-227*, 1985.
6. L. Itti, Ch. Koch, and E. Niebur. A model of saliency-based visual attention for rapid scene analysis. *IEEE Transactions on Pattern Analysis and Machine Intelligence (PAMI), Vol. 20, No. 11, pp. 1254-1259*, 1998.
7. N. Ouerhani and H. Hugli. Real-time visual attention on a massively parallel SIMD architecture. *International Journal of Real Time Imaging, Vol. 9, No. 3, pp. 189-196*, 2003.
8. N. Ouerhani, R. von Wartburg, H. Hugli, and R. Mueri. Empirical validation of the saliency-based model of visual attention. *Electronic Letters on Computer Vision and Image Analysis (ELCVIA), Vol. 3 (1), pp. 13-24*, 2004.
9. D. Parkhurst, K. Law, and E. Niebur. Modeling the role of salience in the allocation of overt visual attention. *Vision Research, Vol. 42, No. 1, pp. 107-123*, 2002.
10. T. Jost, N. Ouerhani, r. von Wartburg, R. Muri, and H. Hugli. Assessing the contribution of color in visual attention. *Computer Vision and Image Understanding Journal (CVIU)*, to appear.
11. N. Ouerhani and H. Hugli. Computing visual attention from scene depth. *International Conference on Pattern Recognition (ICPR'00), IEEE Computer Society Press, Vol. 1, pp. 375-378*, 2000.
12. T. Jost, N. Ouerhani, r. von Wartburg, R. Muri, and H. Hugli. Contribution of depth to visual attention: comparison of a computer model and human. *Early cognitive vision workshop, Isle of Skye, Scotland, 28.5. - 1.6*, 2004.

On the Evolution of Formal Models and Artificial Neural Architectures for Visual Motion Detection

R. Moreno-Díaz jr., A. Quesada-Arencibia, and J.C. Rodríguez-Rodríguez

Instituto Universitario de Ciencias y Tecnologías Cibernéticas,
Universidad de Las Palmas de Gran Canaria,Spain
rmorenoj@dis.ulpgc.es
aquesada@dis.ulpgc.es
jcarlos@ciber.ulpgc.es

Abstract. Motion is a key, basic descriptor of our visual experience of the outside world. The lack of motion perception is a devastating illness that leads to death in animals and seriously impaired behavior in humans. Thus, the study of biological basis of motion detection and analysis and the modelling and artificial implementation of those mechanisms has been a fruitful path of science in the last 60 years. Along this paper, the authors make a review of the main models of motion perception that have emerged since the decade of the 60's stress-ing the underlying biological concepts that have inspired most of them and the traditional architectural concepts imprinted in their functionality and design: formal mathematical analysis, strict geometric patterns of neuron-like processors, selectivity of stimulate etc. Traditional approaches are, then, questioned to include "messy" characteristics of real biological systems such as random distribution of neuron-like processors, non homogeneity of neural architecture, sudden failure of processing units and, in general, non deterministic behavior of the system. As a result is interesting to show that reliability of motion analysis, computational cost and extraction of pure geometrical visual descriptors (size and position of moving objects) besides motion are improved in an implemented model.

1   Motion as an Essential Visual Descriptor

Detection and analysis of movement parameters of objects in motion is one of the main tasks of biological visual systems. Initially not seen as a basic sense (as explained in [1]) motion detection is now considered a fundamental visual dimension and a considerable amount of information has been accumulated in the last 50 years on how biological systems (from insects to vertebrates) cope with the problems of extracting and coding vital information on moving targets. Historically, there has been a classification of seven functional benefits of image motion processing for living systems, namely: (1) encoding of the third dimension, (2) an estimation of time to collision, (3) image segmentation, (4) motion as a proprioceptive sense, (5) motion as a stimulus to drive eye movements, (6)

J. Mira and J.R. Álvarez (Eds.): IWINAC 2005, LNCS 3562, pp. 479–488, 2005.

motion as required for pattern vision and (7) the perception of real moving objects. Thus, biological image motion processing might have a large number of rather different roles to play in vision, at least in higher vertebrates, and can be seen as a source for inspiration when designing artificial visual systems which include motion analysis tasks.

In the last years there has been a huge development not only in our knowledge about the biologic bases of vision but in the formulation of computational theories evolving principles of its functioning and make them applicable in non natural con-texts, in human creations, with two goals in mind, understanding the biological systems and building efficient artificial systems. The interconnection between the two disciplines, Biology and Computer Science, has been decisive: using computational patterns has given rise to new ways of understanding Biology as the same time as new computational patterns have emerged as the result of imitating biological systems and its structures [2]. However this task can be complicated as lot of difficulties arise in understanding how Nature works in most of the cases in such an optimal way.

One of the questions brought out about the perception of motion has to do with the process/mechanism that gives raise to this perception. Is the simply the displacement of a visual image within time?. This implies that identifying specific characteristics in a scene is a prerequisite for the detection of motion. However, although the identification of certain circumstances in the image can have an effect in the perception of motion, it is not necessary as it has been proved in some experiments carried out with different species using periodic and statistic patterns with no prominent characteristics [3].

Is a basic visual dimension as it happens with the perception of color or the stereoscopic vision so that there are basic sensorial processes that extract information about motion or on the contrary is a dimension derived from primitive sensorial processes?

Although the perception of color has always been considered as a fundamental dimension, an "immediate" experience with associated basic sensorial processes that extract information about that perception (there are photoreceptors, specifically the cones, sensitive to color, with three types of cones characterized by having a type of photopigment sensitive to different wavelengths), it is not the same when we talk about the perception of motion: there have been always doubts whether to consider it as a fundamental perception that represents a basic cognitive process or as a characteristic that is reconstructed within our visual system in upper levels [1].

In the 1950's the discovery of techniques that allow to register the activity of individual cells contributed to more evidences. Several experiments found nerve cells sensitive to moving images: records were obtained where the frequency of response of certain cells varied with the modification whether the orientation of the stimulus or its speed, regardless of other parameters like contrast, shape or size [4].

Therefore, motion is a fundamental property of the biologic visual systems that can be isolated from other subsystems using different experimental techniques. The different functions in which it is involved make us think that there are different systems for processing motion that operates in parallel. Each one could carry out different functions and, even more, it is possible that a particular functional application can be carried out by more than one subsystem.

In addition of that, the subsystem of motion plays an important role in the attention and orientation system. However, this influence seems to be reciprocal as some experiments carried out basically with primates show an important effect of the attention in the process of visual information about motion [5]. Taking this studies as the reference point, the authors have proposed a model which reflects the interconnection of both subsystems [6], [7].

2   Essentials of Formal Traditional Motion Detection Models

There are a huge number of theories regarding motion and different types of algorithms that allow to extract characteristics as the orientation and the speed of motion. Under a theoretical point of view, a local mechanism for motion detection must satisfy certain minimum requisites to be able to detect motion in a certain orientation.

Some models characterize in formal terms the processes involved in the detection of motion whereas others deal with the problem under the point of view of the cellular mechanisms that allow that type of processes. Independently of the level of description, the different biological schemes of motion detection can be classified in two main categories: the gradient models and the correlation-type models. Whereas in the gradient models an estimate of the local motion is carried out related with the changes in space and time measured simultaneously in the local intensity of the light of the image, in the schemes of correlation it is carried out evaluating some king of spatio-temporal correlation of a filtered signals of two points of the image present at the retina.

The gradient models emerged from the studies of computer vision whereas the correlation-type detectors of motion were deducted from experimental studies of the observed behavior in the motion detection system of insects.

The gradient models, at least in its mathematical formulation, obtain a measure of the local speed dividing the temporal gradient dI/dt by the space gradient dI/dx of the pattern (x and t are referred to the variables space and time respectively; I makes reference to the intensity of the light), that is

$$V_x = -\frac{dI/dt}{dI/dx}$$

The basic operations of a correlation-type detector are summarized in the figure 1. In its most simple form it operates directly over the intensity distribution present on the retina and executes a multiplication (non-lineal operation) as interaction between the two input channels. These channels are separated a certain distance Dj, called "sample base". This distance determines the spatial

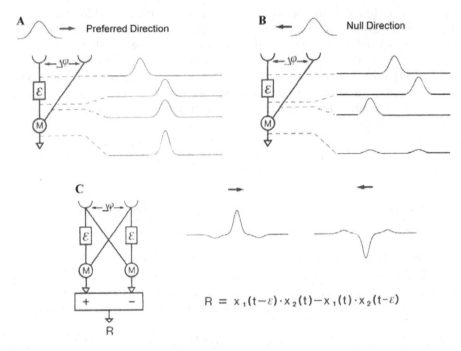

Fig. 1. Motion detection process by a correlation-type detector

resolution of the system and its dependence of the components of the spatial frequency of the stimulus. If this distance is excessively small, the spatial "aliasing" effect can be produced and would confuse the system and if it is too big would produce a loss of sharpness of vision. When the signal in one of the input channels, being activated at first by a moving stimulus, delays a certain time interval, e, the signals in both channels tend to match up in the point M, where both signs are multiplied with the result of a greater amplitude answer (Fig. 1A). Consequently, when the temporal sequence of stimulation is inverted (motion in opposite direction) the temporal displacement between both signals is increased by the delay introduced, e, giving small answers as a result (Fig. 1B). Obviously, the delay introduced and the distance between the two detectors determine the optimal speed of the detector and, consequently, its dynamic range. The combination of the temporal delay and the multiplication between the input channels of the detector is the reason why this type of detectors measure the degree of coincidence of the input signals or, in other words, execute a space-time type correlation.

However, a motion detector like the one shown in figure 1A and B also produces outputs that not only correspond with the result of a motion of the stimulus but are induced by input signals with a high degree of correlation like the variation of the illumination of the environment. To eliminate this phenomenon, a third correlation-type motion detector is shown in figure 1C. This detector is made out of two opposite symmetric units each one with a delay unit and a layer

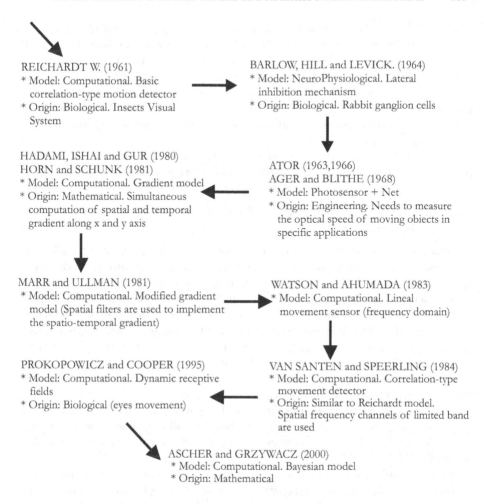

REICHARDT W. (1961)
* Model: Computational. Basic
  correlation-type motion detector
* Origin: Biological. Insects Visual
  System

BARLOW, HILL and LEVICK. (1964)
* Model: NeuroPhysiological. Lateral
  inhibition mechanism
* Origin: Biological. Rabbit ganglion cells

HADAMI, ISHAI and GUR (1980)
HORN and SCHUNK (1981)
* Model: Computational. Gradient model
* Origin: Mathematical. Simultaneous
  computation of spatial and temporal
  gradient along x and y axis

ATOR (1963,1966)
AGER and BLITHE (1968)
* Model: Photosensor + Net
* Origin: Engineering. Needs to measure
  the optical speed of moving obiects in
  specific applications

MARR and ULLMAN (1981)
* Model: Computational. Modified gradient
  model (Spatial filters are used to implement
  the spatio-temporal gradient)

WATSON and AHUMADA (1983)
* Model: Computational. Lineal
  movement sensor (frequency domain)

PROKOPOWICZ and COOPER (1995)
* Model: Computational. Dynamic receptive
  fields
* Origin: Biological (eyes movement)

VAN SANTEN and SPEERLING (1984)
* Model: Computational. Correlation-type
  movement detector
* Origin: Similar to Reichardt model.
  Spatial frequency channels of limited band
  are used

ASCHER and GRZYWACZ (2000)
* Model: Computational. Bayesian model
* Origin: Mathematical

Fig. 2. Summary of the main motion analysis models proposed since the 1960's

where both signs are multiplied. The outputs of both units are subtracted in a
third level so that the answer to motion in opposite direction generates a sign
with the same amplitude but opposite signs. However, this is only true if the
motion detector is mathematically perfect, supposition that is not completely
real given the known properties of the neuronal "hardware". There-fore, a biological motion detector can be not strictly selective to motion, but it is expected
to respond, at least to a certain degree, to the temporal modulation of the light
of a stationary stimulus.

So far we have spoken about two clearly differentiated schemes: the gradient
models and the correlation-type models. However, we can see the motion in a
visual image as a spatio-temporal event that can be represented mathematically
as a Luminance Function $L(x, y, t)$ that depends on two variables: space (x,y)
and time (t). Similarly, this function can be expressed in the frequency domain

were $L(Fx, Fy, Ft)$ is the Fourier transform of $L(x, y, t)$. Since the first processes of visual information codify simultaneously and in an efficient way the spatial and temporal frequency it is interesting to think in terms of space-temporal frequency. Some studies in the 80's allowed, using this approximation, to extract some characteristics of the stimulus used in the experiments. So, for instance, the results of some experiments in which an "apparent" motion of the stimulus is detected, although it doesn't really exist, were explained using a representation in the frequency domain (Fourier).

Figure 2 shows a summary of the main models proposed since the 1960's, period in which the first explicit models about motion analysis are dated.

2.1   Local Motion Detectors and Consequences of the Spatial Integration

Correlation-type individual motion detectors don't give an exact estimation of the lo-cal speed of the pattern. Its response to motion of any pattern, even with constant speed, it is not constant, it is modulated with the passing of time depending on the speed and texture of the pattern. Because of the output signals of a matrix of motion detectors would be out of phase, probably the most simple way of enduring these modulations would consist in carrying out some type of temporal or spatial integration over a sufficiently big group of motion detectors. It is especially interesting the fact that the spatial integration is a quite common mechanism in cells related with motion information processes not only in vertebrates but in invertebrates.

In humans, it is clear the importance of the spatial integration in the perception of motion in several psychophysics studies. Let's illustrate with two examples the responses of individual detectors of motion and the consequences of spatial integration (Fig. 3).

On the visual cortex of the cats, the responses of simple cells sensitive to motion and its directional selectivity to motion of periodic pattern are modulated in the time (Fig. 3C). On the other hand, complex cells combining the outputs of simple cells with antagonistic subfields [8] provide non-modulated responses (Fig. 3D). This match not only the predictions for the responses of local motion detectors but the pre-dictions for the responses of spatiality integrated detectors.

In the visual system of the insects, it has also been possible to investigate the responses of individual detectors of motion and the consequences of spatial integration within the same neuron. So, using anatomical and operational criteria, it has been possible to register the response of a fly cell sensitive to motion and directional selectivity called "HS cell". Since this cells receive inputs from a great matrix of local detectors of motion it is possible to obtain easily the responses of a detector of motion with temporal integration. On the other hand, it is possible to obtain the response of local detectors if we avoid the temporal integration showing the stimulus to the animal through a small slot. The results are shown in Fig. 3A and B respectively. In this way it was demonstrated, as it

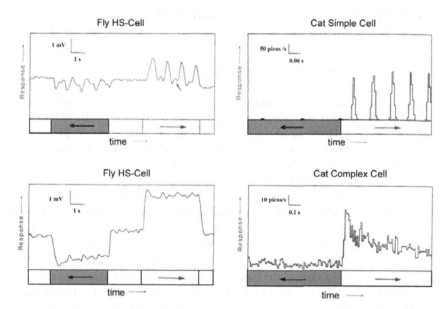

Fig. 3. Local response with spatial integration of two biological motion detectors: the HS cell of the fly and the simple cell in cat

can be predicted from the correlation-type models and ac-cording to the results obtained from the cortical cells of the cat, that the responses of the individual detectors was modulated in time (Fig. 3A) whereas this temporal modulation disappeared as a consequence of the spatial integration (Fig. 3B).

Having reached this stage, it is advisable to notice that the temporal modulations of the responses of the local detectors cannot be predicted using the gradient models, at least in its pure mathematic form.

With this results it is possible to characterize the non-lineal interaction that lies be-hind the detection of motion. If the non-linearity in the motion detection system is a second order one, for instance a multiplication, and no non-lineal process takes place over the visual inputs before and after the detection of motion, the response to the motion of a grille pattern with luminosity following a sinusoid function and at constant speed should contain only the fundamental harmonic and the second harmonic of the temporal frequency of the stimulus and this match the results shown above (Fig. 3A and C).

Unlike the gradient models which represent a pure speed sensor, a correlation-type mechanism of detection of motion does not point out correctly the local motion in the image on the retina in terms of its direction and speed. Also, its outputs depend as well on the structure of the stimulus, for instance on the content of the spatial frequency and the contrast.

The available most relevant biologic experimental data match the predictions carried out from the correlation-type detectors of motion. However, some psychophysics studies carried out in human beings are interpreted from the gradient scheme.

3   A Different Approach: Randomness, Non Homogeneous Architectural Parameters and Multichannel Processing Are Key Tools for Improving Motion Detection Performance

Formal mathematical tools, as seen in previous text, are a powerful way to describe the microstructural behavior of processes that underly in motion detection and analysis. However are of a limited use when trying to implement in a parallel fashion a computational counterpart of a described biological system. In what follows, some results are reviewed from previous material by the authors in which messy concepts like randomness, varying spatial characteristics of neuron-like processors and varying number of neurons are combined to yield a new perspective on the building of visual models. A working vision system for estimating size, location and motion of an object by using a set of randomly distributed receptive fields on a retina has been proposed by the authors [7], [9]. The used approach differs from more conventional ones in which the receptive fields are arranged in a geometric pattern. From the input level, computations are performed in parallel in two different channels: one for purely spatial properties, the other for time-space analysis, and are then used at a subsequent level to yield estimates of the size and center of gravity (CG) of an object and the speed and direction of motion (Fig. 4 and 5). Motion analysis refining is implemented by a lateral interaction (spatial) and memory (temporal) schemes in which direction and speed are used to build a trajectory. The different parameters involved (receptive field RF size, memory weighting function, number of cells) are tested for different speeds and the results compared, yielding new insights on the functioning of the living retina and suggesting ideas for improving the artificial system. A tetra-processor UltraSparc SUN computer was used for simulation and video-outputs in false color show the two-channel activity of the system. A train of input images presenting a moving target was analyzed by the neuron-like processing layers and the results presented as a video movie showing a color coded version of neural activity.

The random distribution of receptive fields obviates the necessity of having a deterministically organized system and it seems that the natural system which has inspired the model does not use a perfectly established network either. Also, taking the number of total cells as a variable parameter is a way of checking the reliability of the system against random loss of computational units.

Seen as a whole, and comparing the location of the CG of the object calculated separately by both channels the results give us some perspective on the usefulness of channel processing in artificial visual systems which could also be of interest in trying to think of a rationale for the same kind of computations in natural perceptual systems. Though an estimate of the position of the object can be calculated from the velocity channel, it is not as good nor as fast in its delivery as it is when presented by the purely spatial CG-size channel, and is greatly dependent on the number of receptive fields that are used. The system could work properly for all descriptors (CG, size, position, speed), in certain

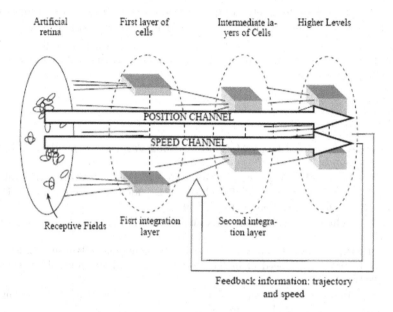

Fig. 4. Representation of a biological based visual motion analysis system

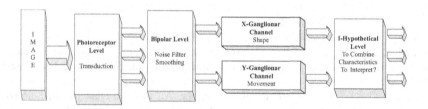

Fig. 5. Specific processing scheme applied to an image reaching our artificial retina

cases, with only the output from the velocity channel, but if precision is needed, then a second channel must be included. In either case, speed and direction of motion can be finely estimated.

From the data obtained in the tests varying the total number of receptive fields one conclusion is the following: a reliable system for computing speed and trajectory of a moving object can be built whose computational complexity and cost (in terms of number of processors, distribution and size of receptive fields) can be controlled to fall within certain desirable ranges. The built-in parallelism of the system allows us to play with those parameters avoiding the increase of the percent error in estimated speed. Thus loss of processors (loss of receptive fields) need not dramatically affect the performance of one subsystem (such as the movement detection and analysis system) provided there is a minimum overlap and the RF sizes are big enough to cope with a range of speeds. A possible extension on which we are currently working, is a system containing several subchannels (for, e.g., effectively computing different speeds or different other descriptors) which might be more reliable and less costly (both in complexity and computational operations) than a single do-it-all channel, even when the number of processors of this last channel could be less than the total sum of them in the former scheme.

References

1. Nakayama, K.: Biological Image Motion Processing: A review. Vision Research, **25**, N°5, (1985) 625–660
2. Cull, P.: Biologically Motivated Computing, Computer Motivated Biology. Wierner's Cy-bernetics:50 years of evolution, Extended Abstracts. Moreno-Díaz, Pichler, Ricciardi, Moreno-Díaz jr Eds (1999)
3. Borst, A., Egelhaaf, M.: Principles of Visual Motion Detection. TINS, **12**, (1989) 297–306
4. Barlow, H. B., Hill, R. M., Levick, W. R.: Retinal ganglion cells responding selectively to direction and speed of image motion in the rabbit. Journal Physiology, **173**, (1964) 377–407
5. Treue, S., Maunsell, J.: "Attentional modulation of visual motion processing in cortical areas MT and MST. Nature, **382**, (1996) 539–541
6. Quesada-Arencibia, A.: Un Sistema Bio-inspirado de análisis y seguimiento visual de movimiento. Implicaciones en Teoría Retinal, Redes Neuronales y Visión Artificial. Tesis Doctoral. Universidad de Las Palmas de Gran Canaria (2001)
7. Quesada-Arencibia, A., Moreno-Díaz jr, R., Alemán-Flores, M.: Biologically Based CAST-Mechanism for Visual Motion Analysis. Springer Lecture Notes in Computer Science, **2178**. Springer-Verlag, Berlin Heidelberg New York (2001) 316–327
8. Hubel, D. H., Wiesel, T. N.: Receptive fields, binocular interaction and functional architec-ture in the cat´s visual cortex. Journal Physiology, London, **160**, (1962) 106–154
9. Rodríguez-Rodríguez, J.C., Quesada-Arencibia, A., Moreno-Díaz jr, R., Leibovic, K.N.: On Parallel Channel Modeling of Retinal Processes. Springer Lecture Notes in Computer Sci-ence, **2809**. Springer-Verlag, Berlin Heidelberg New York (2003) 471–48

Estimation of Fuel Moisture Content Using Neural Networks

D. Riaño[1,2], S.L. Ustin[2], L. Usero[2,3], and M.A. Patricio[3]

[1] Dpto. de Geografía, U. de Alcalá, Colegios 2,
E-28801 Alcalá de Henares, Madrid, Spain
[2] Center for Spatial Technologies and Remote Sensing,
U. California. One Shields Ave. 95616-8617 Davis, CA, USA
[3] Dpto. de Ciencias de la Computación. U. de Alcalá,
Campus Universitario, E-28871 Alcalá de Henares, Madrid, Spain
{driano, slustin, lusero}@ucdavis.edu, miguel.patricio@uah.es

Abstract. Fuel moisture content (FMC) is one of the variables that drive fire danger. Artificial Neural Networks (ANN) were tested to estimate FMC by calculating the two variables implicated, equivalent water thickness (EWT) and dry matter content (DM). DM was estimated for fresh and dry samples, since water masks the DM absorption features on fresh samples. We used the "Leaf Optical Properties Experiment" (LOPEX) database. 60% of the samples were used for the learning process in the network and the remaining ones for validation. EWT and DM on dry samples estimations were as good as other methods tested on the same dataset, such as inversion of radiative transfer models. DM estimations on fresh samples using ANN ($r^2 = 0.86$) improved significantly the results using inversion of radiative transfer models ($r^2 = 0.38$).

1 Introduction

Fuel moisture content (FMC) can be defined as the amount of water per unit dry matter. This variable conditions fire, since the drier the vegetation the easier fires ignite and propagate [1].

FMC is traditionally measured directly through field sampling. This method requires a lot of labour. The use of remote sensing to measure spectral properties of leaves can provide an indirect FMC estimation in order to obtain a comprehensive spatial and temporal distribution. Water stress causes changes in the spectral reflectance and transmittance of leaves [18]. Radiative transfer models show that these spectral measurements are related to equivalent water thickness (EWT), water content per area unit, and dry matter content (DM), matter content per area unit [2], [7] and [12]. FMC is a quotient of these two variables, EWT and DM, that can be estimated independently [16].

Different remote sensing instruments have been used to predict these kind of biochemical variables at different scales measuring their spectral properties. Spectrophotometers that can predict the properties of individual leaves (Figure 1) [2] and [16]. Field spectroradiometers measure the properties of vegetation

J. Mira and J.R. Álvarez (Eds.): IWINAC 2005, LNCS 3562, pp. 489–498, 2005.

Fig. 1. Example of a Spectrophotometer. Perkin-Elmer Lambda-19 Spectrophotometer. Spectral range 250-2500 nm with a 1.00 nm slit width (www.grc.nasa.gov/WWW/epbranch/OpticalProps/lambda.htm)

Fig. 2. Example of a field spectroradiometer. Analytical Spectral Devices FieldSpec ProFR, spectral range 350-2500 nm with a 3-10 nm slit width. (www.asdi.com/products_specifications-FSP.asp)

canopies taken a single spectrum, averaging whatever is seen in their field of view (Figure 2) [16]. Hyperspectral airborne sensors that provide spatial information about the spectral properties of the vegetation (Figure 3) [17] and [19]. Hyperspectral and multispectral satellites that provide repeatable spectral properties of the vegetation at various spatial scales are able to address not only the local fire risk but the regional or global scale risk (Fig. 4) [4], [5], [6] and [20].

Different methods have been used to extract FMC from the spectral measurements. One possible way to estimate FMC is to apply an empirical fitting using linear regression based on band ratios [3], [5], [6], [10] and [14]. This approach

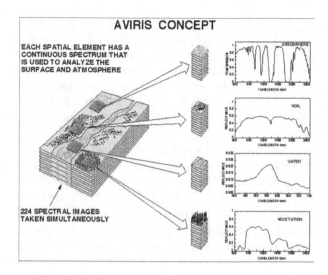

Fig. 3. Example of a hyperspectral airborne sensor. Advanced Visible InfraRed Imaging Spectrometer (AVIRIS), spectral range between 380 -2500 nm with a 10 nm slit width. Spatial resolution, 4 or 20 m, depending on flight altitude. (aviris.jpl.nasa.gov)

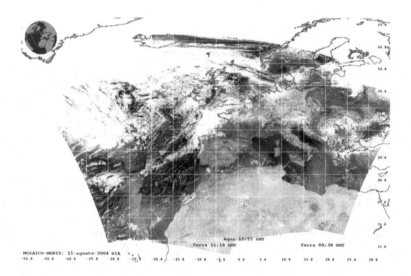

Fig. 4. Example of a hyperspectral satellite. Moderate resolution Imaging Spectrometer (MODIS). 36 bands, spectral range between 405 -14385 nm with a slit width about 20 nm, depending on bands. Spatial resolution, 250 to 1000 m, depending on bands. (www.latuv.uva.es/modisql)

provides a model of known accuracy but needs to be recalibrated in order to be applied to other study areas.

The inversion of radiative transfer models provide a physical estimation of EWT and DM [9], [13], [16] and [20]. This technique is commonly based on information in the entire spectrum instead of band ratios. The physically based method can be extrapolated to other study areas, applying a different model if vegetation type changes. On the other hand, it's use requires intensive computation time, being difficult to apply to an entire image. DM results are more difficult to estimate than EWT, since water in live vegetation masks the DM absorption features [16].

The work presented here explores the estimation of EWT and DM using Artificial Neural Networks (ANN) and compares the results to the inversion of the radiative transfer PROSPECT leaf model [12] and [16]. For the ANN the EWT and DM estimation are not based on the specific absorption features. This approach has been successfully applied to estimate leaf area index [8] and would have the advantage that if we train the network with a wide range of vegetation samples, the method could be applied to an entire image. ANN can be used as a universal approximator [15]. This means that the ANN can approximate any function after a sufficient amount of learning. This is a substantial advantage over traditional statistical prediction models, as the relationship between the data entry and output is highly non-linear with significant, but complex, interactions among both of them.

2   Methods

We used the standard LOPEX dataset produced during an experiment conducted by the Joint Research Center (JRC) of the European Commission (Ispra, Italy) [11]. They measured reflectance and transmittance with a Perkin-Elmer Lambda-19 Spectrophotometer and different biochemical constituents, such as EWT and DM, of a wide variety of species. We selected 49 broad leaf species sampled from 37 different species, collected over an area within 50 km range of the JRC, in Ispra. Each sample contained 5 fresh and 5 near-dry leaves.

We applied ANN to predict EWT and DM from reflectance and transmittance measurements of fresh and dry samples. A total of 2101 measurements were made between 400-2500 nm and linear interpolation was used to select a value every 5 nm, giving a total of 421 reflectance and transmittance values, having a total of 842 input variables. Among the myriad of ANN methods, we chose the feedforward multilayer perceptron trained with the backpropagation learning algorithm. Three different ANN were built:

1. To estimate EWT on fresh leaves.
2. To estimate DM on fresh leaves.
3. To estimate DM on dry leaves.

The JavaNNS (Java Neural Network Simulator) 1.1 was used for ANN development and a specific Java programs for data manipulating. We divide the samples into two data sets: learning and validation data set. Each sample was formed from five leaves, therefore three of them (60 per cent of each sample) were used for the learning data set and the remaining leaves used as validation. A sigmoidal activation function was used in all neurons and a learning rate of 0.2. Since the input and output data are presented to the network in raw format, these required normalizing over the range [0, 1].

There are a few ways to systematically tune the size of the network. In our work, a method of controlled trial and error was used to optimize the network size and make up a network structure. The basic idea of this size optimizing method was to probe different numbers of neurons in the hidden layer with different initial weights. Each time a network was trained, it was tested with the validation set. Through extensive experimentation we finally selected the ANN architecture with a hundred neurons in the hidden layer and one neuron in the output layer (Figure 5).

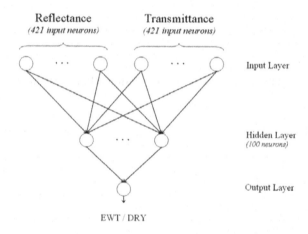

Fig. 5. ANN architecture

3   Results

The estimation of EWT and DM for fresh or dry samples using ANN worked well in all cases (Figure 6 to Figure 11). We compared our results with the estimation of these variables using inversion of radiative transfer models (IRTM) [16] (Table 1). EWT was estimated very accurately with both methods while DM for dry samples was predicted a little bit better by ANN. On the other hand, DM for fresh samples was only accurately estimated by ANN. IRTM worked poorly because DM absorption features are masked by water when samples are fresh [16]. There were complex interactions in the spectra that affect DM that were not captured by the absorption features in the IRTM ($r^2 = 0.38$) but were captured markedly well in the ANN ($r^2 = 0.96$, learning; $r^2 = 0.86$, validation). The learning process included several leaves of the same species, so further research

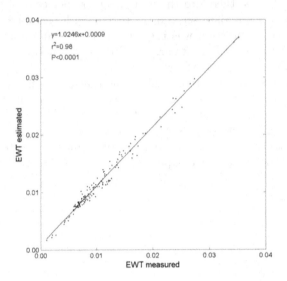

Fig. 6. EWT estimation with data used for learning

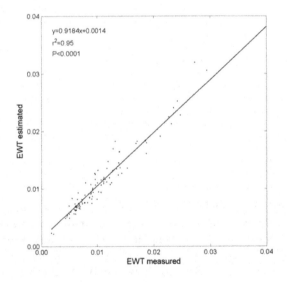

Fig. 7. EWT estimation with data used for validation

is needed to test if ANN will work when species that were not used in the learning process are included in the validation.

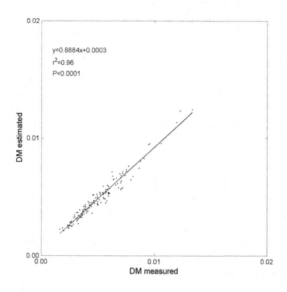

Fig. 8. DM estimation with data used for learning for fresh samples

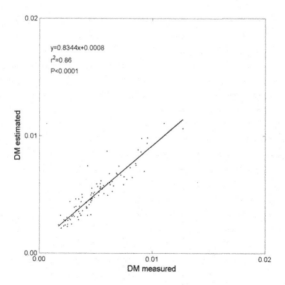

Fig. 9. DM estimation with data used for validation for fresh samples

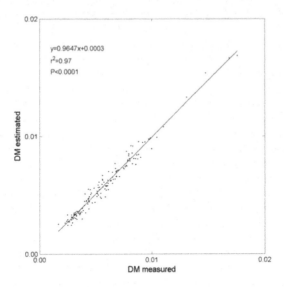

Fig. 10. DM estimation with data used for learning for dry samples

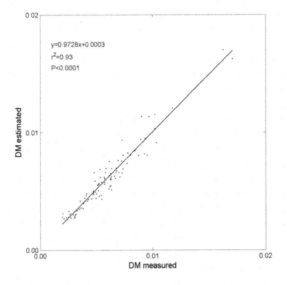

Fig. 11. DM estimation with data used for validation for dry samples

Table 1. Number of samples (n) and correlation (r^2) for the estimation of EWT and DM using IRTM [16] and ANN

	n	EWT	DM (fresh)	DM (dry)
IRTM	245	0.94	0.38	0.85
ANN learning	147	0.98	0.96	0.97
ANN validation	98	0.95	0.86	0.93

Acknowledgements

We thank Brian Hosgood for providing the LOPEX dataset and support.

References

1. Burgan, R.E. & R.C. Rothermel 1984. BEHAVE: Fire Behavior Prediction and Fuel Modeling System. Fuel Subsystem. GTR INT-167, USDA Forest Service, Ogden, Utah.
2. Ceccato, P., N. Gobron, S. Flasse, B. Pinty & S. Tarantola 2002, Designing a spectral index to estimate vegetation water content from remote sensing data: Part 1 - Theoretical approach. Remote Sensing of Environment. 82(2-3): 188-197.
3. Chladil, M.A. & M. Nunez 1995, Assessing grassland moisture and biomass in Tasmania. The application of remote sensing and empirical models for a cloudy environment. International Journal of Wildland Fire. 5: 165-171.
4. Chuvieco, E., I. Aguado, D. Cocero & D. Riaño 2003, Design of an Empirical Index to Estimate Fuel Moisture Content from NOAA-AVHRR Analysis In Forest Fire Danger Studies. International Journal of Remote Sensing. 24(8): 1621-1637.
5. Chuvieco, E., D. Cocero, D. Riaño, P. Martin, J. Martínez-Vega, J. de la Riva & F. Pérez 2004, Combining NDVI and Surface Temperature for the estimation of live fuels moisture content in forest fire danger rating. Remote Sensing of Environment. 92(3): 322-331.
6. Chuvieco, E., D. Riaño, I. Aguado & D. Cocero 2002, Estimation of fuel moisture content from multitemporal analysis of Landsat Thematic Mapper reflectance data: applications in fire danger assessment. International Journal of Remote Sensing. 23(11): 2145-2162.
7. Danson, F.M. & P. Bowyer 2004, Estimating live fuel moisture content from remotely sensed reflectance. Remote Sensing of Enviroment. 92(3): 309-321.
8. Danson, F.M., C.S. Rowland & F. Baret 2003, Training a neural network with a canopy reflectance model to estimate crop leaf area index. International Journal Of Remote Sensing. 24(23): 4891-4905.
9. Fourty, T. & F. Baret 1997, Vegetation water and dry matter contents estimated from top-of-the atmosphere reflectance data: a simulation study. Remote Sensing of Environment. 61: 34-45.
10. Hardy, C.C. & R.E. Burgan 1999, Evaluation of NDVI for monitoring live moisture in three vegetation types of the Western U.S. Photogrammetric Engineering and Remote Sensing. 65: 603-610.

11. Hosgood, B., S. Jacquemoud, G. Andreoli, J. Verdebout, A. Pedrini & G. Schmuck 1994. The JRC Leaf Optical Properties Experiment (LOPEX'93). CL-NA-16095-EN-C, EUROPEAN COMMISSION, Directorate - General XIII, Telecommunications, Information Market and Exploitation of Research, L-2920, Luxembourg.
12. Jacquemoud, S. & F. Baret 1990, Prospect - a Model of Leaf Optical-Properties Spectra. Remote Sensing of Environment. 34(2): 75-91.
13. Jacquemoud, S., S.L. Ustin, J. Verdebout, G. Schmuck, G. Andreoli & B. Hosgood 1996, Estimating Leaf Biochemistry Using the PROSPECT Leaf Optical Properties Model. Remote Sensing of Environment. 56: 194-202.
14. Paltridge, G.W. & J. Barber 1988, Monitoring grassland dryness and fire potential in Australia with NOAA/AVHRR data. Remote Sensing of Environment. 25: 381-394.
15. Poggio, T. & F. Girosi 1989. A theory of networks for approximation and learning. Technical Report AI 1140, MIT, Cambridge, MA.
16. Riaño, D., P. Vaughan, E. Chuvieco, P.J. Zarco-Tejada & S.L. Ustin *in press*, Estimation of fuel moisture content by inversion of radiative transfer models to simulate equivalent water thickness and dry matter content. Analysis at leaf and canopy level. IEEE Transactions on Geoscience and Remote Sensing.
17. Serrano, L., S.L. Ustin, D.A. Roberts, J.A. Gamon & J. Peñuelas 2000, Deriving Water Content of Chaparral Vegetation from AVIRIS Data. Remote Sensing of Environment. 74: 570-581.
18. Ustin, S.L. 2004. Remote Sensing for Natural Resource Management and Environmental Monitoring. John Wiley & Sons, Incorporated.
19. Ustin, S.L. et al. 1998, Estimating canopy water content of chaparral shrubs using optical methods. Remote Sensing of Environment. 65: 280-291.
20. Zarco-Tejada, P.J., C.A. Rueda & S.L. Ustin 2003, Water content estimation in vegetation with MODIS reflectance data and model inversion methods. Remote Sensing of Environment. 85: 109-124.

Adjustment of Surveillance Video Systems by a Performance Evaluation Function

Óscar Pérez, Jesús García, Antonio Berlanga, and José M. Molina**

Universidad Carlos III de Madrid, Departamento de Informática,
Avenida de la Universidad Carlos III, 22 Colmenarejo 28270, Madrid, Spain
oscar.perez.concha@uc3m.es, jgherrer@inf.uc3m.es,
antonio.berlanga@uc3m.es, molina@ia.uc3m.es

Abstract. This paper proposes an evaluation metric for assessing the performance of a video tracking system. This metric is applied to adjust the parameters that regulates the video tracking system in order to improve the system perfomance. Thus, the automated optimization method is based on evolutionary computation techniques. The illustration of the process is carried out using three very different video sequences in which the evaluation function assesses trajectories of airplanes, cars or baggage-trucks in an airport surveillance application.

1 Introduction

The application of video cameras for remote surveillance has increased rapidly in the industry for security purposes. The installation of many cameras produces a great problem to human operators because the incompatibility of a high analysis of received images with the analysis of the whole information provided for the surveillance video camera net. The solution is the automatic analysis of video frames to represent in a simplify way the video information to be presented to the operator. A minimal requirement for automatic video surveillance system is the capacity to track multiple objects or groups of objects in real conditions [1].

The main point of this research consists in the evaluation of surveillance results, defining a metric to measure the quality of a proposed configuration [2]. The truth values from real images are extracted and stored in a file [3] and [4]. To do this, the targets are marked and positioned in each frame with different attributes. Using this metric in an evaluation function, we can apply different techniques to assess suitable parameters and, then, to optimize them. Evolution Strategies (ES) are selected for this problem [5][6][7][8] and [9] because they present high robustness and immunity to local extremes and discontinuities in fitness function. This paper demonstrates that the proposed evaluation function correctly guides the ES optimization in this type of problems. The desired results are reached once an appropriate fitness function has been defined. This allows an automatic adjustment of tracker performance accordingly to all specifications

** Funded by CICYT (TIC2002-04491-C02-02).

J. Mira and J.R. Álvarez (Eds.): IWINAC 2005, LNCS 3562, pp. 499–508, 2005.
© Springer-Verlag Berlin Heidelberg 2005

considered. Furthermore, one of the principal points of this study is that the evaluation and potential optimization are not dependent on the specific type of tracking system used.

In the next section, the whole surveillance system is presented, where specific association problems in this application are analyzed. The third section presents the proposed metric. In section fourth, the system output in several scenarios is presented, indicating the response for complex situations, with real image sequences of representative ground operations. Finally, some conclusions are presented.

2   Surveillance Video System

This section describes the structure of an image-based tracking system.

The system architecture is a coupled tracking system where the detected objects are processed to initiate and maintain tracks. These tracks represent the real targets in the scenario and the system estimates their location and cinematic state. The detected pixels are connected to form image regions referred to as blobs. The association process assigns one or several blobs to each track, while not associated blobs are used to initiate tracks [4].

2.1   Detector and Blobs Extraction

The positioning/tracking algorithm is based on the detection of targets by contrasting with local background, whose statistics are estimated and updated with the video sequence. Then, the pixel level detector is able to extract moving features from background, comparing the difference with a threshold. To illustrate the process, figure 2 depicts the different levels of information interchanged, from the raw images until the tracks.

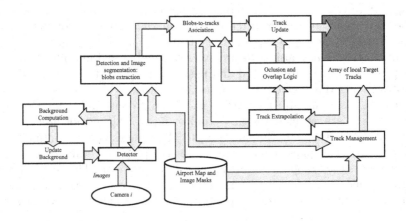

Fig. 1. Structure of video surveillance system

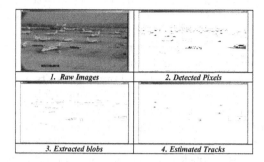

| 1. Raw Images | 2. Detected Pixels |
| 3. Extracted blobs | 4. Estimated Tracks |

Fig. 2. Information levels in the processing chain

Finally, the algorithm for blobs extraction marks with a unique label all detected pixels connected, by means of a clustering and growing regions algorithm [10]. Then, the rectangles which enclose the resulting blobs are built, and their centroids and areas are computed. In order to reduce the number of false detections due to noise, a minimum area, MIN-AREA, is required to form blobs. This parameter is a second data filter which avoids noisy detections from the processing chain.

2.2   Blobs-to-Track Association

The association problem lies in deciding the most proper grouping of blobs and assigning it to each track for each frame processed. Due to image irregularities, shadows, occlusions, etc., a first problem of imperfect image segmentation appears, resulting in multiple blobs generated for a single target. So, the blobs must be re-connected before track assignment and updating. However, when multiple targets move closely, their image regions may overlap. As a result, some targets may appear occluded by other targets or obstacles, and some blobs can be shared by different tracks. For the sake of simplicity, first a rectangular box has been used to represent the target. Around the predicted position, a rectangular box with the estimated target dimensions is defined, (xmin, xmax, ymin, ymax). Then, an outer gate, computed with a parameter defined as a margin, MARGIN-GATE, is defined. It represents a permissible area in which to search more blobs, allowing some freedom to adapt target size and shape. The association algorithm analyses the track-to-blob correspondence. It firsts checks if the blob and the track rectangular gates are compatible (overlap), and marks as conflictive those blobs which are compatible with two or more different tracks. After gating, a grouping algorithm is used to obtain one "pseudoblob" for each track. This pseudoblob will be used to update track state. If there is only one blob associated to the track and the track is not in conflict, the pseudoblob used to update the local track will be this blob. Otherwise, two cases may occur:

1. A conflict situation arises when there are overlapping regions for several targets (conflicting tracks). In this case, the system may discard those blobs

gated by several tracks and extrapolate the affected tracks. However, this policy may be too much restrictive and might degrade tracking accuracy. As a result, it has been left open to design by means of a Boolean parameter named CONFLICT which determines the extrapolation or not of the tracks.

2. When a track is not in conflict, and it has several blobs associated to it, these will be merged on a pseudoblob whose bounding limits are the outer limits of all associated blobs. If the group of compatible blobs is too big and not dense enough, some blobs (those which are further away from the centroid) are removed from the list until density and size constraints are held. The group density is compared with a threshold, MINIMUM-DENSITY, and the pseudo-blob is split back into the original blobs when it is below the threshold.

2.3   Tracks Filtering, Initiation and Deletion

A recursive filter updates centroid position, rectangle bounds and velocity for each track from the sequence of assigned values, by means of a decoupled Kalman filter for each Cartesian coordinate, with a piecewise constant white acceleration model [11]. The acceleration variance that will be evaluated, usually named as "plant-noise", is directly related with tracking accuracy. The predicted rectangular gate, with its search area around, is used for gating. Thus it is important that the filter is "locked" to real trajectory. Otherwise tracks would lose its real blobs and finally drop. So this value must be high enough to allow manoeuvres and projection changes, but not too much, in order to avoid noise. As a result, it is left as an open parameter to be tuned, VARIANCE-ACCEL. Finally, tracking initialization and management takes blobs which are not associated to any previous track. It requires that non-gated blobs extracted in successive frames accomplish certain properties such as a maximum velocity and similar sizes which must be higher than a minimum value established by the parameter MINIMUM-TRACK-AREA. In order to avoid multiple splits of targets, established tracks preclude the initialization of potential tracks in the surrounding areas, using a different margin than the one used in the gating search. This value which allows track initialization is named MARGIN-INITIALIZATION.

3   Evaluation System

The approach used for this work evaluates the detection and tracking system performance using ground truth to provide independent and objective data that can be related to the observations extracted and detected from the video sequence. In each scenario, the ground truth has been extracted frame by frame, selecting the targets and storing the next data for each target:

- Number of analyzed frame
- Track identifier
- Value of the minimum x coordinates of the rectangle that surrounds the target

- Value of the maximum x coordinates of the rectangle that surrounds the target
- Value of the minimum y coordinates of the rectangle that surrounds the target
- Value of the maximum y coordinates of the rectangle that surrounds the target

These ground truth is compared to the real detections by the evaluation system. First of all, the result tracks are checked to see if they match with the ground truth tracks registered in the ground truth table. For example, as we see in the next pictures (figure 3), the real image shows two aircrafts in the parallel taxiways while the tracking system displays three targets. Then, the target which is in the middle of the screen not pass the test and it would be marked as a mismatched track.

Fig. 3. Example of mismatched track

If the test is passed, the evaluation system computes four parameters per target which are classified into 'accurary metrics' and 'continuity metrics':

Accuracy metrics:

- *Overlap-area* (OAP): Overlap Area Percentage between the real and the detected blobs.
- *X-error (E_x) and Y-error (E_y)*: Difference in x and y coordinates between the centers of the ideal blob and the detected blobs.

Continuity metrics:

- *Number of Tracks per target* (NT): It is checked if more than one detected track is matched with the same ideal track. If this happens, the program keeps the detected track which has a bigger overlapped area value, removes the other one and marks the frame with a flag that indicates the number of detected tracks associated to this ideal one.
- *Commutation*(C): A commutation occurs when the identifier of a track matched to an ideal track changes. It typically takes place when the track is lost and recovered later.

Besides these parameters, an evaluation function has been defined, with the objective of extracting a number that measures the quality level of the tracking system. This number is based on the evaluation metrics specified before. Thus, the resultant number is obtained by means of a weighted sum of different terms which are computed target by target:

- Mismatch (M): A counter which stores how many times the ground truth and the tracked object data do not match up (NT is not 1). Furthermore, this counter is normalized by the difference between the last and first frame in which the ideal track disappears and appears (Time of life (T)).
- The three next terms are the total sum of the overlapped areas ($\sum OAP$) and the central errors of x ($\sum E_x$) and y axes ($\sum E_y$). They are normalized by a number which indicates how many times these values are available (there is not continuity problem) in the whole video sequence (DZ).
- The next two elements are two counters:
  - Overmatch-counter (O_c): how many times the ground truth track is matched with more than one tracked object data.
  - Undermatch-counter (U_c): how many times the ground truth track is not matched with any track at all.
- Finally, the last term is the number of commutations in the track under study ($\sum C$). The three last elements are normalized by the same value of normalization as the first one (Time of life, T). It is clear that the lower the Evaluation function, the better the quality of the tracking system. With the objective of minimizing the Evaluation function, the Video Surveillance System has been optimized by ES.

Thus, the evaluation function can be represented as follows:

$$E = \frac{W_1 M}{2\pi} + \frac{W_2 \sum OAP + W_3 \sum E_x + W_4 \sum E_y}{DZ} + \frac{W_5 O_c + W_6 U_c + W_7 \sum C}{T} \tag{1}$$

where $W_{1,2,3,4,5,6,7}$ are the weights for the parameters. Figure 4 depicts the different stages of the Evaluation System in order to have a clear idea of it.

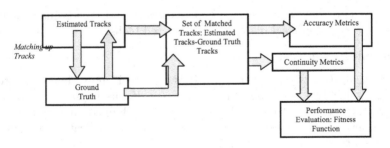

Fig. 4. Evaluation System

4   Perfomance Optimization

This section of the study shows how the analysis of the evaluation system, and its application to ES optimization, improves considerably the performance of a given tracking system. Three parameters explained above are going to be studied in order to see the effects of them in the optimization of the tracking system: the threshold, the minimum area and margin gate. The first example shows the video of the three aircrafts used in the former chapters to explain how the tracking system works. The study is focused on the airplane that moves from the left side to the right side of the screen. The adjusting parameters of the system are randomly selected by the program:

- Threshold: 39
- Minimum Area of blob: 7
- Margin gate: 1.0838

The result of the first execution can be seen in figure 5.

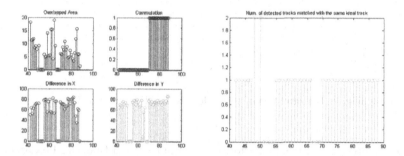

Fig. 5. Performance of example 1 before ES optimization

After using the ES program, the performance of our system improves. The values of the three parameters under study are:

- Threshold: 16.7
- Minimum Area of blob: 3.15
- Margin gate: 10.95

The two first parameters have lower values and the last one is higher. That means, for example, that the criterion for a pixel to be considered as a moving target is less restricted. Then, the sensitivity value and the probability of detection are higher. Moreover, the value of the minimum area that defines a blob is also lower so that much more blobs are considered by the system to likely form future rectangles. And finally, the higher value of margin gate permitted the search of new valuable information around the rectangle to adapt the target size and shape with new information.

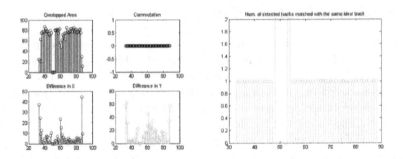

Fig. 6. Performance of example 1 after ES optimization

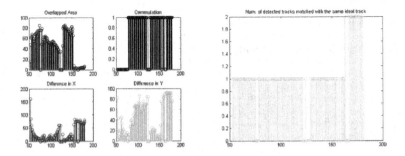

Fig. 7. Performance of example 2 before ES optimization

Thus, the result is a better performance of our system that can be observed in figure 6.

After the optimization of the adjusting parameters, the first estimation track is done earlier (34th frame) than in the previous example (42nd frame) and the track is lost once instead of twice.

The second example takes the second video considered in former chapters and the aircraft that goes from the left side to the right side of the screen as the main target to study. The values of the three parameters on which our study is focus:

- Threshold: 16.7
- Minimum Area of blob: 2
- Margin gate: 2

The surveillance system performance is shown in figure 7.

- Threshold: 38.94
- Minimum Area of blob: 6.9885
- Margin gate: 1.0838

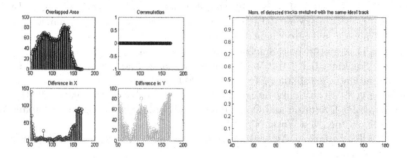

Fig. 8. Performance of example 2 after ES optimization

The surveillance system performance is shown in figure 8.

The new values show the opposite situation that we had in the previous example. The threshold and the minimum area of blob are higher and the margin gate is lower. That means that these values decrease conflicts and interaction among tracks (no commutation), at the same time that the detection probably and the criterion threshold to form a blob are lower.

5   Conclusions

We have presented a novel process to evaluate the performance of a tracking system based on the extraction of information from images filmed by a camera. The ground truth tracks, which have been previously selected and stored by a human operator, are compared to the estimated tracks. The comparison is carried out by means of a set of evaluation metrics which are used to compute a number that represents the quality of the system. Then, the proposed metric has been applied as the argument to the evolutionary strategy (ES) whose function is the optimization of the parameters that rule the tracking system. This process is repeated until the result and the parameters are good enough to assure that the system will do a proper performance. The study tests several videos and shows the improvement of the results for the optimization of three parameters of the tracking system. In future works we will implement the optimization of the whole set of parameters using the results of this paper as valuable background. Furthermore, we plan the evaluation over a high value of videos which present very different number of targets and weather conditions.

References

1. Rosin, P.L., Ioannidis, E.: Evaluation of global image thresholding for change detection, Pattern Recognition Letters, vol. 24, no. 14 (2003) 2345–2356
2. Black, J., Ellis, T., Rosin, P.: A Novel Method for Video Tracking Performance Evaluation, Joint IEEE Int. Workshop on Visual Surveillance and Performance Evaluation of Tracking and Surveillance (VS-PETS) (2003)

3. Piater, J. H., Crowley, J. L.: Multi-Modal Tracking of Interacting Targets Using Gaussian Approximations. IEEE International Workshop on Performance Evaluation of Tracking and Surveillance (PETS) (2001)

4. Pokrajac, D., Latecki, L.J.: Spatiotemporal Blocks-Based Moving Objects entification and Tracking, IEEE Int. W. Visual Surveillance and Performance Evaluation of Tracking and Surveillance (VS-PETS) (2003)

5. Rechenberg, I.: Evolutionsstrategie'94. frommannholzboog, Stuttgart (1994).

6. Schwefel, H.P.: Evolution and Optimum Seeking: The Sixth Generation. John Wiley and Sons, Inc. New York, NY, USA (1995)

7. Bäck, T.: Evolutionary Algorithms in Theory and Practice, Oxford University Press, New York (1996)

8. Bäck, T., Fogel, D.B., Michalewicz, Z.: Evolutionary Computation: Advanced Algorithms and Operators, Institute of Physics, London (2000)

9. Bäck, T., Fogel, D.B., Michalewicz, Z.: Evolutionary Computation: Basic Algorithms and Operators, Institute of Physics, London (2000)

10. Sanka, M., Hlavac, V., Boyle, R.: Image Processing, Analysis and Machine Vision, Brooks/Cole Publishing Company (1999)

11. Blackman, S., Popoli, R.: Design and Analysis of Modern Tracking Systems. Artech House (1999)

Application of Machine Learning Techniques for Simplifying the Association Problem in a Video Surveillance System

Blanca Rodríguez, Óscar Pérez, Jesús García, and José M. Molina**

Universidad Carlos III de Madrid, Departamento de Informática,
Avenida de la Universidad Carlos III, 22 Colmenarejo 28270, Madrid, Spain
mariablanca.rodriguez@uc3m.es, oscar.perez.concha@uc3m.es,
jgherrer@inf.uc3m.es, molina@ia.uc3m.es

Abstract. This paper presents the application of machine learning techniques for acquiring new knowledge in the image tracking process, specifically, in the blobs detection problem, with the objective of improving performance. Data Mining has been applied to the lowest level in the tracking system: blob extraction and detection, in order to decide whether detected blobs correspond to real targets or not. A performance evaluation function has been applied to assess the video surveillance system, with and without Data Mining Filter, and results have been compared.

1 Introduction

Machine learning techniques could be applied to discover new relations among attributes in different domains, this application is named data mining (DM) and it is a part of the knowledge data discovering process (KDD) [1]. The application of data mining techniques to a specific problem takes several perspectives: classification, prediction, optimization, etc. In this work DM techniques will be use to learn a classifier able to determine if a detected surface on an image could be considered as a tentative target or not. This classifier allows avoiding many computations in the association process of the surveillance system. The video surveillance system considered is able of tracking multiple objects or groups of objects in real conditions [2]. The whole system is composed of several processes:

- A predictive process of the image background, usually Gaussian models are applied to estimate variation in the background.
- A detector process of moving targets, detector process works over the previous and actual acquired frames.
- A grouping pixel process, this process groups correlates adjacent detected pixels to conform detected regions. These regions, or blobs, could be defined by a rectangular area or by contour shape.

** Funded by CICYT (TIC2002-04491-C02-02).

J. Mira and J.R. Álvarez (Eds.): IWINAC 2005, LNCS 3562, pp. 509–518, 2005.

– An association process, this process evaluate which detected blob should be considering as belonging to each existing target.
– A tracking system that maintains a track for each existing target. Usually filters are based on Kalman filter.

In this work, we propose the application of DM techniques to add new knowledge into the surveillance system. The surveillance system generates a set of files containing parameters of the detected blobs and, manually, we can generate an identifier to determine if the blob is part of a target or it is just noise. Using this strategy, to add new knowledge, for example the optical flow, the surveillance system is executed with the optimized parameters and the information about the optical flow is recorded for each blob. Then, the DM techniques could use this new information to classify the blobs as a real target (if it is a part of a target) or as false target (if it is only noise). In a previous work [3], an evaluation system has been proposed, and this system will be used to asses the video surveillance system, before and after applying machine learning techniques.

2   Surveillance Video System

This section describes the structure of an image-based tracking system. Figure 1 shows a basic architecture of the system. Specifications and details of this video system have appeared in several publications [4], [5], [6].

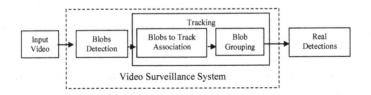

Fig. 1. Basic Architecture of the Video Surveillance System

The system starts capturing the first image, which is used to initialize background estimations. Then, for each new image, tracks are predicted to the capture time. Pixels contrasting with background are detected and blobs related with actual targets are extracted. Background statistics for pixels without detection in this frame are actualized to enable next frame detection. Then, an association process is used to associate one or several blobs to each target. Not associated blobs are used to initiate tracks, and each track is updated with its assigned blobs, while, in parallel, a function deletes the tracks not updated using the last few captures. Since data mining is being applied to the lowest level in the tracking system, that is, to the 'blobs detection' block, this is the only block being briefly described. The positioning/tracking algorithm is based on the detection of targets by contrasting with local background, whose statistics are estimated and updated with the video sequence. Then, the pixel level detector is able to extract moving features from background, comparing the difference with a threshold:

$$Detection(x, y) := [Image(x, y) - Background(x, y)] < THRESHOLD * \sigma \quad (1)$$

being σ the standard deviation of pixel intensity. With a simple iterative process, taking the sequence of previous images, and weighting to give higher significance to most recent frames, the background statistics (mean and variance) are updated. Finally, the algorithm for blobs extraction marks with a unique label all detected pixels connected, by means of a clustering and growing regions algorithm [7], and the rectangles which enclose the resulting blobs are built.

3   Performance Evaluation System

In this section, the evaluation metric proposed in [3], used to assess the quality of the surveillance system, is briefly described. It has been used the typical approach to evaluate the detection and tracking system performance: ground truth is used to provide independent and objective data that can be related to the observations extracted and detected from the video sequence. The ground truth has been extracted frame by frame for each scenario. The targets have been selected and the following data for each target have been stored: number of analyzed frame, track identifier and values of minimun and maximun (x,y) coordinates of the rectangle that surrounds the target.

Ground truth and real detections are compared to by the evaluation system. The evaluation system calculates a number which constitutes the measurement of the quality level for the tracking system. It uses an ideal trajectory as a reference, so the output track should be as similar as possible to this ideal trajectory. With the comparison of the detected trajectories to the ideal one, a group of performance indicators are obtained to analyse the results and determine the quality of our tracking process. The evaluation function is computed by giving a specific weight to each of the next indicators:

- Error in area (in percentage): The difference between the ideal area and the estimated area is computed. If more than one real track corresponds to an ideal trajectory, the best one is selected (although the multiplicity of tracks is annotated as a continuity fault).
- X-Error and Y-Error: The difference among the x and y coordinates of the bounding box of an estimated object and the ground truth.
- Overlap between the real and the detected area of the rectangles (in percentage): The overlap region between the ideal and detected areas is computed and then compared, in percentage, with the original areas. The program takes the lowest value to assess the match between tracking output and ground truth.

– Commutation: The first time the track is estimated, the tracking system
  marks it with an identifier. If this identifier changes in subsequent frames,
  the track is considered a commuted track.
– Number of tracks: It is checked if there is not a single detected track matched
  with the ideal trajectory. Multiple tracks for the same target or lack of tracks
  for a target indicate continuity faults. There are two counters to store how
  many times the ground truth track is matched with more than one tracked
  object data and how many times the ground truth track is not matched with
  any track at all.

With the evaluation function, a number that measures the quality level of
the tracking system is calculated, by means of a weighted sum of different terms
based on the evaluation metrics specified befor. The lower the evaluation function, the better the quality of the tracking system.

4   Data Mining for Blobs Classification

This section describes how Data Mining has been applied to the 'Blobs Detection'
block in the Video Surveillance System. The architecture of the Data Mining-
based Video Surveillance System is shown in figure 2.

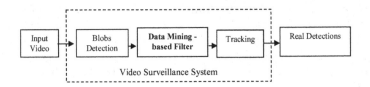

Fig. 2. Architecture of the Data Mining-based Video Surveillance System

The objective of applying Data Mining is to classify detected blobs as "real
targets" or "false targets", removing these false targets to simplify the association
process, and, in this way, improving the whole system. As it has already said, the
detection of targets is based on the intensity gradient in the background image.
But, not all blobs detected are real targets. These false blobs, or false alarms,
may appear because of noise, variation in illumination, etc. It is at this point
where data mining may be applied to remove false alarms without affecting real
targets. The objective of data mining is finding patterns in data in order to make
non-trivial predictions on new data [1]. So, having various characteristics (optical
flow, gradient intensity...) of the detected blobs, the goal is to find patterns that
allow us to decide whether a detected blob corresponds to a real target or not.

The input data take form of a set of examples of blobs. Each instance or
example is characterized by the values of attributes that measure different aspects of the instance. The learning scheme needed in this case is a classification
scheme that takes a set of classified examples from which it is expected to learn

a way of classifying unseen examples. That is, we start from a set of characteristics of blobs together with the decision for each as to whether it is a real
target or not, and the problem is to learn how to classify new blobs as "real
target" or "false target". The output must include a description of a structure
that can be used to classify unknown examples in order that the decision can
be explained. A structural description for the blobs in the form of a decision
tree is the most suitable output here. Nodes in a decision tree involve testing a
particular attribute. Leaf nodes give a classification that applies to all instances
that reach the leaf. To classify an unknown instance, it is routed down the tree
according to the values of the attributes tested in successive nodes, and when
a leaf is reached the instance is classified according to the class assigned to the
leaf. For the blob classification, the algorithm C4.5 has been used [8]. C4.5 is
based in algorithm ID3 [9], both introduced by Quinlan. The basic ideas behind
ID3 are:

- In the decision tree each node corresponds to a non-categorical attribute and
  each arc to a possible value of that attribute. A leaf of the tree specifies the
  expected value of the categorical attribute for the records described by the
  path from the root to that leaf. (This defines what a Decision Tree is).
- In the decision tree at each node should be associated the non-categorical
  attribute which is most informative among the attributes not yet considered
  in the path from the root. (This establishes what a 'good' decision tree is).
- Entropy is used to measure how informative is a node, based in Shannon
  Theory. (This defines what is meant by 'good').

C4.5 is an extension of ID3 that accounts for some practical aspects of real
data, such as unavailable values, continuous attribute value ranges, pruning of
decision trees, rule derivation, and so on. In next subsections, the input data
used to represent this problem and the obtained output data are described.

4.1   Input Attributes

Three scenarios, described in next section, have been used. In all cases, the
training examples are extracted in order to obtain the classifier. These examples
correspond to detected blobs, characterized by several attributes and classified
as 'true target' or 'false target'. Many attributes have been calculated for each
detected blob so a better classification might be done:

1. Intensity Gradient. It is the parameter initially used for detecting blobs. Media, standard deviation, minimum and maximum values of intensity gradient
   inside the blob have been stored.
2. Optical Flow. It calculates the motion of an object, represented as a vector,
   using Horn and Schunck algorithm [10] over consecutive frames. Since it is
   a vector, module and phase may be considered. Media, standard deviation,
   minimum and maximum values of the module and the phase of the optical
   flow inside the blob have been stored.

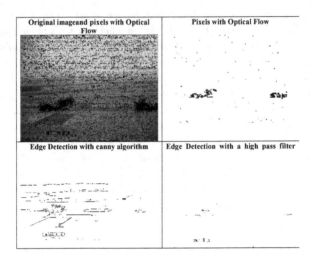

Fig. 3. Optical flow and edge detection

3. Edge Detection. It marks the points in an image at which the intensity changes sharply. Three methods have been used: canny algorithm [11], corner detection and a high-pass filter. In the three cases, the number of pixels of the blob and its surrounding area that correspond to detected edges has been stored.

Examples of Optical flow and edge detection are illustrated in figure 3.

To classify the extracted blobs as 'true target' or 'false target', the ground truth must be used. When the overlap of a detected blob with a real target is superior to a specific value (40%), the blob is classified as a 'true target'; otherwise, it is classified as a 'false target'. This training has been done for the three scenarios we are working with. Figure 4 shows a few training examples.

Intensity Gradient				Optical Flow								Edge Detection			Target?
				Module				Phase							
min	max	μ	σ	min	max	μ	σ	min	max	μ	σ	Canny	Corner	HPF	
16	36	27.75	7.84	2.49	23.79	13.23	6.20	-1.819	-1.01	-1.50	0.21	14	3	16	YES
16	27	21.3	3.95	5.94	13.10	8.90	2.06	-1.68	1.28	-1.48	0.14	5	0	4	YES
2	68	43.99	18.63	0.19	3.20	1.34	0.72	-3.09	3.12	0.32	1.65	0	0	0	NO

Fig. 4. Parameters of detected blobs and their classification

4.2   Output: Trained System and Classifier's Performance

The algorithm C4.5 has been used adjusting the parameter 'confidence factor' to 0.0001 [8] in order to obtain a pruned tree small enough, but also with a small error rate. The trained system obtained, in the form of decision tree, is shown in figure 5.

```
μ|OF⃗| <= 1.38719
|   HPF <= 82
|   |   μ|OF⃗| <= 0.861554: FALSE TARGET (2140.0/237.0)
|   |   μ|OF⃗| > 0.861554
|   |   |   HPF <= 36: FALSE TARGET (727.0/146.0)
|   |   |   HPF > 36: TARGET (102.0/19.0)
|   HPF > 82
|   |   maxΔI <= 37
|   |   |   canny <= 78
|   |   |   |   HPF <= 92: TARGET (4.0/1.0)
|   |   |   |   HPF > 92: FALSE TARGET (26.0)
|   |   |   canny > 78: TARGET (5.0/2.0)
|   |   maxΔI > 37: TARGET (69.0/2.0)
μ|OF⃗| > 1.38719
|   canny <= 0
|   |   σΔI <= 6.33516
|   |   |   μ|OF⃗| <= 2.24537
|   |   |   |   HPF <= 4: FALSE TARGET (50.0/18.0)
|   |   |   |   HPF > 4: TARGET (4.0)
|   |   |   μ|OF⃗| > 2.24537: TARGET (79.0/17.0)
|   |   σΔI > 6.33516: FALSE TARGET (117.0/18.0)
|   canny > 0: TARGET (3153.0/385.0)
```

Fig. 5. Decision tree obtained by algorithm C4.5 for classifying detected blobs as real or false targets

As it can be observed from the decision tree, only 5 out of the 15 attributes are significant; but they cover the three types of parameters:

- Maximum value ($max\Delta I$) and standard deviation value ($\sigma\Delta I$) of the Intensity Gradient. In general, a high $max\Delta I$ means that the detected blob is a true target and a high $\sigma\Delta I$ (probably produced by noise) means that the detected blob is not a true target.
- Mean value of the module of the Optical Flow ($\mu|\boldsymbol{OF}|$). In general, a blob with a high $\mu|\boldsymbol{OF}|$ corresponds to a true target.
- The values corresponding to Edge Detection obtained by canny algorithm and by the high-pass filter (HPF). In general, blobs with high Canny or HPF correspond to true targets.

The algorithm provides as well the classifier's performance in terms of the error rate. It has been executed with cross-validation, with the objective of getting a reliable error estimate. Cross-validation means that part of the instances is used for training and the rest for classification and the process is repeated several times with random samples. The confusion matrix is used by C 4.5 to show how many instances of each class have been assigned to each class:

Table 1. Confusion Matrix

'Target'	'False Target'	classified as
2958	451	'Target'
460	2607	'False Target'

In this case, 2958 blobs have been correctly classified as targets (True Positives, TP), 2607 blobs have been correctly classified as false targets (True Negatives, TN), 451 blobs have been incorrectly classified as false targets (False Negatives, FN) and 460 blobs have been incorrectly classified as true targets (False Positives, FP). The false negatives produce the deletion of true targets, which may have a cost with respect to no applying machine learning. The percentage correct classification (PCC) gives the correctly classified instances:

$$PCC = \frac{TotalTP + TotalTN}{TotalTP + TotalFP + TotalTN + TotalFN} \tag{2}$$

The global PCC for our scenarios is 85.9327, that is 85.9327 % of the detected blobs have been correctly classified.

5   Evaluation of the Data Mining - Based Surveillance Video System

In this section, the Evaluation System described in section 3 is being applied to the Surveillance Video System with and without the Data Mining-based filter, and results are being compared. Firstly, the three scenarios that have been used throughout the whole research are briefly described. They are localized in an airport where several cameras are deployed for surveillance purposes.

- The first scenario presents two aircrafts moving on inner taxiways between airport parking positions. A third aircraft appears at the end, overlapping with one of the other aircraft. Only the aircraft that overlaps is considered.
- In the second scenario, there is an aircraft moving with partial occlusions due to stopped vehicles and aircraft in parking positions in front of the moving object. There are multiple blobs representing a single target that must be re-connected, and at the same time there are four vehicles (vans) moving on parallel roads.
- Finally, in the third scenario, there are three aircrafts moving in parallel taxiways and their images overlap when they cross. The three aircrafts are considered.

	Scenario 1	Scenario 2	Scenario 3		
	Track 0	Track 0	Track 0	Track 1	Track 2
Without DM Filter	33.71	7245.95	8085.81	12226.33	11610.18
With DM Filter	176.02	6642.04	6840.48	9066.24	9140.90
Improvement	-	8.33 %	15.40 %	25.85 %	21.27 %

Fig. 6. Results of the Evaluation Function

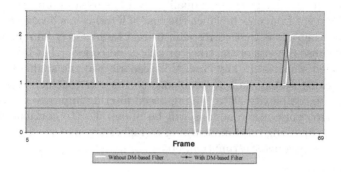

Fig. 7. tracks associated to track 0 in scenario 3 with and without DM-based filter

The complexity of the videos, due to overlaps and occlusions, increases from the first to the third scenario. The Evaluation Function is calculated for each track, and the results are shown in figure 6.

As it was previously explained, the lower the evaluation function, the better the tracking system; so, in four of the five cases, the tracking system is improved. In the only case in which it gets worse is in the simple video, in which the original tracking system had no problems. It gets worse when using the DM Filter because the aircraft is detected one frame later. It is the cost of possibly removing true targets (false negatives), due to the DM filtering (section 4.2). However, in more complex situations the results are better. Next, an example of the most significant of the evaluation metrics in the Evaluation function is given: the number of detected tracks associated to an ideal track. In figure 7 it is shown the number of associated tracks to track 0 in scenario 3:

It can be easily seen how the number of tracks, ideally one, improves with the Data Mining-based filter. Without filtering, the number of tracks associated to track 0 is different from '1' in 18 instances; whilst with filtering, this number reduces to 3, which supposes a decrease of 83.33%. We can see as well, the cost of filtering: during three frames the track is lost, due to the fact that real targets have been removed.

6   Conclusions

The initial surveillance video system has been improved in scenarios with some complexity by applying Data Mining-based filtering. The Data Mining-based

Filter decides whether the blobs extracted by the system correspond to real targets or not. Algorithm C4.5 has been used; this algorithm obtains a classifier in the form of a decision tree from a set of training examples. These training examples consist on detected blobs, characterized by several attributes, based on the following parameters: intensity gradient, optical flow and edge detection. Besides, the training examples must be classified as 'true target' or 'false target', for which, the ground truth, extracted by a human operator, has been used. The result surveillance video system has been evaluated with an evaluation function that measures the quality level. This quality level has been improved in all scenarios tested, except from one, in which the cost of filtering has become manifest. Because of filtering, blobs that correspond to real targets may be removed and this fact may cause the loss of the track or a later detection, what has occurred in the mentioned scenario. In any case, this scenario was the simplest one and the initial tracking system had no problems; so, we can conclude that in scenarios with more complexity Data Mining-based filtering improves the tracking system. In future works some actions will be undertaken to continue this approach, such as, applying machine learning to higher levels of video processing: data association, parameter estimation, etc.

References

1. Witten, I. H., Frank, E.: Data mining : practical machine learning tools and techniques with Java implementations, San Francisco Morgan Kaufmann (2000)
2. Rosin, P.L., Ioannidis, E.: Evaluation of global image thresholding for change detection, Pattern Recognition Letters, vol. 24, no. 14 (2003) 2345–2356
3. Pérez, O., García,J., Berlanga, A., Molina, J.M.: Evolving Parameters of Surveillance Video Systems for Non-Overfitted Learning, Lausanne, Suiza, 7th European Workshop on Evolutionary Computation in Image Analysis and Signal Processing (EvoIASP) (2005)
4. Besada, J. A., Portillo, J., Garca, J., Molina, J.M., Varona, A., Gonzalez, G.: Airport surface surveillance based on video images, FUSION 2001 Conference, Montreal, Canada (2001)
5. Besada, J. A., Portillo, J., Garca, J., Molina, J.M.: Image-Based Automatic Surveillance for Airport Surface, FUSION 2001 Conference, Montreal, Canada (2001)
6. Besada, J. A., Molina, J. M., Garca, J., Berlanga, A., Portillo, J.: Aircraft Identification integrated in an Airport Surface Surveillance Video System, Machine Vision and Applications, Vol 15, No 3 (2004)
7. Sanka, M., Hlavac, V., Bolye, R.: Image Processing, Analysis and Machine Vision, Brooks/Cole Publishing Company (1999)
8. Quinlan, J. R.: C4.5: Programs for machine learning, Morgan Kaufmann (1993)
9. Quinlan, J. R.: Induction of Decision Trees, Machine Learning, vol. 1 (1986)
10. Horn, B. K. P., Schunck, B.G.: Determining Optical Flow, Artificial Intelligence, 17, pp. 185-203 (1981)
11. Canny, J.: A Computational Approach To Edge Detection, IEEE Trans. Pattern Analysis and Machine Intelligence, 8, pp. 679-714 (1986)

A Neurocalibration Model for Autonomous Vehicle Navigation

M.A. Patricio[1], D. Maravall[2], J. Rejón[1], and A. Arroyo[3]

[1] Dpto. Ciencias de la Computación, Universidad de Alcalá, Spain
[2] Dpto. de Inteligencia Artificial, Universidad Politécnica de Madrid, Spain
[3] Dpto. Sistemas Inteligentes Aplicados, Universidad Politécnica de Madrid, Spain
{miguel.patricio, julio.rejon}@uah.es, dmaravall@fi.upm.es,
aarroyo@eui.upm.es

Abstract. The paper evaluates the capability of a neural model to calibrate a digital camera. By calibrate we understand the algorithms that reconstructs the 3D structure of an scene from its corresponding 2D projections in the image plane. The most used 3-D to 2-D geometrical projection models are based in the pin-hole model, a free distortions model. It is based in the correspondence established between the image and the real-world points in function of the parameters obtained from examples of correlation between image pixels and real world pixels. Depending on the sensor used, different kind of chromatic aberrations would appear in the digital image, affecting the brightness or the geometry. To be able to correct these distortions, several theoretical developments based on pin-hole models have been created. The paper proves the validity of applying a neural model to correct the camera aberrations, being unnecessary to calculate any parameters, or any modelling. The calibration of autonomous vehicle navigation system will be used to prove the validity of our model.

1 Introduction

One of the most amazing aspects, when the human perception is studied, is the observer capability determining the 3-D structure of the objects from bidimensional light patterns. First studies focused in the tridimensional analysis from images were done by the photogrammetry scientific community. These studies were retaken by artificial vision scientifics up to develop different approaches for the same problem. As a general rule, all of them use a video camera as sensor which provides information about the 3-D world to an artificial vision system. Our target in 3-D vision must be to explain how to use the 2-D information that we have from the scene to make measures of the 3-D subjacent world.

The most common 3-D to 2-D model used for geometrical projection is the pinhole model or perspective projection model (see Figure 1). In this model, the formation of a digital image is divided in two processes in order to obtain the digital image itself: (1) The scene projection on the image plane (sensor), and

J. Mira and J.R. Álvarez (Eds.): IWINAC 2005, LNCS 3562, pp. 519–528, 2005.
© Springer-Verlag Berlin Heidelberg 2005

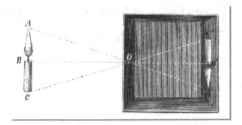

Fig. 1. Pin-hole model: "If the light rays reflected by an illuminated object happen through a tiny hole in a dark box, the image will be projected upside down in a wall inside the box"

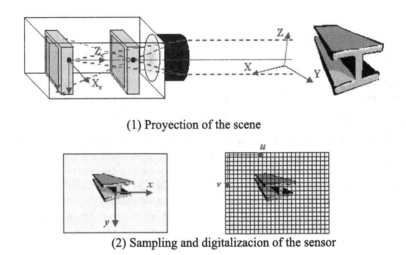

(1) Proyection of the scene

(2) Sampling and digitalizacion of the sensor

Fig. 2. Digital image formation processes

(2) the sample and digitalization of this plane. Each process can be raised as a change in the coordinate system, as depicted in Figure 2.

In this paper we present an alternative to the calibration systems based on physical, geometrical and analytic models. The neural models try to find a correspondence between the sensorial input information and the recognition problem that tries to solve. For that it uses the basic behavior of cells or neurons, with the intention of solving the problem. It is not necessary to know any parameter of the camera or the kind of aberration that can exist.

The paper is divided in two parts. First, the basic principles of the pin-hole model are presented, describing the physical concepts in which it is based and the possible problems involved in itself. Finally, we describe the neural model that has been designed in order to resolve a particular problem as is the calibration system in an autonomous vehicle navigation.

2   Pin-Hole Camera Model

The algorithms that reconstruct the 3-D structure from a scene or calculate the objects' space position need the equations that match the 3-D points with their corresponding 2-D projections. Even though those equations are given by the projection one, it is normal to suppose that the 3-D points can be given by a different coordinates system than the camera one and, in addition, it is necessary to relate the coordinates from a point in the image with their corresponding ones in the projection system given by the camera. We have, therefore, to calculate what is known by the extrinsic parameters (those which relate both 3-D reference systems, see Figure 3) and the intrinsic ones (they relate the reference system of the camera to the image, see Figure 4). This task it is what is known in computer vision as camera calibration process [1] and [2].

The first way to calibrate a camera is from a well-known scene. To make the calibration, a sequence of known points is needed, these points also need to meet some criteria. From these points can then be obtained the extrinsic and intrinsic parameters [3].

We have to bear in mind that the pin-hole model is a free-distortion theoretical model. Depending on the sensor used, aberrations, especially chromatic ones, can appear in the digital image, affecting the brightness or the geometry. A common aberration is the radial distortion. This is a symmetric one on the principal point of the image, which causes straight lines to appear curved in the resulting image as depicted in Figure 5. This causes a displacement of the imaged point from its theoretically correct position. Which effectively is a change in the angle between the incoming ray of light and the incident ray of the imaging sensor. Another kind of camera aberration is the decentring distortions (Figure 6, which are more of an artefact of errors in the assembly and rotational symmetry of the multiple lens components [4] and [5]. Affinity and shear

$$\begin{pmatrix} X_c \\ Y_c \\ Z_c \end{pmatrix} = \begin{pmatrix} r_{11} & r_{12} & r_{13} & t_X \\ r_{21} & r_{22} & r_{23} & t_Y \\ r_{31} & r_{32} & r_{33} & t_Z \end{pmatrix} \begin{pmatrix} X \\ Y \\ Z \\ 1 \end{pmatrix}$$

From the scene to the camera (extrinsic model)

Fig. 3. Extrinsic parameters that mark the camera position and orientation with regard to the scene. They are independent from the type of camera used

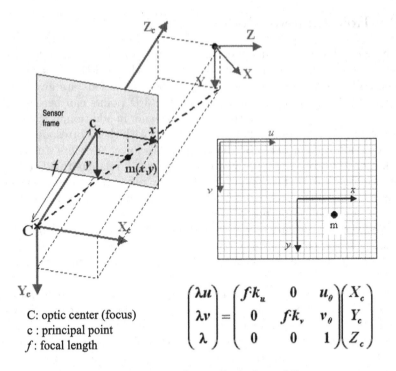

C: optic center (focus)
c : principal point
f: focal length

$$\begin{pmatrix} \lambda u \\ \lambda v \\ \lambda \end{pmatrix} = \begin{pmatrix} f \cdot k_u & 0 & u_\theta \\ 0 & f \cdot k_v & v_\theta \\ 0 & 0 & 1 \end{pmatrix} \begin{pmatrix} X_c \\ Y_c \\ Z_c \end{pmatrix}$$

From sensor to image (Intrinsic model)

Fig. 4. Intrinsic parameters that represent the internal properties of the camera

are aspects of in-plane distortion that are typical characteristics of the capture medium. Affinity is effectively a difference in scales between the x and y axis. And shear describes the non-orthogonallyty between axes. Note that these terms are only applied to one axis, due to the cross axis nature of this distortion. For any digital imaging system, the geometric integrity of the image sensor is very high from geometric consistency inherent in modern semiconductor manufacturing technology, resulting in negligible in-plane distortions. To take into account these aberrations, more complex camera models have been developed which goal is to correct them and obtain the 3-D information from the scene in the most reliable way [6].

Next question we can make ourselves is: which 3-D information can we retrieve when we have more than just one image of the scene?. Even though Euclides and Leonardo Da Vinci observed that the left and right eye project different images of the same scene, it was not until the 19th century when Wheatstone provided the first experimental evidence that showed that the depth is perceived when unlike images are shown to each eye (see Figure 7). Since then, binocular disparity (angular difference between same features images in the left and right eyes) has been considered one of the most important sources in 3-D information recovery [7]. We can start talking now about what it is known in computer vision

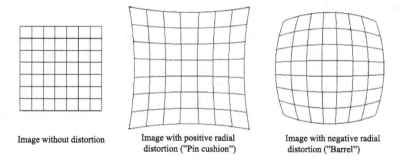

Image without distortion   Image with positive radial distortion ("Pin cushion")   Image with negative radial distortion ("Barrel")

Fig. 5. Radial distortion examples

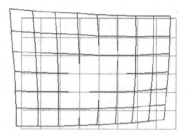

Fig. 6. Decentering distortion

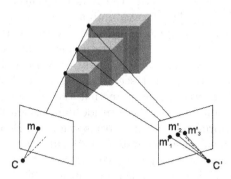

Fig. 7. When the scene is projected on the image plane, depth information about the objects is lost

as stereo vision, that is to say, the capability to infer information about the 3-D structure and distance of a scene obtained from two or more images taken from different points of view.

3   Neurocalibraton Model

This paper presents the development of a neural architecture that simplifies the current calibrating models. In short, the pin-hole model describes the projection of a real world point into an image point by a sequence of reference systems transformations. With the neural network to be designed, we will avoid the calculation complexity of the extrinsic and intrinsic calibrating models, and at the same time, we will take into consideration the possible optical aberrations produced by the sensor used in our experimentation.

In the previous section we presented, one of the more known and used calibration models, as it is the pin-hole model, which establishes the extrinsic and intrinsic parameters. Camera models set up the mathematical relation between the space points and their equivalents in the camera image. As we have seen, the image making process is the projection of a three-dimensional world into a two-dimensional image. Therefore, through the perspective transformation the correspondence between the real world and the image can be obtained. Nevertheless, with the inverse perspective transformation there are more than one possible correspondence (because in the direct transform there is depth information lost) and for each image point there is an infinite set of the three-dimensional environmental points (see Figure 7). Additional information is necessary to obtain a certain space point from the image that can supply the depth lost suffered in the direct transformation. To simplify our problem, we can establish some restrictions to our initial knowledge of the environment, for example, determining that all points in the real world are in the same plane.

Through the autonomous vehicle navigation we will raise a real problem. The complete system is able to control the vehicle actions and take it to a pre-known destination. The vehicle is located into a navigation area controlled by a camera as it is depicted in Figure 8. In order to give the right navigation orders to the vehicle we have to establish the correspondence between the image and the real world points. The scene we are going to work with is a road which is delimited by a set of cones as it seems in Figure 9. With the purpose of simplifying the problem, we will suppose that the road is in one plane and, therefore, we are not going to use the altitude information. We will start form the real world 2-D information and its corresponding 2-D information. Using a differential GPS (Global Position System) with subcentimetrical precision we have taken measurements of the Figure 9 cones.

Pattern election is a classical problem in neural models design used to form our training set. The obtained number should be large enough as well as representative of the problem to be solved. But, what will be the implications of using a reduced sample for designing our neural model?. Let's imagine a quite normal situation when the environment conditions are variable. For example, if we had a camera docked to the car, the correspondence between the real world and image points would change each time the vehicle modifies its position. It would then be necessary a simple system, which, starting from a reduced set of real world points and its corresponding projections into the image plane, would get a calibration of all the camera pixels. Due to environmental changes this

Fig. 8. Vehicle and road images

system needs a single calibrating model, which, from a reduced set of points will be able to obtain a good correspondence between real word and image points. This is an usual task in humanoid locomotion control, where we are trying to track a moving goal. As the robot and the goal are both moving, we need a simple calibration for 3D position measurement [8].

During the execution of a task the vision-system is subject to external influences such as vibrations, thermal expansions, etc. which affect and possibly invalidates the initial calibration. Moreover, it is possible that the parameters of the vision-system, such as the zoom or the focus, are altered intentionally in order to perform specific vision-tasks [9] and [10].

4   Design of the Model for Autonomous Vehicle Navigation

In our experiment we have worked with a reduced set of calibrating points. This relation is the position of 6 out of 12 cones appearing in Figure 9 with it image plane projection:

N.	Latitude(m)	Length(m)	X	Y
1	4462512.89	458960.07	234	436
2	4462510.88	458981.03	129	202
3	4462508.48	459004,79	88	119
4	4462501.99	458999.87	268	115
5	4462503.98	458980.31	374	180
6	4462506.16	458965.70	515	297

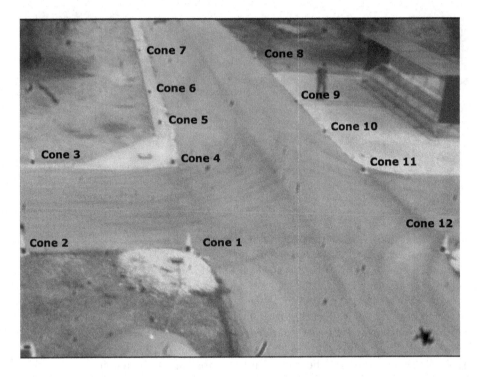

Fig. 9. Image used for calibration

For the neural model design we have used the JavaNSS tool. After an intensive investigation a multilayer perceptron has been proposed as the topological model, using for the learning the backpropagation algorithm. The network topology is depicted in Figure 10.

Since the activating function of this network neurons uses the sigmoidal function between 0 and 1, we will make a normalization process in both input and output data. For each input and output neuron n we apply the following normalization:

$$n^* = \frac{n - n_{min}}{n_{max} - n_{min}} \tag{1}$$

where n_{max} y n_{min} are the maximum and minimum value of the n neuron.

The most exigent and reliable method for the evaluation of any automatic approximator is the *leave-one-out* policy, in which the function –in our case, the feedforward multilayer perceptron- is trained with all the available training samples except one, which is used for the evaluation of the function itself. By repeating the process with all 6 training samples, the average estimation error is an excellent evaluation of its future performance for new calibration. Note that the Medium Square Errors (MSQ) are 0.106 meters for Latitude and 0.130

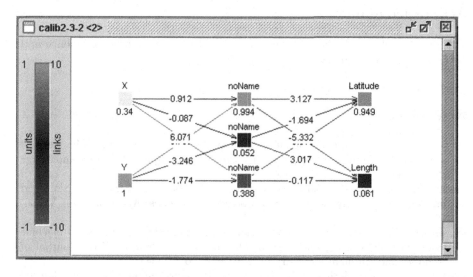

Fig. 10. Calibration model architecture

meters for Length. These results are accurate enough for aiming to autonomous vehicle navigation systems.

N.	Latitude* (m)	Length* (m)	Latitude error	Length error
1	4462512.45	458961.83	0.441668	1.7592848
2	4462511.32	458981.26	0.442144	0.2301248
3	4462508.24	459002,66	0.236234	2.1268832
4	4462502.97	458999.90	0.978057	0.0276320
5	4462503.37	458980.60	0.613003	0.2873744
6	4462506.18	458964.52	0.022140	1.1834904
		MSE	**0.106044**	**0.1307226**

5   Conclusions

The purpose of this paper is to simplify the traditional calibration models using an artificial neural network. This model also considers the possible optical aberrations produced by the camera. We have used an image processing to automatically detect the cone of the images and, then, a backpropagation neural algorithm was used to learn the relationships between the image coordinates and three-dimensional coordinates. This transformation is the main focus of this study. Using these algorithms, the vehicle is able to navigate and reach its objective successfully.

Acknowledgements

This work has been partially supported by the Spanish Ministry of Science and Technology, project DPI2002-04064-CO5-05.

References

1. Yakimovsky, Y., and Cunningham, R., "A system for extracting three-dimensional measurements from a stereo pair of TV cameras", Computer Graphics and Image Processing, 7, 1978, pp. 195–210.
2. Fu, K.S., Gozalez, R.C., and Lee, C.S.G., "Robotics: control, sensing, vision, and intelligence". Mcgraw-Hill Series In Cad/Cam, Robotics, And Computer Vision, 1987.
3. Tsai, R., "A versatile camera calibration technique for high accuracy 3D machine vision metrology using off-the-shelf TV cameras and lenses". IEEE Journal of Robotics and Automation , Vol. 3, No 4, 1987, pp 323–344.
4. Ellum, C.M. and El-Sheimy, N., "Land Based Mobile Mapping Systems", Journal of Photogrammetric Engineering and Remote Sensing, Vol. 68, No. 1, 2002, pp. 13–18.
5. Mikjail, E.M., Bethel, J.S., and McGlone, J.C., "Introduction to Modern Photogrammetry", John Wiley and Sons, Inc. New York, 2001.
6. Weng, J., Cohen P., and Herniou M., "Camera calibration with distortion models and accuracy evaluation", IEEE Transactions on Pattern Analysis and Machine Intelligence, vol. 14, No. 10, 1992, pp. 965–980.
7. Tittle, J. and Todd, J., "Perception of three-dimensional structure". The Handbook of Brain Theory and Neural Networks, MIT Press, 1995, pp 715–718.
8. Kagami S., Okada, K., Inaba, M., and Inoue, H. "Design and implementation of onbody real-time depthmap generation system", In Proc. of IEEE International Conference on Robotics and Automation (ICRA'00), 2000, pp. 1441–1446.
9. Zhang, Z., and Schenk, V., "Self-Maintaining Camera Calibration Over Time" Proc. IEEE Conf. on Computer Vision and Pattern Recognition (CVPR'97), 1997, pp. 231–236.
10. Worrall, A.D., Sullivan, G.D., and Baker, K.D., "A simple, intuitive camera calibration tool for natural images", in Proc. of the 5th British Machine Vision Conference, 1994, pp. 781–790.

Some Remarks on the Application of Artificial Neural Networks to Optical Character Recognition

A. Moratilla and I. Olmeda

Dpto. Ciencias de la Computación, Universidad de Alcalá, Spain
{antonio.moratilla, josei.olmeda}@uah.es

Abstract. In this paper we analyze one important aspect related to handwritten Optical Character Recognition, specifically, we demonstrate that the standard procedure of minimizing the number of misclassified characters could be inconsistent in some applications. To do so, we consider the problem of automatic reading of amounts written in bank cheques and show that the widely used confusion matrix does not provide an appropriate measure of the performance of a particular classifier. We motivate our approach with some examples and suggest a novel procedure, using real data, to improve the performance by considering the true economic costs of the expected misclassification errors.

1 Introduction

Handwritten Optical Character Recognition is one of the most studied problems in the context of Machine Learning. This interest is justified by the enormous amount of important applications where a solution can be implemented as well as because there is a considerable number of techniques that can be applied to the problem, making OCR very appealing both from the practitioner as well as from the researcher point of view. Among the many applications of OCR we can mention the automatic classification of letters using handwritten postal ZIP codes, automatic reading of administrative formularies of fixed structure, car plate automatic detection, recognition of text signs in traffic applications or automatic reading of amounts in bank cheques. From the point of view of the applicable techniques the catalogue is also huge: Artificial Neural Networks, Learning Vector Quantizing, Bayesian Techniques, Hidden Markov Models, or a variety of filtering techniques (such as Wavelets) to cite a few. Moreover, the field has been recently broadened by the consideration of the many "alphabets" employed in the particular application (e.g. Kim and Kim, 2003 [3], for the Chinese, or Battacharya and Chaudry, 2003 [1], for the Bangla).

In this paper we present an application of OCR to the problem of automatic reading of quantities written on bank cheques. This application will allow us to remark some important drawbacks of the "standard" procedure employed that consists on minimizing the number of misclassified digits. We will show that, for

J. Mira and J.R. Álvarez (Eds.): IWINAC 2005, LNCS 3562, pp. 529–537, 2005.

this particular problem, such approach is inconsistent and, consequently, we will propose a possible improvement of the standard procedure.

2   Drawbacks in the "Standard" Approach

It is well known that most of the applications of feedforward neural networks to Optical Character Recognition employ the standard backprop algorithm to find the optimal configuration of the net. The results of the literature are apparently good, since in many cases one finds confusion matrices with misclassification rates well below a 1%. Nevertheless one has to recognize two important facts that question the validity of this general approach. First, it is obvious that the error rate for each of the digits is not the overall error rate for the problem in question: the longer the number one has to recognize, the bigger will be the overall misclassification rate considering the whole number. The second one relates to the problem of the asymmetric information content of each of the digits in a particular application.

To illustrate our arguments, and to motivate the solution proposed, we will begin by presenting some examples in the context of automatic reading of handwritten amounts in bank cheques. To make our results easily replicable, we employ the well known MNIST database, which has been widely applied in the OCR literature (e.g. Milgram et al, 2004 [4]). The MNIST database (available at http://yann.lecun.com/exdb/mnist/) is obtained from the special databases of NIST 1 and 3 which contain handwritten integers between 0 and 9, obtained form 250 individuals. The database is composed of two sets, the first one comprises 60.000 digits, which are usually employed for the training and validation of the particular technique employed; another set of 10.000 digits is reserved for testing. The digits are recorded at a granularity of 28x28 pixels and have been pre-processed using re-dimensioning and normalization based on the moment of the image and a threshold using two levels.

To motivate the method here proposed, we begin with the "standard" procedure of applying a feedforward neural network to classify the handwritten characters into numerals by means of the backprop algorithm. Specifically, we employ a feedforward neural net with 256 binary inputs (corresponding to each of the pixels), 30 hidden units and 10 binary output units for the classification (so that 000000001 corresponds to the prediction of a "9"). The net is trained over 50.000 examples until the validation error, computed along another set of 10.000 examples increases. At this point, training is resumed and the classification of 10.000 unseen examples is done.

To illustrate our first appreciation about the –just- apparent performance of the nets, generally concluded by visual inspection of the confusion matrices (e.g. Battacharya and Chaudry, 2003 [1]), let us consider the confusion matrix for the test set in our example (Table 1). Now assume that we want to automatically read the amount 6.478,59 Eur. written in a cheque bank, note that even though the accuracy is apparently very high, the probability of <u>not</u> making a mistake in that amount is only 50,41% (the product of the conditional probabilities, that

Table 1. Misclassification rates for the testing set

	0	1	2	3	4	5	6	7	8	9
0	**97,24%**	0,09%	0,68%	0,30%	0,00%	0,90%	0,73%	0,19%	0,92%	0,20%
1	0,10%	**96,92%**	1,07%	0,89%	1,53%	0,22%	0,31%	1,26%	1,64%	0,69%
2	0,10%	0,26%	**86,82%**	0,99%	3,16%	0,45%	0,63%	3,60%	1,64%	0,69%
3	0,00%	0,62%	1,36%	**91,68%**	1,12%	2,80%	0,52%	0,78%	1,95%	1,49%
4	0,00%	0,26%	4,55%	0,50%	**82,59%**	0,78%	1,88%	3,02%	1,13%	3,67%
5	0,51%	0,18%	0,39%	2,18%	2,65%	**90,02%**	1,77%	0,10%	1,75%	0,79%
6	1,02%	0,35%	0,29%	0,00%	1,83%	3,14%	**92,90%**	0,10%	1,85%	0,00%
7	0,20%	0,26%	2,23%	1,29%	2,14%	0,34%	0,10%	**89,01%**	0,41%	1,68%
8	0,51%	0,79%	1,65%	1,58%	1,22%	1,01%	1,04%	0,19%	**87,58%**	1,88%
9	0,31%	0,26%	0,97%	0,59%	3,77%	0,34%	0,10%	1,75%	1,13%	**88,90%**

is, 91,56% * 83,61% * 91,32% * 89,60% * 88,73% * 90,70%), so that even with these apparently good results one has almost an equal chance to be "right" than to be "wrong".

Second, we mentioned before that in general applications not all the numbers have the same information content, so that it is biased to consider all the misclassification rates the same. As an example, again consider our application in which the researcher is interested in recognizing the amount written in a cheque. Without loss of generality assume that the cheques are of amounts between 100,00 Eur. and 10.000,00 Eur., also assume that he has to classify 10.000 cheques and that the economic value of the error is measured in terms of the absolute difference between the real amount of the cheque and the predicted amount. Suppose that the researcher has found a "perfect" classifier that makes no mistakes, obviously the error cost of this classifier is 0 Eur. Now assume that the researcher has to choose from another set of classifiers that are slightly worse than the "perfect" classifier but which all of them have exactly the same misclassification rate. For example, suppose that each one of the classifiers make no errors with the exception that, for a particular digit, they have a 90% of probability of recognizing the true digit and a 10% of confusing it with another digit. For example, one classifier might have a 90% of probability of recognizing a "9" when it is a "9" and a 10% of probability of confusing it with an "8", while another might have a 90% of probability of recognizing a "3" when it is a "3" and a 10% of probability of confusing it with an "6".

Even though, in terms of the confusion matrices, all the classifiers have an equal performance, in our particular application not every classifier would be equally valuable. To analyze their relative performance two questions arise. The first, and most obvious one, is what kind of errors are the most relevant in terms of magnitude of the difference between the real and predicted amount. This question has an intuitive answer since one easily recognizes that it would be worse to "see" a "9" when it is a "2" than to "see" a "3". One has to conclude that the bigger the distance between the real and the predicted digit the worse the classifier would be, so that the researcher would choose among classifiers which are biased towards confusing a number with another quite close to it.

The second observation relies on recognizing that not all the digits are equally significant. For example, observe that when writing an amount one rarely employs a "0" to begin a number, so one would write "9.645,34 Eur." instead of

"09.645,34 Eur.". This means that the "0" has less information content than any other number since it always occupies a position relatively less relevant than the others. For similar reasons, the number "5" has less information content than the number "9", since the worse it can happen is to misclassify a "5" as a "0" while it is much worse to classify a "9" as a "0". These two facts intuitively show us that even though a set of classifiers have exactly the same overall misclassification rate they might be quite different in particular applications.

To give an impression of the importance of these two facts, we have simulated the results that one would obtain in the situation similar to the one described above. To see this, we calculate the expected cost of misclassification along our set of 10.000 examples of amounts between 100,00 Eur. and 10.000,00 Eur. by bootstrapping using the conditional probabilities from each of the classifiers of the set described above. Then, we calculate the amount, in euros, of the errors derived. In Table 2 we present the results obtained. First, the costs of the perfect classifiers are represented in the diagonal. Since every digit is correctly classified with 100% confidence, the cost of misclassification is 0 Eur. Now we turn our attention to the cost that, say, a suboptimal classifier would have by confusing a "0" with a "1". This cost (10.244,93 Eur.) is represented in cell (2,1) of the table, while the cost of confusing a "6" with a "7" (96.156,99 Eur.) is represented in cell (8,7). Our first intuition is confirmed by the simulated results: since one can observe that for each one of the columns the amount grows with the distance between the predicted and real digits.

Now, in the last line of the table we have computed the mean amount of making an error in classifying each of the digits. Note that the mean amount of

Table 2. Simulated error cost of a set of suboptimal classifiers. Currency: Euro

	0	1	2	3	4
0	-	101.488,94	190.293,50	334.541,76	432.018,76
1	10.244,93	-	102.854,65	242.034,16	311.531,79
2	17.148,84	119.180,54	-	113.729,35	217.439,70
3	32.397,51	244.366,14	119.593,91	-	108.576,55
4	48.479,20	354.537,45	241.418,86	107.813,51	-
5	48.772,80	476.754,40	293.215,98	189.474,58	82.871,08
6	62.683,44	549.855,70	488.558,44	339.485,91	209.111,04
7	65.052,40	670.441,68	618.025,30	528.642,96	322.980,84
8	87.359,04	755.219,43	614.226,48	610.824,40	457.241,04
9	106.557,39	857.077,12	761.421,29	664.496,22	480.282,75
mean	**47.869,56**	412.892,14	342.960,84	313.104,29	**262.205,36**

	5	6	7	8	9
0	574.812,55	757.700,16	723.564,94	934.527,76	969.709,50
1	427.372,36	617.199,35	706.388,34	713.593,16	976.143,04
2	329.571,66	469.490,44	495.103,35	604.799,64	853.340,95
3	197.027,14	347.190,78	471.538,80	555.461,00	647.441,94
4	103.822,33	256.110,26	335.627,82	482.042,56	555.868,85
5	-	109.497,03	224.590,18	307.918,95	464.573,16
6	95.504,48	-	109.481,54	246.212,30	310.733,49
7	228.364,48	96.156,99	-	108.017,04	222.830,48
8	318.595,17	202.037,30	110.883,44	-	112.295,04
9	494.337,32	357.194,07	262.827,40	112.665,27	-
mean	276.940,75	321.257,64	344.000,58	406.523,77	**511.293,65**

Table 3. Simulated errors' costs in our example. Currency: Euro

	0	1	2	3	4
0	**3.522.579,90**	3.688.126,30	3.880.625,40	4.007.671,60	4.007.435,00
1	3.462.537,20	**3.425.976,90**	3.880.336,30	3.710.309,80	3.855.232,10
2	3.669.532,30	3.372.713,30	**3.626.198,20**	3.677.129,60	3.836.345,40
3	3.427.312,90	3.706.379,90	3.585.130,80	**3.609.577,80**	3.580.478,00
4	3.564.533,30	3.722.341,10	3.802.196,70	3.869.768,40	**3.551.028,30**
5	3.476.497,80	4.052.518,80	3.847.043,30	3.531.399,30	3.684.498,20
6	3.481.483,70	4.081.013,20	4.257.979,30	3.780.995,00	3.800.268,40
7	3.618.668,60	4.088.134,40	4.091.024,90	3.933.851,40	3.778.071,90
8	3.784.587,30	4.361.079,10	4.125.377,00	4.218.901,40	3.936.367,30
9	3.720.595,60	4.278.992,20	4.273.196,10	4.259.486,60	4.134.155,40
mean	**3.572.832,86**	3.877.727,52	3.936.910,80	3.859.909,09	3.816.388,00

	5	6	7	8	9
0	4.192.370,90	4.144.817,40	4.297.987,00	4.419.442,90	4.667.598,70
1	3.909.367,30	4.236.804,70	4.376.207,20	4.366.086,10	4.356.258,80
2	3.833.367,30	4.043.807,20	4.004.849,70	4.144.817,20	4.310.652,50
3	3.555.607,80	3.785.614,50	3.924.564,00	4.198.597,70	4.278.781,80
4	3.801.824,60	3.710.193,80	3.890.579,40	3.852.468,80	4.209.429,80
5	**3.576.906,40**	3.642.085,80	3.865.735,40	3.958.823,60	4.002.366,60
6	3.784.894,60	**3.596.745,80**	3.640.753,80	3.847.567,40	3.857.174,10
7	3.857.712,20	3.724.306,20	**3.555.743,80**	3.717.162,80	3.687.857,90
8	3.705.728,70	3.883.724,10	3.673.293,00	**3.470.576,10**	3.863.739,10
9	3.820.623,60	3.917.022,20	3.585.617,80	3.648.847,30	**3.553.922,70**
mean	**3.803.840,34**	3.868.512,17	3.881.533,11	3.962.438,99	**4.078.778,20**

misclassifying a "0" is 47.869,56 Eur. while the mean amount of misclassifying a "9" is 511.293,65 Eur. This illustrates our second intuition that it is much worse (in fact, ten times worse, in this case) to misclassify a "9" than a "0". Finally note that, as intuitively expected, the central digits (numbers "4"and "5") also lead to lower misclassification costs.

To conclude our illustration, in Table 3 we have repeated a similar experiment but now using the true conditional probabilities obtained in Table 1. Now, in the diagonal we show the actual cost of the classifier without altering the conditional probabilities (so that each of the elements of the table corresponds to a particular realization). In each one of the rest of the cells we compute the corresponding cost by decreasing the probability of being right (predicting a "5" when it is in fact a "5") by a 10% and increasing by this amount the probability of making some kind of error (predicting a "4" when it is in fact a "5" –cell (5,6)- or predicting a "6" when it is in fact a "4" –cell (5,7)-). Note that the even though the differences are less significant that in the simulated extreme case, the conclusions remain he same: it is much better to make an error in predicting "0s" than "9s", it is relatively better to make mistakes in "5s" than in "1s" or "9s" and it is better to predict an "8" when it is in fact a "9" than to predict a "0".

3 Proposed Approach

There are several ways to remedy the problems we have mentioned. Probably the easiest way to improve the accuracy is to employ a different loss function which takes into account the asymmetry in the importance of misclassifications costs and then to employ a modified backprop algorithm to minimize it. Nevertheless, this approach does no take into account that our final objective is to minimize some specific cost associated to the errors made as a consequence of the conditional probabilities of the confusion matrix, and the optimal network in terms of the penalized loss function might not be the one which gives the best results in classifying an arbitrary number.

Alternatively, we suggest employing such loss function to produce confusion matrices which would lead to minimize the expected classification costs specific to the problem. Note that our proposal is dramatically different from the previous in one important aspect: it only employs the penalized function as a mean to calculate confusion matrices which are then used to estimate the expected error of classifying a set of numbers of arbitrary length, and not the numbers (just one digit) of the specific training set.

The procedure here proposed takes two steps. In the first one, we employ the standard to minimize the mean squared misclassification errors, while in the second step we minimize the total cost of misclassifications. To be concrete, let us assume that we have a set of n examples in the training set. In the first step we assume the "standard" procedure which consists on training with backprop until an increase in error along the validation set is found. Assume that we obtain a set of predictions from which we can calculate the corresponding confusion matrix (such as the one reported in table 1). Let $c_{i,j}$ be the conditional probability of classifying digit "j" as an "i" and consider the loss function:

$$error = \sum_{i=1}^{n} \sum_{j=1}^{n} c_{i,j} P(i) |i - j|$$

where

$$P(i) = \begin{cases} \lambda_1, & 1 < \lambda_1 & if & i = 5 \\ \lambda_2, & \lambda_1 < \lambda_2^i & if & i \neq 5, i \neq 0 \\ 1 & & if & i = 0 \end{cases}$$

Note that this function is simply an average of the conditional probabilities out of the main diagonal of the confusion matrix. Also note that it penalizes more heavily the deviations for the non-central digits, gives more weight to important deviations and pays relatively less importance to the misclassification of the number zero[1].

---

[1] One problem with such approach is that the choice of function and the penalty term is completely arbitrary, so that a bad choice might lead to suboptimal results. Alternatively one might directly use the expected misclassification cost to obtain the weights of the network since this is finally the true objective function. Note that this opens the possibility to implicitly consider that probability in the appearance of a particular number in the dataset is not uniform and that its importance in economic terms might differ. Here we only consider the simple case and leave the other possibility for future work.

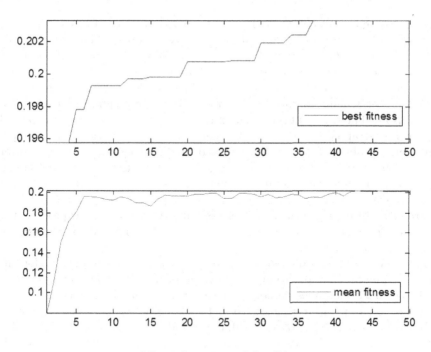

Fig. 1. Iterations of the GA

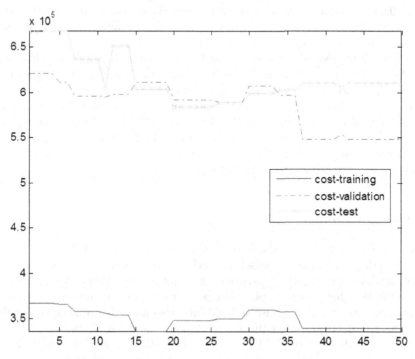

Fig. 2. Misclassification costs

In the second step we try to minimize the expected errors of misclassification derived from the corresponding confusion matrix. To do so, let us now assume that at every iteration, we employ an appropriate optimization procedure to obtain a set of weights that improves the above loss function. In particular, we can employ the simple Genetic Algorithm (Holland, 1976). Then, we make one pass along the training dataset and produce the corresponding new confusion matrix. Note that this confusion matrix can be used to estimate the expected error by bootstrapping from the implied conditional probabilities and by simply comparing, in absolute terms, the real and predicted figures. Note that after that we compute the expected cost of misclassification using the corresponding confusion matrices for the training and validation sets, that is $c_{tra}(i,j)$, $c_{val}(i,j)$ we can employ an increase in the later as the stopping criterion to prevent from overtraining. After the optimal network is found we employ it to produce the final confusion matrix for the test set which is finally employed to estimate the final cost of misclassification.

The approach proposed is extremely computationally demanding and, for this reason, we present an application on a "toy" problem to illustrate the procedure. We employ 6.000 examples randomly chosen from the database to configure our training, validation and testing set (so that each of the sets has 2.000 examples). To compute the expected training, validation and testing error costs we employ 1.000 random amounts between 100 and 10.000Eur. (no decimals). The GA is run using a population of 20 individuals, with crossover and mutation rates of 0.5 and 0.00001, respectively. The number of iterations was set to 50 since, as we shall see, the algorithm converges quite rapidly. The parameters in our penalty function were $\lambda_1 = 1.2$ and $\lambda_2^i = 0.7 + 0.5 * log(|5-i|) + 1)$.

In figure 1 we plot the fitness values for the best individual as well as the mean fitness for the whole population. Note that the algorithm converges quite fast. In figure 2 we present the total expected costs (in euros) of the misclassification along the training, testing and validation sets. Note that the results are promising: for the net chosen to minimize the expected validation costs, we find improvement in the expected testing cost of about a 9%. This means a reduction, in economic terms, of about 60.000 Eur. since the total testing expected error reduces from 668.029 Eur. to 610.130 Eur.

4   Conclusions

In this paper we have proposed a novel method, in the context of Optical Character Recognition, to reduce the misclassification error of neural network classifiers in problem of recognizing the handwritten amounts of bank cheques. Our method relies in the fact that, in this application, neither all the characters have the same information content nor the misclassification errors have the same economic impact. The approach proposed effectively exploits these two facts by employing a modified penalty function, which is optimized through heuristic procedures, and by using the conditional probabilities implicit in the confusion matrix to estimate the expected error of misclassifications. Even though our results are purely preliminary, they suggest several extensions to improve network accuracy in OCR problems.

References

1. Battacharya, U and B.B. Chaudry (2003): A Majority Voting Scheme for Multiresolution Recognition of Handprinted Numerals, in Proceedings of the Seventh International Conference on Document Analysis and Recognition ICDAR 2003, IEEE Press.
2. Holland, J. (1976): *Adaptation in Natural and Artificial Systems*, The MIT Press.
3. Kim; I-J and J-H Kim (2003): 'Statistical character structure modeling and its application to handwritten Chinese character recognition', *IEEE Transactions on Pattern Analysis and Machine Intelligence*, Volume: 25 , Issue: 11 , 1422 – 1436
4. Milgram, J.; M Cheriet and R. Sabourin (2004): 'Speeding Up the Decision Making of Support Vector Classifiers', *Frontiers in Handwriting Recognition, IWFHR-9 2004*. 26-29 Oct. 2004. pp. 57 – 62

Using Fuzzy Clustering Technique for Function Approximation to Approximate ECG Signals

A. Guillén, I. Rojas, E. Ros, and L.J. Herrera

Department of Computer Architecture and Computer Technology,
Universidad de Granada, Spain

Abstract. Radial Basis Function Neural Networks (RBFNN) has been applied successfully to solve function approximation problems. In the design of an RBFNN, it is required a first initialization step for the centers of the RBFs. Clustering algorithms have been used to initialize the centers, but these algorithms were not designed for this task but rather for classification problems. The initialization of the centers is a very important issue that affects significantly the approximation accuracy. Because of this, the CFA (Clustering for Function Approximation) algorithm has been developed to make a better placement of the centers. This algorithm performed very good in comparison with other clustering algorithms used for this task. But it still may be improved taking into account different aspects, such as the way the partition of the input data is done, the complex migration process, the algorithm's speed, the existence of some parameters that need to be set in a concrete order to obtain good solutions, and the convergence guaranty. In this paper, it is proposed an improved version of this algorithm that solves some problems that affected its predecessor. The approximation of ECG signals is not straightforward since it has low and high frequency components in different intervals of a heart stroke. Furthermore, each interval (P wave, the QRS complex, T wave) is related with the behaviour of specific parts of the heart. The new algorithm has been tested using the ECG signal as the target function to be approximated obtaining very small approximation errors when it is compared with the traditional clustering technique that were used for the centers initialization task. The approximation of the ECG signal can be useful in the diagnosis of certain diseases such as Paroxysmal Atrial Fibrillation (PAF).

1   Introduction

Designing an RBF Neural Network (RBFNN) to approximate a function from a set of input-output data pairs, is a common solution since this kind of networks are able to approximate any function [5, 10]. Formally, a function approximation problem can be formulated as, given a set of observations $\{(\boldsymbol{x}_k; y_k); k = 1, ..., n\}$ with $y_k = F(\boldsymbol{x}_k) \in \mathbb{R}$ and $\boldsymbol{x}_k \in \mathbb{R}^d$, it is desired to obtain a function $\mathcal{G}$ so $y_k = \mathcal{G}(\boldsymbol{x}_k) \in \mathbb{R}$ with $\boldsymbol{x}_k \in \mathbb{R}^d$. Once this function is learned, it will be possible to generate new outputs from input data that were not specified in the original data set.

J. Mira and J.R. Álvarez (Eds.): IWINAC 2005, LNCS 3562, pp. 538–547, 2005.

The initialization of the centers of RBFs is the first step to design an RBFNN. This task has been solved traditionally using clustering algorithms [8] [11]. Clustering techniques have been applied to classification problems [7], where the task to solve is how to organize observed data into meaningful structures. In classification problems, the input data has to be assigned to a pre-defined set of labels, thus, if a label is not assigned correctly, the error will be greatly increased. In the functional approximation problem, a continuous interval of real numbers is defined to be the output of the input data. Thus, if the generated output value is near the real output, the error does not increase too much.

In this context, a new clustering algorithm for functional approximation problems was designed in our research group: Clustering for Functional Approximation (CFA)[6].The CFA algorithm uses the information provided by the function output in order to make a better placement of the centers of the RBFs. This algorithm provides better results in comparison with traditional clustering algorithms but it has several elements that can be improved.

In this paper, a new algorithm is proposed, solving all the problems presented in the CFA algorithm using fuzzy logic techniques, and improving results, as it will be shown in the experiments section. In this section, the target function will be the ECG signal of a human person. The approximation an ECG signal is not straightforward since it has low and high frequency components in different intervals of a heart stroke. Furthermore, each interval (P wave, the QRS complex, T wave) is related with specific parts of the heart. Therefore, diverse pathologies are diagnosed studying in detail specific intervals within the ECG traces, for instance the P wave is related with the atrial activity while ventricular fibrillation can be easily detected observing the evolution of the QRS complexes.

For certain diagnosis applications the shape of a certain interval of the ECG is of interest. Some approaches extract templates of a certain area by averaging the activity in this area over a number of heart strokes [9] or by extracting morphological features of these intervals [13, 4] (such as the amplitude of the P wave, the integral of the P wave). These techniques require the definition of a P wave interval and the accurate measurement of different characteristics. But the previous methods are simplified ways to characterize the heart activity in a certain interval, which can be done more accurately with sophisticated function approximation methods.

2   RBFNN Description

A RBFNN $\mathcal{F}$ with fixed structure to approximate an unknown function F with n entries and one output starting from a set of values $\{(\boldsymbol{x}_k; y_k); k = 1, ..., n\}$ with $y_k = F(\boldsymbol{x}_k) \in \mathbb{R}$ and $\boldsymbol{x}_k \in \mathbb{R}^d$, has a set of parameters that have to be optimized:

$$\mathcal{F}(\boldsymbol{x}_k; C, R, \Omega) = \sum_{j=1}^{m} \phi(\boldsymbol{x}_k; \boldsymbol{c}_j, r_j) \cdot \Omega_j \tag{1}$$

where $C = \{c_1, ..., c_m\}$ is the set of RBF centers, $R = \{r_1, ..., r_m\}$ is the set of values for each RBF radius, $\Omega = \{\Omega_1, ..., \Omega_m\}$ is the set of weights and $\phi(x_k; c_j, r_j)$ represents an RBF. The activation function most commonly used for classification and regression problems is the Gaussian function because it is continuous, differentiable, it provides a softer output and improves the interpolation capabilities [2, 12]. The procedure to design an RBFNN for functional approximation problem is shown below:

1. Initialize RBF centers c_j
2. Initialize the radius r_j for each RBF
3. Calculate the optimum value for the weights Ω_j

The first step is accomplished by applying clustering algorithms, the new algorithm proposed in this paper will initialize the centers, providing better results than other clustering algorithms used for this task.

3   Clustering for Function Approximation Algorithm: CFA

This algorithm uses the information provided by the objective function output in such a way that the algorithm will place more centers where the variability of the output is higher instead of where there are more input vectors.

To fulfill this task, the CFA algorithm defines a set $O = \{o_1, ..., o_m\}$ that represents a hypothetic output for each center. This value will be obtained as a weighted mean of the output of the input vectors belonging to a center.

CFA defines an objective function that has to be minimized in order to converge to a solution:

$$\frac{\sum_{j=1}^{m} \sum_{x_k \in C_j} \|x_k - c_j\|^2 \omega_{kj}}{\sum_{j=1}^{m} \sum_{x_k \in C_j} \omega_{kj}} \qquad (2)$$

where ω_{kj} weights the influence of each input vector in the final position a center. The bigger the distance between the expected output of a center and the real output of an input vector is, the bigger the influence in the final result will be. The calculation of w is obtained by:

$$\omega_{kj} = \frac{|F(x_k) - o_j|}{\max_{i=1}^{n} \{F(x_i)\} - \min_{i=1}^{n} \{F(x_i)\}} + \vartheta_{\min}, \quad \vartheta_{\min} > 0. \qquad (3)$$

The first addend in this expression calculates a normalized distance (in the interval [0,1]) between $F(x_k)$ and o_j, the second addend is a minimum contribution threshold. The smaller $\vartheta_{\min}$ becomes, the more the centers are forced to be in areas where the output is more variable.

The CFA algorithm is structured in three basic steps: Partition of the data, centers and estimated output updating and a migration step.

The partition is performed as it is done in Hard C-means [3], thus, a Voronoi partition of the data is obtained. Once the input vectors are partitionated, the centers and their estimated outputs have to be updated, this process is done iteratively using the equations shown below:

$$
c_j = \frac{\sum\limits_{x_k \in C_j} x_k \omega_{kj}}{\sum\limits_{x_k \in C_j} \omega_{kj}} \quad o_j = \frac{\sum\limits_{x_k \in C_j} F(x_k)\omega_{kj}}{\sum\limits_{x_k \in C_j} \omega_{kj}} . \tag{4}
$$

The algorithm, to update centers and estimated outputs, has an internal loop that iterates until the total distortion of the partition is not decreased significantly.

The algorithm has a migration step that moves centers allocated in input zones where the target function is stable, to zones where the output variability is higher. The idea of a migration step was introduced in [14] as an extension of Hard C-means.

CFA tries to find an optimal vector quantization where each center makes an equal contribution to the total distortion [5]. This means that the migration step will iterate, moving centers that make a small contribution to the error to the areas where centers make a bigger contribution.

3.1   Flaws in CFA

CFA has some flaws that can be improved, making the algorithm more robust and efficient and providing better results.

The first disadvantage of CFA is the way the partition of the data is made. CFA makes a hard partition of the data where an input vector can belong uniquely to a center, this is because it is based on the Hard C-means algorithm. When Fuzzy C-means [1] was developed, it demonstrated how a fuzzy partition of the data could perform better than a hard partition. For the functional approximation problem, it is more logical to apply a fuzzy partition of the data because an input vector can activate several neurons with a certain degree of activation, in the same way an input vector can belong to several centers in a fuzzy partition.

The second problem is the setting of a parameter which influences critically the results that can be obtained. The parameter is ϑ_{min}, the minimum contribution threshold. The smaller this parameter becomes, the slower the algorithm becomes and the convergence becomes less warranted. The need of a human expert to set this parameter with a right value is crucial when it is desired to apply the algorithm to different functions, because a wrong value, will make the algorithm provide bad results.

The third problem of CFA is the iterative process to converge to the solution. The convergence is not demonstrated because it is presented as a weighted version of Hard C-means, but the equations proposed do not warrant the convergence of the algorithm. The iterative method is quite inefficient because it has to iterate many times on each iteration of the main body of the algorithm.

The last problem CFA presents is the migration process. This migration step is quite complex and makes the algorithm run very slow. It is based on a distortion function that require as many iterations as centers, and adds randomness to the algorithm making it not too robust.

4   Improved CFA Algorithm: ICFA

Let's introduce the new elements in comparison with CFA, and let's see the reasons why this new elements are introduced.

4.1   Input Data Partition

As it was commented before, for the functional approximation problem, is better to use a fuzzy partition, but CFA uses a hard partition of the data. In ICFA, in the same way as it is done in Fuzzy C-means, a fuzzy partition of the data is used, thus, an input vector belongs to several centers at a time with a certain membership degree.

4.2   Parameter w

In CFA, the estimated output of a center is calculated using a parameter w (3). The calculation of w implies the election of a minimum contribution value ($\vartheta_{\min}$) that will affect in a serious way the performance and the computing time of the algorithm.

In order to avoid the establishment of a parameter, ICFA removes this threshold, and the difference between the expected output of a center and the real output of the input data is not normalized. Thus, the calculation of w is done by:

$$w_{kj} = \frac{|F(\boldsymbol{x}_k) - o_j|}{\max\limits_{i=1}^{n}\{F(\boldsymbol{x}_i)\} - \min\limits_{i=1}^{n}\{F(\boldsymbol{x}_i)\}} \tag{5}$$

where $F(\boldsymbol{x})$ is the function output and o_j is the estimated output of $\boldsymbol{c}_j$.

4.3   Objective Function and Iterative Process

In order to make the centers closer to the areas where the target function is more variable, a change in the similarity criteria used in the clustering process it is needed. In Fuzzy C-means, the similarity criteria is the euclidean distance. Proceeding this way, only the coordinates of the input vectors are used, thus, the values of the membership matrix U for a given center will be small for the input vectors that are far from that center, and the values will be big if the input vector is close to that center. For the functional approximation problem, this is not always true because, given a center, its associated cluster can own many input vectors even if they are far from this center but they have the same output values.

To consider these situations, the parameter w is introduced (5) to modify the values of the distance between a center and an input vector. w will measure the difference between the estimated output of a center and the output value of an input vector. The smaller w is, the more the distance between the center and the vector will be reduced. This distance is calculated now by modifying the norm in the euclidean distance:

$$d_{kj} = \|\boldsymbol{x}_k - \boldsymbol{c}_j\|^2 \cdot w_{kj}^2. \tag{6}$$

The objective function to be minimize is redefined as:

$$J_h(U, C, W) = \sum_{k=1}^{n} \sum_{i=1}^{m} u_{ik}^h \|\boldsymbol{x}_k - \boldsymbol{c}_i\|^2 w_{ik}^2 \tag{7}$$

where $w_{ik} = |Y_k - o_i|$. This function is minimized applying the LS method, obtaining the following equations that will converge to the solution:

$$u_{ik} = \left(\sum_{j=1}^{m} \left(\frac{d_{ik}}{d_{jk}} \right)^{\frac{2}{h-1}} \right)^{-1} \qquad c_i = \frac{\sum_{k=1}^{n} u_{ik}^h \boldsymbol{x}_k w_{ik}^2}{\sum_{k=1}^{n} u_{ik}^h w_{ik}^2}$$

$$o_i = \frac{\sum_{k=1}^{n} u_{ik}^h Y_k d_{ik}^2}{\sum_{k=1}^{n} u_{ik}^h d_{ik}^2} \tag{8}$$

where d_{ij} is the weighted euclidean distance between center i and input vector j, and $h > 1$ is a parameter that allow us to control how fuzzy will be the partition and usually is equal to 2.

These equations are the equivalence of the ones defined for CFA (4) where the centers and their expected outputs are updated. These equations are obtained applying Lagrange multipliers and calculating the respect derivatives of the function, so convergence is warranted, unlike in CFA. ICFA, requires only one step of updating, being much more efficient than CFA where an internal loop is required on each iteration of the algorithm to update the centers and the outputs.

4.4   Migration Step

As in CFA, a migration step is incorporated to the algorithm. CFA's migration iterates many times until each center contributes equally to the error of the function defined to be minimized. On each iteration, all centers are considered to be migrated, making the algorithm inefficient and, since it adds random decisions, the migration will affect directly to the robustness of the final results.

ICFA only makes one iteration and instead of considering all centers to be migrated, it performs a pre-selection of the centers to be migrated. The distortion of a center is the contribution to the error of the function to be minimized. To decide what centers will be migrated, it is used a fuzzy rule that selects centers that have a distortion value above the average.By doing this, centers that do not add a significant error to the objective function are excluded because their placement is correct and they do not need help from other center. The idea is that if two centers introduce a big error, putting them together can decrease the total error.

There is a fixed criteria to choose the centers to be migrated, in opposite to CFA where a random component was introduced at this point. The center to be migrated will be the one that has assigned the smallest value of distortion. The destination of the migration will be the center that has the biggest value of distortion. The repartition of the input vectors between those two it is like in CFA. If the error is smaller than the one before the migration step, the migration is accepted, otherwise is rejected.

4.5   ICFA General Scheme

Once all the elements that compose the algorithm have been described, the general scheme that ICFA follows is:

Do
    Calculate the weighted distance between C_i and X using w
    Calculate the new U_i, C_i using U_i and O_i using C_i
    Migrate
While$(\text{abs}(C_{i-1}\text{-}C_i<threshold)$

In ICFA, the start point is not a random initialization of matrix U as in Fuzzy C-means. In the new algorithm, centers will be distributed uniformly through the input data space and their estimated outputs will be equal to the difference between the maximum and the minimum value of the output function. Proceeding like this, the robustness of the algorithm is only affected by the random component added in the migration with the simulated annealing.

5   Experimental Results

The first experiment consists in approximate a normal person ECG signals. To compare the results provided by the different algorithms, it will be used the normalized root mean squared error (NRMSE).

The radii of the RBFs were calculated using the k-neighbors algorithm with k=1. The weights were calculated optimally by solving a linear equation system.

The data set used for this experiment consist in the record of the signals P,Q,R,S,T obtained by using the electrocardiogram technique. The data is provided by PhysioNet at its Prediction Challenge Database. The signal used belongs to the first minute of record "n01".

Table 1 shows the errors when approximating the first pulse of ECG (Fig. 1) using the ICFA, Fuzzy C-means and CFA algorithms. In Fig. 1 are represented

Table 1. Mean and Standard Deviation of the approximation error (NRMSE) for function the *ECG* signal (Training Pulse) before and after local search algorithm

Clusters	FCM	CFA	ICFA	Clusters	FCM	CFA	ICFA
4	0.966(0.009)	0.948(0.010)	0.449(0)	4	0.238(0.036)	0.495(0.236)	0.160(0)
5	0.938(3E-4)	0.914(0.001)	0.446(0)	5	0.880(0.001)	0.204(0.009)	0.066(0)
6	0.929(1E-4)	0.883(0.023)	0.437(0)	6	0.515(0.383)	0.154(0.029)	0.068(0)
7	0.915(020)	0.871(0.020)	0.422(0)	7	0.694(0.351)	0.101(0.044)	0.055(0)
8	0.914(0.001)	0.846(0.010)	0.336(0)	8	0.657(0.334)	0.067(0.029)	0.050(0)
9	0.893(3E-4)	0.838(0.017)	0.506(0)	9	0.658(0.353)	0.077(0.030)	0.044(0)

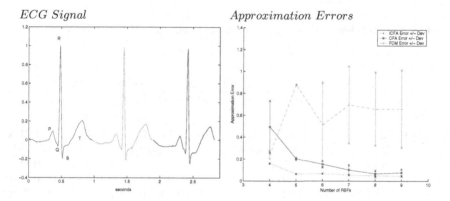

Fig. 1. *ECG* signal, in red, the first heartbeat of training, in green, the test heartbeat. Mean and Standard Deviation of the approximation error (NRMSE) after local search algorithm

graphically the results shown in Table 1. In this table are represented the results before and after applying the Levenberg-Mardquardt local search algorithm. As it is shown, the new algorithm proposed makes a better placement of the centers providing good results even before applying the local search algorithm.

The results clearly show the improvement in performance of ICFA in comparison with CFA and its predecessors, not improving only the results, but the robustness. The ICFA algorithm provides the best approximation in comparison with the others clustering algorithms using only 5 centers. This means that it identifies the 5 low and high frequency components in different intervals of a heart stroke (P wave, the QRS complex, T wave). In Figure 2 this is illustrated, this figure shows the placement of the centers after the execution and the local search algorithm using ICFA. It is clear how it places one center on each signal.

The approximation of the signals is tested with the next pulse on the *ECG* obtaining the results shown in Table 2, in this table we can appreciate how the RBFNN generated using the new algorithm still makes a good approximation of the signal.

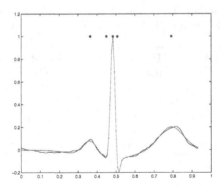

Fig. 2. *ECG* signal and its approximation. Each star represents one center

Table 2. Mean and Standard Deviation of the approximation error (NRMSE) for the *ECG* signal (Test Pulse) before and after local search algorithm

Clusters	FCM	CFA	ICFA	Clusters	FCM	CFA	ICFA
4	0.972(0.)	0.954(0.008)	0.458(0)	4	0.254(0.042)	0.513(0.243)	0.184(0)
5	0.946(3E-4)	0.925(0.001)	0.462(0)	5	0.894(0.001)	0.209(0.015)	0.114(0)
6	0.936(1E-4)	0.895(0.023)	0.457(0)	6	0.525(0.383)	0.172(0.021)	0.100(0)
7	0.924(020)	0.883(0.044)	0.432(0)	7	0.702(0.351)	0.121(0.039)	0.083(0)
8	0.923(0.001)	0.857(0.010)	0.352(0)	8	0.665(0.334)	0.083(0.013)	0.085(0)
9	0.907(3E-4)	0.848(0.020)	0.507(0)	9	0.665(0.353)	0.088(0.016)	0.075(0)

6   Conclusions

RBFNNs provides good results when they are used for functional approximation problems. The CFA algorithm was designed in order to make the right initialization of the centers for the RBFs improving the results provided by the clustering algorithms that were used traditionally for this task. CFA had some mistakes and disadvantages that could be improved. In this paper, a new algorithm which fix all the problems in CFA is proposed. This new algorithm performs much better than its predecessor.

If we use function approximation to characterize ECG traces with a restricted number of centres. These centres (their position with respect to the QRS complex) and their specific characteristics (amplitude, sigma) embed automatically features that can be of interest for diagnosis of heart diseases. The evolution of these features along a ECG trace can be of interest for instance to predict heart crisis such as atrial or ventricular fibrillation which is part of our future work.

Acknowledgements. This work has been partially supported by the Spanish CICYT Project TIN2004-01419.

References

1. J. C. Bezdek. *Pattern Recognition with Fuzzy Objective Function Algorithms.* Plenum, Nueva York, 1981.
2. Adrian G. Bors. Introduction of the Radial Basis Function (RBF) networks. *On-Line Symposium for Electronics Engineers.*
3. R.O. Duda and P.E. Hart. *Pattern classification and scene analysis.* New York:wiley, 1973.
4. P. Kastner G. Schreier and W. Marko. An automatic ECG processing algorithm to identify patients prone to paroxysmal atrial fibrillation. *Computers in Cardiology*, 3:133–136, 2001.
5. A. Gersho. Asymptotically Optimal Block Quantization. *IEEE Transanctions on Information Theory*, 25(4):373–380, July 1979.
6. J. González, I. Rojas, H. Pomares, J. Ortega, and A. Prieto. A new Clustering Technique for Function Aproximation. *IEEE Transactions on Neural Networks*, 13(1):132–142, January 2002.
7. J. A. Hartigan. *Clustering Algorithms.* New York:wiley, 1975.
8. N. B. Karayannis and G.W. Mi. Growing radial basis neural networks: Merging supervised and unsupervised learning with network growth techniques.
9. C. Heneghan P. de Chanzal. Automated Assessment of Atrial Fibrillation. *Computers in Cardiology*, pages 117–120, 2001.
10. J. Park and J. W. Sandberg. Universal approximation using radial basis functions network. *Neural Computation*, 3:246–257, 1991.
11. Y. Cai Q. Zhu and L. Liu. A global learning algorithm for a RBF network.
12. I. Rojas, M. Anguita, A. Prieto, and O. Valenzuela. Analysis of the operators involved in the definition of the implication functions and in the fuzzy inference proccess. *Int. J. Approximate Reasoning*, 19:367–389, 1998.
13. E. Ros, S. Mota, F.J. Fernndez, F.J. Toro, and J.L. Bernier. ECG Characterization of Paroxysmal Atrial Fibrillation: Parameter Extraction and Automatic Diagnosis Algorithm. *Computers in Biology and Medicine*, 34(8):679–696, 2004.
14. M. Russo and G. Patan. Improving the LBG Algorithm. *Lecture Notes in Computer Science*, 1606:621–630, 1999.

Information Retrieval and Classification with Wavelets and Support Vector Machines

S. Dormido-Canto[1], J. Vega[2], J. Sánchez[1], and G. Farias[1]

[1] Dpto. de Informática y Automática -
E.T.S.I. Informática - U.N.E.D. Madrid 28040, Spain
{sebas, jsanchez}@dia.uned.es, gfarias@bec.uned.es
[2] Asociación EURATOM/CIEMAT, Madrid 28040, Spain
jesus.vega@ciemat.es

Abstract. Since fusion plasma experiment generates hundreds of signals. In analyzing these signals it is important to have automatic mechanisms for searching similarities and retrieving of specific data in the waveform database. Wavelet transform (WT) is a transformation that allows to map signals to spaces of lower dimensionality, that is, a smoothed and compressed version of the original signal. Support vector machine (SVM) is a very effective method for general purpose pattern recognition. Given a set of input vectors which belong to two different classes, the SVM maps the inputs into a high-dimensional feature space through some non-linear mapping, where an optimal separating hyperplane is constructed. This hyperplane minimizes the risk of misclassification and it is determined by a subset of points of the two classes, named support vectors (SV). In this work, the combined use of WT and SVM is proposed for searching and retrieving similar waveforms in the TJ-II database. In a first stage, plasma signals will be preprocessed by WT in order to reduce the dimensionality of the problem and to extract their main features. In the next stage, and using the new smoothed signals produced by the WT, SVM will be applied to show up the efficency of the proposed method to deal with the problem of sorting out thousands of fusion plasma signals.

1   Introduction

Databases in nuclear fusion experiments are made up of thousands of signals. For this reason, data analysis must be simplified by developing automatic mechanisms for fast search and retrieval of specific data in the waveform database. In particular, a method for finding similar waveforms would be very helpful.

In [1] a method is proposed to find similar time sequences using *Discrete Fourier Transformation* (DFT) to reduce the dimensionality of the feature vectors, that is, to minimize the computation time for indexing and comparing signals. In [2], the previous DFT based method is used to search similar phenomena in waveform databases but just it is applied with slowly varying signals. However, the DFT has difficulties when used with fast varying waveforms since

J. Mira and J.R. Álvarez (Eds.): IWINAC 2005, LNCS 3562, pp. 548–557, 2005.

time information is lost when transforming to the frequency domain and non-stationary or transitory characteristics can not be detected. The *Short Time Fourier Transform* (STFT) can obtain the non-stationary characteristics using an analysis window. However, the precision is determined by the analysis window that is the same for all frequencies. *Wavelets* (WT) offers an efficient alternative to data processing and provides many advantages: 1) data compression, 2) computing efficiency, and 3) simultaneous time and frequency representation. Because of these characteristics, wavelets have a growing impact on signal processing applications [3, 4].

Support Vector Machines (SVM) is a very effective method for general purpose pattern recognition [5, 6, 7]. In a few words, given a set of input vectors which belong to two different classes, the SVM maps the inputs into a high-dimensional feature space through some non-linear mapping, where an optimal separating hyperplane is constructed in order to minimize the risk of misclassification. The hyperplane is determined by a subset of points of the two classes, named *Support Vectors* (SV). Several methods had been proposed to cope with multi-category classification [8].

In this work, preliminary results are shown when using WT techniques for characterizing the signals and SVM as the technique for pattern recognition and information retrieval. The proposed method has been applied to the TJ-II stellarator database. The TJ-II is a stellarator device [9] (heliac type, $B(0) \leq 1.2T$, $R(0) = 1.5m$, $\langle a \rangle \leq 0.22m$) located at CIEMAT (Madrid, Spain) that can explore a wide rotational transform range ($0.9 \leq iota/2\pi \leq 2.2$). TJ-II plasmas are produced and heated with ECRH (2 gyrotrons, 300 kW each, 53.2 GHz, 2nd harmonic, X-mode polarization) and NBI (300 kW). At present, 928 digitization channels are available for experimental measurements in the TJ-II.

2   Wavelet Transform

WT are basis functions used in representing data or other functions. Wavelet algorithms process data at different resolutions or decomposition levels in contrast with DFT where only frequency components are considered. The construction of the first orthonormal system by Haar [10] is an important milestone since the Haar basis is still a foundation of modern wavelet theory. In this work, the use of the Haar wavelets in the problem of extracting characteristic of the plasma signals will be considered.

The motivation for using the WT is to have a decomposition method that is fast to compute and requires little data storage for each signal. The Haar wavelet is chosen for many advantages: (1) it allows good approximation with a subset of coefficients, (2) it can be computed quickly and easily, requiring linear time in the length of the signal and simple coding, and (3) it preserves Euclidean distance. Concrete mathematical foundations can be found in [11]. In the WT with Haar base, there are two kinds of functions called *approximation function and difference function*. The approximation function generates a sequence of the averages between two adjacent values of the input sequence, that is, the sampled signal. The difference function generates a sequence of the differences between two consecutive data in the current approximation sequence. These functions are applied recursively until the number of the elements in the difference sequence is one. That is, the ith approximation sequence A_i and difference sequence D_i are defined as follow:

$$A_i = \{ \frac{A_{i-1}(1)+A_{i-1}(2)}{2}, \frac{A_{i-1}(3)+A_{i-1}(4)}{2}, \ldots, \frac{A_{i-1}(m-1)+A_{i-1}(m)}{2} \}$$

$$D_i = \{ \frac{A_{i-1}(1)-A_{i-1}(2)}{2}, \frac{A_{i-1}(3)-A_{i-1}(4)}{2}, \ldots, \frac{A_{i-1}(m-1)-A_{i-1}(m)}{2} \}$$

where $A_i(j)$ is the j-th element in the sequence A_i and m is the number of the elements in the sequence A_{i-1}. Next, a brief example of the Haar transformation of a discrete sequence $\overrightarrow{X} = \{9, 7, 4, 8, 5, 3, 8, 8\}$ is shown (Table 2).

Table 1. Example of the Haar transformation

Approximation	Averages	Coefficients (Differences)
8	{9, 7, 4, 8, 5, 3, 8, 8}	-
4	{8, 6, 4, 8}	{1, -2, 1, 0}
2	{7, 6}	{1, -2}
1	{6.5}	{0.5}

Approximation 8 is the full resolution of the discrete sequence. In approximation 4, {8, 6, 4, 8} are obtained by taking the average of {9, 7}, {4, 8}, {5, 3} and {8, 8} at resolution 8 respectively. The coefficients {1, -2, 1, 0} at resolution 4 are the differences of {9, 7}, {4, 8}, {5, 3} and {8, 8} divided by two respectively. This process is continued until an approximation of 1 is reached. The Haar transform $H(\overrightarrow{x}) = \{c, d_0^0 \, d_0^1, d_1^1, d_0^2, d_1^2, d_2^2, d_3^2\}$ is obtained which is composed of the last average value 6.5 and the coefficients found on the right most column. It should be pointed out that c is the overall average value of the whole time sequence.

The reason of using Haar transform to replace DFT is based on several evidences. The first reason is on the pruning power. The nature of the Euclidean distance preserved by Haar transform and DFT are different. In DFT, comparison of two time sequences is based on their low frequency components, where most energy is presumed to be concentrated on. On the other hand, the comparison of Haar coefficients is matching a gradually refined resolution of the two time sequences. Another reason is the complexity consideration. The complexity of Haar transform is $O(n)$ whilst $O(n \log n)$ computation is required for DFT. Both impose restriction on the length of time sequences which must be an integral power of 2. Although these computations are all involved in pre-processing stage, the complexity of the transformation can be a concern especially when the database is large, as happens in our case. Another advantage of using WT is the multi-resolution representation of signals since it has the time-frequency localization property. Thus, WT is able to give locations in both time and frequency.

Therefore, wavelet representations of signals bear more information than that of DFT, in which only frequencies are considered. While DFT extracts the lower harmonics which represent the general shape of a time sequence, WT encodes a coarser resolution of the original time sequence with its preceding coefficients.

Fig. 1 shows the WT is applied to the original signals in order to compute a few coefficients for each signal in a fast way.

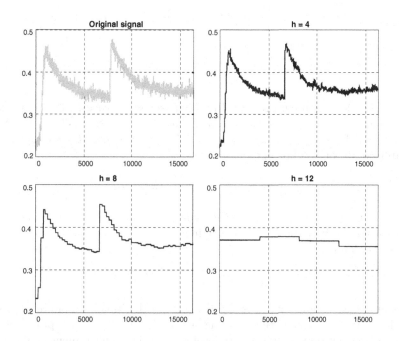

Fig. 1. Original signal and its Wavelet transform approximations with three different decomposition levels (h=4, 8, 12)

3   Support Vector Machines for Classification

The support vector machine (SVM) is a universal constructive learning procedure based on the statistical learning theory [5]. The SVM maps input data into a high-dimensional space using a non-linear function. Once input data are mapped into the high-dimensional space, linear functions with constraints on complexity (i.e., hyperplanes) are used to discriminate the inputs, and a quadratic optimization problem must be solved to determine the parameters of these functions. Nevertheless for high-dimensional feature spaces, the large number of parameters makes this problem intractable. For this reason, duality theory of optimization is used in SVM to make the estimation of parameters in the high-dimensional feature space computationally affordable. The linear approximation function corresponding to the solution of the dual problem is given in the kernel representation and it is called the optimal separating hyperplane. The solution in the kernel representation is written as a weighted sum of the support vectors, that is, a subset of the training data. Let's explain how to obtain the optimal separating hyperplane.

A separating hyperplane is a linear function that is capable of separating the training data without error. Consider the problem of separating the set of training vectors belonging to two separate classes (a binary classifier),

$$\{(x_1, y_1), \ldots, (x_n, y_n)\}, \quad x \in \mathbb{R}, \quad y \in \{+1, -1\}$$

with a hyperplane decision function $D(x)$,

$$D(x) = <w, x>$$

where $< \cdot, \cdot >$ denotes inner product. In linearly separable cases, SVM constructs a hyperplane which separates the training data without error. The hyperplane is constructed by finding another vector w and a parameter b that minimizes $\|w\|^2$ and satisfies the following conditions:

$$y_i = [<w, x> +b] \geq 1, \quad i = 1, \ldots, n$$

where w is a normal weight vector to the hyperplane, $|b|/\|w\|$ is the perpendicular distance from the hyperplane to the origin, and $\|w\|^2$ is the Euclidean norm of w. After the determination of w and b, a given vector x can be classified by:

$$sign(<w, x> +b) \ . \tag{1}$$

In non-linearly separable cases, SVM can map the input vectors into a high dimensional feature space. By selecting a non-linear mapping a priori, SVM constructs and optimal separating hyperplane in this higher dimensional space. A kernel function $K(x, x')$ performs the non-linear mapping into feature space [7], and the original constrains are the same. In this way, the evaluation of the inner products among the vectors in a high-dimensional feature space is done indirectly via the evaluation of the kernel between support vectors and vectors in the input space (Fig. 2).

This provides a way of addressing the technical problem of evaluating inner products in a high-dimensional feature space. Examples of kernel functions are shown in Table 2.

Linear support vector machine is applied to this feature space and then the decision function is given by Eq. 2:

$$f(x) = sign(\sum_{i \in SV} \alpha_i y_i K(x_i, x) + b) \ . \tag{2}$$

where the coefficients α_i and b are determined by maximizing the following Langrangian expression:

$$\sum_{i=1}^{n} \alpha_i - \frac{1}{2} \sum_{i=1}^{n} \sum_{j=1}^{n} \alpha_i \alpha_j y_i y_j K(x_i, x_j) \ , \quad \text{where: } \alpha_i \geq 0 \text{ and } \sum_{i=1}^{n} \alpha_i y_i = 0$$

A positive or negative value from Eq. 1 or Eq. 2 indicates that the vector x belongs or not to class 1. The data samples for which the are nonzero are the support vectors. The parameter b is given by:

$$b = y_s \sum_{i \in SV} \alpha_i y_i K(x_s, x_i)$$

where (x_s, y_s) is any one of the support vectors.

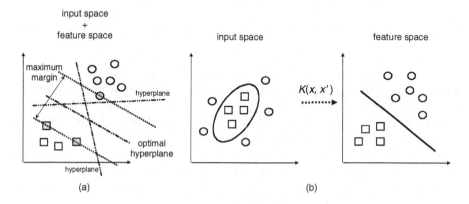

(a)   (b)

Fig. 2. The idea of SVMs: map the training data into a higher-dimensional feature space via K, and construct a separating hyperplane with maximum range there. This yields a nonlinear decision boundary in input space. By the use of kernel functions, it is possible to compute the separating hyperplane without explicity carrying out the map into the feature space. (a) Linearly separable case. (b) Non-linearly separable case

Table 2. Kernel functions extensively used

Kernel Function	Description
Inner product	$K(x, x') = <x, x'>$
Polynomial of degree d	$K(x, x') = (<x, x'> +1)^d$
Gaussian Radial Basis Function	$K(x, x') = \exp\{-\|x - x'\|^2 / 2\sigma^2\}$
Exponential Gaussian Radial Basis Function	$K(x, x') = \exp\{-\sqrt{\|x - x'\|^2} / 2\sigma^2\}$

4 Performance Evaluation

Some preliminary results of our pattern classification approach based on wavelets and SVM are presented in this Section. We have focused the attention in showing the method validity instead of looking for a specific application. Our proof was based on classifying and recognizing temporal evolution signals from the TJ-II database. It is accomplished in a two-step process. A first step provides signal conditioning (Fig. 3), to ensure the same sampling period and number of samples.

This requirement is a consequence of the fact that signals could have been collected with different acquisition parameters. A second step is devoted to perform,

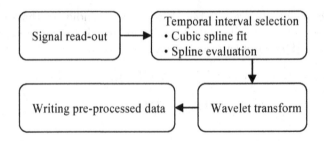

Fig. 3. Signal conditioning data flow

firstly, the learning process with SVM and some of the pre-processed data. Secondly, classification tasks are carried out. All processes have been programmed from the MATLAB software package. In order to evaluate the approach, two experiments have been carried out to classify signals stored in the TJ-II database. These signals belong to one of the classes shown in Table 3.

Table 3. Classes of signals of the TJ-II database

Classes	Description
BOL5	Bolometer signal
ECE7	Electron ciclotron emission
RX306	Soft x-ray
ACTON275	Espectroscopic signal (CV)
HALFAC3	Hα
Densidad2	Line averaged electron density

In the first stage of our approach, the signals are preprocessed in both of our experiments by Haar transform (with a decomposition level of 8) to reduce the dimensionality of the problem. In the second stage, the test signals are classified using SVM.

The method applied is one versus the rest, that allows to get multi-class classifiers. For that reason, we construct a set of binary classifiers as it is explained in Section III. Each classifier is trained to separate one class from the rest, and to combine them by doing the multi-class classification according to the maximal output before applying the sign function (Eq. 1). Next, two experiments are shown to demonstrate the viability of the proposed approach.

In the first experiment, 4 classes have been considered: ECE7, BOL5, RX306, and Densidad2. The training set is composed by 40 signals and the test set by 32 signals obtained from the TJ-II database.

The Fig. 4 displays the positive support vectors for each class using a linear kernel, the training signal corresponding to the original signal in TJ-II, and the wavelet approach which is the signal resampled to 16384 samples after the wavelet transform.

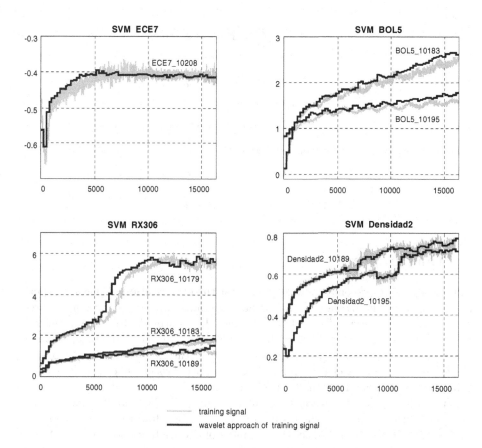

Fig. 4. Positive support vector for every class in the experiment 1

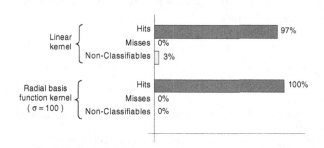

Fig. 5. Results of the experiment 1

The percentage of hits, misses, and non-classifiable signals are illustrated in Fig. 5.

In a second experiment, the training and test sets are composed by 60 and 48 signals and the number of classes was 6, respectively. Fig. 6 shows the results.

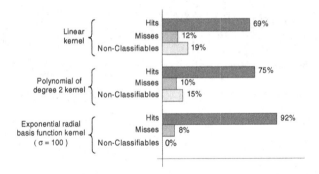

Fig. 6. Results of the experiment 2

5   Conclusions

In this paper, we present a new approach for the classification of plasma experiments signals. The method proposed here contains two processing stages, preprocessing of the original signals by Wavelet Transform (WT) and multi-class classification by Support Vector Machines (SVM). In the first stage, wavelet transformations are applied to signals to reduce the number of dimensions of the feature vectors. After that, a SVM-based multi-class classifier is constructed using the preprocessed signals as input space.

From observation of several experiments, our WT+SVM method is very viable and efficient time, and the results seem promising. However, we have further work to do. We have to finish the development of a Matlab toolbox for WT+SVM processing and to include new relevant features in the SVM inputs to improve the technique, even developing new kernel functions. We have also to make a better preprocessing of the input signals and to study the performance of other generic and self-custom kernels.

References

1. Radiei, D., Mendelzon, A.: Efficient Retrieval of Similar Time Sequences Using DFT. Proc. 5th Inter. Conf. on Foundations of Data Organization (FODO'98), (1998), 249-257
2. Nakanishi, H., Hochin, T., Kojima, M. and LABCOM group: Search and Retrieval Methods of Similar Plasma Waveforms. 4th IAEA TCM on Control, Data Acquisition, and Remote Participation for Fusion Research. San Diego. July 21-23, (2003)
3. Mallat, S.: A Wavelet Tour of Signal Processing. 2 Edition, Academic Press, (2001)
4. Vetterli, M.: Wavelets, Approximation and Compression. IEEE Signal Processing Magazine, (2000), 59-73
5. Vapnik, V.: The Nature of Statistical Learning Theory. Springer, (1995)
6. Vapnik, V.: Statistical Learning Theory. John Wiley & Sons, INC, (1998)

7. Duda, R.O., Hart, P.E., Stork, D.G.: Pattern Classification. 2 Edition, John Wiley, (2001)
8. Sebald, J.D., Bucklew, J.A.: Support Vector Machines and the Multiple Hypothesis Test Problems. IEEE Trans. on Signal Processing, vol. 49, no. 11, (2001), 2865-2872
9. Alejaldre, C. et al.: Plasma Phys. Controlled Fusion 41, 1 (1999), pp. A539
10. Haar A.: Theorie der orthogonalen funktionen-systeme. Mathematische Annalen, (1910) 69:331-371
11. Grochening, K., Madych, W.R.: Multiresolution analysis, Haar bases, and self-similar tilings of rn. IEEE Trans. of Information Theory, vol. 38, no 2, (1992), 556-568

A New Approach to Clustering and Object Detection with Independent Component Analysis

Ingo R. Keck[1], Salua Nassabay[2],
Carlos G. Puntonet[2], and Elmar W. Lang[1]

[1] Institute of Biophysics, Neuro- and Bioinformatics Group,
University of Regensburg, D-93040 Regensburg, Germany
{ingo.keck, elmar.lang}@biologie.uni-regensburg.de
[2] Departamento de Arquitectura y Tecnologia de Computadores,
Universidad de Granada/ESII, E-18071 Granada, Spain
{salua, carlos}@atc.ugr.es

Abstract. It has previously been suggested that the visual cortex performs a data analysis similar to independent component analysis (ICA). Following this idea we show that an incomplete ICA, applied after filtering, can be used to detect objects in natural scenes. Based on this we show that an incomplete ICA can be used to efficiently cluster independent components. We further apply this algorithm to toy data and a real-world fMRI data example and show that this approach to clustering offers a wide variety of applications.

1 Introduction

Clustering of data seems to occur very naturally during data processing in the human brain. We are used to look at our surrounding and perceive things there as "objects" without further thinking why perception actually performs some kind of clustering: all what we can see of one object is clustered to form the image of that object, e.g. all we can see of a chair is clustered to form the image of the chair in our minds.

At the same time the problem of finding data samples which belong together in huge data spaces is an old problem in data analysis. This is especially true with applications of independent component analysis (ICA) in brain research. i.e. to functional magnetic resonance imaging (fMRI) (see [1] for a recent review). While ICA has the power to detect activity in the brain which other methods fail to notice [2], the amount of data any researcher has to analyze manually is a formidable task and forms a severe problem.

Bell *et al* [3] showed that the filters which are produced by analyzing natural scenes with ICA are similar to filters which are found in the visual cortex of animals. This leads to the idea that the visual cortex indeed applies some kind of ICA to learn these filters. But what happens if another ICA is applied

J. Mira and J.R. Álvarez (Eds.): IWINAC 2005, LNCS 3562, pp. 558–566, 2005.

to the outputs, i.e. after filtering of natural scenes? In this article we concentrate on this question and show that an incomplete ICA will automatically perform clustering based on the appearance of the independent components in the mixtures.

2   Theory

In this section we will show that clustering with ICA is related to dimension reduction of the data space. Based on this idea we develop a clustering algorithm.

2.1   Incomplete Independent Component Analysis

ICA can be used to solve the "blind source separation" (BSS) problem. It tries to separate a mixture of originally statistically independent components based on higher order statistical properties of these components:

Let $s_1(t), \ldots, s_{(}t)$ be m independent signals with unit variance for simplicity, represented by a vector $s(t) = (s_1(t), \ldots, s_m(t))^T$, where T denotes the transpose. Let the mixing matrix $\mathbf{A}$ generate n linear mixtures $x(t) = (x_1(t), \ldots, x_n(t))^T$ from these source signals according to:

$$x(t) = \mathbf{A}s(t) \tag{1}$$

(Note that each column of the mixing matrix $\mathbf{A}$ represents the contribution of one source to each mixture.) Assume that only the mixtures $x(t)$ can be observed. Then ICA is the task to recover the original sources $s(t)$ along with the mixing matrix $\mathbf{A}$. For the complete case $n = m$ many algorithms exist to tackle this problem, e.g. Infomax (based on entropy maximization [4]) and FastICA (based on negentropy using fix-point iteration [5]), just to mention some of the most popular ones. The other cases like the more difficult overcomplete ($n < m$) and the more trivial undercomplete ($n > m$) case have also been widely studied in the literature, see e.g. [6,7].

In this paper we concentrate on the incomplete case: What will happen if we try to extract deliberately fewer sources than can be extracted from the mixtures $x(t)$? We do not want to extract a subset of all independent sources, instead we try to cluster all sources into fewer components than could be extracted in principle. In this way the incomplete case differs from the overcomplete case.

At the same time the incomplete case obviously makes a dimension reduction of the data set necessary. A common way to keep the loss of information minimal is to apply a principal component analysis to the mixtures $x(t)$ and to do the dimension reduction based on the eigenvectors e_i corresponding to the smallest eigenvalues λ_i of the data covariance matrix $\mathbf{C}$ [8]. This is also a basic preprocessing step ("whitening") for many ICA algorithms (and can be done quite efficiently with neural networks), as it reduces the degrees of freedom in the space of the solutions by removing all second order correlations of the data and setting the variances to unity:

$$\tilde{x} = \mathbf{E}\Lambda^{-\frac{1}{2}}\mathbf{E}^T x, \tag{2}$$

where $\mathbf{E}$ is the orthogonal matrix of eigenvectors of the covariance matrix of x, with $\mathbf{C}(x) = \mathrm{E}((x - \mathrm{E}(x))(x - \mathrm{E}(x))^T)$, and $\mathbf{\Lambda}$ the diagonal matrix of its eigenvalues.

This dimension reduction will cluster the independent components (IC) $s_i(t)$ based on their presence in the mixing matrix $\mathbf{A}$, as the covariance matrix of x depends on $\mathbf{A}$: [9]

$$\mathrm{E}(xx^T) = \mathrm{E}(\mathbf{A}ss^T\mathbf{A}^T) \tag{3}$$
$$= \mathbf{A}\mathrm{E}(ss^T)\mathbf{A}^T \tag{4}$$
$$= \mathbf{A}\mathbf{A}^T \tag{5}$$

If two columns in the mixing matrix $\mathbf{A}$ are almost identical up to a linear factor, i.e. are linearly dependent, this means that the two sources represented by those columns are almost identically represented (up to a linear factor) in the mixtures. A matrix with two linearly dependent columns does not have full rank, hence will have at least one zero eigenvalue due to its restricted dimension.

This also holds for the transpose $\mathbf{A}^T$ of the matrix $\mathbf{A}$ as the transpose has the same dimensionality as the original matrix, as well as for the product of both matrixes $\mathbf{A}\mathbf{A}^T$.

Setting this close-to-zero eigenvalue to zero in the course of a dimension reduction will thus combine two almost identical columns of $\mathbf{A}$ to a single one. This means that components that appear to be similar to each other in most of the mixtures will be clustered together into new components by the dimension reduction with PCA.

Another possibility is to use only parts of the original data set. This also will cause the ICA to form clusters of independent components with similar columns in the mixing matrix in the reduced data set.

2.2   Clustering with Incomplete ICA

For ICA the literature on clustering so far is based on the comparison of the independent components themselves. To name just a few published algorithms for this problem: the tree-dependent ICA [10] and the topographic ICA [11], which have also been applied to fMRI data [12].

As shown in the section before, clustering based on the columns of the mixing matrix comes naturally to incomplete ICA. However, normally ICA is applied to find the basic independent components of the original data. So a second step is necessary to get from the clusters to their separate parts: The idea is to compare different ICA runs with a different level of dimension reduction applied beforehand. First a complete ICA is performed extracting the maximal number of independent components (ICs) from the data set. In a second run, an incomplete ICA is performed on a reduced data set which resulted from a dimension reduction during PCA pre-processing.

The independent components of the complete ICA without dimension reduction are then compared with the IC of several incomplete ICA runs. Independent components which form part of the components of the incomplete ICA are then grouped into the cluster which is represented by the IC of the incomplete ICA at hand. Hence the ICs of any incomplete ICA form sort of prototype ICs of the clusters formed by ICs from the complete set. This leads immediately to an algorithm for clustering by incomplete ICA:

1. apply a standard ICA to the data set without dimension reduction: ICA1
2. apply a standard ICA to the data set with dimension reduction: ICA2
3. find the independent components in ICA1 that are similar to some forms of components in ICA2 for a further analysis of the independent components.

3   Examples

In this section we demonstrate that incomplete ICA will cluster the response of a filter to a set of images in a way that the results can be used to detect objects. Then we demonstrate with a toy example that the clustering with incomplete ICA outperforms a standard *k-means* clustering algorithm. Finally we apply the presented clustering algorithm to a real world fMRI example.

3.1   Object Detection

Incomplete ICA can be used to detect objects in small movies: while each object typically consists of many separate independent components, these ICs will appear together in the result of the incomplete ICA because they also appear together in the mixtures. To demonstrate this we took 4 pictures of a palm tree in slightly different angles. This mimics the view of a passing observer (figure 1). Then a filter for vertical edges was applied (see figure 2 on the left side for an example). After this an incomplete ICA was used to separate the resulting filter responses of the four images.

Fig. 1. The four black and white images that were used for the object detection test. each time the palm tree is in the centre of the image, but the angle of the view differs slightly

As written before we can expect that this ICA will show us at least one IC with the highest values for that part of the image that represents an object within the filter responses in the image. At the same time this IC should have the overall highest values in his column of the mixing matrix as it should be the IC that consistently appears in the images.

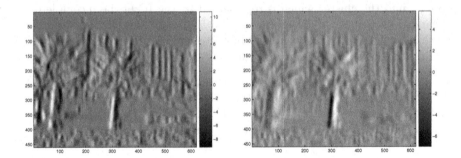

Fig. 2. Left side: The filter response for one of the images. As expected the filter for vertical edges will have the highest results for the trunks of the trees and the buildings in the background. Right side: The component with the highest value in the incomplete ICA. The trunk of the palm tree was detected as object

Figure 2 shows on the right side the IC that showed the highest values in the columns of the mixing matrix. The trunk of the palm tree in the middle here clearly shows the highest values and thus can be marked as "object".

Combined with the results of other filters that are able to detect other parts of objects, the clustering feature of an incomplete ICA could be used to build a sophisticated object detection system.

3.2   Toy Data

To test the quality of the clustering we choose to build a toy data set. 64 sources were created where each of the sources represents a circle in a square lattice of 100×100 lattice points. The mixing matrix was initialized randomly and then modified so that two sets of circles – one representing the letter "A" in the upper left corner, the other representing the letter "C" in the lower right corner – appeared together in the mixtures by setting the columns of these sets to the same values, with random differences of up to 5%. Figure 3 shows one of these mixtures on the left side.

K-Means Analysis. A standard *k-means* clustering analysis was performed to cluster the columns of the mixing matrix **A**. Figure 3 shows on the right the mean overall error of 100 *k-means* analysis runs for 3 up to 63 clusters. It can be seen that in this case this statistic gives no hint on the number of clusters in the data set.

In figure 4 the mean number of wrong circles is plotted, on the left side for the A-class circles, on the right side for the C-class circles. While the k-means analysis obviously clusters all the A-class circles in one cluster up to an overall number of searched-for clusters of 10, it fails to do so with an average error of almost 1 not-A circle. For more than 20 clusters this error disappears, but at the same time A-class circles appear in other clusters. The results for the C-class circles are practically the same as can be seen on the right side of the figure.

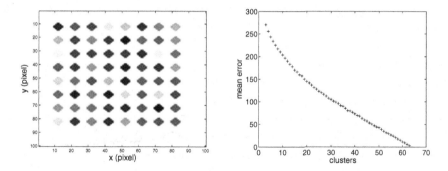

Fig. 3. The figure on the left shows one mixture of the toy data set. Note the almost undetectable A-set in the upper left corner and the C-set in the lower right corner. The figure on the right shows the overall mean error against the number of clusters that were used for the k-means analysis

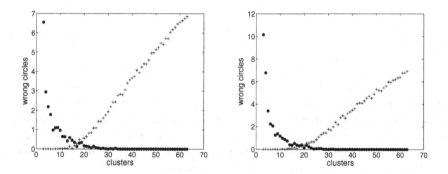

Fig. 4. In the left figure the number of wrong circles in the A-cluster (dots) and the number of A-circles in the wrong cluster (crosses) are plotted against the number of clusters. The right figure shows the same for the C-Cluster. It can be seen that within this data set the k-means analysis fails to cluster the right circles without any error, independent of the number of clusters

Incomplete ICA. For this analysis the *FastICA* algorithm [5] was used. 3 up to 63 components (data reduction via a preceding PCA or via a random pick of the mixtures from the original data) were estimated in 100 runs for each analysis. As the ICA first had to de-mix the mixture of circles, a simple de-noising of the resulting components was necessary (every pixel with a level of 70% was counted as activated).

On the left side of figure 5 the plot for the number of falsely classified circles shows that the algorithm using PCA for dimension reduction worked well for 3 up to 31 used components, thus being remarkably stable. For more than 52 components the ICA separated the single circles of the A-Cluster, as can be expected due to the setting of the data. The incomplete ICA also was able to

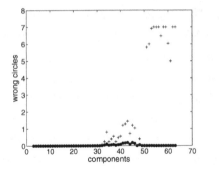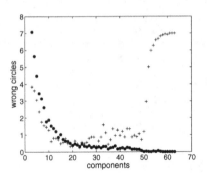

Fig. 5. In the plot on the left side the number of wrong circles in the A-cluster (dots) and the number of A-circles in the wrong cluster (crosses) are plotted against the number of components for clustering with the incomplete ICA using the PCA dimension reduction. The right side shows the result for the dimension reduction by randomly picking mixtures of the data set

find the right column of the mixing matrix for the class-A circles for all used components in every run.

The right side of figure 5 shows the results for the algorithm using a random subset of the original data set for dimension reduction. Obviously the algorithm here had problems in clustering the correct circles together. Also the ICA here was not always successful in finding the correct column of the mixing matrix for the A-class circles. It seems that the simple dimension reduction via the random picking of only some of the original mixtures erases too much of the information, so that the algorithm can not work properly anymore.

3.3   fMRI Data

We also applied the clustering algorithm to functional magnetic resonance data (fMRI) of a modified *Wisconsin Card Sorting Test* of one subject. This data set, consisting of 467 scans, was created at the institute for medicine at the Research Center Jülich, Germany. It has been preprocessed to remove motion artifacts, normalized and was filtered with a gaussian filter to increase the signal to noise ratio. Spatial ICA was used for the analysis so that the independent components correspond to activation maps and the columns of the mixing matrix correspond to the time courses of this activation. [13].

For the clustering with incomplete ICA the data was first reduced via PCA to 400 dimensions, so that almost no information was lost. Then the (spatial) ICA1 was calculated using the *extended Infomax* algorithm [4]. After this step the (spatial) ICA2 was calculated, with the dimension of the data reduced to 20. The 20 independent components of ICA2 were manually compared to a cluster of activations found in the data set using a general linear model analysis. Figure 6 (left) shows the component that was found to correspond to this cluster. Then all independent components of ICA1 were searched automatically for similar

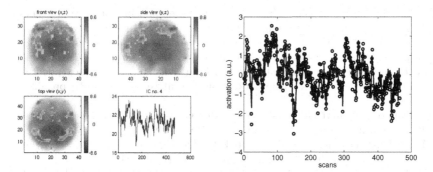

Fig. 6. Left side: This cluster of activations was found in ICA2. The positions of activity match the activations found in a general linear model analysis of the same data. Right side: comparison of the time course of the cluster from ICA2 (solid line) with the sum of the time courses of its components from ICA1 (circles)

activations. The time courses of these components differ in detail, while their sum closely follows the time course of the cluster component of ICA2 (figure 6, right side).

4   Conclusion

Based on the idea that an incomplete ICA will cluster independent components based on their distribution in the mixtures, thus allowing a basic form of object detection, we were able to show that this feature of ICA can be used to develop a promising new kind of clustering algorithm. We also showed that this algorithm was able to out-perform *k-means* in a toy example and can successfully be applied to a real world fMRI data problem.

Acknowledgment

The authors would like to thank Karsten Specht and Gereon Fink from the Research Center Jülich for the fMRI data set. This work was supported by the German ministry of education and research BMBF (project ModKog) and by the project TEC 2004-0696 (SESIBONN).

References

1. V.D. Calhoun, T. Adali, L.K. Hansen, J. Larsen, J.J. Pekar, "ICA of Functional MRI Data: An Overview", in *Fourth International Symposium on Independent Component Analysis and Blind Source Separation*, pp. 281–288, 2003
2. I.R. Keck, F.J. Theis, P. Gruber, E.W. Lang, K. Specht, C.G. Puntonet, "3D Spatial Analysis of fMRI Data on a Word Perception Task", in Springer Lecture Notes LNCS 3195, Proceedings of the ICA 2004 Granada, pp. 977–984, 2004.

3. A.J. Bell, T.J. Sejnowski, "The 'Independent Components' of Natural Scenes are Edge Filters", Vision Research **37**(23), pp. 3327-3338, 1997.
4. A.J. Bell, T.J. Sejnowski, "An information-maximisation approach to blind separation and blind deconvolution", Neural Computation, **7**(6), pp. 1129–1159 (1995).
5. A. Hyvärinnen, "Fast and Robust Fixed-Point Algorithms for Independent Component Analysis", IEEE Transactions on Neural Networks **10**(3), pp. 626–634, 1999.
6. S. Amari, "Natural Gradient Learning for Over- and Under-Complete Bases in ICA", Neural Computation **11**, pp. 1875–1883 (1999).
7. F.J. Theis, A. Jung, C.G. Puntonet, E.W. Lang, "Linear geometric ICA: Fundamentals and algorithms", Neural Computation, **15**, pp. 419–439, 2003
8. A. Hyvärinnen, E. Oja, "Independent Component Analysis: Algorithms and Applications", Neural Networks, **13**(4-5), pp. 411–430, 2000.
9. A. Belouchrani, K. Abed-Meraim, J.-F. Cardoso, E. Moulines, "A Blind Source Separation Technique Using Second-Order Statistics", IEEE Transactions on Signal Processing, **45**(2), pp. 434–444, 1997
10. F.R. Bach, M.I. Jordan, "Beyond independent components: Trees and Clusters", Journal of Machine Learning Research **4**, pp. 1205–1233, 2003.
11. A. Hyvärinnen, P. Hoyer, "Topographic independent component analysis", Neural Computation **13**, pp. 1527–1558, 2001.
12. A. Meyer-Bäse, F.J. Theis, O. Lange, C.G. Puntonet, "Tree-Dependent and Topographic Independent Component Analysis for fMRI Analysis", in Springer Lecture Notes LNCS 3195, Proceedings of the ICA 2004 Granada, pp. 782–789, 2004.
13. M.J. McKeown, T.J. Sejnowski, "Analysis of fmri data by blind separation into independent spatial components", Human Brain Mapping **6**, 160–188, 1998.

Bispectra Analysis-Based VAD for Robust Speech Recognition

J.M. Górriz, C.G. Puntonet, J. Ramírez, and J.C. Segura

E.T.S.I.I., Universidad de Granada,
C/Periodista Daniel Saucedo, 18071 Granada, Spain
gorriz@ugr.es

Abstract. A robust and effective voice activity detection (VAD) algorithm is proposed for improving speech recognition performance in noisy environments. The approach is based on filtering the input channel to avoid high energy noisy components and then the determination of the speech/non-speech bispectra by means of third order autocumulants. This algorithm differs from many others in the way the decision rule is formulated (detection tests) and the domain used in this approach. Clear improvements in speech/non-speech discrimination accuracy demonstrate the effectiveness of the proposed VAD. It is shown that application of statistical detection test leads to a better separation of the speech and noise distributions, thus allowing a more effective discrimination and a tradeoff between complexity and performance. The algorithm also incorporates a previous noise reduction block improving the accuracy in detecting speech and non-speech. The experimental analysis carried out on the AURORA databases and tasks provides an extensive performance evaluation together with an exhaustive comparison to the standard VADs such as ITU G.729, GSM AMR and ETSI AFE for distributed speech recognition (DSR), and other recently reported VADs.

1 Introduction

Speech/non-speech detection is an unsolved problem in speech processing and affects numerous applications including robust speech recognition [1], discontinuous transmission [2, 3], real-time speech transmission on the Internet [4] or combined noise reduction and echo cancellation schemes in the context of telephony [5]. The speech/non-speech classification task is not as trivial as it appears, and most of the VAD algorithms fail when the level of background noise increases. During the last decade, numerous researchers have developed different strategies for detecting speech on a noisy signal [6, 7] and have evaluated the influence of the VAD effectiveness on the performance of speech processing systems [8]. Most of them have focussed on the development of robust algorithms with special attention on the derivation and study of noise robust features and decision rules [9, 10, 11]. The different approaches include those based on energy thresholds [9], pitch detection [12], spectrum analysis [11], zero-crossing rate [3],

J. Mira and J.R. Álvarez (Eds.): IWINAC 2005, LNCS 3562, pp. 567–576, 2005.

periodicity measure [13], higher order statistics in the LPC residual domain [14] or combinations of different features [3, 2].

This paper explores a new alternative towards improving speech detection robustness in adverse environments and the performance of speech recognition systems. The proposed VAD proposes a noise reduction block that precedes the VAD, and uses Bispectra of third order cumulants to formulate a robust decision rule. The rest of the paper is organized as follows. Section II reviews the theoretical background on Bispectra analysis and shows the proposed signal model. Section III analyzes the motivations for the proposed algorithm by comparing the speech/non-speech distributions for our decision function based on bispectra and when noise reduction is optionally applied. Section IV describes the experimental framework considered for the evaluation of the proposed endpoint detection algorithm. Finally, section V summarizes the conclusions of this work.

2   Model Assumptions

Let $\{x(t)\}$ denote the discrete time measurements at the sensor. Consider the set of stochastic variables y_k, $k = 0, \pm 1 \ldots \pm M$ obtained from the shift of the input signal $\{x(t)\}$:

$$\mathbf{y}_k(t) = \mathbf{x}(t + k \cdot \tau) \tag{1}$$

where $k \cdot \tau$ is the differential delay (or advance) between the samples. This provides a new set of $2 \cdot m + 1$ variables by selecting $n = 1 \ldots N$ samples of the input signal. It can be represented using the associated Toeplitz matrix:

$$T_{x(t_0)} = \begin{pmatrix} y_{-M}(t_0) & \cdots & y_{-m}(t_N) \\ y_{-M+1}(t_0) & \cdots & y_{-M+1}(t_N) \\ \cdots & \cdots & \cdots \\ y_M(t_0) & \cdots & y_M(t_N) \end{pmatrix} \tag{2}$$

Using this model the speech-non speech detection can be described by using two essential hypothesis(re-ordering indexes):

$$H_o = \begin{pmatrix} \mathbf{y}_0 = n_0 \\ \mathbf{y}_{\pm 1} = n_{\pm 1} \\ \cdots \\ \mathbf{y}_{\pm M} = n_{\pm M} \end{pmatrix} \tag{3}$$

$$H_1 = \begin{pmatrix} \mathbf{y}_0 = s_0 + n_0 \\ \mathbf{y}_{\pm 1} = s_{\pm 1} + n_{\pm 1} \\ \cdots \\ \mathbf{y}_{\pm M} = s_{\pm M} + n_{\pm M} \end{pmatrix} \tag{4}$$

where s_k's/n_k's are the speech (see section /refsec:speech) /non-speech (any kind of additive background noise i.e. gaussian) signals, related themselves with some differential parameter. All the process involved are assumed to be jointly stationary and zero-mean. Consider the third order cumulant function $C_{y_k y_l}$ defined as:

$$C_{y_k y_l} \equiv E[y_0 y_k y_l] \tag{5}$$

and the two-dimensional discrete Fourier transform (DFT) of $C_{y_k y_l}$, the bispectrum function:

$$\mathcal{C}_{y_k y_l}(\omega_1, \omega_2) = \sum_{k=-\infty}^{\infty} \sum_{l=-\infty}^{\infty} C_{y_k y_l} \cdot \exp(-j(\omega_1 k + \omega_2 l))) \tag{6}$$

2.1 A Model for Speech / Non Speech

The voice detection is achieved applying biespectrum function to the set of new variables detailed in the previous section. Then the essential difference between speech (s_k) and non-speech (n_k) (i.e. noise) will be modelled in terms of the value of the spectral frequency coefficients. We also assume that the noise sequences (n_k) are statistically independent of s_k with vanishing biespectra. Of course the third order cumulant sequences of all process satisfy the summability conditions retailed in [15].

The sequence of cumulants of the voice speech is modelled as a sum of coherent sine waves:

$$C_{y_k y_l} = \sum_{n,m=1}^{K} a_{nm} cos[kn\omega_0^1 + lm\omega_0^2] \tag{7}$$

where a_{nm} is amplitude, $K \times K$ is the number of sinusoids and ω is the fundamental frequency in each dimension. It follows from [14] that a_{mn} is related to the energy of the signal $\mathcal{E}_s = E\{s^2\}$. The VAD proposed in the later reference only works with the coefficients in the sequence of cumulants and is more restrictive in the model of voice speech. Thus the Bispectra associated to this sequence is the DTF of equation 7 which consist in a set of Dirac´s deltas in each excitation frequency $n\omega_0^1, m\omega_0^2$. Our algorithm will detect any high frequency peak on this domain matching with voice speech frames, that is under the above assumptions and hypotheses, it follows that on H_0, $\mathcal{C}_{y_k y_l}(\omega_1, \omega_2) \equiv \mathcal{C}_{n_k n_l}(\omega_1, \omega_2) \simeq 0$ and on H_1 $\mathcal{C}_{y_k y_l}(\omega_1, \omega_2) \equiv \mathcal{C}_{s_k s_l}(\omega_1, \omega_2) \neq 0$. Since $s_k(t) = s(t + k \cdot \tau)$ where $k = 0, \pm 1 \ldots \pm M$, we get:

$$\mathcal{C}_{s_k s_l}(\omega_1, \omega_2) = \mathcal{F}\{E[s(t + k \cdot \tau)s(t + l \cdot \tau)s(t)]\} \tag{8}$$

The estimation of the bispectrum is deep discussed in [16] and many others, where conditions for consistency are given. The estimate is said to be (asymptotically) consistent if the squared deviation goes to zero, as the number of samples tends to infinity.

2.2 Detection Tests for Voice Activity

The decision of our algorithm is based on statistical tests including the Generalized Likelihood ratio tests (GLRT) [17] and the Central χ^2-distributed test statistic under H_O [18]. We will call them GLRT and χ^2 tests. The tests are based on some asymptotic distributions and computer simulations in [19] show that the χ^2 tests require larger data sets to achieve a consistent theoretical asymptotic distribution.

GRLT: Consider the complete domain in biespectrum frequency for $0 \leq \omega_{n,m} \leq 2\pi$ and define P uniformly distributed points in this grid (m, n), called coarse grid. Define the fine grid of L points as the L nearest frequency pairs to coarse grid points. We have that $2M + 1 = P \cdot L$. If we reorder the components of the set of L Bispectrum estimates $\hat{\mathcal{C}}(n_l, m_l)$ where $l = 1, \ldots, L$, on the fine grid around the bifrequency pair into a L vector β_{ml} where $m = 1, \ldots P$ indexes the coarse grid [17] and define P-vectors $\phi_i(\beta_{1i}, \ldots, \beta_{Pi})$, $i = 1, \ldots L$; the generalized likelihood ratio test for the above discussed hypothesis testing problem:

$$H_0 : \mu = \mu_n \quad against \quad H_1 : \eta \equiv \mu^T \sigma^{-1} \mu > \mu_n^T \sigma_n^{-1} \mu_n \tag{9}$$

where $\mu = 1/L \sum_{i=1}^{L} \phi_i$ and $\sigma = 1/L \sum_{i=1}^{L} (\phi_i - \mu)(\phi_i - \mu)^T$ are the maximum likelihood gaussian estimates of vector $\mathcal{C} = (\mathcal{C}_{\mathbf{y}_k \mathbf{y}_l}(m_1, n_1) \ldots \mathcal{C}_{\mathbf{y}_k \mathbf{y}_l}(m_P, n_P))$, leads to the activity voice detection if:

$$\eta > \eta_0 \tag{10}$$

where η_0 is a constant determined by a certain significance level, i.e. the probability of false alarm. Note that:

1. We have supposed independence between signal s_k and additive noise n_k [1] thus:

$$\mu = \mu_n + \mu_s; \quad \sigma = \sigma_n + \sigma_s \tag{11}$$

2. The right hand side of H_1 hypothesis must be estimated in each frame (it's a-priori unknown). In our algorithm the approach is based on the information in the previous non-speech detected intervals.

The statistic considered here η is distributed as a central $F_{2P,2(L-P)}$ under the null hypothesis. Therefore a Neyman-Pearson test can be designed for a significance level α.

χ^2 **tests:** In this section we consider the χ^2_{2L} distributed test statistic[18]:

$$\eta = \sum_{m,n} 2M^{-1} |\Gamma_{\mathbf{y}_k \mathbf{y}_l}(m, n)|^2 \tag{12}$$

where $\Gamma_{\mathbf{y}_k \mathbf{y}_l}(m, n) = \frac{|\hat{\mathcal{C}}_{\mathbf{y}_k \mathbf{y}_l}(n,m)|}{[S_{\mathbf{y}_0}(m) S_{\mathbf{y}_k}(n) S_{\mathbf{y}_l}(m+n)]^{0.5}}$ which is asymptotically distributed as $\chi^2_{2L}(0)$ where L denotes the number of points in interior of the principal

---

[1] Observe that now we do not assume that n_k $k = 0 \ldots \pm M$ are gaussian

domain. The Neyman-Pearson test for a significant level (false-alarm probability) α turns out to be:

$$H_1 \quad if \quad \eta > \eta_\alpha \tag{13}$$

where η_α is determined from tables of the central χ^2 distribution. Note that the denominator of $\Gamma_{\mathbf{y}_k \mathbf{y}_l}(m, n)$ is unknown a priori so they must be estimated as the bispectrum function (that is calculate $\hat{\mathcal{C}}_{\mathbf{y}_k \mathbf{y}_l}(n, m)$). This requires a larger data set as we mentioned above in this section.

2.3   Noise Reduction Block

Almost any VAD can be improved just placing a noise reduction block in the data channel before it. The noise reduction block for high energy noisy peaks, consists of four stages and was first developed in [20]:

i) Spectrum smoothing. The power spectrum is averaged over two consecutive frames and two adjacent spectral bands.

ii) Noise estimation. The noise spectrum $N_e(m, l)$ is updated by means of a 1st order IIR filter on the smoothed spectrum $X_s(m, l)$, that is, $N_e(m, l) = \lambda N_e(m, l - 1) + (1 - \lambda)X_s(m, l)$ where $\lambda = 0.99$ and $m = 0, 1, ..., NFFT/2$.

iii) Wiener Filter (WF) design. First, the clean signal $S(m, l)$ is estimated by combining smoothing and spectral subtraction and then, the WF $H(m, l)$ is designed. The filter $H(m, l)$ is smoothed in order to eliminate rapid changes between neighbor frequencies that may often cause musical noise. Thus, the variance of the residual noise is reduced and consequently, the robustness when detecting non-speech is enhanced. The smoothing is performed by truncating the impulse response of the corresponding causal FIR filter to 17 taps using a Hanning window. With this operation performed in the time domain, the frequency response of the Wiener filter is smoothed and the performance of the VAD is improved.

iv) Frequency domain filtering. The smoothed filter H_s is applied in the frequency domain to obtain the de-noised spectrum $Y(m, l) = H_s(m, l)X(m, l)$.

Fig. 1 shows the operation of the proposed VAD on an utterance of the Spanish SpeechDat-Car (SDC) database [21]. The phonetic transcription is: ["siete", "θinko", "dos", "uno", "otSo", "seis"]. Fig 1(b) shows the value of η versus time. Observe how assuming η_0 the initial value of the magnitude η over the first frame (noise), we can achieve a good VAD decision. It is clearly shown how the detection tests yield improved speech/non-speech discrimination of fricative sounds by giving complementary information. The VAD performs an advanced detection of beginnings and delayed detection of word endings which, in part, makes a hang-over unnecessary. In Fig 2 we display the differences between noise and voice in general and in figure we settle these differences in the evaluation of η on speech and non-speech frames.

According to [20], using a noise reduction block previous to endpoint detection together with a long-term measure of the noise parameters, reports important benefits for detecting speech in noise since misclassification errors are significantly reduced.

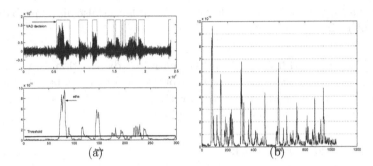

Fig. 1. Operation of the VAD on an utterance of Spanish SDC database. (a) Evaluation of η and VAD Decision. (b) Evaluation of the test hypothesis on an example utterance of the Spanish SpeechDat-Car (SDC) database [21]

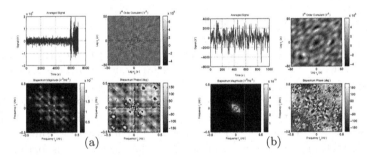

Fig. 2. Different Features allowing voice activity detection. (a) Features of Voice Speech Signal. (b) Features of non Speech Signal

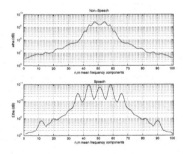

Fig. 3. Speech/non-Speech η values for Speech-Non Speech Frames

3   Experimental Framework

Several experiments are commonly conducted to evaluate the performance of VAD algorithms. The analysis is mainly focussed on the determination of the error probabilities or classification errors at different SNR levels [11] vs. our VAD operation point, The work about the influence of the VAD decision on the performance of speech processing systems [8] is on the way. Subjective performance tests have also been considered for the evaluation of VADs working in combination with speech coders [22]. The experimental framework and the objective performance tests conducted to evaluate the proposed algorithm are partially showed for space reasons (we only show the results on AURORA-3 database)in this section.

The ROC curves are frequently used to completely describe the VAD error rate. The AURORA subset of the original Spanish SpeechDat-Car (SDC) database [21] was used in this analysis. This database contains 4914 recordings using close-talking and distant microphones from more than 160 speakers. The files are categorized into three noisy conditions: quiet, low noisy and highly noisy conditions, which represent different driving conditions with average SNR values between 25dB, and 5dB. The non-speech hit rate (HR0) and the false alarm rate (FAR0= 100-HR1) were determined in each noise condition being the actual speech frames and actual speech pauses determined by hand-labelling the database on the close-talking microphone. These noisy signals represent the most probable application scenarios for telecommunication terminals (suburban train, babble, car, exhibition hall, restaurant, street, airport and train station).

In table 1 shows the averaged ROC curves of the proposed VAD (BiSpectra based-VAD) and other frequently referred algorithms [9, 10, 11, 6] for recordings from the distant microphone in quiet, low and high noisy conditions. The working points of the G.729, AMR and AFE VADs are also included. The results show improvements in detection accuracy over standard VADs and over a representative set VAD algorithms [9, 10, 11, 6]. It can be concluded from these results that:

i) The working point of the G.729 VAD shifts to the right in the ROC space with decreasing SNR.

ii) AMR1 works on a low false alarm rate point of the ROC space but exhibits poor non-speech hit rate.

iii) AMR2 yields clear advantages over G.729 and AMR1 exhibiting important reduction of the false alarm rate when compared to G.729 and increased non-speech hit rate over AMR1.

iv) The VAD used in the AFE for noise estimation yields good non-speech detection accuracy but works on a high false alarm rate point on the ROC space.

Table 1. Average speech/non-speech hit rates for SNRs between $25dB$ and $5dB$. Comparison of the proposed BSVAD to standard and recently reported VADs

(%)	G.729	AMR1	AMR2	AFE (WF)	AFE (FD)
HR0	55.798	51.565	57.627	69.07	33.987
HR1	88.065	98.257	97.618	85.437	99.750
(%)	Woo	Li	Marzinzik	Sohn	χ^2/GLRT
HR0	62.17	57.03	51.21	66.200	66.520/68.048
HR1	94.53	88.323	94.273	88.614	85.192/90.536

It suffers from rapid performance degradation when the driving conditions get noisier. On the other hand, the VAD used in the AFE for FD has been planned to be conservative since it is only used in the DSR standard for that purpose. Thus, it exhibits poor non-speech detection accuracy working on a low false alarm rate point of the ROC space.

v) The proposed VAD also works with lower false alarm rate and higher non-speech hit rate when compared to the Sohn's [6], Woo's [9], Li's [10] and Marzinzik's [11] algorithms in poor SNR scenarios. The BSVAD works robustly as noise level increases.

The benefits are especially important over G.729, which is used along with a speech codec for discontinuous transmission, and over the Li's algorithm, that is based on an optimum linear filter for edge detection. The proposed VAD also improves Marzinzik's VAD that tracks the power spectral envelopes, and the Sohn's VAD, that formulates the decision rule by means of a statistical likelihood ratio test.

It is worthwhile mentioning that the experiments described above yields a first measure of the performance of the VAD. Other measures of VAD performance that have been reported are the clipping errors [22]. These measures provide valuable information about the performance of the VAD and can be used for optimizing its operation. Our analysis does not distinguish between the frames that are being classified and assesses the hit-rates and false alarm rates for a first performance evaluation of the proposed VAD. On the other hand, the speech recognition experiments conducted later on the AURORA databases will be a direct measure of the quality of the VAD and the application it was designed for. Clipping errors are evaluated indirectly by the speech recognition system since there is a high probability of a deletion error to occur when part of the word is lost after frame-dropping.

These results clearly demonstrate that there is no optimal VAD for all the applications. Each VAD is developed and optimized for specific purposes. Hence, the evaluation has to be conducted according to the specific goal of the VAD. Frequently, VADs avoid loosing speech periods leading to an extremely conservative behavior in detecting speech pauses (for instance, the AMR1 VAD). Thus, in order to correctly describe the VAD performance, both parameters have to be considered.

4   Conclusions

This paper presented a new VAD for improving speech detection robustness in noisy environments. The approach is based on higher order Spectra Analysis employing noise reduction techniques and order statistic filters for the formulation of the decision rule. The VAD performs an advanced detection of beginnings and delayed detection of word endings which, in part, avoids having to include additional hangover schemes. As a result, it leads to clear improvements in speech/non-speech discrimination especially when the SNR drops. With this and other innovations, the proposed algorithm outperformed G.729, AMR and AFE standard VADs as well as recently reported approaches for endpoint detection. We think that it also will improve the recognition rate when it was considered as part of a complete speech recognition system.

References

1. L. Karray and A. Martin, "Towards improving speech detection robustness for speech recognition in adverse environments," *Speech Communitation*, no. 3, pp. 261–276, 2003.
2. ETSI, "Voice activity detector (VAD) for Adaptive Multi-Rate (AMR) speech traffic channels," *ETSI EN 301 708 Recommendation*, 1999.
3. ITU, "A silence compression scheme for G.729 optimized for terminals conforming to recommendation V.70," *ITU-T Recommendation G.729-Annex B*, 1996.
4. A. Sangwan, M. C. Chiranth, H. S. Jamadagni, R. Sah, R. V. Prasad, and V. Gaurav, "VAD techniques for real-time speech transmission on the Internet," in *IEEE International Conference on High-Speed Networks and Multimedia Communications*, 2002, pp. 46–50.
5. S. Gustafsson, R. Martin, P. Jax, and P. Vary, "A psychoacoustic approach to combined acoustic echo cancellation and noise reduction," *IEEE Transactions on Speech and Audio Processing*, vol. 10, no. 5, pp. 245–256, 2002.
6. J. Sohn, N. S. Kim, and W. Sung, "A statistical model-based voice activity detection," *IEEE Signal Processing Letters*, vol. 16, no. 1, pp. 1–3, 1999.
7. Y. D. Cho and A. Kondoz, "Analysis and improvement of a statistical model-based voice activity detector," *IEEE Signal Processing Letters*, vol. 8, no. 10, pp. 276–278, 2001.
8. R. L. Bouquin-Jeannes and G. Faucon, "Study of a voice activity detector and its influence on a noise reduction system," *Speech Communication*, vol. 16, pp. 245–254, 1995.
9. K. Woo, T. Yang, K. Park, and C. Lee, "Robust voice activity detection algorithm for estimating noise spectrum," *Electronics Letters*, vol. 36, no. 2, pp. 180–181, 2000.
10. Q. Li, J. Zheng, A. Tsai, and Q. Zhou, "Robust endpoint detection and energy normalization for real-time speech and speaker recognition," *IEEE Transactions on Speech and Audio Processing*, vol. 10, no. 3, pp. 146–157, 2002.
11. M. Marzinzik and B. Kollmeier, "Speech pause detection for noise spectrum estimation by tracking power envelope dynamics," *IEEE Transactions on Speech and Audio Processing*, vol. 10, no. 6, pp. 341–351, 2002.
12. R. Chengalvarayan, "Robust energy normalization using speech/non-speech discriminator for German connected digit recognition," in *Proc. of EUROSPEECH 1999*, Budapest, Hungary, Sept. 1999, pp. 61–64.
13. R. Tucker, "Voice activity detection using a periodicity measure," *IEE Proceedings, Communications, Speech and Vision*, vol. 139, no. 4, pp. 377–380, 1992.
14. E. Nemer, R. Goubran, and S. Mahmoud, "Robust voice activity detection using higher-order statistics in the lpc residual domain," *IEEE Trans. Speech and Audio Processing*, vol. 9, no. 3, pp. 217–231, 2001.
15. C. Nikias and A. Petropulu, *Higher Order Spectra Analysis: a Nonlinear Signal Processing Framework*. Prentice Hall, 1993.

16. D. Brillinger and M. Rossenblatt, *Spectral Analysis of Time Series.* Wiley, 1975, ch. Asymptotic theory of estimates of kth order spectra.
17. T. Subba-Rao, "A test for linearity of stationary time series," *Journal of Time Series Analisys,* vol. 1, pp. 145–158, 1982.
18. J. Hinich, "Testing for gaussianity and linearity of a stationary time series," *Journal of Time Series Analisys,* vol. 3, pp. 169–176, 1982.
19. J. Tugnait, "Two channel tests fro common non-gaussian signal detection," *IEE Proceedings-F,* vol. 140, pp. 343–349, 1993.
20. J. Ramí´ýrez, J. Segura, C. Bení´ýtez, A. delaTorre, and A. Rubio, "An effective subband osf-based vad with noise reduction for robust speech recognition," *In press IEEE Transactions on Speech and Audio Processing,* vol. X, no. X, pp. X–X, 2004.
21. A. Moreno, L. Borge, D. Christoph, R. Gael, C. Khalid, E. Stephan, and A. Jeffrey, "SpeechDat-Car: A Large Speech Database for Automotive Environments," in *Proceedings of the II LREC Conference,* 2000.
22. A. Benyassine, E. Shlomot, H. Su, D. Massaloux, C. Lamblin, and J. Petit, "ITUT Recommendation G.729 Annex B: A silence compression scheme for use with G.729 optimized for V.70 digital simultaneous voice and data applications," *IEEE Communications Magazine,* vol. 35, no. 9, pp. 64–73, 1997.

On-line Training of Neural Networks: A Sliding Window Approach for the Levenberg-Marquardt Algorithm

Fernando Morgado Dias[1], Ana Antunes[1], José Vieira[2], and Alexandre Manuel Mota[3]

[1] Escola Superior de Tecnologia de Setúbal do Instituto Politécnico de Setúbal,
Departamento de Engenharia Electrotécnica, Campus do IPS,
Estefanilha, 2914-508 Setúbal, Portugal Tel: +351 265 790000
{aantunes, fmdias}@est.ips.pt
[2] Escola Superior de Tecnologia de Castelo Branco,
Departamento de Engenharia Electrotécnica, Av. Empresário,
6000 Castelo Branco, Portugal Tel: +351 272 330300
zevieira@est.ipcb.pt
[3] Departamento de Electrónica e Telecomunicações,
Universidade de Aveiro, 3810 - 193 Aveiro,
Portugal, Tel: +351 234 370383
alex@det.ua.pt

Abstract. In the Neural Network universe, the Backpropagation and the Levenberg-Marquardt are the most used algorithms, being almost consensual that the latter is the most effective one. Unfortunately for this algorithm it has not been possible to develop a true iterative version for on-line use due to the necessity to implement the Hessian matrix and compute the trust region. To overcome the difficulties in implementing the iterative version, a batch sliding window with Early Stopping is proposed, which uses a hybrid Direct/Specialized evaluation procedure. The final solution is tested with a real system.

1 Introduction

Many algorithms have been used in the field of Artificial Neural Networks (ANNs). Among these the Backpropagation and the Levenberg-Marquardt are the most used, being almost consensual that the latter is the most effective one. Its use has though been mostly restricted to the off-line training because of the difficulties to implement a true iterative version. The on-line versions are very useful for identifying time varying systems and to build black box approaches for identification. These difficulties to implement an iterative version come from computing the derivatives for the Hessian matrix, inverting this matrix and computing the region for which the approximation contained in the calculation of the Hessian matrix is valid (the trust region). In the present work a different approach is suggested: the use of the Levenberg-Marquardt algorithm on-line in a batch version with sliding window and Early Stopping, allowing the use of the Levenberg-Marquardt algorithm as in the off-line approaches. The implementation also uses a hybrid Direct/Specialized evaluation procedure and the final solution is tested with a real system composed of a reduced scale prototype kiln.

J. Mira and J.R. Álvarez (Eds.): IWINAC 2005, LNCS 3562, pp. 577–585, 2005.

2   Review of the Algorithm

In this section, a short review of the Levenberg-Marquardt algorithm is done to enable easier perception of the problems found in the on-line implementation. Equation 1 shows the updating rule for the algorithm where x_k is the current iteration, $v(x)$ is the error between the output obtained and the pretended output, $J(x_k)$ is the Jacobian of the system at iteration k and $2.J^T(x_k).J(x_k) + \mu_k I$ is the Hessian matrix approximation used where I is the identity matrix and μ_k is a value (that can be changed in each iteration) that makes the approximation positive definite and therefore allowing its inversion.

$$\triangle x_k = -\left[2.J^T(x_k).J(x_k) + \mu_k I\right]^{-1}.2.J^T(x_k).v(x_k) \qquad (1)$$

The Levenberg-Marquardt algorithm is due to the independent work of both authors in [1] and [2].

The parameter μ_k is the key of the algorithm since it is responsible for stability (when assuring that the Hessian can be inverted) and speed of convergence. It is therefore worth to take a closer look on how to calculate this value.

The modification of the Hessian matrix will only be valid in a neighbourhood of the current iteration. This corresponds to search for the correct update of the next iteration x_{k+1} but restricting this search to $|x - x_k| \leqslant \delta_k$.

There is a relationship between δ_k and μ_k since raising μ_k makes the neighbourhood δ_k diminish [3]. As an exact expression to relate these two parameters is not available, many solutions have been developed.

The one used in the present work was proposed by Fletcher [3] and uses the following expression:

$$r_k = \frac{V_N(x_k) - V_N(x_k + p_k)}{V_N(x_k) - L_k(x_k + p_k)} \qquad (2)$$

to obtain a measure of the quality of the approximation. Here V_N is the function to be minimized, L_k is the estimate of that value calculated from the Taylor series of second order and p_k is the search direction, in the present situation, the search direction given by the Levenberg-Marquardt algorithm.

The value of r_k is used in the determination of μ_k according to the following algorithm:

1-Choose the initial values of x_0 e μ_0.
2-Calculate the search direction p_k.
3-If $r_k > 0.75$ then set $\mu_k = \mu_k/2$.
4-If $r_k < 0.25$ then set $\mu_k = 2.\mu_k$.
5-If $V_N(x_k + p_k) < V_N(x_k)$ then the new iteration is accepted.
6-If the stopping condition for the training is not met, return to step 2.

3   On-line Version

As pointed out before, the difficulties come from computing the derivatives for the Hessian matrix, inverting this matrix and computing the trust region, the region for which the approximation contained in the calculation of the Hessian matrix is valid.

In the literature, some attempts to build on-line versions can be found, namely the work done by Ngia [4] developing a modified iterative Levenberg-Marquardt algorithm which includes the calculation of the trust region and the work in [5] which implements a Levenberg-Marquardt algorithm in sliding window mode for Radial Basis Functions.

3.1   A Double Sliding Window Approach with Early Stopping

The current work is an evolution of the one presented in [6] where an on-line version of the Levenberg-Marquardt algorithm was implemented using a sliding window with Early Stopping and static test set. In the present work two sliding windows are used, one for the training set and another for the evaluation set with all the data being collected on-line. As in the previous work, the Early Stopping technique [7], [8] is used to avoid the overfitting problem because it is almost mandatory to employ a technique to avoid overtraining when dealing with systems that are subject to noise. The Early Stopping technique was chosen over other techniques that could have been used (like Regularization and Prunning techniques) because it has less computational burden.

The use of two sliding windows will introduce some difficulties since both data sets will be changing during training and evaluation phases. For these two windows it is necessary to decide their relative position. In order to be able to perform Early Stopping in a valid way, it was decided to place the windows in a way that the new samples will go into the test window and the samples that are removed from the test set will go in to the training set according to figure 1.

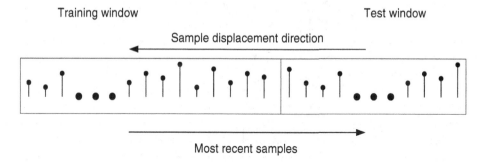

Fig. 1. Relative position of the training and test sets

If the inverse relative position of the two windows was used, the samples would be part of the test set after they have been part of the training set and so the objective of evaluating the generalization ability would be somehow faked.

In order to save some of the time necessary to collect all the samples needed to fill both the test and training window, the training is started after some data has been collected but before the windows are both filled. The test window always keeps the same number of samples, while the training window is growing in the initial stage. The choice of maintaining the test window always with the same number of points was taken with the objectives of maintaining this window as stable as possible (since it is responsible for producing the numerical evaluation of the models) and assuming the use of a minimal test window that should not be shortened.

The windows may not change in each training iteration since all the time between samplings is used for training which may permit several training epochs before a new sample is collected. But each time the composition of the windows is changed, the test and training errors will probably be subjected to an immediate change that might be interpreted as an overtraining situation. The Early Stopping technique is here used in conjunction with a measure of the best model that is retained for control. Each time there is a change in the windows, the values of the best models (direct and inverse) must be re-evaluated because the previous ones, obtained over a different test set, are no longer valid for a direct comparison.

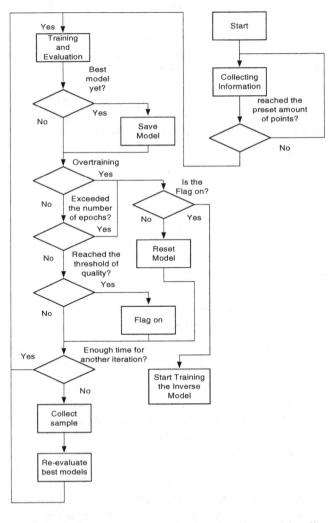

Fig. 2. Block diagram for the identification of a direct model on-line

The procedure used for the identification of the direct model on-line is represented in figure 2.

As was already explained, training starts when a predefined amount of points have been collected. After each epoch the ANN is evaluated with a test set. The value of the Mean Square Error (MSE) obtained is used to perform Early Stopping and to retain the best models.

The conditions for overtraining and the maximum number of epochs are then verified. If they are true, the Flag, which indicates that the threshold of quality has been reached, will also be verified and if it is on, the training of the inverse model starts, otherwise the models will be reset since new models need to be prepared. Resetting here means that the model's weights are replaced by random values between -1 and 1 as in used in the initial models.

After testing the conditions for overtraining and the maximum number of epochs, if they are both false, the predefined threshold of quality will also be tested and if it has been reached the variable Flag will be set to on. In either case the remaining time of the sampling period is tested to decide if a new epoch is to be performed or if a new sample is to be collected and training is to be performed with this new sample included in the sliding window.

Each time a new sample is collected both the direct and inverse models must be re-evaluated with the new test set and the information about the best models updated.

The procedure for the inverse model is very similar and almost the same block diagram could be used to represent it. The on-line training goes on switching from direct to inverse model each time a new model is produced. The main difference between the procedure for direct and inverse model lies in the evaluation step. While the direct model is evaluated with a simple test set, the inverse model is evaluated with a control simulation corresponding to the hybrid Direct/Specialized approach for generating inverse models [9].

During the on-line training the NNSYSID [10] and NNCTRL [11] toolboxes for MATLAB were used.

4   Test System

The test system chosen is a reduced scale prototype kiln for the ceramic industry which is non-linear and will be working under measurement noise because of the type B thermocouple used.

The system is composed of a kiln, electronics for signal conditioning, power electronics module and a Data Logger from Hewlett Packard HP34970A to interface with a Personal Computer (PC) connected as can be seen in figure 3.

Through the Data Logger bi-directional real-time information is passed: control signal supplied by the controller and temperature data for the controller. The temperature data is obtained using a thermocouple. The power module receives a signal from the Data Logger, with the resolution of 12 bits (0 to 4.095V imposed by the Data Logger) which comes from the controller implemented in the Personal Computer, and converts this signal in a Pulse Width Modulation (PWM) signal of 220V applied to the heating element.

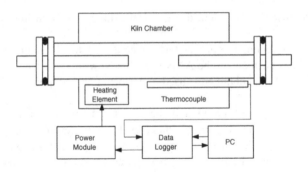

Fig. 3. Block diagram of the system

Fig. 4. Picture of the kiln and electronics

The signal conversion is implemented using a sawtooth wave generated by a set of three modules: zero-crossing detector, binary 8 bit counter and D/A converter. The sawtooth signal is then compared with the input signal generating a PWM type signal.

The PWM signal is applied to a power amplifier stage that produces the output signal. The signal used to heat the kiln produced this way is not continuous, but since the kiln has integrator behaviour this does not affect the functioning.

The Data Logger is used as the interface between the PC and the rest of the system. Since the Data Logger can be programmed using a protocol called Standard Commands for Programmable Instruments (SCPI), a set of functions have been developed to provide MATLAB with the capability to communicate through the RS-232C port to the Data Logger.

Using the HP34902A (16 analog inputs) and HP34907A (digital inputs and outputs and two Digital to Analog Converters) modules together with the developed functions, it is possible to read and write values, analog or digital, from MATLAB. A picture of the system can be seen in figure 4. The kiln is in the centre and at the lower half are the prototypes of the electronic modules.

5   Results

Figure 5 and 6 show the results obtained in a training situation followed by control using the Direct Inverse Control (DIC) and Internal Model Control (IMC) strategies.

The test sequence is composed of 100 points, the sliding window used for training has a maximum of 200 samples and training starts after 240 samples have been collected. It should be noted that comparing these values with the ones used in [6], the windows are now smaller to compensate for the increasing of computational effort needed in this approach.

The training is also started later since the samples for the test window must all be collected before training starts in order to permit the direct comparison of the test errors through all the procedure.

Both direct and inverse models were one hidden layer models with 6 neurons on the hidden layer and one linear output neuron. The direct model has as inputs the past two samples of both the output of the system and the control signal.

The sampling period used was 150 seconds, which allowed performing several epochs of training between each control iteration.

During the initial phase of collecting data a PI was used in order to keep the system operating within the range of interest. The PI parameters are Kp=0.01 and Ki=0.01. After this initial phase the PI is replaced by the DIC or the IMC controller, using the direct and inverse models trained on-line.

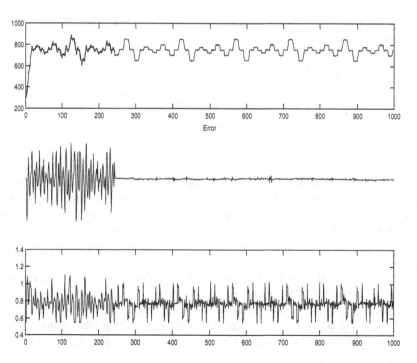

Fig. 5. On-line identification and control. The first part of the control is performed by a PI and the second by DIC

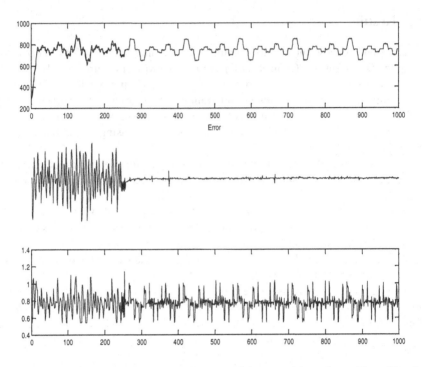

Fig. 6. On-line identification and control. The first part of the control is performed by a PI and the second by IMC

Table 1. Mean Square Error for the two control strategies used

Control Type	Sampls 300 to 1000
DIC	0,46
IMC	0,65

In both situations the first inverse model is ready at sample 243, that is only 2 samples after the training has started. After the 240 samples have been collected it only took one sampling period to complete the training of the direct model and another sampling period to complete the inverse model, even though the Matlab code was running on a personal computer with a Celeron processor at 466MHz using 64Mbytes of memory.

The quality of the control obtained in both situations is resumed in therms of Mean Square error in table 1. As can be seen, the error value for both solutions is very small (less than one degree per sample) and of the same order.

Although it should not be expected that the error of the DIC solution be smaller than the one for the IMC, it should be noted that a disturbance can have here a more important role than the control strategy.

6   Conclusion

This paper presents on-line identification and control using the Levenberg-Marquardt algorithm in a batch version with two sliding windows and Early Stopping.

The problems pointed out in section 3 to perform Early Stopping under a changing sliding window for the training set were not critical and a good choice of the parameters for identification of the overtraining situation and for the maximum number of iterations for each attempt to create a model were sufficient to obtain reasonable models to perform IMC control.

It should be noted that during the time presented in figures 5 and 6 the quality of control is increasing as the number of samples advances and better models are produced.

The PI is here just used to maintain the system in the operating range while data is being collected and is disconnected as soon as the ANN models are ready.

The sliding window approach with Early Stopping solves the problems for using the Levenberg-Marquardt algorithm on-line due to the difficulty of creating a true iterative version which includes the computation of the trust region.

As shown here, even for a noisy system, for which overtraining is a real problem it is possible to create models on-line of acceptable quality as can be concluded from the values presented in table 1.

References

1. K. Levenberg, "A method for the solution of certain problems in least squares," *Quart. Appl. Math.*, vol. 2, pp. 164–168, 1944.
2. D. Marquardt, "An algorithm for least -squares estimation of nonlinear parameters," *SIAM J. Appl. Math.*, vol. 11, pp. 431–441, 1963.
3. M. Nørgaard, O. Ravn, N. K. Poulsen, and L. K. Hansen, *Neural Networks for Modelling and Control of Dynamic Systems*, Springer, 2000.
4. Lester S. H. Ngia, *System Modeling Using Basis Functions and Application to Echo Cancelation*, Ph.D. thesis, Department of Signals and Systems School of Electrical and Computer Engineering, Chalmers University of Technology, 2000.
5. P. Ferreira, E. Faria, and A. Ruano, "Neural network models in greenhouse air temperature prediction," *Neurocomputing*, vol. 43, no. 1-4, pp. 51–75, 2002.
6. Fernando Morgado Dias, Ana Antunes, José Vieira, and Alexandre Manuel Mota, "Implementing the levenberg-marquardt algorithm on-line: A sliding window approach with early stopping," *2nd IFAC Workshop on Advanced Fuzzy/Neural Control*, 2004.
7. N. Morgan and H. Bourlard, "Generalization and parameter estimation in feedforward nets: Some experiments.," *Advances in Neural Information Processing Systems, Ed. D.Touretzsky, Morgan Kaufmann*, pp. 630–637, 1990.
8. Jonas Sjöberg, *Non-Linear System Identification with Neural Networks*, Ph.D. thesis, Dept. of Electrical Engineering, Linköping University, Suécia, 1995.
9. Fernando Morgado Dias, Ana Antunes, and Alexandre Mota, "A new hybrid direct/specialized approach for generating inverse neural models," *WSEAS Transactions on Systems*, vol. 3, Issue 4, pp. 1521–1529, 2004.
10. M. Nørgaard, "Neural network based system identification toolbox for use with matlab, version 1.1, technical report," Tech. Rep., Technical University of Denmark, 1996.
11. M. Nørgaard, "Neural network based control system design toolkit for use with matlab, version 1.1, technical report," Tech. Rep., Technical University of Denmark, 1996.

Boosting Parallel Perceptrons for Label Noise Reduction in Classification Problems

Iván Cantador and José R. Dorronsoro*

Dpto. de Ingeniería Informática and Instituto de Ingeniería del Conocimiento,
Universidad Autónoma de Madrid, 28049 Madrid, Spain

Abstract. Boosting combines an ensemble of weak learners to construct a new weighted classifier that is often more accurate than any of its components. The construction of such learners, whose training sets depend on the performance of the previous members of the ensemble, is carried out by successively focusing on those patterns harder to classify. This fact deteriorates boosting's results when dealing with malicious noise as, for instance, mislabeled training examples. In order to detect and avoid those noisy examples during the learning process, we propose the use of Parallel Perceptrons. Among other things, these novel machines allow to naturally define margins for hidden unit activations. We shall use these margins to detect which patterns may have an incorrect label and also which are safe, in the sense of being well represented in the training sample by many other similar patterns. As candidates for being noisy examples we shall reduce the weights of the former ones, and as a support for the overall detection procedure we shall augment the weights of the latter ones.

1 Introduction

The key idea of boosting methods is to iteratively construct a set $\{h_t\}$ of weak learners that are progressively focused on the "most difficult" patterns of a training sample in order to finally combine them in a final averaged hypothesis $H(X) = \sum_t \alpha_t h_t(X)$. More precisely [7], a starting uniform distribution $D_0 = \{d_0(X)\}$ is progressively updated (see section 3 for more details) to

$$d_{t+1}(X) = \frac{1}{Z_t} d_t(X) e^{-\alpha_t y_X h_t(X)}, \tag{1}$$

where Z_t is a probability normalization constant, $y_X = \pm 1$ is the class label associated to X and the averaging constant $\alpha_t > 0$ is related to the training error ϵ_t of h_t. If a pattern X is incorrectly classified after iteration t, we have $y_X h_t(X) < 0$ and, therefore, boosting iteratively focuses on the incorrectly classified patterns.

When label noisy dataset classification problems (i.e., problems where some pattern labels are incorrect) are considered, this property has two consequences.

* With partial support of Spain's CICyT, TIC 01–572, TIN 2004–07676.

J. Mira and J.R. Álvarez (Eds.): IWINAC 2005, LNCS 3562, pp. 586–593, 2005.

On the one hand, while boosting has been used with great success in several applications and over various data sets [2], it has also been shown [3, 4] that it may not yield such good results when applied to noisy datasets. In fact, assume that a given pattern has label noise, that is, although clearly being a member of one class, its label corresponds to the alternate class. Such a label noisy pattern is likely to be repeatedly misclassified by the successive hypotheses which, in turn, will increase its sampling probability, causing boosting to hopelessly concentrate on it and progressively deteriorate the final hypothesis.

Furthermore, mislabeled training examples will tend to have high probabilities as the boosting process advances, which implies that after few iterations, most of the training examples with high weights should correspond to mislabeled patterns, and gives a good motivation to use boosting as a noise filter, as done for instance in [8], where it is proposed that after a number N of rounds, the examples with the highest weights are removed from the training sample. This should allow a more efficient subsequent learning, but also has problems of its own as, for instance, the definition of the exact percentage of presumably noisy examples to be filtered (or, more generally, how to choose a "high enough" weight), or the appropriate choosing of the number N of boosting rounds.

On the other hand, a second issue to be considered on any boosting strategy is the fact that the probabilities of correctly classified examples are progressively diminished. Intuitively, this is a good idea, because the examples that are incorrectly predicted by previous classifiers are chosen more often than examples that were correctly predicted, and boosting will attempt to produce a new classifier more able to handle correctly those patterns for which the current ensemble performance is poor. However, in the presence of medium to high levels of noisy examples, the weak learners may have many difficulties to obtain good separation frontiers, as there may be not enough correctly classified examples in the training samples to do so. In particular, they may be not be able to distinguish between the true mislabeled examples and the incorrectly classified (but well labeled) ones. In other words, it may be sensible to keep an adequate representation of correctly classified patterns in boosting's training sets, as they should make noisy examples to be more easily found. Thus, a "reverse boosting" strategy of keeping well classified patterns, i.e., the "safe" ones, while diminishing label noisy ones may allow a reasonable learning procedure in the label noisy setting. The problem, of course, is how to detect good patterns (even if they may be somewhat redundant) and, more importantly, how to detect noisy patterns. If done properly, an adequate weighting of boosting's exponent should dismiss the latter patterns while keeping the former.

We shall follow this general approach in this work, using Parallel Perceptrons (PPs; [1]) or, more precisely, their activation margins, to detect simultaneously good and noisy patterns. As we shall see in section 2, PPs, a variant of the classical committee machines, not only learn "best" perceptrons but also stabilize their outputs by learning optimal hidden unit activation margins. These margins shall be used in section 3 to classify training patterns in the just mentioned safe and noisy categories and a third one, borderline patterns, somehow in between of the other ones. This, in turn, can be used to adjust boosting's probability updates. We will do so here by changing the exponent in (1) to $\alpha_t R(X) y_X h_t(X)$, where the $R(X)$ factor will be -1 for safe patterns (increasing their probability) and noisy ones (decreasing now it), and 0 for the bordeline ones (leaving their probability essentially unchanged). The resulting approach will be favorably compared in section 4 with other options such as boosting and bagging of PP and also of standard multilayer perceptrons (MLPs). Finally, the paper will close with a brief summary section and a discussion of further work.

2   Parallel Perceptron Training

Parallel Perceptrons (PP) have the same structure of the well-known committee machines (CM) [6]. These are made up of an odd number H of standard perceptrons P_i with ± 1 outputs and they have a single one–dimensional output that is simply the sign of the sum of these perceptrons' outputs (that is, the sign of the overall perceptron vote count). More precisely, if perceptron i has a weight W_i, its output for a given D–dimensional input pattern X is $P_i(X) = s(W_i \cdot X) = s(act_i(X))$, where $s(\cdot)$ denotes the sign function and $act_i(X) = W_i \cdot X$ is the activation of perceptron i due to X (we may assume $x_D = 1$ for bias purposes). The output of the CM is then $h(X) = s\left(\sum_{i=1}^{H} P_i(X)\right)$. We will assume that each input X has an associated ± 1 label y_X and take the output $h(X)$ as correct if $y_X h(X) = 1$. If this is not the case, i.e., whenever $y_X h(X) = -1$, classical CM training tries to change the smallest number of perceptron outputs so that X could then be correctly classified (see [6], chap. 6). On the other hand, PP training changes the weights of all incorrect perceptrons, i.e. those P_i verifying $y_X P_i(X) = -1$. In both cases, the well-known Rosenblatt's rule

$$W_i := W_i + \eta \ y_X X \tag{2}$$

is applied, with η a learning rate. However, the main difference between CM and PP training is the margin stabilization procedure of the latter. More precisely, when a pattern X is correctly classified, PP training also applies a margin–based output stabilization procedure to those perceptrons for which $0 < y_X act_i(X) < \gamma$. That is, it tries to keep their activations $act_i(X)$ away from zero, so that a small perturbation of X does not cause P_i to change its output over X. To get activations far away from zero, Rosenblatt's rule is again applied

$$W_i := W_i + \mu \ \eta \ y_X X \tag{3}$$

when $0 < y_X act_i(X) < \gamma$. The new parameter μ weights the importance we give to clear margins. Moreover, the value of the margin γ is dynamically adjusted: after a pattern X is processed correctly, γ is increased to $\gamma + 0.25\eta$ if for all correct perceptrons we have $y_X act_i(X) > \gamma$, while we decrease γ to $\gamma - 0.75\eta$ if $0 < y_X act_i(X) < \gamma$ for at least one correct perceptron. The updates (2) and (3)

can be seen as a kind of gradient descent with respect to the following criterion
function

$$J(\mathcal{W}) = - \sum_{\{X:y_X h(X)=-1\}} \left(\sum_{\{i:y_X W_i \cdot X < 0\}} y_X W_i \cdot X \right) +$$

$$\mu \sum_{\{X:y_X h(X)=1\}} \left(\sum_{\{i:0 < y_X W_i \cdot X < \gamma\}} (\gamma - y_X W_i \cdot X) \right)$$

$$= J_1(\mathcal{W}) + \mu J_2(\mathcal{W}), \tag{4}$$

where we have written $\mathcal{W} = (W_1, \ldots, W_H)$. J_1 is basically a classification error
measure (it is 0 if all X are correctly classified), while J_2 can be seen as a
regularization term with μ being the regularization's weight (it is 0 if all correct
activations are above the margin γ). PP training can be performed either on line
or in batch mode. Since we will use PPs in a boosting framework, we shall use
the second procedure. Moreover, notice that for the margin to be meaningful,
weights have to be normalized somehow; we will make their euclidean norms to
be 1 after each batch pass.

In spite of their very simple structure, PPs do have a universal approximation
property and, as shown in [1], they provide results in classification and regression
problems quite close to those offered by C4.5 decision trees and only slightly
weaker that those of standard multilayer perceptrons (MLPs).

3   Label Noise Reduction Through Parallel Perceptron Boosting

We first discuss how PP's activation margins can be used to detect safe, label
noisy and borderline patterns. More precisely, as just described, PPs adaptively
adjust these margins, making them to converge to a final value γ that ensures stable PP outputs. Thus, if for a pattern X its i–th perceptron activation verifies
$|act_i(X)| > \gamma$, $s(act_i(X))$ is likely to remain unchanged after small perturba-
tions of X. Given the voting outputs of PP, we will accordingly take a pattern
as safe if for $\lfloor H/2 \rfloor$ perceptrons P_i (i.e., their majority) we have $y_X act_i(X) > \gamma$,
as such an X is likely to be also correctly classified later on. Similarly, if for
$\lfloor H/2 \rfloor$ perceptrons we have $y_X act_i(X) < -\gamma$, X is likely to remain wrongly
classified, and we will take it to be label noisy. As borderline patterns we will
simply take the remaining X. We shall use the notations S_t, N_t and B_t for the
safe, noisy and borderline training sets at iteration t. As an example, in figure 1 we show how the safe (squared points) and label noisy (crossed points)
patterns of a 2–dimensional XOR problem with 10% of noise level have been
detected using a 3–perceptron PP. It shows that almost all label noisy and safe
patterns that are quite likely to remain stable in further trainings have been
selected.

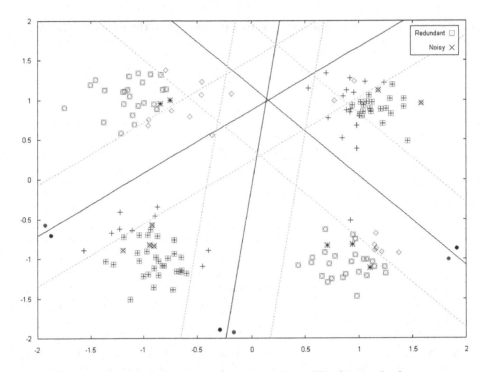

Fig. 1. Safe and mislabeled patterns detection using a PP of 3 standard perceptrons on a 2–dimensional XOR problem with 10% of noise level

Let us see how to use this categorization in a boosting–like setting. Recall that boosting's probability updates are given by the rule (1), where $Z_t = \sum_X d_t(X)$,

$$\alpha_t = \frac{1}{2} \ln \left(\frac{1 - \epsilon_t}{\epsilon_t} \right),$$

and ϵ_t is the iteration error with respect to d_t, i.e.,

$$\epsilon_t = \sum_{\{X \,:\, y_X h_t(X) = -1\}} d_t(X).$$

We introduce safe, noisy and borderline into the boosting process through a pattern dependent factor $R(X)$ in the boosting probability actualization procedure as follows

$$d_{t+1}(X) = \frac{1}{Z_t'} d_t(X) e^{-\alpha_t R(X) y_X h_t(X)}, \tag{5}$$

with Z_t' again a normalization constant. Notice that we recover standard boosting by setting $R(X) = 1$. Now, we want to diminish the influence of label noisy patterns $X \in N_t$, so we put $R(X) = -1$, which should make $d_{t+1}(X) < d_t(X)$, diminishing therefore their importance at the $t+1$ iteration. Moreover, setting also $R(X) = -1$ for safe patterns, we now have $\alpha_t R(X) y_X h_t(X) < 0$, as $y_X h_t(X) > 0$

Table 1. Accuracies for the PP, MLP bagging and boosting procedures over 3 synthetic datasets with 0%, 5%, 10%, 20% and 30% of noise level in the training samples (best in bold face, second best in italics)

Dataset	Noise level	PP bagging	PP boosting	PP NR boosting	MLP bagging	MLP boosting
twonorm	0 %	96.597	96.267	*96.800*	96.733	**97.100**
	5 %	93.600	89.400	**95.933**	*95.700*	93.000
	10 %	90.900	84.600	**95.733**	*94.033*	90.733
	20 %	83.800	75.233	**93.367**	*90.900*	81.967
	30 %	73.800	68.467	**90.167**	*85.200*	72.833
threenorm	0 %	79.367	75.933	78.233	*82.933*	**83.533**
	5 %	77.033	73.700	78.233	**81.367**	*80.100*
	10 %	74.867	71.367	*77.700*	**79.067**	77.300
	20 %	70.100	65.467	**75.000**	*74.733*	71.133
	30 %	66.467	60.967	**71.233**	*69.600*	63.033
ringnorm	0 %	63.900	60.800	64.067	*75.900*	**78.233**
	5 %	61.533	59.100	63.400	**75.400**	*75.300*
	10 %	61.133	59.033	62.100	**72.200**	*71.433*
	20 %	57.967	56.367	60.600	**67.967**	*64.433*
	30 %	56.667	54.000	*59.233*	**61.700**	57.300

for them; hence, $d_{t+1}(X) > d_t(X)$, as desired. Finally, for borderline patterns $X \in B_t$ we shall take $R(X) = 0$, that except for changes on the normalization constant, should give $d_{t+1}(X) \simeq d_t(X)$. Notice that except for borderline patterns, the proposed procedure, which we call NR boosting, comes to essentially being a "reversed" boosting, as it gives bigger emphasis to correct patterns and smaller to the wrong ones, just the other way around to what boosting does. We shall numerically compare next NR boosting against standard boosting and bagging of PPs and of the much stronger learners given by MLPs.

4   Experiments

In order to have a better control of the noise added, we have used in our experiments three well known synthetically generated datasets of size 300, the twonorm, threenorm and ringnorm datasets, also used in other boosting experiments [2]. We briefly recall their description. They are all 20–dimensional, 2–class problems. In twonorm each class is drawn from a multivariate normal distribution with unit covariance matrix. Class #1 has mean $(a, a, \ldots, a)$ and class #2 has mean $(-a, -a, \ldots, -a)$ where $a = 2/\sqrt{20}$. In threenorm, class #1 is now drawn from two unit covariance normals, one with mean $(a, a, \ldots, a)$ and the other with mean $(-a, -a, \ldots, -a)$. Class #2 is drawn from a unit covariance normal with mean $(a, -a, a, -a, \ldots, -a)$. Here $a = 2/\sqrt{20}$ too. Finally, in ringnorm, class #1 follows a normal distribution with mean 0 and covariance matrix 4 times the identity and class #2 is a unit covariance normal and mean

$(a, a, \ldots, a)$ where now $a = 1/\sqrt{20}$. The twonorm and threenorm problems are clearly the easier ones, as they are essentially linearly separable, although there is a greater normal overlapping in threenorm, that gives it a much higher optimal Bayes error probability. Ringnorm is more difficult: the more concentrated inner second normal is quite close to the mean of the wider first normal. In particular, the Bayes boundary is basically a circle, quite difficult to learn with hyperplane based methods such as PPs (this is also the case [5] with other simple methods, such as nearest neighbors, linear discriminants or learning vector quantization).

We will compare the results of the PP procedure described in section 3 with those of standard bagging and boosting. These two will also be applied to MLPs. PP training has been carried out as a batch procedure. In all examples we have used 3 perceptrons and parameters $\gamma = 0.05$ and $\eta = 10^{-3}$. As proposed in [1], the η rate does not change if the training error diminishes, but is decreased to 0.9η if it augments. Training epochs have been 250 in all cases; thus the training error evolution has not been taken into account to stop the training procedure. Anyway, this error has an overall decreasing behavior. The MLPs, each with a single hidden layer of 3 units and a learning rate value of $\eta = 10^{-3}$, were trained during 2000 epochs.

In all cases the number of boosting rounds was 10 and we have used 10–times 10–fold cross validation. That is, the overall data set has been randomly split in 10 subsets, 9 of which have been combined to obtain the initial training set. To ensure an appropriate representation of both classes in all the samples, stratified sampling has been used. The final PPs and MLPs' behaviors have been computed on the remaining, unchanged subset, that we keep for testing purposes.

All training samples were artificially corrupted with different levels of classification noise: 5%, 10%, 20% and 30%. Table 1 gives the overall accuracies for the five construction procedures (best values are in bold face, second best in italics). They are also graphically represented in figure 2. In all cases MLPs give best results in absence of noise, with PPs being close seconds for twonorm and

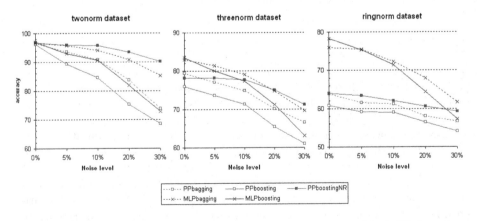

Fig. 2. Graphical comparative results of the accuracies given in table 1. NR boosting accuracies decrease quite slowly; it is thus the more noise robust method

threenorm, but not so for the more difficult ringnorm. When noise increases, NR boosting gives best results in twonorm for all noise levels and for the 20% and 30% levels in threenorm; it is second best in the 10% level. It cannot overcome the large head start of MLPs in ringnorm, although is second best for the 30% noise level. On the other hand, NR boosting is clearly the most robust method. For instance, their 30% noise accuracies are just 6.63 points below the noise free ones in twonorm, 7.00 points in threenorm and 4.83 in ringnorm. For MLP bagging (the more robust MLP procedure), its drops are 11.53, 13.39 and 14.20 respectively. This behavior is clearly seen in figure 2.

5   Conclusions and Further Work

We have shown how the concept of activation margin that arises very naturally on PP training can be used to provide a robust approach to label noise reduction. This is done adding an extra factor $R(X)$ to boosting's exponential probability update, with values $R(X) = -1$ for safe and label noisy patterns and $R(X) = 0$ for the borderline ones. The assignment of a pattern to each category is done through its activation margins. The resulting procedure has been successfully tested on three well-known synthetic datasets, artificially corrupted with different levels of classification noise. NR boosting has been shown to have a very good overall performance and it is the most robust method, as its accuracies decrease very slowly and it gives the smallest overall drop. Further work will concentrate on the pursuit of a more general approach to malicious noise detection using the ideas of redundant and label noisy patterns categorizations.

References

1. P. Auer, H. Burgsteiner, W. Maass, *Reducing Communication for Distributed Learning in Neural Networks*, Proceedings of ICANN'2002, Lecture Notes in Computer Science 2415 (2002), 123–128.
2. L. Breiman, *Bias, variance and arcing classifiers*, Tech. Report 460, Department of Statistics, University of California, Berkeley, 1996.
3. T. Dietterich, *An experimental comparison of three methods for constructing ensembles of decision trees: Bagging, boosting, and randomization*, Machine Learning, 40 (2000) 139–158.
4. Y. Freund, R.E. Schapire, *Experiments with a New Boosting Algorithm*, Proceedimgs of the 13th International Conference on Machine Learning (1996), 148–156.
5. D. Meyer, F. Leisch, K. Hornik, *The support vector machine under test*, Neurocomputing, 55 (2003), 169–186.
6. N. Nilsson, **The Mathematical Foundations of Learning Machines**, Morgan Kaufmann, 1990.
7. R.E. Schapire, Y. Freund, P. Bartlett, W.S. Lee, *Boosting the margin: a new explanation for the effectiveness of voting methods*, Annals of Statistics, 26 (1998), 1651–1686.
8. S. Verbaeten, A.V. Assche. *Ensemble methods for noise elimination in classification problems*. In Fourth International Workshop on Multiple Classifier Systems (2003), 317-325.

On the Connection Between the Human Visual System and Independent Component Analysis

Susana Hornillo-Mellado[1], Rubén Martín-Clemente[1],
Carlos G. Puntonet[2], and José I. Acha[1]

[1] Dpto. de Teoría de la Señal y Comunicaciones,
Escuela Superior de Ingenieros, Universidad de Sevilla,
Avda. de los Descubrimientos s/n, 41092-Sevilla, Spain
{susanah, ruben}@us.es
[2] Dpto. de Arquitectura y Tecnología de Computadores,
E.T.S. de Ingeniería Informática, Universidad de Granada,
c/Periodista Daniel Saucedo s/n, 18071-Granada, Spain
carlos@atc.ugr.es

Abstract. The Human Visual System (HVS) presents some properties that are shared by the results obtained when Independent Component Analysis (ICA) is applied to natural images. Particularly, the special appearance of the ICA bases (they look like "edges") and the sparse distribution of the independent components have been linked to the receptive fields of the simple neurons in the visual cortex and their way to codify the visual information, respectively. Nevertheless, no theoretical study has been provided so far to explain that behaviour of ICA. The objective of this paper is to analyze, both mathematical and experimentally, the results obtained when ICA is applied to natural images in order to supply a theoretical basis for the connection between ICA and the HVS.

1 Introduction

The primary visual cortex (V1) is the maximum responsible of the way in which we perceive the world. It uses the luminous information detected by the photoreceptors in the retina (cones and sticks) to generate a topographical representation of the environment [7][13].

In V1, the response of a each neuron is associated to the intensity pattern of a very small part or *patch* of the visual field. In addition, this dependence is determined by a pattern of excitation and inhibition called *receptive field*: for example, the so-called *simple cells* of V1 respond to *oriented* image structures, meaning that their receptive fields match with "edges" [7]. The model usually accepted for the response of V1 neurons is the linear one [10]:

$$s_j = \sum_{x,y} w_j(x,y) I(x,y), \text{ for any neuron } j \tag{1}$$

J. Mira and J.R. Álvarez (Eds.): IWINAC 2005, LNCS 3562, pp. 594–603, 2005.

where (x, y) denotes spatial position, $w_j(x, y)$ model the receptive field of the neuron j in question, and $I(x, y)$ is one patch of the natural scene perceived by the eye.

Several researchers have argued that this representation of natural images in V1 is characterized by two important features. First, the visual cortex uses a factorial code to represent the visual environment, i.e., the responses s_j are *sparsely distributed*. This means that only a small number of neurons are activated at the same time. Second, the neurons in V1 use an strategy based on the principle of *redundancy reduction*, i.e., neurons should encode information in such a way as to minimize statistical dependencies among them [1][6][13][14].

Recent works about visual perception have attempted to explain image representation in area V1 in terms of a probabilistic model of the structure of natural scenes, based on the theory of *probabilistic inference*. This theory assumes that a given state of the environment can be inferred from a particular state of activity of the neurons [13]. Many models described in the literature have used a simple way to model natural images in which each portion of a scene is described in terms of a linear superposition of *features* or *basis functions* [2][13]:

$$I(x, y) = \sum_i s_i \, a_i(x, y) + \nu(x, y) \tag{2}$$

where $a_i(x, y)$ are the features and the coefficients s_i determine how strong each feature is present in the image. The variable ν represents Gaussian noise (i.i.d.) and is included to model structures in the images that may not be well represented by the basis functions. In addition to this generative model for images, another thing is needed to infer a solution to this problem: a prior distribution over the coefficients s_i, which has to be designed to enforce sparseness in the representation [13].

This model of generation of natural images was first linked with *Independent Component Analysis* (ICA) by Bell and Sejnowski [2] (more results can be found in [10][11]), a mathematical procedure for analyzing empirical data. Applying the theory of probabilistic inference, the idea was that ICA could provide a good model of image representation in V1. This hypothesis was based on several interesting characteristics that bore a strong resemblance with some properties of the HVS mentioned previously. First, ICA assumes that each particular signal of an ensemble of signals is a superposition of elementary components that occur independently of each other, recalling the behaviour supposed for V1 neurons. Furthermore, when ICA is applied to natural images, the independent components obtained turn to be *sparse*, just like the distribution assumed for the responses of V1 neurons. Moreover, ICA bases appear to be like "edges", similarly to the receptive fields of simple cells. The objective of this paper is to provide a mathematical basis for this properties of ICA, that are also useful in many fields of digital image processing like texture segmentation, edge detection [8], digital watermarking [9] or elimination of noise [11].

The paper is structured as follows. In section 2 we will analyze mathematically the behaviour of ICA when it is applied to natural images and we will go more deeply into its connection with the HVS. Section 3 is dedicated to show some experiments. We will finish with some conclusions in section 4.

2   Mathematical Analysis of the Connection Between ICA and the HVS

Independent Component Analysis (ICA) is an emergent technique for studying empirical data [3][4][11]. It involves a mathematical procedure that transforms a number of zero-mean *observed variables* $x_1, ..., x_N$ into a number of "as statistically independent as possible" variables $s_1, ..., s_N$ called *independent components*. In linear ICA, this transformation reads:

$$s_i = \sum_{n=1}^{N} b_{in} x_n, \ \forall i = 1, ..., N \tag{3}$$

The inverse model is given by:

$$x_i = \sum_{n=1}^{N} a_{in} s_n, \ \forall i = 1, ..., N \tag{4}$$

Here a_{in} and b_{in} $(i, n = 1, ..., N)$ are real coefficients. In matrix form,

$$\mathbf{S} = \mathbf{B}\,\mathbf{X} \Longleftrightarrow \mathbf{X} = \mathbf{A}\,\mathbf{S} \tag{5}$$

where $\mathbf{B} = (b_{ij})$ and $\mathbf{A} = \mathbf{B}^{-1} = (a_{ij})$.

Based on this mathematical representation of the data and on the results obtained by Barlow [1] and Field [6], Bell and Sejnowski [2] hypothesized that ICA could provide a good model of the behaviour of the simple cells localized in the primary visual cortex (see also the works published by Hyvärinen *et al* [10][11] and Olshausen *et al* [14]). More precisely, they showed that when ICA is applied to natural images (i.e., variables x_i contain the pixels of the particular image) the results obtained presented some interesting resemblance with the behaviour of the V1 neurons hypothesized by Barlow and Field. First of all, the independent components, s_i, were sparsely distributed, modelling the neurons' responses. Secondly, the ICA bases (columns of matrix $\mathbf{A}$) appeared to be like "edges", recalling the structure of the receptive fields of simple cells. Our justification for these results is as follows.

First of all, among the different ways to carry out the independent component analysis of empirical data, we will focus on the maximization of high-order cumulants, like skewness and kurtosis, which is a very popular criterion that is inspired in the Central Limit Theorem [4][11][12].

To apply ICA to an image $\mathbf{I}$, we divide it into N_2 patches of $k_1 \times k_2$ pixels each. These patches are stacked into the columns of matrix $\mathbf{X}$, obtaining $\mathbf{X} = [\mathbf{x}_1|\mathbf{x}_2|...|\mathbf{x}_{N_2}]$ where $\mathbf{x}_j = [x_{1j}, x_{2j}, ..., x_{N_1 j}]^T$ contains the j-th patch, with $j = 1, 2, ..., N_2$ and $N_1 = k_1 k_2$. We consider each row of matrix $\mathbf{X}$ as a realization of a random process.

Let us define $\mathbf{y} = \mathbf{b}^T \mathbf{X} = [y_1, y_2, ..., y_{N_2}]$, where $y_i = \mathbf{b}^T \mathbf{x}_i$. Our objective is to find the vector $\mathbf{b}$ that solves the following constrained optimization problem:

$$\max_{\mathbf{b}} J_p(\mathbf{y}) = \frac{1}{N_2} \sum_{j=1}^{N_2} y_j^p \text{ subject to } \frac{1}{N_2} \sum_{j=1}^{N_2} y_j^2 = 1 \tag{6}$$

with $p > 2$. The restriction is necessary to avoid the solution $\mathbf{b} \to \infty$. When $p = 3$, $J_p(\mathbf{y})$ is an estimation of the skewness. If $p = 4$, $J_p(\mathbf{y})$ leads to the kurtosis.

The Lagrangian $L(\mathbf{y}, \lambda)$ is given by:

$$L(\mathbf{y}, \lambda) = J_p(\mathbf{y}) - \lambda \left(\sum_{j=1}^{N_2} y_j^2 - N_2 \right) \tag{7}$$

where λ is the Lagrange multiplier. It is shown in the Appendix that the stationary points of $L(\mathbf{y}, \lambda)$ are the solutions to:

$$\frac{\partial}{\partial \mathbf{b}} L(\mathbf{y}, \lambda) = \frac{1}{N_2} \left(p \mathbf{X} \mathbf{y}^{p-1} - 2\lambda \mathbf{X} \mathbf{y} \right) = 0$$

$$\frac{\partial}{\partial \lambda} L(\mathbf{y}, \lambda) = \frac{1}{N_2} \|\mathbf{y}\|_2^2 - 1 = 0 \tag{8}$$

where $\frac{\partial}{\partial \mathbf{b}} L(\mathbf{y}, \lambda)$ is a $N_1 \times 1$ vector whose k-th component is $\frac{\partial}{\partial b_k} L(\mathbf{y}, \lambda)$, and

$$\mathbf{y}^k \stackrel{def}{=} \begin{bmatrix} y_1^k \\ y_2^k \\ ... \\ y_{N_2}^k \end{bmatrix} \tag{9}$$

One class of solution is obtained when $p \mathbf{y}^{p-1} = 2\lambda \mathbf{y}$, i.e., either $y_j = 0$ or:

$$y_j = \left(\frac{2\lambda}{p} \right)^{\frac{1}{p-2}} \tag{10}$$

for each $j = 1, 2, ..., N_2$, with the restriction $\|\mathbf{y}\|_2 = \sqrt{N_2}$. In this case, *sparse* solutions are possible. Specifically, it can be easily shown that (6) attains a global maximum when one and only one of the y_j is different from zero.

Another class of solution is obtained when vector $p \mathbf{y}^{p-1} - 2\lambda \mathbf{y}$ is **orthogonal** to every row of $\mathbf{X}$ (then, it is said that $p \mathbf{y}^{p-1} - 2\lambda \mathbf{y}$ belongs to the kernel or null-space of $\mathbf{X}$). Mathematically:

$$p \sum_{n=1}^{N_2} x_{kn} y_n^{p-1} - 2\lambda \sum_{n=1}^{N_2} x_{kn} y_n = 0, \ \forall k = 1, 2, ..., N_1 \tag{11}$$

Again, we have find a solution when :

$$\lambda = \frac{p \sum_{n=1}^{N_2} x_{kn} y_n^{p-1}}{2 \sum_{n=1}^{N_2} x_{kn} y_n} \tag{12}$$

Fig. 1. The "Lena" image (256×256)

for every row $k = 1, 2, ..., N_1$ of **X**. So, we will have that:

$$\frac{\sum_{n=1}^{N_2} x_{kn} y_n^{p-1}}{\sum_{n=1}^{N_2} x_{kn} y_n} = \frac{\sum_{n=1}^{N_2} x_{jn} y_n^{p-1}}{\sum_{n=1}^{N_2} x_{jn} y_n}, \forall k \neq j \tag{13}$$

For N_2 sufficiently large, this solution is equivalent to

$$\frac{E\{x_{kn} y_n^{p-1}\}}{E\{x_{kn} y_n\}} = \frac{E\{x_{jn} y_n^{p-1}\}}{E\{x_{jn} y_n\}}, \forall k \neq j \tag{14}$$

Also in this case, sparse solutions are possible.

3   Example

Let us take the natural grey-scale image showed in Fig. 1. We divide it into 8×8 patches to obtain the matrix of observations and apply ICA (for example, using FastIca algorithm [5]).

In Fig. 2 we show some of the rows of matrix the **S** corresponding to the "Lena" image, which have, as we shown in the previous section, an sparse distribution.

As we introduced before, there is a striking resemblance between this result and the conclusions obtained by Barlow [1], who hypothesized that the primary visual cortex of mammals uses factorial code (i.e., independent components) to represent the visual environment. Such a code should be sparse in order to reduce the redundancy of the sensory signal, which would appear to constitute a sensible coding strategy for making the most use of the limited resources of the optic nerve [13]. This hypothesis was confirmed by Field [6], who also justified a relation between the receptive fields of V1 neurons and the Gabor filters [11][13].

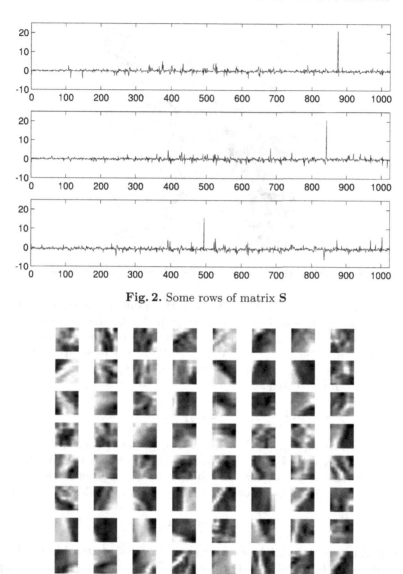

Fig. 2. Some rows of matrix **S**

Fig. 3. ICA bases obtained for the "Lena" image, that has been divided into 8 × 8 patches

Moreover, the $(k_1 k_2 \times k_1 k_2)$ matrix **A** has a very particular structure. If we arrange its columns into $(k_1 \times k_2)$ blocks and represent them as images, we observe that most of them look like *edges* with different orientations, as shown in Fig. 3. These images are usually called *ICA bases* and would play the role of the *features* in the generative model of natural images (2). Besides, these ICA bases (more specifically matrix **B** = **A**$^{-1}$) have been linked to the mentioned receptive fields of the simple cells localized in the primary visual cortex [2][11].

Fig. 4. Patches of the image that are similar to the ICA bases

Nevertheless, all these similarities between the results obtained when ICA is applied to natural images and the behaviour of the V1 neurons, have been mere observations so far. The purpose of this paper is to provide a mathematical basis for this connection between ICA and the HVS. Let us return to the ICA model (5). Since $\mathbf{X} = \mathbf{A}\mathbf{S}$, the i-th column of $\mathbf{X}$, $\mathbf{x}_i = [x_{1i}, x_{2i}, ..., x_{N_1 i}]^T$, $i = 1, 2, ..., N_2$, can be written as:

$$\mathbf{x}_i = \mathbf{A}\,\mathbf{s}_i = s_{1i}\,\mathbf{a}_1 + s_{2i}\,\mathbf{a}_2 + ... + s_{N_1 i}\,\mathbf{a}_{N_1} \tag{15}$$

where $\mathbf{s}_i = [s_{1i}, s_{2i}, ..., s_{N_1 i},]^T$, $i = 1, 2, ..., N_2$, is the i-th column of $\mathbf{S}$ and $\mathbf{a}_k = [a_{1k}, a_{2k}, ..., a_{N_1 k},]^T$, $k = 1, 2, ..., N_1$, is the k-th column of $\mathbf{A}$. As shown in the previous Section, $\mathbf{s}_i$ has a sparse distribution, so that most of its elements are negligible (e.g., see Fig. 2). It follows that there even exist indexes j for which

$$\mathbf{x}_i = \mathbf{A}\,\mathbf{s}_i \cong s_{ji}\,\mathbf{a}_j \tag{16}$$

It means that most of the ICA bases are, except for a scale factor, like patches of the original image. In practice, as illustrated in Fig. 3, the ICA bases are *edges* that already exist in the image (e.g., the locations of the edges shown in Fig. 3 are highlighted in Fig. 4).

To illustrate that when the ICA bases are edges (6) is maximum, let us do the following experiment. We build two matrices of bases, **A1**, obtained from edges of the "Lena" image, and **A2**, obtained from patches taken randomly from the same image. We calculate the matrices **S1** and **S2** using **A1** and **A2**, respectively, and obtain both the skewness and the kurtosis of these two groups of independent components. Skewness is a particular case of $J_p(\mathbf{y})$ when $p = 3$ and kurtosis is very similar to $J_4(\mathbf{y})$ (see (6)). In Fig. 5 are shown the results obtained. Both the skewness and the kurtosis are greater when the ICA bases are edges than when they are any other kind of pattern.

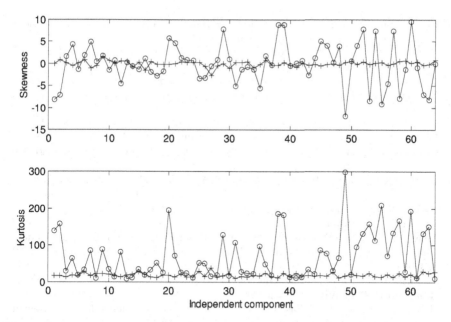

Fig. 5. Skewness and kurtosis of the independent components **y** when bases are like edges (o) and when bases are like any other pattern (+)

4   Conclusions

The Human Visual System (HVS) present some properties that have been connected to the mathematical technique called Independent Component Analysis (ICA), when it is applied to natural images. Particularly, the ICA bases appears to be like "edges", just like the receptive fields of a certain kind of neurons of the primary visual cortex (V1), the simple cells. On the other hand, the sparse distribution of the independent components have been linked with the responses of V1 neurons, also sparsely distributed. Nevertheless, no theoretical study has been provided so far to explain that special behaviour of ICA when it is applied to natural images. In this paper we have analyzed this question, both mathematical and experimentally, in order to give a theoretical basis for the connection between ICA and the HVS.

Applications arise in digital image processing like texture segmentation, edge detection [8], digital watermarking [9] or elimination of noise [11].

References

1. Barlow, H.: Sensory Communication. Possible Principles Underlying the Transformation of Sensory Messages. MIT Press (1961) 217–234
2. Bell, A., Sejnowski, T.: The "Independent Components" of Natural Scenes are Edge Filters. Vision Research, vol. 37 (23) (1997) 3327–3338

3. Cao, X., Liu, R.: General Approach to Blind Source Separation. IEEE Transactions on Signal Processing, vol. 44 (3) (1996) 562–571
4. Cichocki, A., Amari, S. I.: Adaptive Blind Signal and Image Processing. John Willey & Sons (2002)
5. Fastica algorithm available at http://www.cis.hut.fi/projects/ica/fastica/
6. Field, F.: What is the Goal of Sensory Coding? Neural Computation, vol. 6 (1994) 559–601
7. Guyton, A. C., Hall, J. E.: Textbook of Medical Physiology, Saunders, 2000.
8. Hornillo-Mellado, S., Martín-Clemente, R., Puntonet, C. G., Acha, J. I.: Application of Independent Component Analysis to Edge Detection. Proc. World Automation Congress (WAC 2004). Sevilla, Spain (2004)
9. Hornillo-Mellado, S., Martín-Clemente, R., Acha, J. I., Puntonet, C. G.: Application of Independent Component Analysis to Edge Detection and Watermarking. Lecture Notes in Computer Science, vol. 2 (2003) 273–280
10. Hyvärinen, A., Hoyer, P. O., Hurri, J.: Extensions of ICA as Models of Natural Images and Visual Processing. Proc. International Symposium on Independent Component Analysis and Blind Signal Separation (ICA 2003). Nara, Japan (2003) 963–974
11. Hyvärinen, A., Karhunen, J., Oja, E.: Independent Component Analysis. Wiley-Interscience, John Wiley & Sons (2001)
12. Hyvärinen, A., Oja, E.: A Fast Fixed-Point Algorithm for Independent Component Analysis. Neural Computation, vol.6 (1997) 1484–1492
13. Olshausen, B. A.: Principles of Image Representation in Visual Cortex. The Visual Neurosciences, L.M. Chalupa, J.S. Werner, Eds. MIT Press (2003) 1603–1615
14. Olshausen, B. A., Field, D. J.: Sparse Coding with an Overcomplete Basis Set: A Strategy Employed by V1? Vision Research, vol. 37 (23) (1997) 3311–3325

Appendix: Proof of eqn. (8)

The derivative of (7) with respect to k-th element of $\mathbf{b}$, b_k, is given by:

$$\frac{\partial}{\partial b_k} L(\mathbf{y}, \lambda) = \frac{\partial}{\partial b_k} J_p(\mathbf{y}) - \lambda \frac{\partial}{\partial b_k} \left(\sum_{j=1}^{N_2} y_j^2 - N_2 \right) \tag{17}$$

The first term is given by:

$$\frac{\partial}{\partial b_k} J_p(\mathbf{y}) = \frac{1}{N_2} \frac{\partial}{\partial b_k} \sum_{j=1}^{N_2} y_j^p = \frac{1}{N_2} \frac{\partial}{\partial b_k} \sum_{j=1}^{N_2} \left(\mathbf{b}^T \mathbf{x}_j \right)^p = \frac{p}{N_2} \sum_{j=1}^{N_2} y^{p-1} x_{kj} \tag{18}$$

$$= \frac{p}{N_2} [x_{k1}, x_{k2}, ..., x_{kN_1}] \mathbf{y}^{p-1}$$

since $y_j = \mathbf{b}^T \mathbf{x}_j = \sum_{j=1}^{N_1} b_k x_{kj}$, where x_{kj} is the entry (k, j) of matrix $\mathbf{X}$, and

$$\mathbf{y}^k \overset{def}{=} \begin{bmatrix} y_1^k \\ y_2^k \\ ... \\ y_{N_2}^k \end{bmatrix} \tag{19}$$

Secondly,

$$\lambda \frac{\partial}{\partial b_k} \left(\sum_{j=1}^{N_2} y_j{}^2 - N \right) = 2\lambda \sum_{j=1}^{N_2} y_j \frac{\partial}{\partial b_k} y_j = 2\lambda \sum_{j=1}^{N_2} y_j x_{kj} \tag{20}$$

In matrix form, we have:

$$\frac{\partial}{\partial \mathbf{b}} L\left(\mathbf{y}, \lambda\right) = \frac{1}{N_2} \left(p\mathbf{X}\mathbf{y}^{p-1} - 2\lambda\mathbf{X}\mathbf{y} \right) \tag{21}$$

where $\frac{\partial}{\partial \mathbf{b}} L(\mathbf{y}, \lambda)$ is a $N_1 \times 1$ vector and $\frac{\partial}{\partial b_k} L(\mathbf{y}, \lambda)$ is its k-th component. On the other hand:

$$\frac{\partial}{\partial \lambda} L\left(\mathbf{y}, \lambda\right) = \sum_{j=1}^{N_2} y_j^2 - N_2 = \|\mathbf{y}\|_2^2 - N_2 \tag{22}$$

Image Classifier for the TJ-II Thomson Scattering Diagnostic: Evaluation with a Feed Forward Neural Network

G. Farias[1], R. Dormido[1], M. Santos[2], and N. Duro[1]

[1] Dpto. de Informática y Automática -
E.T.S.I. Informática - U.N.E.D. Madrid 28040, Spain
gfarias@bec.uned.es, {raquel, nduro}@dia.uned.es
[2] Dpto. de Arquitectura de Computadores y Automática -
Facultad de CC. Fisicas - Universidad Complutense, Madrid 28040, Spain
msantos@dacya.ucm.es

Abstract. There are two big stages to implement in a signal classification process: features extraction and signal classification. The present work shows up the development of an automated classifier based on the use of the Wavelet Transform to extract signal characteristics, and Neural Networks (Feed Forward type) to obtain decision rules. The classifier has been applied to the nuclear fusion environment (TJ-II stellarator), specifically to the Thomson Scattering diagnostic, which is devoted to measure density and temperature radial profiles. The aim of this work is to achieve an automated profile reconstruction from raw data without human intervention. Raw data processing depends on the image pattern obtained in the measurement and, therefore, an image classifier is required. The method reduces the 221.760 original features to only 900, being the success mean rate over 90%. This classifier has been programmed in MATLAB.

1 Introduction

The TJ-II is a medium-size stellarator (heliac type) [1] located at CIEMAT (Spain). The Thomson Scattering (TS) in plasmas consists in the re-emission of incident radiation (from very powerful lasers) by free electrons. Electron velocity distribution generates a spectral broadening of the scattered light (by Doppler effect) related to the electronic temperature. The total number of scattered photons is proportional to the electronic density.

Every laser shot produces a bi-dimensional image from which is possible to obtain radial profiles of temperature and density. Only a restricted number of pattern images appear in the TJ-II. They represent different physical situations related to either the plasma heating or the system calibration. Depending on the pattern obtained, data are processed in different ways. Therefore, to perform an automated data analysis, a computerized classification system must provide the kind of pattern obtained in order to execute the proper analysis routines.

J. Mira and J.R. Álvarez (Eds.): IWINAC 2005, LNCS 3562, pp. 604–612, 2005.

As in any classification process, Thomson Scattering images need to be pre-processed in a suitable way [2]. Most of the analyses try to extract either unique or common signal features, thereby allowing identification of patterns that reflect similar experimental conditions [3, 4].

The present work shows up the development of an automated classifier (programmed in MATLAB) made up of two phases. The first one (feature extraction) uses the Wavelet Transform, and the second one (classification) makes use of Multilayer Neural Networks (Feed Forward type).

1.1   Image Patterns

The TJ-II Thomson Scattering images can be grouped under five different classes (Fig. 1).

Table 1 shows a brief description corresponding to every pattern.

As it can be seen in Fig. 1, all the patterns except BKGND correspond to images with, at least, four important features: an empty zone in the middle, two central vertical components, and a thin line on the right. Without giving details about the physical meaning of these characteristics, the differences among the patterns are consequence of the light intensity: very high in the central

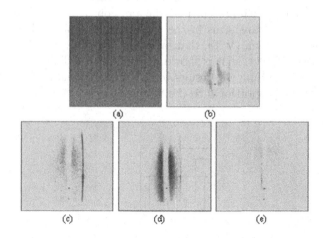

Fig. 1. Image patterns: (a) BKGND (b) COFF (c) ECH (d) NBI (e) STRAY

Table 1. Description of TJ-II Thomson Scattering patterns

Pattern	Description
BKGND	CCD Camera background
COFF	Reached cut off density for plasmas heated by Electron Cyclotron Resonant Heating
ECH	Plasmas heated by Electron Cyclotron Resonant Heating
NBI	Plasmas heated by Neutral Beam Injectors
STRAY	Measurement of parasitic light without plasma

components for the NBI case, low for the ECH case (although with a very intense thin line), central components grouped at the bottom for the COFF case, and practically null for the STRAY case.

2   Procedures for Data Mining and Information Retrieval

In this section effective feature extraction and classification methods for images are briefly illustrated. Firstly, a short review of the Wavelet Transform and its application to the signals is presented. Secondly, the Neural Networks technique used in the signal classification process is described. Finally, training and validation procedures of the classification method are commented.

2.1   Wavelet Transform

In many cases the signals present pseudo-periodic behaviour, oscillating around a fixed frequency. The most widely used analysis tool for periodic signals is the Fourier Transform, which allows to study time depended signals in the frequency domain. However, there are many other signals that can present a non-periodical behaviour, whose principal features must be obtained from a temporal analysis. For these signals, the Fourier Transform is unsuitable.

To analyze signals which present periodic and non-periodic behaviour is necessary to make use of transforms in the time-frequency plane. For such a reason the Wavelet Transform (WT), that overcomes the drawbacks of the Fourier Transform, is used. In fact, as it is shown in Fig. 2, it is a Time-Scale approach, which allows understanding the results in the time-frequency plane. Note that the scale is inversely proportional to the frequency.

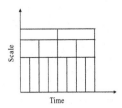

Fig. 2. Time-Frequency relation of the Wavelet analysis

Wavelet Transform Processing. The Wavelet Transform compares the original signal with the so-called Mother Wavelet. The Mother Wavelet is a wavelet prototype function, which can be modified to scale and to shift the signal as needed. Fig. 3 displays two types of Mother Wavelet function belonging to the Daubechies and Haar types. The correlation coefficients can be obtained from the comparison between the different Mother Wavelet functions and the original signal. From these coefficients it is possible to reconstruct the original signal using the inverse of the Wavelet Transform.

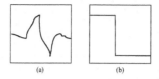

Fig. 3. Mother Wavelets, (a) Daubechies 2 and (b) Haar

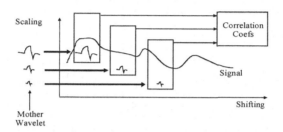

Fig. 4. Wavelet Transform processing

Fig. 4 shows the described process.

Once the Wavelet Transform is performed for all-possible scales (or levels), many characteristics of the signal are obtained. However, to select the most interesting scales and shifts for a concrete signal is a very difficult task. For such a reason, it is very common to analyze the signal by the Discrete Wavelet Transform. This transform consists in choosing only the scales and shifts based on powers of two. Thanks to this choosing, the redundant information is minimized, and so the computational load is substantially cut down [5, 6, 7].

Application of the Wavelet Transform to Images. Analysis of bidimensional signals is getting great improvements by using Wavelet based methods. For this problem Wavelet analysis technique makes use of regions with variable size. This technique allows not only to analyze regions of considerable size where information associated to low frequencies can be found (nearly homogeneous regions), but also small regions where information related to high frequencies is contained (vertices regions, edges or colours changes).

It is possible to characterize an image as a series of approximations and sets of finer details. The Wavelet Transform provides such a representation. The WT descomposition is multi-scale: it consists of a set of Approximation coefficients and three sets of Detail coefficients (Horizontal, Vertical and Diagonals). The Approximation coefficients represent coarse image information (they contain the most part of the image's energy), while the Details are close to zero, but the information they represent can be relevant in a particular context.

Fig. 5 illustrates this point. It has been obtained applying the WT to the image of a signal belonging to COFF class, using a Mother Wavelet Haar at level 2.

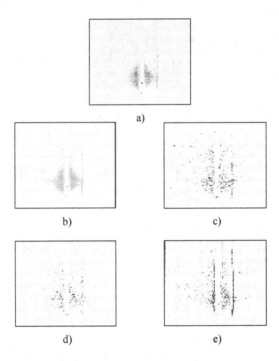

Fig. 5. DWT application to a signal of class COFF: (a) Original Signal, (b) Approximation, (c) Horizontal Detail, (d) Diagonal Detail, and (e) Vertical Detail

The family of Mother Wavelet functions used plays an important role in the final results. Its choice comes from experimentation with the work signals. Other important properties to be considered are the features of the wavelet coefficients and the decomposition level of the transform.

In relation to the TS signals, it has been found [8] that the best coefficient to characterize the images is the Vertical Detail, when the selected Mother Wavelet is the Haar at level 4. When applying the mentioned Wavelet Transform to the signals of the TS, the attributes are reduced from 221.760 to 900. So, the obtained attributes with the Wavelet Transform represent the 0.4% of the complete original signal.

2.2   Neural Networks: Feed Forward Multilayer

Neuronal Networks (NN) have been used successfully in great number of problems and applications. There is a diversity of types of NN, each one with a different structure according to the intentions of designer. However, in every case is possible to find the basic elements that define them. The NNs consist of elements of processing called neurons, which are grouped in layers and connected by synapses with an associate weight [9, 10].

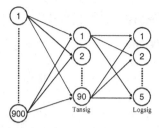

Fig. 6. Structure of the proposed NN

In the present work, a NN Feed Forward has been used. One of the possibilities of this type of NN is to use it for the supervised learning, where it is necessary to train the NN indicating to the input layer the attributes of a signal and the wished values to the output layer.

Fig. 6 shows the NN that has generated the best results. Note that the NN has an input layer of 900 attributes which come from the previous processing stage (generated by WT). The hidden layer used 90 with functions of activation Tansig, whereas the output layer has 5 neurons with functions of activation Logsig. After the training of the NN, every signal is associated with its class through of the activation from its output neuron and resetting the remaining ones.

The functions of activation are defined in Eq.1 and Eq.2.

$$Tansig(n) = \frac{2}{1 + \exp^{-2n}} - 1 \ . \tag{1}$$

$$Logsig(n) = \frac{1}{1 + \exp^{-n}} \ . \tag{2}$$

2.3   Train, Classify, and Testing Process

The training processing of the NN implies to obtain a set of weight through a Back-Propagation algorithm, which produces the minimal error between the values of the output layer and the wished values.

The classification process implies to use the trained NN to decide whether a signal presented to the NN belongs to a class or another.

To validate the efficiency of the classifier, the signals were divided randomly in training and testing sets. In this case, the training set was composed of 60% of the all signals. Later on, testing set was used to compare the obtained results with the wished values. To obtain an average performance, the procedure was realized for 100 different training and testing sets.

3   The Classifier

To implement the previous ideas in the present work, an application named Thomson Scattering Classifier has been designed in MATLAB [7, 11]. This ap-

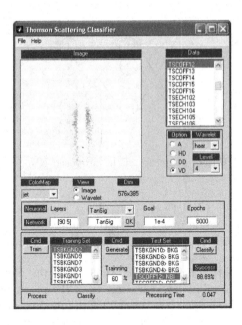

Fig. 7. Thomson Scattering Signals Classifier

plication allows to manipulate a set of labeled signals, whose main function is to evaluate the performance of the different classifiers. These classifiers can be obtained easily by modifying some parameters as: the Mother Wavelet, the decomposition level, the number of the NN layers, the activation functions, etc.

Fig. 7 presents the graphical user interface of the application.

3.1   Description of the Application

A brief description of the capabilities and available options of the developed classifier is given in the following sections.

Signal Image. The image of the signals can be displayed in the Image window at the left side. To select the signal to be displayed, an item of the Data list has to be clicked.

Wavelet Transform Configuration. In the application, it is possible to specify the different parameters associated with the WT. So, if to set up a decomposition level for the signal is wanted, there is to select an option from the popup Level menu. In this application, it is also possible to define the Mother Wavelet and in addition the set of obtained wavelet coefficients, that is, the Approximation (A) or the Detail (D).

Wavelet Transform Image. Once the type of the WT has been selected, the application displays the Wavelet Transform image of a particular signal. For this purpose, it is necessary to select the signal from the Data list and then to click on the Wavelet option in the View section.

Random Generation of Training and Testing Sets. When pressing the Generate button, two sets of signals are randomly obtained for training and testing purposes. The proportion of signals that compose the training set is defined by the user.

Neuronal Network Parameters. The application allows to set up the NN parameters to specify: the number of layers, the number of neurons in every layer, the functions of activation, the required goal, and the training epochs.

Neuronal Network Training. After the application has generated the training and testing sets, it is necessary to train the NN. For this aim, it is only necessary to press the Train button. The NN is now ready to classify, only just if the required goal has been reached.

Testing Signals Classification. To evaluate the testing set, it is necessary to press the Classify button. So, the classifier will make the predictions for every signal of the testing set. Automatically, the classifier will also compare the obtained results with the labels of each one of the signals, identifying therefore, the percentage of hits.

4   Results and Conclusions

Thomson Scattering Classifier allows to do many different kind of classifiers, due to fact that NN or WT parameters can be changed according to user requirements.

We have selected a Mother Wavelet Haar at level 4 with vertical details, while the selected Neuronal Network has the same structure that the NN proposed in Fig. 6.

To test the designed classifier, many experiments were made. It is necessary to indicate that the number of signals available at the moment to do the experiments was 46.

We made 100 experiments, where every experiment was composed of two signal groups, that is, training and testing set, which were randomly generated. Fig. 8 shows the results for each one of the classes, being the average percentage of hits of 90.89%.

The previous classifier, using Wavelet Transform and Neuronal Network, constitutes an alternative for automatic classification of Thomson Scattering signals.

The development of Thomson Scattering Classifier to experiment with the described techniques, allows to observe the effect of each one in the classification, to reduce evaluation time, and to search satisfactory parameters.

It is necessary to consider that probably better results can be obtained if knowledge about the problem context is added. However the presented results are a start point for new analysis.

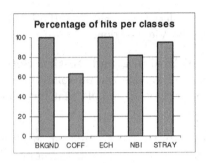

Fig. 8. Results for Each One of the Classes

Finally, it is necessary to emphasize that the training of the classifier has been done by a limited set of signals. So if the number of signal increase, it will be possible to improve the Percentage of hits obtained.

References

1. Alejaldre, C. et al.: Plasma Phys. Controlled Fusion 41, 1 (1999), pag. A539
2. Duda, R., Hort, P., Stork, D.: Pattern Classification, Second Edition, A Wiley-Interscience Publication (2001)
3. Nakanishi, H.; Hochin, T.; Kojima, M. and LABCOM group: Search and Retrieval Methods of Similar Plasma Waveforms. 4th IAEA TCM on Control, Data Acquisition, and Remote Participation for Fusion Research. San Diego, USA (July 21-23, 2003). (To appear in Fusion Engineering and Design)
4. Dormido-Canto, S., Farias, G., Dormido, R., Vega, J., Snchez, J., Santos, M. and The TJ-II Team: TJ-II Wave Forms Analysis with Wavelets and Support Vector Machines. Review Scientific Instruments, vol. 75, no. 10, (2004), 4254-4257
5. Daubechies, I.: Ten Lectures on Wavelets, SIAM (1992)
6. Mallat, S.: A Wavelt Tour of signal Processing, 2 Edition, Academia Press (2001)
7. Misiti, M., Oppenheim, G., Poggi, J., Misita, Y.: Wavelet Toolbox User's Guide (V.1). The MathWorks, Inc. (1995-98)
8. Farias, G., Santos, M., Marrn, J. L., Dormido-Canto, S.: Determinacin de Parmetros de la Transformada Wavelets para la Clasificacin de Seales del Diagnstico Scattering Thomson. XXV Jornadas de Automtica, Ciudad Real (Espaa), ISBN: 84-688-7460-4 (2004)
9. Hilera, J.R., Martnez, V.J.: Redes Neuronales Artificiales. Fundamentos, modelos y aplicaciones. Ed. Rama (1995)
10. Freeman, J., Skapura, D. : Redes Neuronales, Algoritmos, aplicaciones y tcnicas de programacin. Addison-Wesley/Diaz de Santos (1993)
11. Demuth, H., Beale, M.: Neural Network Toolbox User's Guide (V.3). The Math-Works Inc. (1992-1998)

Computerized Adaptive Tests and Item Response Theory on a Distance Education Platform

Pedro Salcedo, M. Angélica Pinninghoff, and Ricardo Contreras

Research and Educational Informatics Department,
Informatics Engineering and Computer Science Department,
Universidad de Concepción, Chile
psalcedo@udec.cl

Abstract. This article presents how Computerized Adaptive Tests and Item Response Theory are modified for using in a Distance Education Platform, MISTRAL, showing the advantages of using this technique, the way in which the knowledge acquisition is accomplished, how it links to student profile and how the students and materials are evaluated.

Keywords: Distance Education Platform, Computerized Adaptive Tests, Item Response Theory, Adaptive Systems.

1 Introduction

The use of tests for evaluation is a widely employed technique in education. Traditional test design and management depends on whether or not the test is oriented to a group or to a single person. Group oriented tests are cheaper than individual tests in both time and resources, and present the advantage that every member in the group under examination is faced with exactly the same constraints (time, environment, etc.). As a trade off, these kind of test must contain as many difficulty levels as knowledge levels may exist in the group, while individual tests contain more selective material that is tailored to a given student.

Group tests can impart undesirable consequences, such as loss of motivation in students with high knowledge level or disappointment in students with lower knowledge levels. At the beginning of the 70s, some studies suggested the use of more flexible tests. Lord [4] established the theoretical structure of a massive test which was individually adapted: *The basic idea of an adaptive test is to imitate a judicious examiner's behavior*, i.e., if a particular question is not correctly answered, the next one should be easier than the first; although testing in this manner only became possible in the early 80s by using cheaper and more powerful computers. This era gave birth to Computerized Adaptive Tests (CAT). A CAT is a test managed by computer in which each item is introduced and the decision to stop are dynamically imposed based on the students answers and his/her estimated knowledge level.

J. Mira and J.R. Álvarez (Eds.): IWINAC 2005, LNCS 3562, pp. 613–621, 2005.

First efforts in designing and applying computerized adaptive strategies come from the 70s as a result of Item Response Theory (IRT) a theoretical modelling development [4, 5], that represents the mathematical basis of any adaptive strategy.

MISTRAL [9, 10, 11] is software developed to support distance learning. By incorporating different Artificial Intelligence techniques on Adaptive Hypermedia Systems [1] it is aimed at offering an interesting alternative as a tool for distance learning because of the rapid development on the internet and the World Wide Web.

Although it can be argued that there are many solutions that use the same techniques, the novelty here is that MISTRAL takes into account things such as adapting teaching strategies to different learning styles, adaptive communication tools, the student profile and the ability to choose the teaching strategy to be used. Among MISTRAL tools, a module has been developed that allows generation of activities and CATs [6, 7] which also partially automate the process of distance evaluation and update the student's profile. All of this is presented in this study.

2   The Process of Evaluating in MISTRAL

On whatever educational platform, the process of evaluation is one of the more complex steps. In MISTRAL, the capability to adapt to a student issue has a direct connection with results obtained at that stage. This process, in part, depends on the sequence of activities the system proposes to the student in the learning process (Learning Strategy), which must take into account elements such as the student's profile and the course model. The course model establishes the order, links and difficulty level for each node of the coursework, which are supplied by the course instructor as will be explained later.

2.1   Student Profiling in MISTRAL

In MISTRAL, profiling the student is the process devoted to identify student knowledge by a specific point in the learning process, starting with an individually configurable knowledge level supplied by the instructor or by the student.

The initial configuration is composed of two parts:

1. Suppositions about the students' knowledge are entered the first time the user activates the system. This user profile is supplied by the course instructor, based on an anticipated student knowledge level and it serves as the starting point for the system use and evaluation.

2. Student profiles with respect to psychosociological features are divided into:

- Suppositions about student learning style, which is measured by using the Learning Styles Inventory [3], and
- Psychosocial features, supplied by the user through a simple questionaire the first time the application is used by that student.

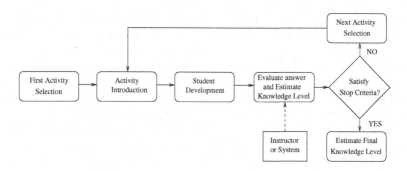

Fig. 1. Evaluation Algorithm

MISTRAL user profile maintenance involves:

- Automatic evaluation through adaptive tests, True/False tests and multiple choice tests, all of these update the user profile.
- Instructor evaluation is permitted through use of portfolio technique [2], for activities that cannot be automatically evaluated. Portfolios are virtual spaces on the server hard drive where activities are stored.

In general terms, the algorithm for updating the Student's profile is based on IRT and CAT, as shown in figure 1.

Profile updating in MISTRAL, considers the following steps:

1. To estimate the input knowledge level which is instructor configurable, at the time the student is registered in the software, which is his/her presumed knowledge level in that course.
2. Then, to determine the first question, the updated profile is used. Questions are stored in the Database, ordered by difficulty level.
3. Once the student has completed the first item or he/she has answered the first question of a CAT, evaluation occurs and the student profile is updated. The knowledge level in our model is determined by means of a Bayesian Method described in [6, 7].
4. When the stop criteria have been satisfied (the last activity or a certain percentage for the objective has been reached), then the process is finished or the next problem is selected. The stop criteria is a combination of a minimum/maximum number of questions and a minimum value to reach for each "content" (contents of the material covered on the activity/test). Besides the expected value for knowledge (θ), a minimum and a maximum number of questions or activities are established.

2.2   The Knowledge in MISTRAL

When constructing a course (Course Model) the first action to carry out is to establish the general objective, a unique objective for each course which will guide the coursework.

To accomplish this general objective, specific objectives must be established, i.e., goals that can be reached in the short time. Linked to these specific objectives, a score for content is used, which is, the minimum knowledge units included in each course. This learning content or coursework is reflected in goals reached for each specific objective. Each goal is reachable through a series of different problems that must be solved by students and stored for revision using the portfolio technique. These activities are evaluated to determine if the students know (or not) the contents, of the material, and consequently, if course objectives have been reached.

2.3   The Bayesian Approach

A Bayesian Net (BN) [8] is an acyclic and directed graph in which nodes represent variables and arcs are causal influence relationships among them. Parameters used for representing uncertainty are conditioned probabilities for each node, given their parent node states; i.e., if net variables are $\{X_i, i = 1, \ldots, n\}$ and $pa(X_i)$ represents the set that contains X_i parents, then the net parameters are $\{P(X_i/pa(X_i)), i = 1, \ldots, n\}$. This set defines the compound probability distribution through:

$$P(X_1, \ldots, X_n) = \prod_{i=1}^{n} P(X_i/pa(X_i))$$

Henceforth, to define a BN we must specify: a set of variables $X_1, \ldots, X_n$, a set of links among these variables in such a way that the resulting net is a directed and acyclic graph, and for each variable its probability, conditioned to its parent set, i.e., $\{P(X_i/pa(X_i)), i = 1, \ldots, n\}$.

The following example can help establish these ideas: The simplest nontrivial BN in MISTRAL contains two variables, called C and A_1 (concept and activity respectively), and one arc going from the first, to the second, as shown in figure 2.

Here C represents the student knowledge about a concept and A_1 represents his/her capability to solve a certain activity (or question if the activity is a test) related to that concept. In other words, if a student knows a concept C, which implies a causal influence concerning the fact that he/she is capable to solve A_1, which is represented through the arc that appears in figure 2. The meanings of the values in the nodes are:

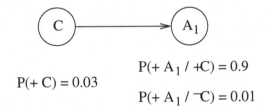

Fig. 2. Two nodes Bayesian network

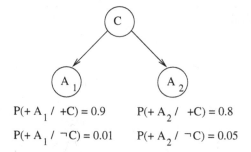

P(+ A$_1$ / +C) = 0.9   P(+ A$_2$ / +C) = 0.8

P(+ A$_1$ / ¬C) = 0.01   P(+ A$_2$ / ¬C) = 0.05

Fig. 3. Three nodes Bayesian network

- $P(+C) = 0.3$ indicates that 30 per cent of students in the study group know the concept C. $P(+A_1/+C) = 0.9$ indicates that the 90 per cent of students that know the concept C successfully accomplish activity A_1. $P(+A_1/\neg C) = 0.01$ indicates that only 1 per cent of students that do not know concept C are capable of successfully solving the activity A_1.

Knowing this information the *a priori* probability can be calculated for any student that successfully accomplish an activity A_1,

$$P(+A_1) = P(+A_1/+C) \cdot P(+C) + P(+A_1/\neg C) \cdot P(\neg C) = 0.277$$
$$P(\neg A_1) = P(\neg A_1/+C) \cdot P(+C) + P(\neg A_1/\neg C) \cdot P(\neg C) = 0.723$$

And the probability of knowing C concept is given by Bayes Theorem:

$$P^*(C) = P(C/A_1) = \frac{P(C) \cdot P(A_1/C)}{P(A_1)}$$
$$P^*(+C) = P(+C/+A_1) = \frac{P(+C) \cdot P(+A_1/+C)}{P(+A_1)} = \frac{0.3 \cdot 0.9}{0.277} = 0.97473$$

If we add a new activity, A_2, the BN is as shown in figure 3.

In this case, Bayes theorem let us calculate the probability of knowing C, given A_1 and A_2 successfully solved.

$$P^*(C) = P(C/A_1, A_2) = \frac{P(C) \cdot P(A_1, A_2/C)}{P(A_1, A_2)}$$

In MISTRAL, knowledge is modelled from a Bayesian point of view. To do that, it is necessary to define the three basic elements that constitute a bayesian net: variables, links between variables and parameters. It allows us to get a structure that encapsulates the data mentioned in the previous paragraph, with variables defined through the General Objective, Specific Objectives, Contents and Activities; links represent existing relationships between variables that take into account the causal influence and, finally, parameters indicate the influence weights that each child node has on its parent node.

The resulting bayesian net (see figure 4) is used in student profiling. Using this net it is possible to detect the level of the student for each knowledge variable

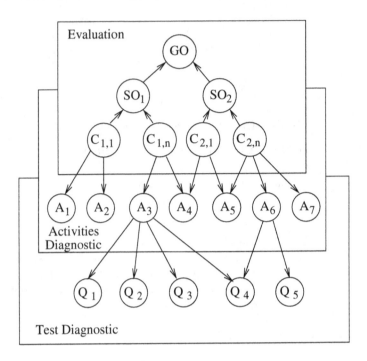

Fig. 4. Bayesian net for knowledge modelling

at any moment and determine the sequence of activities most appropriated to complete the coursework.

Two important stages of the evaluation process are incorporated into this model as shown in figure 4:

Student Diagnostic Process: The stage, in which the set of contents the student knows (or doesn't know) is created, based on evidence nodes, which in this case are called activities nodes.

Evaluation Process: This stage is based on the previous diagnostic (contents the student knows, or doesn't know) and measures the degree of knowledge the student has reached with respect to different levels. Specifically content level in Specific Objectives and also the General Objective for the course.

In practical terms, the activities the students must carry out are evaluated by a human examiner through the portfolio technique or automatically by the system, depending on the nature of the activity. Results are received in the software as an input for the student diagnostic, giving an estimate of the students' knowledge. If, as a result of the diagnostic it is assumed that the student knows the content, then the system goes to the next section of the coursework; otherwise, MISTRAL will propose a series of activities to reinforce the unassimilated knowledge, taking into account the student profile.

This is analogous to a classic evaluation in any CAT. The difference here is that evaluation can be made by a human or by the software.

2.4   The Computerized Adaptive Test in MISTRAL

Activities MISTRAL proposes are related to the students profile and the goals of the coursework. These activities may be diverse, exercises, reading, report generation, discussion participation, and so on. With a corresponding weight (dependant on the coursework), different difficulty levels, and a characteristic activity curve, to foresee the probability the student will successfully accomplish the activity and if, in fact, he/she has the necessary knowledge. As an example Table 1 shows part of an Object Orientation course.

In this context, there is an additional activity with the same features but it uses a different treatment. It is the Computerized Adaptive Test. In the domain of activities that can be proposed, a CAT is one of the more effective tools to evaluate the student behavior and if the instructor can establish a complete itemset, results are improved.

Generating a net similar to that used in the General Evaluation Process, a new student profile is acquired. In this new net evaluation and diagnostic steps are also supported, but here CAT is responsible for the diagnostic process, in which, only items and concepts are required. Once the test is finished, the platform, using the evaluating stage, is utilized to estimate the knowledge level the student has reached for each specific objective and for the general objective of the course.

CAT is responsible for the diagnostic processes requiring only questions and concepts. Once the test is finished, and during the evaluation stage MISTRAL considers each objective which will update the student's profile. The corresponding test algorithm is similar to the classic CATs presented. MISTRAL contains the following basic elements: Answer model for each question, grading method, question database, question selection method and stop criteria.

2.5   Answer Model Associated to Item in MISTRAL

It can be said that the Item Response Theory is a conceptual framework that is based on basic measurement concepts using statistical and mathematical tools, to try to find a theoretical description to explain empirical data behavior that is obtained by applying a psychometric device. Parameters that the model estimates, allow the evaluation of the technical quality for each item separately, and for the device as a whole, estimating the level of each individual under examination for the specific topic. In MISTRAL, an item is considered an activity that can be modelled through a mathematical function called Activity Characteristic Curve (for item in IRT).

In IRT the item play a fundamental role; in MISTRAL this dichotomous variable is the activity and what we are fundamentally interested in is whether the student answered yes or no and not in what the total score was. For each knowledge level θ there exists a probability to correctly answer the question, represented by $P(U = 1/\theta)$; $P(\theta)$ for short. This function is called characteristic activity curve (item). Typically the graphics for this function approach zero for

Table 1. Part of the contents and activities for an Object Oriented Course

Cont.	Activities	c	a	b
1	Software Quality			
1.1	Read text 1.1 "Introduction". Underline relevant topics and put on your portfolio as Act1.1	0	0.2	1
1.2	Read text 1.2 "Software Quality". Underline relevant topics and put on your portfolio as Act1.5	0	0.2	1
...	...	...	...	...
4	The UML Language			
4.1	Read document 4.1.2 and underline items you think are important. Put on your portfolio as Act4.1	0	0.2	1
4.1, 4.3	Study Power Point slides 4.1.1 of Jesus Garcia (Murcia University). Write an abstract and put on your portfolio as Act4.2 (using Word)	5	1.2	2

small values of θ and approach one for higher values. In MISTRAL, the following three parametric formula have been chosen:

$$P_i(\theta) = c_i + (1 - c_i)\frac{1}{1+e^{-1.7a_i(\theta-b_i)}}$$

Where, $P_i(\theta)$: the probability that a randomly chosen student has the knowledge θ to successfully answer item i.

a_i : is a discriminating index for item i

b_i : difficulty level for item i.

c_i : guess probability factor for item i

In Table 1, above, c is a guess probability factor, a is a discriminating index, b is a difficulty level.

MISTRAL can be tested at http://152.74.11.209/pide/principal.htm.

3   Conclusions

MISTRAL has been developed to consider different course materials when evaluating status. The primary difficulty is determining parameters linked to each activity and the time resources necessary to develop a complete course, typically about one year. In spite of that, the instructors' evaluation has been very favorable, because it is possible to connect evaluation and distance education processes in a clear way. Besides that, MISTRAL allows the instructor to select the most appropriate activity for a specific knowledge level for each student, and because of this, a student having prior knowledge may require less time than a student without prior training.

The general purpose of this paper is to show the potential capabilities of combining BN and IRT in support systems for distance teaching (learning) development. In particular, we have shown the use of BN in student profiling and IRT in adaptive processes.

MISTRAL incorporates these AI techniques with the proposed adaptation and evaluation mechanisms. Emphasis in testing is directed towards different coursework for students having different profiles. This is the basis of our work and clearly shows its advantages and limitations.

In summary, the global objective of our work has been to study distance teaching support systems from the point of view of knowledge engineering, as knowledge based systems, and to search for whatever AI techniques will improve results.

References

1. Brusilovsky, P.: Methods and techniques of adaptive hypermedia, in Brusilovsky, P., Kobsa, A., and Vassileva J. (eds): Adaptive Hypertext and Hypermedia. Dordrecht: Kluwer Academic Publishers (1998) 1-43
2. Danielson, C., and Abrutyn, L.: An Introduction to Using Portfolios in the Classroom. Alexandria, Virginia: Association for Supervision and Curriculum Development (1997)
3. Kolb, D., and Rubin, I., and McIntyre, J.: Organizational Psychology. Prentice Hall, Englewood Cliffs (1984)
4. Lord, F. M.: Some test theory for tailored testing, in W. H. Holtzman (Ed), Computer assisted instruction, testing and guidance, New York, Harper and Row (1970) 139-183
5. Martínez, Rosario Arias.: Psicometría: Teoría de los Tests Psicológicos y Educativos. Editorial Síntesis S.A., Madrid (1996)
6. Millán, E., Muñoz, A., Pérez, J.L., Triguero, F.: GITE: Intelligent generation of tests, in Lecture Notes in Computer Science, Vol. 1108. Proceedings of the 3rd International Conference CALISCE'96. Springer Verlag. Berlin (1996)
7. Millán, E., Pérez, J.L., Triguero, F.: Using bayesian networks to build and handle the student model in exercise based domains, in Lecture Notes in Computer Science, Vol. 1142. Proceedings of the 4th International Conference ITS'98. Springer Verlag. Berlin (1998)
8. Pearl, J.: Probabilistic Reasoning in Expert Systems: Networks of Plausible Inference. San Francisco. Morgan Kaufmann Publishers, Inc. (1988)
9. Salcedo, P., Labraña, C., Farran, Y.: Una plataforma inteligente de educación a distancia que incorpora la adaptabilidad de estrategias de enseñanza al perfil, estilos de aprendizaje y conocimiento de los alumnos, in XXVIII Latin-American Conference on Informatics (CLEI 2002). November 25-29, Uruguay (2002)
10. Salcedo, P., Farran, Y., Mardones E.: MISTRAL: An Intelligent platform for distance education incorporating teaching strategies which adapt to students learning styles, in World Conference on e-learning in Corporate, Government, Healthcare and Higher Education. (2002)
11. Salcedo, P, Pinninghoff, M.A., Contreras R.: MISTRAL: A Knowledge-Based System for Distance Education that Incorporates Neural Networks Techniques for Teaching Decisions, in Artificial Neural Nets Problem Solving Methods, Lectures Notes in Computer Science, Vol. 2687, Springer-Verlag. Berlin (2003)

Stochastic Vs Deterministic Traffic Simulator. Comparative Study for Its Use Within a Traffic Light Cycles Optimization Architecture

Javier Sánchez Medina, Manuel Galán Moreno, and Enrique Rubio Royo

Innovation Center for Information Society – C.I.C.E.I.,
University of Las Palmas de Gran Canaria,
Campus de Tafira s/n, Las Palmas de Gran Canaria, Spain
jsanchez@polaris.ulpgc.es, mgalan@cicei.com, rubio@cicei.com
http://www.cicei.com

Abstract. Last year we presented at the CEC2004 conference a novel architecture for traffic light cycles optimization. The heart of this architecture is a Traffic Simulator used as the evaluation tool (fitness function) within the Genetic Algorithm. Initially we allowed the simulator to have a random behavior. Although the results from this sort of simulation were consistent, it was necessary to run a huge amount of simulations before we could get a significant value for the fitness of each individual of the population . So we assumed some simplifications to be able to use a deterministic simulator instead of the stochastic one. In this paper we will confirm that it was the right decision; we will show that there is a strong linear correlation between the results of both simulators. Hence we show that the fitness ranking obtained by the deterministic simulator is as good as the obtained with the stochastic one.

1 Introduction

The traffic affair is not only a comfort factor for every major city in the world, but also a very important social and economical problem when not correctly managed. The progressive overload process that traffic infrastructures are suffering forces us to search for new solutions. In the vast majority of the cases it is not viable to extend traffic infrastructures due to costs, lack of available space, and environmental impacts. So it is a must to optimize existing infrastructures in order to obtain from them the very best service they can provide.

One of the most relevant problems in traffic optimization is the traffic light cycles optimization. In [1] it is demonstrated that the traffic light cycles have a strong influence in traffic flow results. This is the reason why we decided to deal with this problem.

In our group we presented in CEC2004 ([2]) a new architecture for optimizing the traffic light cycles in a traffic network with several intersections. Through other work in the same year ([3]) – we demonstrated that this architecture was such a scalable one, performing optimizations with networks from 4 to 80 intersections.

J. Mira and J.R. Álvarez (Eds.): IWINAC 2005, LNCS 3562, pp. 622–631, 2005.

The rest of this article is organized as follows. In the next subsection, we comment on the State of the Art about traffic optimization. In the section 2 we briefly explain our system architecture. In section 3 we explain some concepts related to traffic simulation and our simulator in detail. In section 4 we present the comparative study of the stochastic simulator versus deterministic simulator. Finally, section 5 includes our conclusions and some future work ideas.

1.1   State of the Art

There many works on traffic optimization. In this subsection we pretend to give some outstanding examples. In [4] an "ad hoc" architecture is used to optimize a 9 intersections traffic network. It uses Genetic Algorithms as an optimization technique running on a single machine. The CORSIM model is used within the evaluation function of the Genetic Algorithm. In this work scalability is not solved. Authors recognize that it is a customized non scalable system.

In [5] it is proposed the concept of optimal green time algorithm, which reduces average vehicle waiting time while improving average vehicle speed using fuzzy rules and neural networks. This work only considers a very small amount of traffic signals — two near intersections — in the cycle optimization.

In [6], a cycle-based isolated intersection is controlled applying efficient optimal control techniques based on linear systems control theory for alleviating the recurrent isolated intersection congestion. Again this work deals with very small scale traffic networks — one intersection.

In [7] the authors presented a neural network approach for optimal light timing. The approach is based on a neural network (or other function approximator) serving as the basis for the control law, with the weight estimation occurring in closed-loop mode via the simultaneous perturbation stochastic approximation (SPSA) algorithm. The training process of the NN is fed exclusively with real data. So, it would only be useful in systems with an on-line data acquisition module installed.

In [8], Dr. Tveit, senior researcher with *SINTEF* (Norway), explains that a common cycle time for a set of intersections is a worse approach than a distributed and individual one. We do agree with him. In our system every intersection has independent cycles.

In [9] a real-time local optimization of one intersection technique is proposed. It is based on fuzzy logic. Although an adaptive optimization may be very interesting – we realize this in [2] – we believe that a global optimization is a more effective approach to the problem.

Finally, we cite [10]. In this paper, Petri Nets are applied to provide a modular representation of urban traffic systems regulated by signalized intersections. An interesting feature of this model consists in the possibility of representing the offsets among different traffic light cycles as embedded in the structure of the model itself. Even though it is a very interesting work, the authors only optimize the coordination among traffic light timings. Our cycle optimization is a complete flexible one because we implicitly optimize not only traffic light offsets but also green times.

2   Architecture of Our System

The architecture of our system comprises three items, namely a Genetic Algorithm, a Cellular Automata based Traffic Simulator inside the evaluation routine of the Genetic Algorithm, and a Beowulf Cluster as MIMD multicomputer. We briefly will explain the Genetic Algorithm and the Beowulf cluster in this section, and the Traffic Simulator in detail in the next one. To know more about this architecture you may consult [2].

2.1   Genetic Algorithm Description

In this subsection we are to succinctly describe the genetic algorithm we are employing. Again, for a more profound description have a look at the [2] paper.

We optimize the traffic light cycles for all the intersections in a traffic network. Every cycle is represented by an array of integers. Every integer indicates which traffic light is open at every cycle step for every intersection.

We have chosen a Truncation and Elitism combination as selection strategy. It means that at every generation a little group of individuals — the best two individuals in our case — is cloned to the next generation. The remainder of the next generation is created by crossovering the individuals in a best fitness subset – usually a 66 percent of the whole population. We have tested many selection strategies but, so far, this one seems to have better results.

About the Crossover and Mutation operators, we have used a standard two points crossover and a variable mutation probability within the mutation operator.

Finally, for the evaluation we use the Mean Elapsed Time (M.E.T.). This is the average elapsed time (iterations) since a new vehicle arrives at the network until it leaves. We calculated the fitness simply as the inverse of the evaluation value since this is a minimization problem.

2.2   Beowulf Cluster Description

The Architecture of our system is based on a five Pentium IV node Cluster Beowulf, due to its very interesting price/performance relationship and the possibility of employing Open Software on it. On the other hand, this is a very scalable MIMD computer, a very desirable feature in order to solve all sorts — and scales — of traffic problems. The chosen operating system is Red Hat 9.

3   The Traffic Simulator

Traffic Simulation is known to be a very difficult task. There are two different sorts of traffic models. The first one is the macroscopic model set. They are based on Fluid Dynamics since they consider traffic flow as a continuous fluid. On the other hand, we have microscopic approaches. In them, traffic is considered as a set of discrete particles following some rules. In the last decade there is a common belief about the better performance of Microscopic approaches to do Traffic

Modeling. One such widely used approach is the Cellular Automata Model. Scientific literature is plenty of macroscopic approaches for traffic modeling. In the 50s appeared some "first order" continuum theories of highway traffic. In the 70s and later some other "second order" models were developed in order to correct the formers' deficiencies. In [11] "second order" models are questioned due to some serious problems, i.e. a negative flow predictions and negative speeds under certain conditions.

Nowadays the microscopic simulators are widely used. Their main drawback is that they have a lower performance than the macroscopic ones, except in the case of the Cellular Automata. Cellular Automata and macroscopic simulators have similar computing times.

3.1   The Cellular Automata as Inspiring Model

The Cellular Automata Simulators are based on the Cellular Automata Theory developed by John Von Neumann [12] at the end of the forties at the Logic of Computers Group of the University of Michigan. Cellular Automata are discrete dynamical systems whose behavior is specified in terms of local relation. Space is sampled into a grid, with each cell containing a few bits of data. As time advances each cell decides its next state depending on the neighbors state and following a small set of rules.

In the Cellular Automata model not only space is sampled into a set of points, but also time and speed are sampled too. Time becomes iterations. It sets a relationship between time and iterations, e. g. $1(sec.) \equiv 1(iteration.$ Consequently, speed turns into "cells over iterations".

In [13] we can find a well described list of microscopic models and a comparative study of them. Although conclusions are not definitive, this work seems to demonstrate that models using less parameters have better performance.

We have developed a traffic model based on the SK model ([14]) and the SchCh model ([15]). The SchCh model is a combination of a highway traffic model — Nagel-Schreckenberg [16] — and a very simple city traffic model — Biham-Middleton-Levine [17]. The SK model adds the "smooth braking" for avoiding abrupt speed changes. We decided to base our model in the SK model due to its better results for all the tests shown in [13].

3.2   Our Improved Cellular Automata Model

Based on the Cellular Automata Model we have developed a non-linear model for simulating the traffic behavior. The basic structure is the same like the one used in the Cellular Automata. However, in our case we add two new levels of complexity by the creation of two abstractions, "paths" and "vehicles".

"paths" – are overlapping subsets included in the Cellular Automata set. There is one "path" for every origin-destination pair. To do this, every one has a collection of positions and, for each one of them, an array of permitted next positions. In figure 1 we pretend to illustrate this idea.

"Vehicles" is an array of objects, each one of them having the following properties:

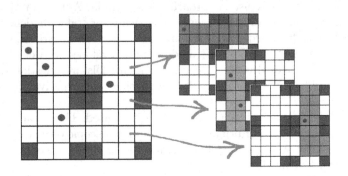

Fig. 1. Paths in our Improved Cellular Automata Model

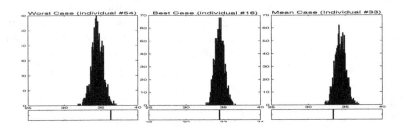

Fig. 2. Three Chromosome Histograms for the Huge Network

1. Position: the Cellular automata where it is situated. Note that every cell may be occupied by one and only one vehicle.
2. Speed: the current speed of the vehicle. It means the number of cells it moves over every simulator step.
3. Path: Every vehicle is univocally related to a Path in our model.

In our model these are the rules applied to every vehicle.

1. A vehicle ought to accelerate up to the maximum speed allowed. If it has no obstacle in its way (other vehicle, or a red traffic light), it will accelerate at a pace of 1 point per iteration, every iteration.
2. If a vehicle can reach an occupied position, it will reduce its speed and will occupy the free position just behind the preceding.
3. If a vehicle has a red traffic light in front of, it will stop.
4. Smooth Braking: Once the vehicle position is updated, then the vehicle speed is also updated. To do this, the number of free positions from current position forward is taken into account.
5. Multi-lanning: When a vehicle is trying to move on, or update its speed, it is allowed to consider positions on other parallel lanes.

By this means we can have lots of different path vehicles running in the same network. This model may be seen as a set of N_{paths} traditional Cellular Automata networks working in parallel over the same physical points.

3.3   The Stochastic Version vs Deterministic Version

In our traffic simulator there are three key items susceptible of making the simulator a Stochastic or a Deterministic one.

1. The cells updating order. In the Stochastic version the order of the cells to be updated is chosen at random. In the Deterministic version we have written a recursively function to calculate the dependencies graph. With it we prepare a non-deadlocking updating order at the beginning of the simulation.
2. The new vehicle creation time. In the Stochastic version every input has a new vehicle arrival probability. So the new vehicle creation time depends on a random variable. In the Deterministic case we create a new vehicle at a fixed period proportional to the traffic flow at every input.
3. The acceleration probability. In the Deterministic case when updating every vehicle speed, if it has free way and is under the maximum allowed speed it will always accelerate. However, in the Stochastic case, there is an accelerating probability – usually bigger than 0.7. So it could accelerate or not.

4   Stochastic Deterministic Comparison

4.1   Tests Performed

We used three different traffic network scales for this test. Their statistics are shown in the table 1.

Table 1. "Points" means the number of cells of the respective network, sampled at a rate of a sample every 7 m approximately — the minimal distance required in a traffic jam. "T.Lights" means the number of traffic lights optimized. "Intersections", "Inputs" and "Outputs" mean the number of intersections, inputs and outputs of the respective network. Finally "Chromosome Size" means the number of bytes that every chromosome includes

Scale	Points	T. Lights	Intersections	Inputs	Outputs	Chromosome Size
Small	80	16	4	6	8	96 bytes
Big	202	24	12	14	14	192 bytes
Huge	248	40	20	18	18	320 bytes

For every network scale we have launched 1000 stochastic simulations and a deterministic one. The size of the population is 100 individuals. Every simulation runs through 1500 iterations.

4.2   Results

In this part we present the results of the tests performed. In the first picture – figure 2 – we display three individual histograms resulting from the "huge case" test. They are the worst case, the best case and the mean case concerning the difference between deterministic and stochastic values obtained. Note that for every individual there is a histogram on the top representing the stochastic simulator outputs and a mark at the bottom representing the deterministic simulator value.

In this picture one may see that the stochastic simulator follows a unimodal Gaussian distribution. This fact plus the low dispersion also observed mean that it is a stochastically converging evaluation. Hence, we may trust that the arithmetic mean of the statistics obtained from the stochastic simulator significantly represent the fitness value of each individual in the population.

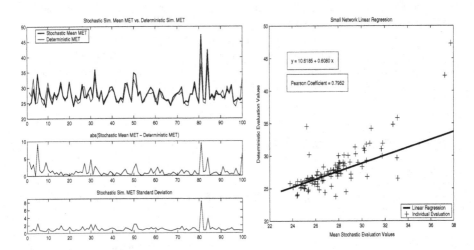

Fig. 3. Small Network MET Comparison   **Fig. 4.** Small Network Linear Regression

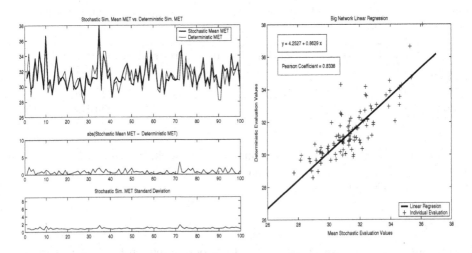

Fig. 5. Big Network MET Comparison   **Fig. 6.** Big Network Linear Regression

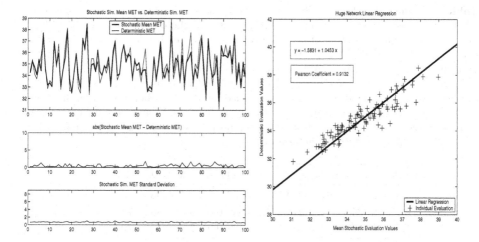

Fig. 7. Huge Network MET Comparison   **Fig. 8.** Huge Network Linear Regression

In second place we present a set of six pictures – figures 3 and 4 for the small network; figures 5 and 6 for the big network and 7 and 8 for the huge network. In figures 3, 5 and 7 we have represented:

1. At the first row we draw the Mean Elapsed Time from both simulators. Note that in the stochastic case there are represented the average values of all the executions.
2. At the second row it is represented the absolute value of the difference between the mean values of the stochastic simulator and the deterministic simulator values.
3. Finally it is plotted the standard deviation of the stochastic simulator values.

From this pictures one may observe that the two main plots – mean stochastic values and deterministic values – are highly dependent.

In figures 4, 6 and 8 we plot the linear regression function and the individual evaluation values. It is obvious that there is a strong linear regression correlating the mean stochastic values with the deterministic ones – shown in the next paragraph. In this sense, the "huge case" is the best fitted with a linear regression.

Finally, we present the table 2 with some other interesting statistics. Note that for every network scale we have two 100 element arrays, one for the stochastic simulator values and one for the deterministic one. For clarifying purposes we will represent the stochastic average values as x and the deterministic values as y. So, every $x[i]$ element means the average value of 1000 executions of the stochastic simulator, and every $y[i]$ element means the deterministic evaluation value, both for the i chromosome.

$$\rho = \frac{\sum_{i=1}^{n}(x_i - \overline{x})(y_i - \overline{y})}{\sqrt{\sum_{i=1}^{n}(x_i - \overline{x})^2 \sum_{i=1}^{n}(y_i - \overline{y})^2}} \tag{1}$$

Table 2. In the first column we have the network scale of the test. In the second one, labeled as $\bar{y}$, we have the average evaluation value in the deterministic case. $Std(y)$ means the standard deviation of every individual in the stochastic case. The following two columns – $\bar{x}$ and $Std(x)$ – mean the average of the standard deviation of every individual in the stochastic case and the standard deviation respectively. In the fifth column we have the Pearson Correlation Coefficient calculated as equation 1. The sixth column displays the Mean Euclidean Distance – labeled as M.E.D. – between the individual evaluations and the regression function. Finally we have the Mean Computational Cost Ratio – labeled as M.C.C.R. – what is just the ratio between the execution time of the stochastic simulations over the deterministic time

Scale	$\bar{y}$	$Std(y)$	$\bar{x}$	$Std(x)$	ρ	M.E.D.	M.C.C.R.
Small	27.6250	2.6149	27.9701	1.0832	0.7952	2.4595	63.90
Big	31.3756	1.6793	31.4305	0.8630	0.8338	1.6786	74.49
Huge	34.8757	1.5745	34.8786	0.6929	0.9132	1.7214	70.87

In this table we want to remark that for every network scale there is a Pearson Correlation Coefficient always over 0.7. It confirms that there is a strong linear correlation between x and y. The last column is also very interesting. It shows the high computational cost of using the stochastic simulator.

From figures 3, 4, 5, 6, 7 and 8 and the table 2 it is clear that we have a strong linear correlation between values obtained from both simulators. So it is acceptable to use the deterministic one for arranging the individuals of the population in order of fitness value.

5   Conclusions

In a previous work we developed a new architecture for traffic light cycles optimization. The heart of this architecture was a traffic simulator that we use as evaluation device.

In that work we decided to use a deterministic simulator instead of a stochastic one. Although traffic is known to be an intrinsic stochastic process, it was a must to take this decision. It was extremely overwhelming to run all the stochastic simulations needed to guide properly the genetic algorithm in comparison with the only one simulation needed for the deterministic evaluation case.

In this paper we have founded that decision on numerical facts. Firstly we confirm that the stochastic simulator is a suitable – convergent – statistical process to compare with. Secondly we demonstrate that the deterministic simulator outputs are highly linearly correlated with the stochastic ones. So our deterministic simulator can arrange the population ranking in order of fitness at least as the stochastic simulator, but with a remarkable lower computing power.

As future work we will try to apply this system to a real world case and see what happens. In further time we will improve this architecture trying to adapt it to more dynamical situations.

References

1. Brockfeld, E., Barlovic, R., Schadschneider, A., Schreckenberg, M.: Optimizing Traffic Lights in a Cellular Automaton Model for City Traffic. http://arxiv.org/ps/cond-mat/0107056. (2001)
2. Sanchez, J., Galan, M., Rubio, E.: Genetic Algorithms and Cellular Automata: A New Architecture for Traffic Light Cycles Optimization. Proceedings of The Congress on Evolutionary Computation 2004 (CEC2004) (2004) Volume II, 1668–1674.
3. Sanchez, J., Galan, M., Rubio, E.: Genetic Algorithms and Cellular Automata for Traffic Light Cycles Optimization. Scalability Study. To appear in the Proceedings of The Optimization and Design in Industry congress (OPTDES IV) (2004)
4. Rouphail, N., Park, B., Sacks,J.: Direct Signal Timing Optimization: Strategy Development and Results. In XI Pan American Conference in Traffic and Transportation Engineering, Gramado, Brazil (2000)
5. You Sik Hong, JongSoo Kim, Jeong Kwangson, ChongKug Park: Estimation of optimal green time simulation using fuzzy neural network. Fuzzy Systems Conference Proceedings, 1999. FUZZ-IEEE '99. 1999 IEEE International, vol.2. (1999) 761 – 766.
6. Wann-Ming Wey: Applications of linear systems controller to a cycle-based traffic signal control. Intelligent Transportation Systems, 2001. Proceedings. 2001 IEEE , 25-29 Aug. 2001, 179 – 184.
7. Spall, J.C.; Chin, D.C.: A model-free approach to optimal signal light timing for system-wide traffic control. Decision and Control, 1994., Proceedings of the 33rd IEEE Conference on, vol.2. (1994) 1868 – 1875,
8. Tveit, O.: Common cycle time - a stregth or barrier in traffic light signalling. Traffic Engineering and Control (TEC) Magazine, (2003) 44(1) 19–21
9. GiYoung, L.; JeongJin, K.;YouSik, H.: The optimization of traffic signal light using artificial intelligence. Fuzzy Systems, 2001. The 10th IEEE International Conference on ,Volume: 3 , 2-5 Dec.(2001) 1279 – 1282
10. Di Febbraro, A.; Giglio, D.; Sacco, N.: On applying Petri nets to determine optimal offsets for coordinated traffic light timings. Intelligent Transportation Systems, 2002. Proceedings. The IEEE 5th International Conference on, (2002) 773 – 778.
11. Daganzo,C.F.: Requiem for second order fluid approximations of traffic flow. Transportation Research B, (1995)
12. von Neumann,J.: The General and Logical Theory of Automata. John von Neumann—Collected Works, Volume V, A. H. Taub (ed.), Macmillan, New York, (1963) 288–328.
13. Brockfeld, E.; Khne, R. D.; Wagner, P.: Towards Benchmarking Microscopic Traffic Flow Models Networks for Mobility, International Symposium, Volume I, (2002) 321–331.
14. Krauss, S., Wagner, P., and Gawron, C.: Metastable states in a microscopic model of traffic flow, Phys. Rev. E55 5597 – 5605 (1997)
15. Schadschneider, A ; Chowdhury, D ; Brockfeld, E ; Klauck, K ; Santen, L ; Zittartz, J: A new cellular automata model for city traffic. "Traffic and Granular Flow '99: Social, Traffic, and Granular Dynamics". Springer, Berlin (1999)
16. Nagel, K., Schreckenberg, M.: A Cellular Automaton Model for Freeway Traffic. Journal de Physique I France, **332** (1992) 2221–2229.
17. Biham, O., Middleton, A. A., Levine, D.: Phys. Rev. A **46** (1992) 6124.

Author Index

Lecture Notes in Computer Science

For information about Vols. 1–3453

please contact your bookseller or Springer